W9-ADY-942

# NUCLEOSIDE ANTIBIOTICS

# NUCLEOSIDE ANTIBIOTICS

R. J. SUHADOLNIK

*Research Laboratories*
*Department of Bio-organic Chemistry*
*Albert Einstein Medical Center*
*Philadelphia*

WILEY-INTERSCIENCE, A DIVISION OF JOHN WILEY & SONS, INC.

NEW YORK • LONDON • SYDNEY • TORONTO

Library of Congress Catalogue Card Number: 73–115655

ISBN 471 0 83543 9

Printed in the United States of America

10 9 8 7 6 5 4 3 2 1

# Preface

The nucleoside antibiotics represent a diverse group of biological compounds structurally related to the purine and pyrimidine nucleosides and or nucleotides found in the cell. The nucleoside antibiotics have been useful as models for conformational studies, mass spectrometry, nmr, and ORD measurements. They have also been equally important as biochemical tools in cellular reactions, as illustrated in those studies where they have aided in the elucidation of the complex steps involved in reading the genetic message on the ribosomes for protein synthesis, RNA synthesis, DNA synthesis, regulation of purine and pyrimidine nucleotide synthesis, mechanisms of enzymatic reactions, subcellular organization, intermediary metabolism, and cell wall biosynthesis. The close structural relationship of the nucleoside antibiotics to purine and pyrimidine nucleosides and nucleotides have made them useful as structural analogs and inhibitors. Finally, the studies that have been made on the biosynthesis of eleven of the nucleoside antibiotics add another role for preformed purine or pyrimidine nucleosides and nucleotides in cellular reactions.

This book attempts to provide the first clear, comprehensive, and up-to-date review of the nucleoside antibiotics. It is my hope that the contents will serve as a useful reference for teaching and research. The extraordinary growth of the research related to the nucleoside antibiotics is vividly illustrated by a comparison of the material available for this review with that of recent reviews (Cohen, *Progress in Nucleic Acid Research and Molecular Biology*, Academic Press, New York, 1966; Fox, Watanabe and Bloch, *Nucleoside Antibiotics*, Academic Press, New York, 1966; Korzyski, Kowszyk-Gindifer, and Kurylowicz, *Antibiotics*, Vols. I and II, Pergamon Press, New York 1967; Gottlieb and Shaw, *The Antibiotics*, Vols. I and II, Springer-Verlag, New York, 1967; Umezawa, *Recent Advances in Chemistry and Biochemistry of Antibiotics*, Nissin Tosho Insatsu Co., Ltd., Tokyo, 1964; Umezawa, *Index of Antibiotics From Actinomycetes*, University Park Press, 1970 ;Rinehart, *The Neomycins and Related Antibiotics*, 1964; John Wiley & Sons, Inc.; Roy-Burman, Recent Results in Cancer Research, **25**, Springer Verlag, 1970).

The structural features of the nucleoside antibiotics provided a convenient basis for grouping these compounds into eleven chapters in this book. In general, the first five chapters deal with those nucleosides in which structural modifications occur in which ribose is replaced by a 3′-deoxyribose, ketohexoses, D-arabinose, or 4- or 5-aminohexuronic sugars. With aristeromycin (Sect. 6.1), the oxygen in the ribofuranosyl moiety has been changed to a methylene group. The remaining chapters are concerned with those nucleosides in which structural modifications occur in the aglycone moiety, such as the substitution of a *s*-triazine, a pyrrolopyrimidine, a pyrazolopyrimidine, a maleimide, or an isoguanine ring for the normal purine or pyrimidine base. For consistency, each of the 35 nucleosides reviewed has the following general outline: Introduction; Discovery, Isolation, and Production; Physical and Chemical Properties; Structural Elucidation; Chemical Synthesis; Synthesis of Analogs; Inhibition of Growth; Biosynthesis; Biochemical Properties; Summary; References.

Seven nucleosides discussed in this book have not exhibited any antibiotic properties. They are 3′-acetamido-3′-deoxyadenosine, spongosine, arabinosyluridine, nebularine, crotonoside, pseudouridine, and oxoformycin. They are included since they represent naturally occurring nucleosides which are closely related to the nucleoside antibiotics that have been isolated and studied. Septicidin, a $N^6$-glycosyl adenine antibiotic, is reviewed since it consists of the adenine chromophore and a unique seven-carbon sugar.

Finally, with the exception of very few references, I have carefully read all the original publications cited in this book. Indeed, the writing of this book was made possible by the reprints, preprints, and personal communications from those scientists who have contributed to the studies on the nucleoside antibiotics. Therefore, I am extremely grateful to those principal investigators throughout the world who have sent me their material and who have generously given time to read, criticize, and correct the chapters in this book. However, I assume responsibility for any shortcomings in the material cited.

*January 1970*
*Philadelphia, Pennsylvania*

R. J. SUHADOLNIK

# Acknowledgments

I would especially like to acknowledge the following colleagues for their valuable editorial comments, suggestions and criticisms: Drs. G. Acs, M. Agranoff, M. Atkinson, C. Baugh, L. L. Bennett, Jr., T. Beppu, G. Brown, A. Cerami, S. S. Cohen, J. M. Clark, Jr., C. Coutsogeorgopoulos, E. F. Elstner, J. J. Fox, L. Flexner, J. Florini, S. Frederiksen, P. Goldman, A. Guarino, T. Haskell, L. J. Hanka, F. Henderson, M. Hori, M. Ikehara, K. Isono, A. Kaplan, T. Kishi, H. Klenow, A. Kornberg, J. Lucas-Lenard, G. A. LePage, K. Moldave, J. Montgomery, R. E. Monro, H. Moyed, J. Nagyvary, R. Parks, S. Pestka, K. V. Rao, E. Reich, R. Robins, F. Schabel, Jr., Y. Shealy, H. Shigeura, C. Smith, S. Suzuki, F. Sorm, P. T'so, I. Tsukada, T. Uematsu, H. Umezawa, J. Vesely, M. von Saltza, D. C. Ward, T. Webb, B. Weisblum, R. Williams, H. Yonehara.

I also wish to express my sincerest appreciation to Miss Nancy L. Reichenbach, Mrs. Carol Cook, Miss Alyce Miller, Donald Chernoff and David Purdy for their valuable assistance and patience in helping me to prepare this book.

R. J. S.

# Contents

# NUCLEOSIDE ANTIBIOTICS

CHAPTER 1

# 3′-Deoxypurine Nucleosides

## 1.1. PUROMYCIN 3

## 1.2. CORDYCEPIN (3′-DEOXYADENOSINE) 50



## 1.3. 3′-AMINO-3′-DEOXYADENOSINE 76



## 1.4. 3′-ACETAMIDO-3′-DEOXYADENOSINE 86

---

## ABBREVIATIONS

5′-Adenylylmethylene-diphosphonate: AMP-PCP; 5′guanylylmethylene-diphosphonate: GMP-PCP; formylmethionine: F-met; methionine: met; acetylmethionine: acetyl-met; transfer RNA: tRNA; trinucleoside diphosphates: AUG, GUG, UUG, CUG; formylatable species of tRNA are represented by $tRNA_F$; nonformylatable species of tRNA by $tRNA_M$.

Structural changes at C-3′ of purine nucleosides result in nucleosides with marked changes in their biological properties. To date five 3′-deoxyadenine nucleoside antibiotics have been isolated from the *Streptomyces* and fungi that are extremely toxic to bacteria, animal cells in culture, tumors, and viruses. They are puromycin, cordycepin (3′-deoxyadenosine), 3′-amino-3′-deoxyadenosine, homocitrullylaminoadenosine, and lysylamino adenosine. The 3′-deoxynucleoside (3′-acetamido-3′-deoxyadenosine) is not toxic to bacteria or tumor cells. These nucleosides have been excellent biochemical tools for studying numerous reactions with partially purified enzymes or cellular processes. This chapter reviews the chemical and biochemical properties of these six naturally occurring adenine nucleosides.

## 1.1. PUROMYCIN

### INTRODUCTION

Puromycin, 6-dimethylamino-9-[3-(*p*-methoxy-L-β-phenylalanylamino)-3-deoxy-β-D-ribofuranosyl]purine (Fig. 1.1*A*), is a nucleoside antibiotic elaborated by *Streptomyces alboniger*. The structure and total chemical synthesis have been reported. Puromycin is structurally similar to the 3′-terminus of aminoacyl-tRNA (Fig. 1.1*B*). Puromycin has been a valuable biochemical tool in molecular biology for the elucidation of the mechanisms involved in reading the genetic message by which the nascent polypeptide chain on the peptidyl site is transferred to the aminoacyl-tRNA on the

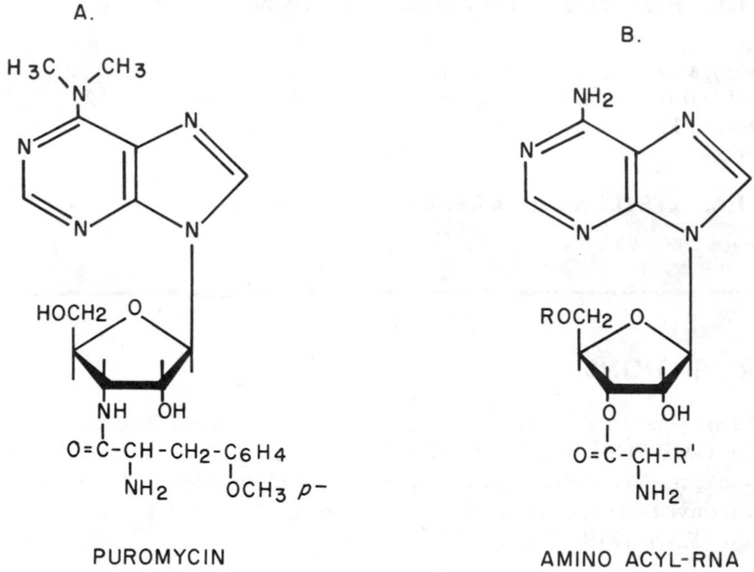

PUROMYCIN

AMINO ACYL-RNA

FIGURE 1.1 (*A*) *Structure of puromycin.* (*B*) *3′-Terminus of aminoacyl-tRNA. R = tRNA; R′ = H; alkyl group of amino acids.*

aminoacyl site on the ribosomes. The use of puromycin in the elucidation of initiation and elongation as related to protein synthesis will be discussed. Puromycin inhibits a wide spectrum of organisms. The final enzymatic reaction in the biosynthesis of puromycin has been reported.

## DISCOVERY, PRODUCTION, AND ISOLATION

Puromycin (achromycin or stylomycin) (Fig. 1*A*) was isolated from the culture filtrates of *S. alboniger* (ATCC, 12,462) by Porter et al. (1952). The medium reported by Porter et al. (1956) to produce puromycin was modified by Szumski and Goodman (1957). The composition of the medium is as follows: 6% corn steep liquor (50% solids), 4% corn starch, 0.7% calcium carbonate, and 1% lard oil (to prevent foaming) (Szumski and Goodman, 1957). The pH was 6.0–8.5. The addition of the purinyl radical (uric acid, adenine, hypoxanthine, or xanthine) increased the yield of puromycin from 627 to 479 μg/ml. Seed flasks (100 ml medium/500 ml flask) were inoculated with spores from an agar slant (suspended in water). The flasks were maintained on an incubator shaker at 27°C for 2 days. A 2.5% inoculum was used to inoculate larger volumes of the same medium (pH 7.0) (Porter et al., 1956). Puromycin was isolated from the fermentation medium 70 hr after inoculation by adjustment of the pH to 4.0–4.5, filtration, readjustment of the pH to 9.0–9.5, and extraction with 1-butanol. The butanol phase was extracted with distilled water (pH 1.5–2.0). The acid–water was concentrated *in vacuo*, and puromycin crystallized in the cold. It was recrystallized by dissolving in water at 40°C, pH 2–4 in a 10% solution. Hydrochloric acid was added to make a 1 *N* solution. Puromycin hydrochloride crystallized on standing. The free base has been isolated by addition of NaOH to pH 7.0 to the hydrochloride salt (Porter et al., 1956; Szumski and Goodman, 1957).

## PHYSICAL AND CHEMICAL PROPERTIES

The molecular formula for puromycin is $C_{22}H_{29}N_7O_5$; mp 175.5–177°C; $[\alpha]^2_D$ −11° (in ethanol); $\lambda_{max}$ 267.5 m$\mu$ ($\epsilon$ = 19,500) in 0.1 *N* HCl; $\lambda_{max}$ 275 m$\mu$ ($\epsilon$ = 20,300) in 0.1 *N* NaOH. It is hydrolyzed to 6-dimethylaminopurine, *O*-methyl-L-tyrosine and 3-amino-3-deoxyribose in acid (Waller et al., 1953). The mass spectra of puromycin have been reported by Eggers et al. (1966). The fragmentation pattern of puromycin is in agreement with the structure assigned and can be compared with the mass spectra of the purine nucleoside analogs cordycepin, 3′-amino-3′-deoxyadenosine, and 3′-acetamido-3′-deoxyadenosine (see pp. 55, 77, and 88).

## STRUCTURAL ELUCIDATION AND CHEMICAL SYNTHESIS

The partial structure of puromycin was reported by Waller et al. (1953.) When puromycin (**1**) was treated with ethanolic HCl, three compounds were formed (Fig. 1.2): 6-dimethylaminopurine (**2**), *O*-methyl-L-tyrosine (**3**), and an aminopentose (**4**) that consumed periodic acid (Waller et al., 1953). The amino pentose was shown to be identical with synthetic 3-amino-3-deoxyribose synthesized by Baker and Schaub (1954) and Baker et al. (1955a). When puromycin was treated with phenyl isothiocyanate and sodium methoxide, aminonucleoside (**5**) was isolated (Baker et al., 1955b). This compound (**5**) consumed 1 mole of periodic acid, which established the furanoid structure for the 3-amino-3-deoxyribose moiety. Since puromycin failed to consume periodic acid (Waller et al., 1953), the carboxyl group of *O*-methyl-L-tyrosine was covalently linked to the amino group on carbon-3′ of the pentose. The total chemical synthesis of 3-amino-3-deoxy-D-ribose made possible the synthesis of the aminonucleoside (Baker et al., 1955c). When the titanium chloride complex of the amino sugar was treated with the chloromercuri derivative of 2-methylmercapto-6-dimethylaminopurine, followed by desulfurization with Raney nickel and de-*O*-benzoylation, the *N*-acetyl derivative of the aminonucleoside was obtained. This derivative was then converted to the aminonucleoside (**5**), which was converted to puromycin (Baker et al., 1955b). The glycosidic bond of puromycin has the $\beta$ configuration (Baker et al., 1954; Baker and Joseph, 1955).

The syntheses of various analogs of puromycin with varying amino acids substituted for *O*-methyl-L-tyrosine have been described (Montgomery and Thomas, 1962; Nathans and Neidle, 1963; Symons et al., 1969). At 3 × $10^{-4}$

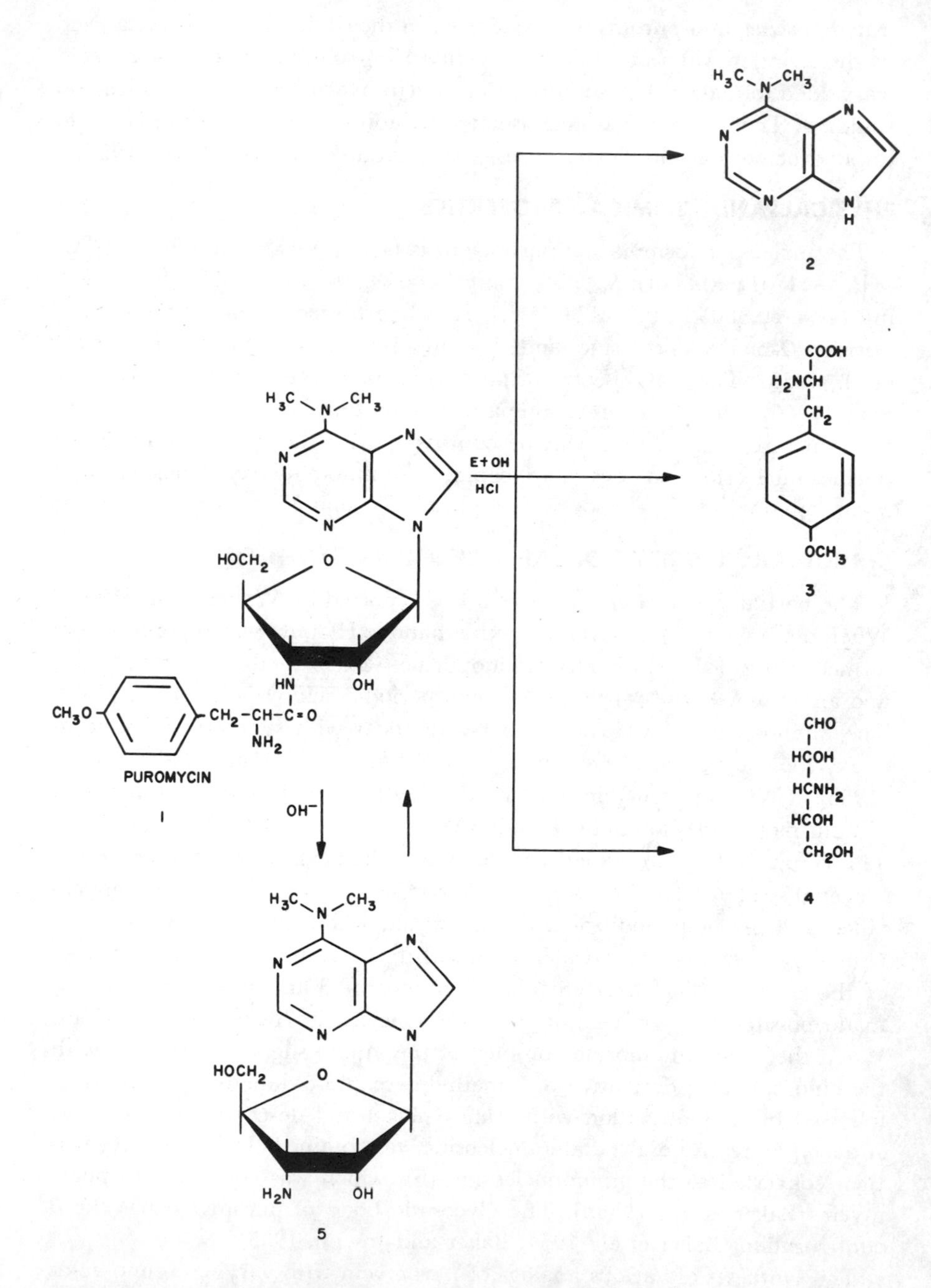


FIGURE 1.2 *Hydrolysis of puromycin. (From Waller et al., 1953; Baker et al., 1955b.)*

*M*, L-phenylalanine, *S*-benzyl-L-cysteine, and L-tyrosine were respectively, 99, 78, and 62% as inhibitory as puromycin (Symons et al., 1969). It is of interest that the X-ray structure of puromycin dihydrochloride reveals that the $N^6,N^6$-dimethyladenine and the *p*-methoxyl group in tyrosine form alternate stacks such that the *p*-methoxyl group underlies the adenine ring. This type stacking suggests a hydrophobic interaction between the *p*-methoxyl group and the adenine ring and might explain the function of puromycin as an inhibitor of protein synthesis when this analog interacts with the CCA end of tRNA. These findings may explain the lower activity of demethoxypuromycin (Sundaralingum, private communication). Carbon-14, tritium-labeled, and $^{32}P$-labeled puromycin have also been synthesized (Allen and Zamecnik, 1962; Shelton and Clark, 1967; Smith et al., 1965). Fisher, Lee, and Goodman have prepared a number of puromycin analogs (1970).

## INHIBITION OF GROWTH

Puromycin is a strong inhibitor of gram-positive organisms, but weakly active against the gram-negatives and acid-fast bacilli. The gram-positive organisms are generally more sensitive. It is also effective *in vivo* against *Trypanosoma equiperdum*, *Entamoeba histolytica*, oxyurids, tapeworms, a glioblastoma cultivated in chick embryos, a mammary adenocarcinoma of $C_3H$ mice, viruses, and HeLa cells (Hewitt et al., 1954, 1955). The unusual broad spectrum of inhibition of puromycin may probably be attributed to the fact that puromycin has been shown to be an inhibitor of peptide synthesis.

White and White (1964) reported that puromycin is bacteriostatic and not bacteriocidal. However, the concentration of puromycin added to bacterial cultures is important.

Since puromycin is an effective inhibitor of peptide synthesis, it is also very toxic in higher animals. The 50% lethal dose for mice by the intravenous route is 335 mg/kg, while that supplied intraperitoneally is 580 mg/kg (Sherman et al., 1954/1955). Intravenous injections of 25 mg/kg in cats cause a drop in blood pressure. In rats, intraperitoneal doses of 25 mg/kg caused the animals to lose weight and evoked weakness. At 100 mg/kg, there was renal and bone marrow impairment.

## BIOSYNTHESIS

Pogell and his co-workers reported on the biosynthesis of 2-aminoribose-5-phosphate and 2-aminolyxose-5-phosphate with cell-free extracts from *S. alboniger* (Rebello et al., 1969). They reported that D-ribose-5-phosphate is enzymatically converted to 2-amino-2-deoxy-D-ribose-5-phosphate and 2-amino-2-deoxy-D-lyxose-5-phosphate. The specificity for the carbon donor for aminopentose phosphate synthesis with the dialyzed crude supernatant of *S. alboniger* was shown to be ribose-5-phosphate. When ribose was the substrate, ATP was an essential energy source. Glucose-6-phosphate, fructose-6-phosphate, and ribose were not precursors for amino pentose biosynthesis. Linearity of product formation with time was observed with the partially purified enzyme. A very active phosphoriboisomerase is present in the cell-free extracts. Evidence for the presence of two phosphorylated pentoses was established by paper chromatography following enzyme incubation. Rebello et al. (1969) concluded that the aminopentose phosphates formed from ribose-5-phosphate were 2-amino-2-deoxyribose-5-phosphate and 2-amino-2-deoxylyxose 5-phosphate. The significance of 2-aminopentose-5-phosphates in extracts of *S. alboniger* as related to the biosynthesis of puromycin is not known. The 2-aminoribose may be an important precursor for the biosynthesis of the 3-aminoribose moiety of puromycin. Additional *in vitro* studies are necessary to establish this biosynthetic relationship.

Rao et al. (1969) have also isolated and partially purified the enzyme from sonicated extracts of *S. alboniger* that catalyzed the enzymatic methylation of *O*-demethylpuromycin. No cofactors were needed. The physical and chemical properties of this methylated puromycin established unequivocally that *O*-methylation of *O*-demethylpuromycin had occurred. Pattabiraman and Pogell (1969) have isolated and identified *O*-demethylpuromycin as a contaminant of crystalline puromycin. These data strongly suggest that the final step in puromycin biosynthesis is the methylation of the tyrosine moiety of *O*-demethylpuromycin.

## BIOCHEMICAL PROPERTIES

Since puromycin has been used in a number of cellular reactions, this section on the biochemical properties of this nucleoside antibiotic will be divided into sub-headings applicable to each study. Lipmann (1969), Ono et al. (1969), Vazquez and Monro (1968), Monro (1969), Pestka (1970c), Coutsogeorgopoulos (1970), Nathans (1967), and Lengyel and Söll (1969) have recently published and reviewed polypeptide chain elongations in protein synthesis and the role that puromycin and other antibiotics have played in understanding the mechanisms involved in reading the genetic message. For additional information on the mechanism of protein synthesis, the reader should consult volume **34** of the Cold Spring Harbor Symposia, 1970, The Mechanism of Protein Synthesis.

### I. Site of Action of Puromycin

#### 1. *Puromycin: A Biochemical Tool for Studying Protein Synthesis*

Puromycin has been used extensively in the study of the mechanism of protein biosynthesis in both mammalian and bacterial cell-free ribosomal and non-ribosomal systems. The results from these studies indicate that puromycin releases only those peptides from ribosomes that are bound to the so-called donor site, or peptide binding site (Heintz et al., 1966). Peptidyl-tRNA on the acceptor site (aminoacyl-tRNA binding site), will not react with puromycin. The terms "donor" and "acceptor" reflect the idea that the tRNA on the donor site "donates" its peptidyl moiety to the acceptor, aminoacyl-tRNA, that is bound on the acceptor site. Hence, growth of the polypeptide is from amino to carboxyl end.

Yarmolinsky and de la Haba (1959) were the first to recognize the close structural similarity between puromycin and the aminoacyl end of aminoacyl-tRNA. They showed that the linkage of the carboxyl group of amino acids to the 2′ or 3′ hydroxyl group of adenosine at the 3′-terminus of tRNA is structurally similar to the *p*-methoxytyrosine in which the carboxyl group is covalently bound to the 3′-amino group of puromycin. They reported that puromycin inhibited peptide synthesis in rat liver preparations by blocking the transfer of $^{14}$C-leucine from $^{14}$C-leucyl-tRNA to the nascent peptide bound to the ribosome. Nathans and Lipmann (1961) subsequently confirmed these results using a cell-free preparation from *E. coli*. Gardner et al. (1962) and Nathans and Neidle (1963) showed that the inhibition of protein synthesis by puromycin was not amino acid specific. Both polylysine and polyphenylalanine synthesis were inhibited. A subsequent paper by Morris et al. (1963) revealed several characteristics of the mode of action of puromycin in hemoglobin biosynthesis. They showed that the amount of TCA-insoluble polypeptide formed was reduced in the presence of puromycin. The release of $^{14}$C-labeled peptide from the ribosomes by puromycin occurred in the presence of the supernatant enzyme fraction with no measurable breakdown of ribosomes. When puromycin was added to intact cells, there was a release of polypeptide that contained valine in the *N*-terminal residues. These results showed that puromycin released incomplete globin chains. One molecule of puromycin was covalently bound for each *N*-terminal valine.

Allen and Zamecnik (1962) were the first to suggest that the amino group of the *p*-methoxyphenylalanyl moiety of puromycin attacked the *C*-terminal acyl group of the growing peptide chain with the displacement of tRNA. Additional proof for the release of shorter, acid-soluble and alcohol-soluble peptides by puromycin was supplied by Nirenberg et al. (1962) and Nathans et al. (1963). Gilbert (1963) similarly showed that puromycin released polypeptides from peptidyl-tRNA bound to the 50-S subunit by reacting with the ester bond of the peptidyl-tRNA. The composition of the peptidyl–puromycin reaction was determined by studying the puromycin reaction in the polyadenylate-directed synthesis of polylysine with $^{32}$P-$\beta$-cyanoethylphosphate 5′-puromycin (Smith et al., 1965). Each of the peptides released to the supernatant contained one molecule of puromycin. Hydrolysis of dilysyl-puromycin with trypsin resulted in the isolation of lysyl-lysine. Lysyl-puromycin was not a product released to the supernatant. These studies clearly showed that the carboxyl end of the endogenous peptidyl-tRNA reacted with the amino group of puromycin to form peptidyl-puromycin.* Thus, puromycin inhibits protein synthesis by substituting for an aminoacyl-tRNA.

Heintz et al. (1968) used the puromycin reaction in their reticulocyte ribosome complex to study the mechanism of protein synthesis. Using poly U as the messenger, they observed that phenylalanyl-tRNA was bound to two different sites on the reticulocyte ribosomes. When phenylalanyl-tRNA was located on the donor site, it reacted directly to form phenylalanyl-puromycin in the presence of transfer enzyme 2 or transferase II. When phenylalanyl-tRNA was bound to the acceptor site, GTP and possibly the transferase I were required to complete the peptide–puromycin reaction in addition to the transferase II. Cycloheximide, a protein-inhibiting antibiotic, blocked the puromycin reaction. Increasing concentrations of transferase II did not reverse this inhibition. However, preincubation of transferase II with ribosomes before the addition of puromycin allowed the necessary reactions to occur on the ribosome that prevented the normal inhibition by cycloheximide.

To understand the mechanism of peptide bond formation, it has been necessary to recognize the function of the enzymes and cofactors involved. The use of puromycin in the design of many of these experiments will be described here. Two soluble protein fractions have been reported as necessary for the synthesis of peptides in the systems from rabbit reticulocytes (Arlinghaus et al., 1964), rat liver (Skogerson and Moldave, 1968a, 1968b, 1968c), and yeast (Richter and Klink, 1967). Three soluble factors appear to be involved in the bacterial systems (Lucas-Lenard and Lipmann, 1966, 1967). The soluble protein factors obtained from rat liver have been designated by Moldave as aminoacyltransferase I and II. Transferase I is the soluble aminoacyl-tRNA protein binding factor; GTP is required for this binding (Ibuki and Moldave, 1968). Transferase II (peptidyl-tRNA translocase or the translocation factor) is the soluble protein factor (with GTP) required for translocation of messenger RNA and peptidyl-tRNA from the aminoacyl site to the peptidyl site on the ribosomes (Skogerson and Moldave 1968a, 1968b; Pestka, 1968; Schneider et al., 1968). Galasinski and Moldave (1969) have now reported on the purification of aminoacyltransferase II from rat liver. The specific activities are up to 1,000 fold higher than the "pH 5 supernatant" fraction. They estimated the molecular weight to be 60,000 to 65,000.

Using the rat liver system, Skogerson and Moldave (1968a, 1968b, 1968c) reported that puromycin rapidly reacted with endogenous peptidyl-tRNA bound to the peptidyl site on ribosomes without requiring either GTP,

* Ribosomal peptidyl transferase also catalyzes the formation of an ester bond as shown by the formation of N-formyl-methionyl-ψ-hydroxypuromycin (Fahnestock, S., Neumann, H., Shashoua, V., and Rich, A., *Biochemistry*, **9**, 2477 (1970)). Chloramphenicol and gougerotin inhibit ester formation.

aminoacyltransferase I or aminoacyltransferase II. On the other hand, if the peptidyl-tRNA were bound to the aminoacyl site, the puromycin reaction specifically required transferase II and GTP. Peptidyl-tRNA was isolated on the aminoacyl site following peptide bond formation between the endogenous peptidyl-tRNA on the peptidyl site and the aminoacyl-tRNA on the aminoacyl site. The conversion of peptidyl-tRNA from this non-puromycin aminoacyl site to the puromycin reactive peptidyl site is termed "translocation." McKeehan and Hardesty (1969a) and Lin et al. (1969) have now characterized the highly purified aminoacyl-tRNA binding enzyme from rabbit reticulocytes. This enzyme catalyzes the GTP-dependent binding of phenylalanyl-tRNA to ribosomes in the presence of messinger RNA. GTP hydrolysis is required for this reaction. GMP-PCP gives less than 10% of the activity of GTP for enzymatic binding (Lin et al., 1969). McKeehan and Hardesty (1970) reported that binding enzyme (T-I) and GTP promote binding of aminoacyl-tRNA to the mRNA-reticulocyte ribosome complex at 0°C without GTP hydrolysis. This ribosomal bound aminoacyl-tRNA cannot accept a peptide from peptidyl-tRNA on the donor site. However, when the temperature is increased to 37°C, peptide bond formation occurs.

Fusidic acid (Fig. 1.3), cycloheximide, and erythromycin, aided by puromycin, are known to block translocation in the highly coordinated sequence of reactions required for protein synthesis on polynucleotide templates (Pestka, 1968, 1969c, 1970b; Igarashi et al., 1969; McKeehan and Hardesty, 1969b; Tanaka et al., 1968; Kinoshita et al., 1968; Tanaka et al., 1969a, 1969b). Lin et al. (1968) have also shown that bottromycin $A_2$, another known inhibitor of protein synthesis, inhibits the puromycin reaction enhanced by G factor and GTP, indicating the inhibition of peptidyl-tRNA translocation. Tanaka et al., (1969a) have shown that bottromycin $A_2$ does not inhibit the ribosome dependent GTPase reaction of G factor. Their data indicate that bottromycin $A_2$ may be interacting with ribosomes. This was shown by the reversal of the inhibition of bottromycin $A_2$ by increasing the concentration of ribosomes (Lin and Tanaka, 1968). Of extreme interest are the recent findings of Tanaka et al. (private communication) in which they have shown that bottromycin $A_2$ and fusidic acid do not inhibit dipeptide synthesis, but show a marked inhibition of tripeptide synthesis. These data may be explained by the fact that fusidic acid inhibits the dissociation of the ribosome-factor-G-GDP complex such that a second round of GTP can not be used (Bodley, 1970). Therefore, puromycin-peptide bond formation involves at least two steps: (1) movement of peptidyl-tRNA on the ribosome and (2) formation of a peptidyl-puromycin bond (Brot et al., 1968; Skogerson and Moldave, 1968c; Igarashi et al., 1969).

Ravel et al. (1968) and Lucas-Lenard and Haenni (1968) reported that factor T (subfractions, $T_u$ and $T_s$) from *E. coli* (analogous to aminoacyl transferase I from rat liver, Skogerson and Moldave, 1968b; Skoultchi et al., 1968; Ono et al., 1969) and GTP were required for the binding of aminoacyl-tRNA

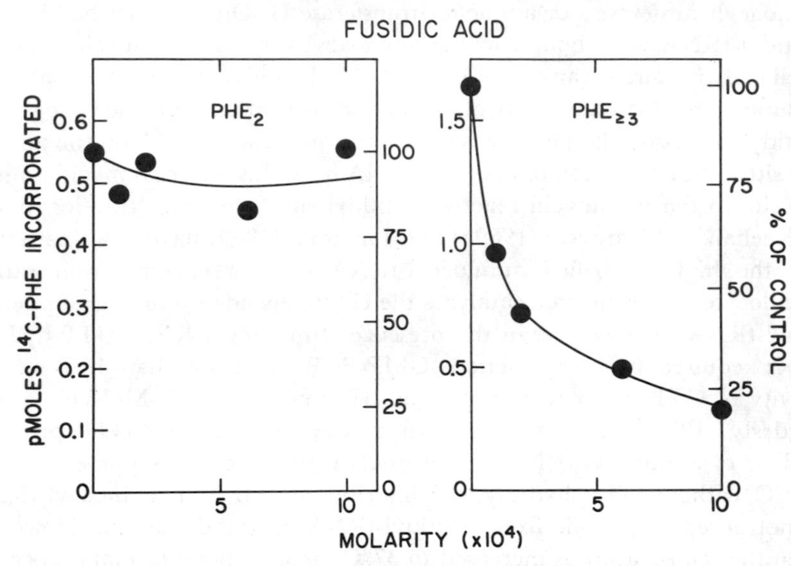


FIGURE 1.3. *Effect of fusidic acid (sodium salt) on di- and oligophenylalanine (Phe* $\geq$ *3) synthesis as a function of antibiotic concentration. Each 0.050-ml reaction mixture contained the following components: 0.05 M Tris-acetate, pH* 7.2; 0.05 *M KCl;* 0.005 *M* $MgCl_2$; 0.016 *M putrescine;* 0.004 *M spermidine;* 21 *nmoles of base residues of poly U* (0.21 $A_{260}$ *unit);* 6.8 *pmoles of* [$^{14}$C]*Phe-tRNA* (0.26 $A_{260}$ *unit);* 0.38 $A_{260}$ *unit of ribosomes;* $5 \times 10^{-5}$ *M GTP; and* 0.6 *μg of factor G. Incubations were performed at 37° for 5 min. Diphenylalanine formation is shown in the left panel and oligo-phenylalanine (Phe* $\geq$ *3) synthesis in the right panel. Fusidic acid concentration is given on the abscissa. Data are expressed as pmoles of* [$^{14}$C]*Phenylalanine incorporated into di- or oligophenylalanine (left ordinate); and as percentage of control value without antibiotic (right ordinate).* [$^{14}$C]*Phe-tRNA was added last to start the reactions. (From Pestka, 1970b).*

to the acceptor site of ribosomes. Peptide bond formation occurred when the peptidyl site carried a peptidyl-tRNA or an aminoacyl-tRNA with a free or blocked α-amino group on the amino acid (Skogerson and Moldave, 1968a; Lucas-Lenard and Haenni, 1968). When the peptide reaction is completed, the peptidyl-tRNA is located on the acceptor site. In order for protein synthesis to resume, GTP and transfer factor G are necessary for the translocation of the newly synthesized peptidyl-tRNA. Upon completion of this reaction, the aminoacyl site is open for entry for the next aminoacyl-tRNA.

More recently, Haenni and Lucas-Lenard (1968) have elaborated further on the T-dependent binding of aminoacyl-tRNA to ribosomes. The system utilized *N*-acetylphenylalanyl-tRNA as the initiator in the poly U directed synthesis of polyphenylalanine with ribosomes from *E. coli*. Factors T and G and GTP were necessary for *N*-acetylphenylalanyl peptide synthesis. Once translocation and charging of the aminoacyl site on the ribosomes with aminoacyl-tRNA occurred, none of factors T, G or GTP were required for peptide bond formation. Maden et al. (1968) also reported that GTP was not needed for peptide bond formation with charged ribosomes. When the GTP analog (GMP-PCP) and factor T bring about the binding of aminoacyl-tRNA to ribosomes (prebound with *N*-acetylphenylalanyl-tRNA), peptide bond formation does not occur; yet, *N*-acetylphenylalanyl-puromycin is formed despite the presence of phenylalanyl-tRNA on the ribosomes. The observation that phenylalanyl-tRNA binding occurs with GMP-PCP, but peptide bond formation is prevented while the puromycin reaction continues, leads to the speculation that GTP is necessary for a critical alignment of aminoacyl-tRNA on the ribosomes such that the aminoacyladenosine end of tRNA is available for peptide bond formation on the 50S subunits. This proper attachment apparently does not occur when GMP-PCP replaces GTP. According to these data, it may be that puromycin does not occupy the acceptor site, but is bound in site S (Fig. 1.4) and is close enough to undergo peptide formation with the peptidyl-tRNA located on the donor site. It is interesting to note that Coutsogeorgopoulos (1967b) suggested that puromycin does not enter the aminoacyl-RNA-ribosome-mRNA complex by way of the "amino acid" site (acceptor site), but by way of another site (site S) where the aminoacyladenylyl terminus and the peptidyl-adenylyl terminus interact to form a peptide bond. His conclusion was based on the

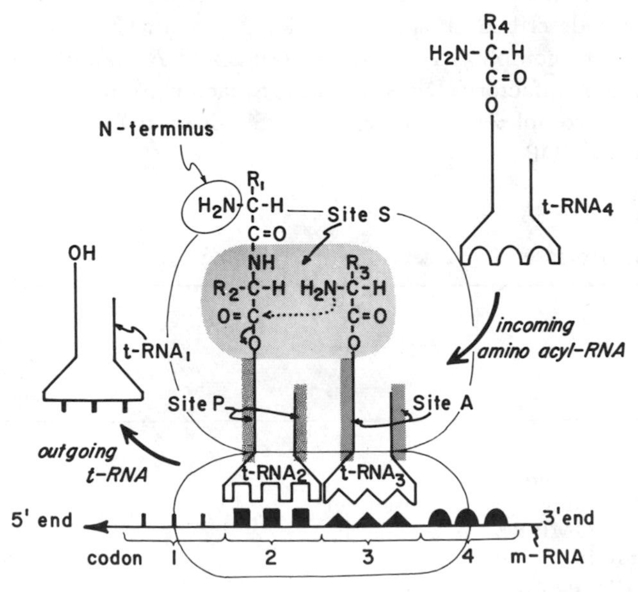


FIGURE 1.4. *Schematic diagram of peptide-bond formation on the ribosome.* (*From Coutsogeorgopoulos, 1967.*)

fact that the inhibition caused by puromycin was not reversed with increasing amounts of phenylalanyl-tRNA. Finally, Haenni and Lucas-Lenard (1968) described experiments which showed that after one peptide (*N*-acetyldiphenylalanyl-tRNA) was formed, factor G and GTP were required for translocation and then incubation with T factor and phenylalanyl-tRNA was necessary before the peptide chain could be lengthened to form *N*-acetyltriphenylalanyl-tRNA. Puromycin could replace phenylalanyl-tRNA and form *N*-acetyldiphenylalanyl-puromycin (Table 1.1). The role of factor G in protein synthesis has also been demonstrated by Pestka and Felicetti. Oligophenylalanine formation, but not diphenylalanine, requires factor G (Fig. 1.5). An *E. coli* mutant with an altered factor G showed that factor G is needed for polymerization but not for peptide bond synthesis (Table 1.2). Weissbach et al. (1970) have shown that factor $T_s$ catalyzes the release of GDP from $T_u$-GDP to continue the cycle for peptide chain elongation as follows:

$$T_s + T_u\text{-GDP} \rightleftharpoons T_s\text{-}T_u + \text{GDP}$$

$$T_u\text{-}T_s + \text{GTP} \rightleftharpoons T_u\text{-GTP} + T_s$$

$$T_u\text{-GTP} + \text{AA-tRNA} \rightleftharpoons \text{AA-tRNA-}T_u\text{-GTP}$$

$$\text{AA-tRNA-}T_u\text{-GTP} \xrightarrow[\text{messenger}]{\text{ribosome}} (\text{AA-tRNA-messenger-ribosome}) + T_u\text{-GDP} + \text{Pi}$$

GDP can be phosphorylated to GTP in the $T_u$-GDP complex by PEP and PEP kinase. Waterson, Beaud, and Lengyel (*Nature*, in press; private communication) described a similar cycle; $S_3$ factor-GTP-aminoacyl-tRNA form a ternary complex (I) on the ribosomes of *B. stearothermophilus*. GTP is cleaved and $S_3$-factor-GDP is released; $S_1$ factor promotes the reformation of the ternary complex (I). Factors $T_s$ and $S_1$ appear to function by promoting the release of GDP.

**Table 1.1**

Puromycin Release of *N*-Acetyldiphenylalanine

| Additions | *N*-Acetyldiphenylalanyl-puromycin released ($^{14}$C or $^3$H), $\mu\mu$moles |
|---|---|
| Puromycin | 1.8 |
| G + GTP + puromycin | 5.4 |
| G + GTP + fusidic acid + puromycin | 2.0 |
| G + GTP + GMP-PCP + puromycin | 4.1 |
| G + puromycin | 2.0 |
| GTP + puromycin | 1.6 |

From Haenni and Lucas-Lenard, 1968.

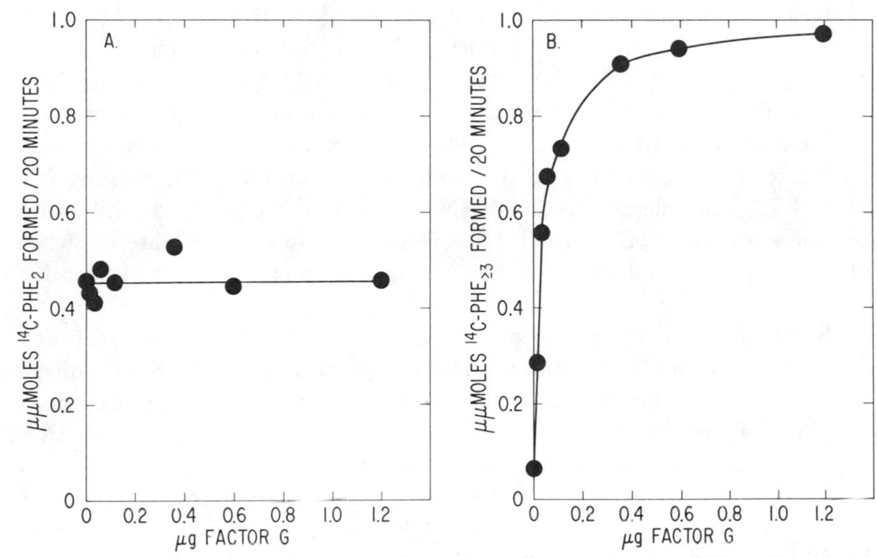


FIGURE 1.5. *Di- and oligophenylalanine formation as a function of factor G concentration. (A) Dipyhenylalanine formation; (B) Oligophenylalanine formation (From Pestka, 1968).*

**Table 1.2**

Reaction with Puromycin of Polyphenylalanine-Charged Ribosomes from Wild-Type and Mutant[a]

| | $^{14}C$ Phenylalanine-charged ribosomes | | | |
|---|---|---|---|---|
| | Wild-type | | Mutant | |
| Samples | μμ-moles | % release | μμ-moles | % release |
| Control | 2.6 | 0 | 1.82 | 0 |
| Puromycin ($10^{-3}$ *M*) | 1.24 | 48 | 0.66 | 64 |

From Felicetti et al., 1969.
[a] The incubation mixture contained, in a final volume of 0.1 ml, either 100 μg of polyphenylalanine-charged ribosomes from wild-type or 200 μg from mutant.

Brot et al. (1969) have now described experiments demonstrating that factor G and GTP, GDP, or dGTP are bound to the ribosomes. This binding of nucleotides to the ribosome–factor G complex does not appear to require hydrolysis of GTP, since GMP–PCP is more effective in binding than is GTP, even though hydrolysis of GTP did occur. The findings are closely related to the report of Skogerson and Moldave (1968b).

The conditions necessary for the release of tRNA during peptide chain elongation have recently been reported (Lucas-Lenard and Haenni, 1969; Igarashi and Kaji, 1969). They showed that the tRNA bearing the initiator amino acid or peptide is not released from the ribosomes following peptide bond formation until the translocation step mediated by the factor G and GTP takes place. Fusidic acid prevents the release. GMP-PCP cannot replace GTP. The release of donor tRNA appears to be coupled with translocation since factor G and GTP are insufficient to cause release of tRNA after *N*-acetylphenylalanyl-puromycin formation, a case where translocation is not occurring.

The formation of a peptide bond between CCA-met$_F$ (fragment reaction) and puromycin in the presence of methanol and isolated 50-S subunits is shown in Figure 1.6*a*. Monro and Vazquez (1967) showed that antibiotic inhibitors of protein synthesis blocked the reaction shown in Figure 1.6*a*. Similarly, alcohol plus the 50-S subunits catalyzed the transfer of F-met-tRNA$_F$ to puromycin (Fig. 1.6*b*) (Monro et al., 1968). Most recently, Monro (1969) provided experimental evidence for peptide bond formation with the 50-S subunits and methanol. CCA-met$_F$ was bound to the donor site and aminoacyl-tRNA was bound to the acceptor site (Fig. 1.6*c*). Monro showed that recycling occurs and a mixture of di-, tri-, and tetrapeptidyl-tRNA species formed. The peptides were of random amino acid sequence. The diagrammatic representation of peptide bond formation with the 50-S subunits and alcohol (Figs. 1.6*a*,*b*,*c*) can be compared with Figures 1.4, 1.15, which show the requirements for chain elongation.

The effect of pH on the puromycin reaction with the poly U dependent phenylalanine incorporation is also important (Cathey and Klebanoff,

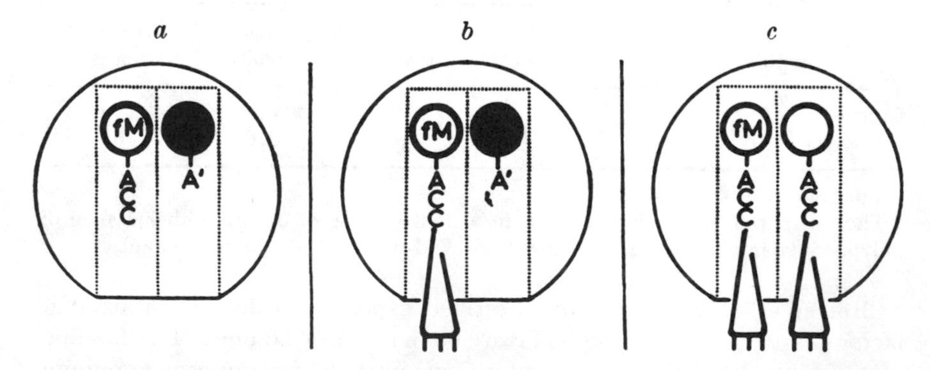


FIGURE 1.6. *Systems for the study of peptide bond formation with isolated 50S subunits in presence of alcohol. The substrate-ribosome complex is depicted at the moment before peptidyl transfer.* (*a*) *CCA-met-*$_F$ *and puromycin,* ●*-A′;* (*b*) *fMet-tRNA*$_F$ *and puromycin,* ●*-A′; c, fMet-tRNA*$_F$ *and aminoacyl-tRNA.* (*From Monro, 1969.*)

1967). At pH 5.2 puromycin does not inhibit the peptidyl transfer reaction with the ribosome–mRNA complex from *E. coli* or *L. acidophilus*. Since the ionizable groups of puromycin have p$K$ values of 7.3 and 3.7, it appears that protonation of the amino group of puromycin diminishes the inhibitory effect of this analog on polyphenylalanine synthesis. These findings suggest that the uncharged $\alpha$-amino group of puromycin is needed for inhibition. The aminonucleoside from puromycin was not inhibitory.

A new technique was introduced by Maden et al. (1968) to study the puromycin reaction in protein synthesis with polyphenylalanine-charged *E. coli* ribosomes. The technique was based on differential solubilities of substrate–product in *m*-cresol. The role of the 50-S subunit, without added supernatant and GTP, was also studied (Fig. 1.7). In general, the puromycin reaction with the 50-S ribosomes was similar to the 70-S system. The evidence presented strongly suggested that peptidyl transferase is an integral part of the 50-S ribosomal subunit (Vazquez and Monro, 1968). Mukundan et al. (1968) also showed that while F-met-tRNA$_F$ is bound initially to the 30-S particle, peptide bond formation, as determined by the puromycin reaction, occurs on the 50-S particle. They reported that F-met-tRNA$_F$ bound by the initiation factors ($f_1$ and $f_2$) reacted with puromycin only after the addition of the 50-S particle. AMP–PCP and GMP–PCP were able to replace GTP in the binding reaction. Rudland et al. (1969) concluded that the initiation factors recognize F-met-tRNA$_F$. Chloramphenicol inhibited the reaction of charged 50-S subunits with puromycin. Four molar urea or 0.5% sodium

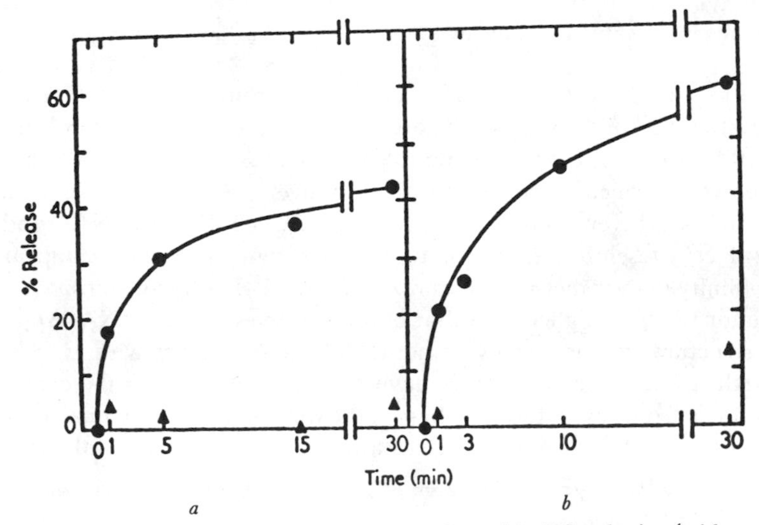


FIGURE 1.7. *Time-course of puromycin reaction with 50S subunits (without added supernatant or GTP). (a) Pre-incubated 30 min, 30° before adding puromycin, (b) no pre-incubation.* (●-●) *Standard conditions;* (▲) *no puromycin. (From Maden et al., 1968.)*

dodecyl sulfate or 2% formaldehyde (16 hr at 0°C) completely inhibited the puromycin reaction (Table 1.3). Removal of the urea partially reversed the inhibition of the puromycin reaction. These data show that native ribosomes are required for the peptide reaction.

**Table 1.3**

Effects of Urea and Sodium Dodecyl Sulfate (SDS) on the Puromycin Reaction

| Addition | Release of polyphenylalanine, % | |
|---|---|---|
| | 1 min | 30 min |
| Expt. 1 | | |
| (a) None | 43 | 73(13)[a] |
| (b) 4 *M*-urea, immediately before cresol | | 80 |
| (c) 4 *M*-urea | 4 | 13(10) |
| (d) 4 *M*-urea, then dialyzed | 23 | 45(18) |
| Expt. 2 | | |
| (a) None | | 73 |
| (b) 2% SDS, immediately before cresol | | 72 |
| (c) 0.5% SDS, pre-inc. 1 min, 30°C | | 26 |
| (d) 0.5% SDS, pre-inc. 10 min, 30°C | | 10 |
| (e) 2% SDS, pre-inc. 1 min, 30°C | | 9(6) |

From Maden et al., 1968.
[a] Figures in parentheses represent estimates for release in absence of puromycin.

Caskey et al. (1968) have also used the puromycin reaction recently to study the dissimilar responses to mammalian and bacterial tRNA fractions to mRNA-codons. In the presence of AUG and *E. coli* ribosomes, methionyl-puromycin formed at a faster rate with liver met-$tRNA_1$ than with liver met-$tRNA_2$. Liver met-$tRNA_1$ resembles *E. coli* F-met-$tRNA_F$ and liver met-$tRNA_2$ resembles *E. coli* met-$tRNA_M$. From these data it appears that the affinity of liver met-$tRNA_1$ and *E. coli* F-met-$tRNA_F$ for ribosomal sites of initiator tRNA is higher than that of other species of met-tRNA. Apparently the structural features of liver met-$tRNA_1$ that interact with *E. coli* transformylase and *E. coli* ribosomes have been retained during the evolution of mammals from lower forms. These data were not taken as final proof that liver met-$tRNA_1$ functions *in vivo* as an initiator of protein synthesis.

### 2. Effect of Puromycin on Transfer of Amino Acids on Protein Acceptors in the Absence of Ribosomes

Kaji (1968) reported on the characteristics of a soluble amino acid-incorporating system from *E. coli* independent of ribosmes. Kaji also reported

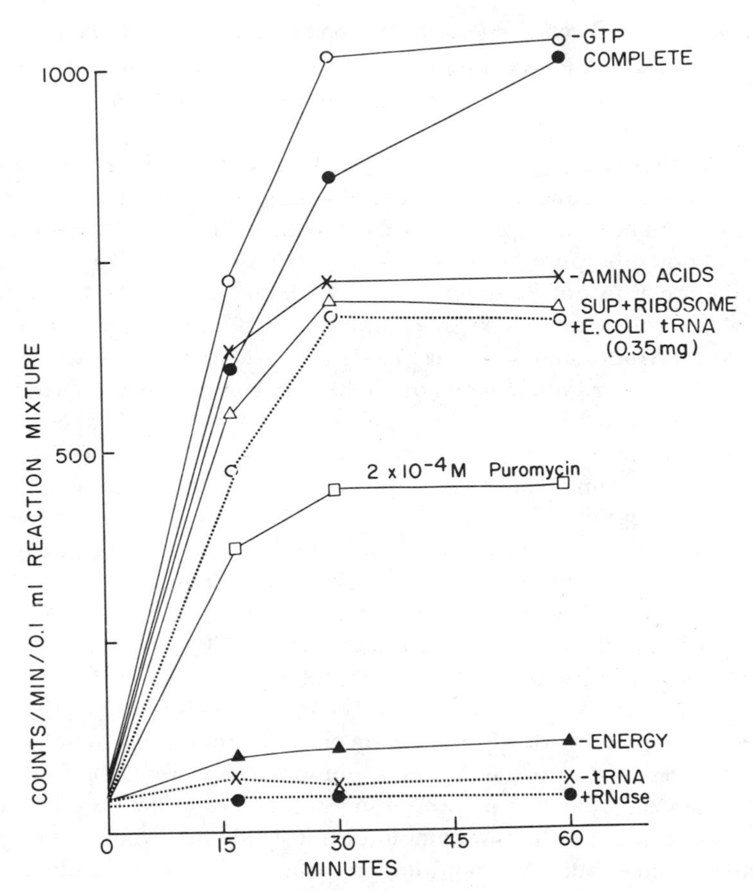


FIGURE 1.8. *Time-course of arginine-$^{14}C$ incorporation into protein by Fraction II (nonribosomal)* (●-●) *as well as by the ribosomal system* (△-△) *(From Kaji, 1968.)*

that rat liver contains a system that incorporates amino acids into a preformed protein acceptor. Both the *E. coli* and the rat liver-soluble amino acid-incorporating non-ribosomal systems are sensitive to puromycin (Fig. 1.8). The incorporation 15 min after the start of the reaction was inhibited only 30% by puromycin. However, when the reaction was allowed to proceed for 60 min, the incorporation of arginine was inhibited more than 50% (□—□). It appears that the initial rate of incorporation of arginine is less sensitive to puromycin than the incorporation that takes place later in the soluble amino acid incorporating system from rat liver. No explanation was given for these observations. There is no requirement for GTP (○—○). The elimination of GTP actually stimulates the incorporation of arginine when compared to the complete system. (●—●) (Fig. 1.8).

### 3. *Inhibition of the Puromycin Reaction by Inhibitors of Protein Synthesis*

A number of studies have been reported in which the puromycin reaction is blocked by the addition of compounds that are known inhibitors of protein synthesis.

Nathans (1964), Traut and Monro (1964), and Rychlík (1966) showed that chlortetracycline and chloramphenicol inhibited the puromycin reaction. When chloramphenicol is present in the protein-synthesizing system, polysome breakdown by puromycin is prevented (Flessel, 1968; Cannon, 1968) (see section on Effect of Puromycin on Polyribosomes, p. 33). This inhibition of the puromycin reaction by chloramphenicol might be explained by the studies of Monro and Vazquez (1967) and Weber and DeMoss (1969). Monro and Vazquez stated that chloramphenicol may act by interfering with substrate binding at the peptidyl site. Weber and DeMoss (1969) examined the kinetics of the puromycin-induced release of peptides from tRNA with and without chloramphenicol using washed *E. coli* ribosomes with radioactive nascent peptides. Twenty-five per cent of the peptide was released by puromycin. Chloramphenicol blocked only 50% of this reaction; therefore, some of the peptides react with puromycin even in the presence of chloramphenicol. The additional puromycin reaction was completely blocked by chloramphenicol when supernatant fraction and GTP were added. Weber and DeMoss concluded that chloramphenicol does not inhibit the peptidyl transferase reaction as determined by the puromycin-induced release of peptides from peptidyl-tRNA, but is involved in inhibiting some step in peptide synthesis prior to this reaction. Similar results were obtained with a cell-free system from *E. coli* by Coutsogeorgopoulos (1969), who showed that, under certain conditions, chloramphenicol inhibited the formation of polyphenylalanine, but it did not interfere with the formation of phenylalanine-containing oligopeptides. Pestka (1969a, 1969b, 1970c) has suggested that chloramphenicol may inhibit protein synthesis by interfering with the binding of the aminoacyl end of charged tRNA to ribosomes. The effect of chloramphenicol on binding of [$^{3}$H]-phenylalanyl-oligonucleotide (Phe-oligonucleotide) and $Mg^{2+}$(20m*M*) to ribosomes is shown in Table 1.4 (Pestka 1969a). It is also possible that chloramphenicol inhibits the binding of peptidyl-tRNA to ribosomes. Pestka (1970a) (Fig. 1.9) has also shown that chloramphenicol is a competitive inhibitor of puromycin in the formation of *N*-acetyl-phenylalanyl-puromycin from *N*-acetyl-phenylalanyl-tRNA and puromycin on ribosomes. The results of Pestka (1969a, 1970a, 1970b, 1970c) directly support the suggestion made by Coutsogeorgopoulos (1966, 1967a, 1967c) that chloramphenicol and other antibiotics exert their action by blocking the binding of the aminoacyl-adenylyl terminus of charged tRNA to ribosomes (Pestka 1970a). Pestka has now compared the effect of these antibiotics on the inhibition of binding of aminoacyl-oligonucleotides to

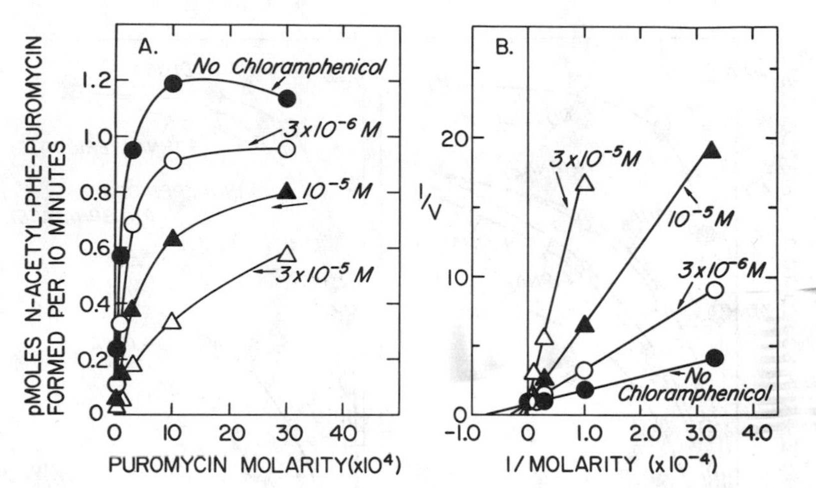


FIGURE 1.9. *A Effect of chloramphenicol on the rate of N-acetyl-phenylalanyl-puromycin formation as a function of puromycin concentration. Each 0.050-ml reaction mixture contained the following components: 0.05 M Tris-acetate, pH 7.2; 0.08 M KCl, 0.14 M ammonium chloride; 0.02 M magnesium chloride; 2.3 $A_{260}$ units of ribosomes; 1.8 pmoles of N-Ac[$^{14}$C]Phe-tRNA (0.14 $A_{260}$ unit); puromycin concentration as given on the abscissa; 0, $3 \times 10^{-6}$, $10^{-5}$, or $3 \times 10^{-5}$ M chloramphenicol where indicated. Reactions were performed at 37° for 10 min. B. Lineweaver-Burk plot of the data in Fig. 9A.* ●, *no chloramphenicol;* ●, *$3 \times 10^{-6}$ M chloramphenicol;* ▲, *$10^{-5}$ M chloramphenicol;* ▲, *$3 \times 10^{-5}$ M chloramphenicol (From Pestka, 1970a.)*

ribosomes with that of the inhibition of binding of intact, aminoacyl-tRNA to ribosomes (Pestka, 1969b; 1969c; 1970b). While there is no inhibition of binding of the intact, aminoacyl-tRNA to ribosomes (Pestka, 1970b), there is complete inhibition of binding of the aminoacyl-oligonucleotides (i.e., GCACCA-phenylalanine; CACCA-leucine; UACCA-leucine; CACCA-methionine; CACCA-valine) (Pestka 1969b, 1970c, private communication). Pestka proposes that the primary effect of these antibiotics as inhibitors of protein synthesis is not observed until the aminoacyl-tRNA is hydrolyzed to the aminoacyl-oligonucleotide. Apparently, the intact, aminoacyl-tRNA has several points of contact with the ribosomes or else the CCA-end of tRNA forms an unstable complex with ribosomes which masks the inhibitory effect of the antibiotics. Therefore, the inhibition of binding to ribosomes is not observed *in vitro* until assays are performed with the aminoacyl-oligonucleotides. Pestka (1970d) has now reported studies on the binding of enzymatically and periodate altered tRNA molecules to 70-S ribosomes and the 30-S subunits. Removal of the 3′-hydroxyl, pA, pCpA and pCpCpA enzymatically resulted in altered tRNA molecules that were capable of binding to 70-S ribosomes (Fig. 1.10A). Their binding

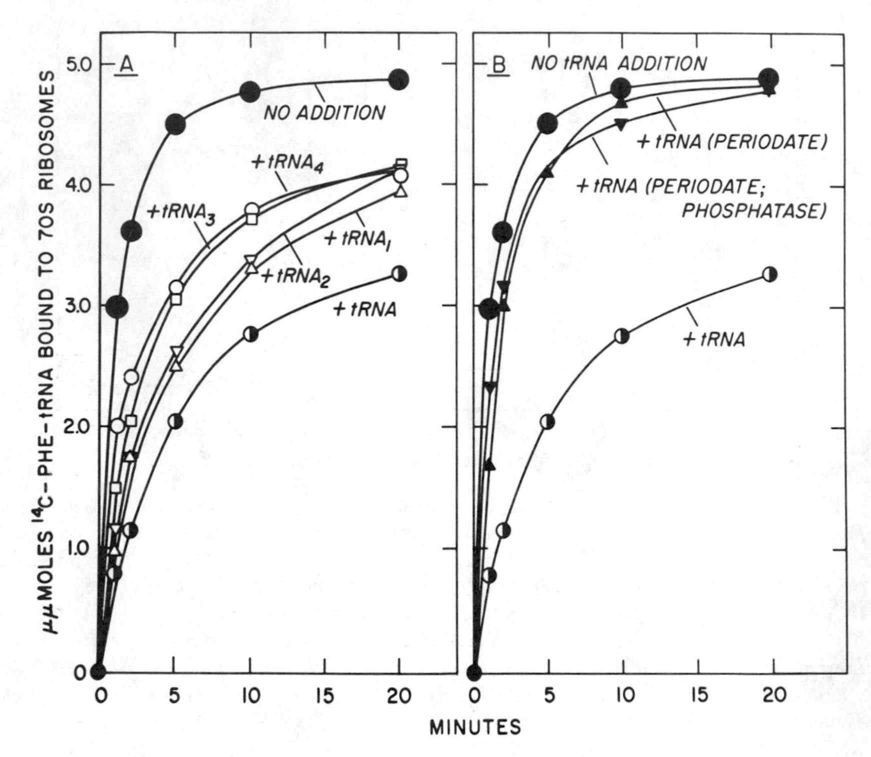


FIGURE 1.10. *Binding of [$^{14}$C] Phe-tRNA to 70S ribosomes in the presence of intact and modified tRNA preparations. Binding of [$^{14}$C] Phe-tRNA to 70S ribosomes was done in* 0.050-*ml reaction mixtures containing the following components:* 5.7 *pmoles of [$^{14}$C] Phe-tRNA* (0.18 $A_{260}$ *unit);* 1.0 $A_{260}$ *unit of the various tRNA preparations; 0.02 M magnesium acetate,* 0.05 *M potassium acetate;* 0.05 *M Tris-acetate, pH* 7.2; 23 *nmoles of base residues of poly U; and* 2.3 $A_{260}$ *units of* 70*S ribosomes. Incubations were performed at* 24°. *A solution of ribosomes and poly U was added last to start the reaction. The deacylated tRNA was treated with snake venom phosphodiesterase or sodium periodate. The tRNA$_1$ was incubated with venom phosphodiesterase for* 15 *min; tRNA$_2$ for* 30 *min; tRNA$_3$ for* 120 *min; tRNA$_4$ for* 420 *min. This treatment resulted in the progressive removal of the terminal pCpCpA-end of deacylated tRNA. tRNA (periodate; phosphatase) refers to periodate oxidized tRNA which was dephosphorylated following periodate oxidation. The data is presented in two similar panels (Fig. A and B) simply to avoid the confusion of too many curves on one graph. For reference two curves are repeated in each panel, namely, no addition (●) and tRNA addition (◑). ●, no additions; ◑, unfractionated tRNA; △, tRNA$_1$; ▽, tRNA$_2$; □, tRNA$_3$; ○, tRNA$_4$; ▲, periodate oxidized tRNA; ▼, tRNA oxidized by periodate, followed by removal of terminal phosphates by alkaline phosphatase (From Pestka,* 1970*d).*

to 30-S subunits, when compared to intact deacylated tRNA, was unimpaired. Periodate oxidation of phenyalanyl-charged tRNA or lysyl-charged tRNA caused a marked reduction in binding to 70-S ribosomes

**Table 1.4.**

Effect of Chloramphenicol on Binding of [$^3$H]phe-Oligonucleotide to Ribosomes[a]

| Chloramphenicol molarity | Per cent of control | *p*moles bound to ribosomes |
|---|---|---|
| 0 | 100 | 0.44 |
| $10^{-6}$ | 79 | 0.35 |
| $10^{-5}$ | 51 | 0.23 |
| $10^{-4}$ | 26 | 0.12 |
| $10^{-3}$ | 15 | 0.07 |

From Pestka, 1969a.

[a] Binding of [$^3$H]Phe-oligonucleotide to ribosomes was determined in 0.050-ml reaction mixtures containing the following components: 0.05 *M* Tris-acetate, pH 7.2; 0.05 *M* potassium acetate; 0.3 *M* ammonium chloride; 0.02 *M* magnesium acetate; 43 nmoles of base residues of poly U; and 5.1 $A_{260}$ units of unfractionated *E. coli* B tRNA. Chloramphenicol was added to the reactions prior to the addition of 4.3 pmoles of [$^3$H]Phe-oligonucleotide (5330 cpm), which was added last to begin the reaction. Each reaction contained 6 $A_{260}$ units of 70S ribosomes. Incubations were performed at 24°C for 10 min and assayed as described by Pestka (1969a). After the reaction, phenylalanine can be recovered as Phe-oligonucleotide and the amino acid is not transferred to a trichloroacetic acid precipitable state.

(Fig. 1.10B). Pestka suggests that periodate oxidizes one or more internal residues of certain tRNA species.

The experimental data obtained by Pestka (1969b; 1970c) in his studies on the inhibition of protein synthesis by use of the phenylalanyl-oligonucleotide reaction (20 m*M* $Mg^{2+}$), strongly suggest that many antibiotics interfere with the binding of phenylalanyl-oligonucleotide to ribosomes by blocking the attachment of the aminoacyl end of aminoacyl-tRNA to the 50-S subunit of ribosomes. Pestka suggests that this may indeed be the manner in which amicetin, gougerotin, sparsomycin, chloramphenicol, vernamycin A, D-WIN-5094, PA114A, streptogramin, tylosin, and spiramycin III exert their mode of action. Puromycin also inhibits the binding of phenylalanyl-oligonucleotide to human ribosomes (Fig. 1.11). It does not form phenylalanyl-puromycin. Pestka (1969b) emphasized that his findings support the idea that the binding of aminoacyl-tRNA to ribosomes is probably a multistep process and many antibiotics may interfere with the specific areas of ribosomes where portions of tRNA molecules are bound. As discussed in Chapter 4 (Sect. 4.1, Gougerotin), there are two additional experiments describing the effects of gougerotin in the binding of intact, charged phenylalanyl-tRNA to ribosomes and or oligopeptide bond formation. They are as

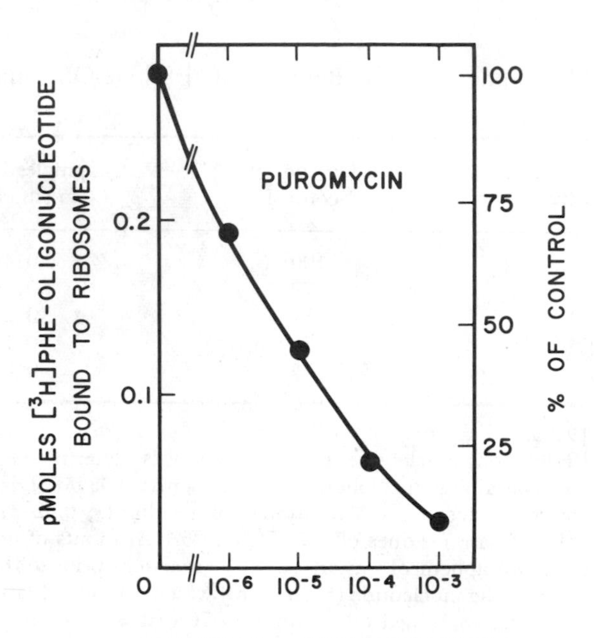


FIGURE 1.11. *Binding of phenylalanyl-oligonucleotide to ammonium chloride washed ribosomes isolated from human placenta. Each* 0.050-*ml reaction mixture contained the following components:* 0.050 *M Tris-chloride, pH* 7.2*;* 0.4 *M KCl;* 0.06 *M* $NH_4Cl$*;* 0.04 *M magnesium acetate;* 4.1 $A_{260}$ *units of ribosomes;* 3.0 *pmoles* [$^3H$] *phenylalanyl-oligonucleotide; reaction*-20 *min*, 24° (*From Pestka, private communication*).

follows: (*1*) gougerotin (4.5 × $10^{-4}$ *M*) has been shown to stimulate the binding of phenylalanyl-tRNA to ribosomes ($Mg^{2+}$ 10 m*M*) (Yukioka and Morisawa, 1969a, 1969b, 1969c), and (*2*) although gougerotin ($10^{-5}$ *M*) blocks polypeptide synthesis, oligophenylalanine synthesis still takes place (Coutsogeorgopoulos, 1969, 1970) (see Chap. 4, Table 4.4). Casjens and Morris (1965) and Gottesman (1967) reported that gougerotin, a nucleoside similar in structure to puromycin, exerted a competitive inhibition upon the puromycin reaction. Their studies suggested that the locus of action of gougerotin was the peptidyl transferase. Clark and Chang (1965) also reported that gougerotin prevented the release of peptidyl materials from peptidyl-tRNA on the ribosome complex. These results suggest that gougerotin inhibits protein synthesis by a mechanism that differs from puromycin. Although both puromycin and gougerotin appear to be structural analogs of aminoacyl-tRNA, the former, but not the latter, can accept the growing peptidyl chain from peptidyl-tRNA. Therefore, gougerotin appears to act as a nonfunctional analog of aminoacyl-tRNA.

Goldberg and Mitsugi (1967) in their studies on the sulfur-containing antibiotic sparsomycin reported that it blocked the puromycin-induced release of polyphenylalanine from ribosomes. It does not affect the binding of polypeptidyl-tRNA to ribosomes. Gougerotin and chloramphenicol, however, inhibited the puromycin reaction at levels almost 200-fold greater than that required for sparsomycin. The kinetic data suggest that sparsomycin is a competitive inhibitor of puromycin, while gougerotin and chloramphenicol act as "mixed" type inhibitors. Although the question of whether or not the two antibiotics act at the same site has not been answered, Monro et al. (1969) presented data to show that sparsomycin competes with puromycin for reaction with the bound peptidyl donor substrate. Once the sparsomycin-induced complex is formed in the fragment reaction with CACCA-leucylacetate, puromycin can not react to form *N*-acetylleucyl-puromycin (step 2; Fig. 1.12). It appears that sparsomycin forms an inert complex between a peptidyl donor and a 50-S ribosomal subunit that is unreactive and non-interconvertible (step 2; Fig. 1.12). Monro et al. (1969) proposed this hypothetical scheme to explain the results obtained with puromycin. The observation of Herner, Goldberg, and Cohen (1969) that sparsomycin stimulates the poly U-directed binding of *N*-acetylphenyl-alanyl-tRNA is consistent with the findings that sparsomycin also stimulates the AUG-directed binding of *N*-formylmethionyl-$tRNA_F$ to *E. coli* ribosomes (Anderson, Clark, and Monro, unpublished results). The effect of sparsomycin on the puromycin reaction is shown in Table 1.5. It appears that the

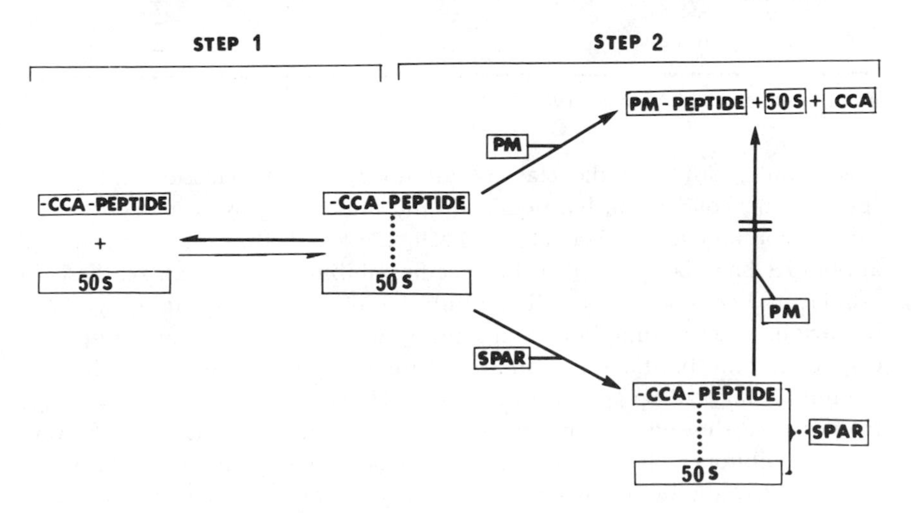


FIGURE 1.12. *Hypothetical scheme to explain inhibition of protein synthesis by puromycin (PM) and sparsomycin (SPAR) (From Monro et al.,* 1969.)

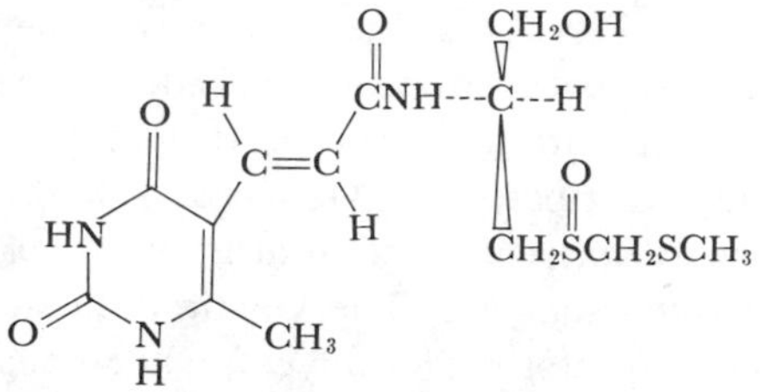

SPARSOMYCIN

primary site of action of sparsomycin is to inhibit peptide bond formation (Jayaraman and Goldberg, 1968; Herner et al., 1969; Igarashi et al., 1969; Monro et al,, 1969) by blocking the binding of the aminoacyl-adenylterminus of charged tRNA to ribosomes (Pestka, 1969b, 1970a, 1970b). Although sparsomycin inhibits protein synthesis, it is not an aminoacyl nucleoside. Wiley and MacKellar (1970) have now reported the structure of this pyrimidine analog as shown.

**Table 1.5**

Sparsomycin Inhibition of Polyphenylalanine Release by Puromycin

| | % Inhibition of release at | |
|---|---|---|
| Sparsomycin, *M* | 5 min | 1 min |
| $8.8 \times 10^{-5}$ | 38 | 52 |
| $2.6 \times 10^{-4}$ | 53 | — |
| $7.9 \times 10^{-4}$ | 80 | — |

From Goldberg and Mitsugi, 1967.

A detailed study of the class of aminoacylaminonucleoside antibiotics (puromycin, gougerotin, blasticidin S, and amicetin) has been reported by Coutsogeorgopoulos (1967a, 1967b, 1969, 1970). All these aminonucleoside antibiotics have been found to be specific inhibitors of protein synthesis in cellular and cell-free systems. The inhibition of polyphenylalanine synthesis by various selective inhibitors of protein synthesis shows that chloramphenicol, gougerotin, blasticidin S, and amicetin are much stronger inhibitors in the initial stages of polypeptide formation. However, incubations at a lower temperature showed that puromycin, gougerotin, and blasticidin S were better inhibitors. The recent studies of Coutsogeorgopoulos (1969, 1970) with gougerotin, blasticidin S, amicetin, and chloramphenicol show that these antibiotics do not inhibit the formation of phenylalanine-containing oligopeptides at concentrations at which they block polyphenylalanine formation (see Chap. 4, Tables 4.4 and 4.5 and discussion on p. 185). From these findings, Coutsogeorgopoulos suggests that modifications are necessary in the current scheme which visualizes these antibiotics as specific inhibitors of the peptidyl transfer reaction (peptide bond formation per se). It may be that these antibiotics are involved in and interfere with steps in protein synthesis which are distinct from the formation of the peptide bond per se. From these new data, it is now becoming obvious that the formation of peptidyl-puromycin in the protein-synthesizing system may be an oversimplication of protein synthesis in that the possible role of substeps in peptide bond formation have not been revealed.

Similar studies using the fragment reaction showed that chloramphenicol, lincomycin, celesticetin, streptogramin A, amicetin, gougerotin, and sparsomycin inhibited the puromycin reaction (Monro and Vazquez, 1967).

Erythromycin is another inhibitor of protein synthesis and of the puromycin reaction. Erythromycin is a macrolide antibiotic that inhibits the transfer of amino acids from aminoacyl-tRNA into polypeptide (Taubman et al., 1963). More recently, Oleinick and Corcoran (1969) reported that erythromycin binds to sensitive and resistant ribosomes from *B. subtilis*. The binding of this macrolide antibiotic involves the 50-S subunit. The evidence to date indicates that erythromycin is bound to the donor site that is normally occupied by polypeptidyl-tRNA. When nascent peptides are stripped from ribosomes by puromycin, the binding of erythromycin increases (Corcoran, unpublished results). It is interesting to note that erythromycin does not inhibit tripeptide synthesis, but does inhibit tetrapeptide synthesis (Tanaka, private communication). Additional experiments now show that erythromycin, like fusidic acid, inhibits translocation through factor G (Cundliffe and McQuillen, 1967; Igarashi et al., 1969; Albrecht et al., 1970). Igarashi et al. (1969) used puromycin under two different experimental conditions (condition A—involves only peptide bond formation; condition B—involves peptide bond formation and translocation).

Cohen et al. (1969a, 1969b) have shown that the stability of the initiation complex is decreased by the interaction of pactamycin with *E. coli* ribosomes to which *N*-acetylphenylalanyl-tRNA is bound. Pactamycin affects the initiation complex on the 30-S ribosomal subunit but does not interfere with the puromycin reaction. Pactamycin causes the release of prebound *N*-acetyl-L-phenylalanyl-tRNA from ribosomes (Fig. 1.13*B*). The pactamycin induced release of prebound *N*-acetyl-L-phenylalanyl-tRNA from the ribosomes is slower than that due to puromycin and is not associated with deacylation (Fig. 1.13*A*). When puromycin and pactamycin are both present, there is a slight increase in the release of radioactivity from the ribosome (Fig. 1.13*B*), while the extent of deacylation is less (Fig. 1.13*A*). This decrease in the puromycin deacylation caused by pactamycin apparently may be attributed to its action on *N*-acetyl-L-phenylalanyl-tRNA binding to ribosomes.

### 4. The Use of the Puromycin Reaction and the Initiation of Protein Synthesis

The puromycin reaction has been used to show that the methionyl-tRNA species that can be formylated (F-met-tRNA$_F$) binds to the peptidyl-tRNA site on the ribosome, while the other methionyl-tRNA (met-tRNA$_M$) species binds to the aminoacyl-tRNA site (Bretscher and Marcker, 1966). The F-met-tRNA$_F$ binds to the donor site and acts as an initiator of protein synthesis even if the methionine is not formylated. Binding to the ribosome occurs with AUG and GUG. This indicates a lack of specificity in the first nucleotide residue (at the 5′-OH end) (Clark and Marcker, 1966). F-met-

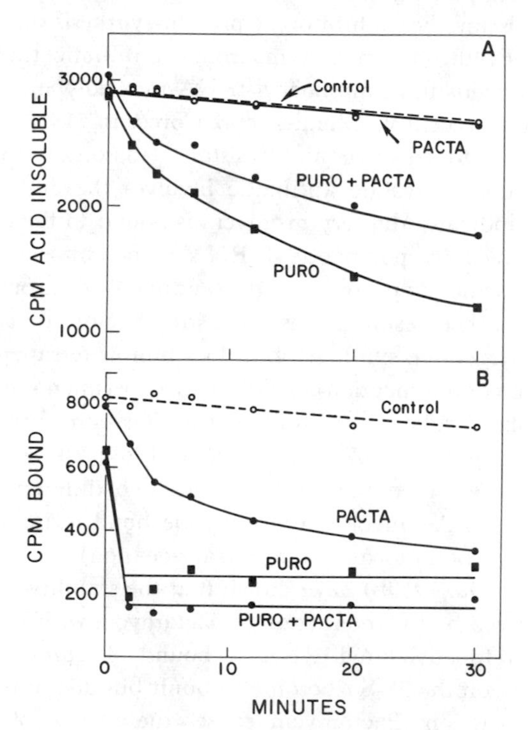


FIGURE 1.13. *Effect of pactamycin and puromycin on N-acetyl-*L*-phenylalanyl-tRNA prebound to ribosomes. Ribosomes were pre-incubated with* [$^{14}$C] *N-acetyl-*L*-phenylalanyl-tRNA for* 10 *min in* 1.5-*ml reaction mixtures. After the pre-incubation, aliquots* (100 *μl*) *were removed for determination of* (1) *binding to ribosomes and* (2) *radioactivity remaining that was precipitable by cold* 5% *trichloroacetic acid* (*zero time*); *pactamycin* ($2 \times 10^{-5}$ *M*) *and/or puromycin* ($4 \times 10^{-4}$ *M*) *were added and incubation was continued for the indicated times when* 100-*μl aliquots were removed for assay. Pactamycin, PACTA; puromycin, PURO.* (*From Cohen et al.*, 1969*a*.)

tRNA$_F$ in the presence of the initiation complex (formed by GMP-PCP, mRNA and initiation factor), while not reacting with puromycin, is bound to the ribosomes (Ohta et al., 1967; Mukundan et al., 1968). Apparently the initiator tRNA is not bound to the peptidyl site on the ribosome under these conditions. However, F-met-tRNA$_F$ must be bound initially to the aminoacyl site on the ribosome since tetracycline (an antibiotic inhibiting the binding of aminoacyl-tRNA to the acceptor site) (Gottesman, 1967) blocks F-met-tRNA$_F$ binding to the ribosomes (Sarkar and Thach, 1968). F-met-tRNA$_F$ in the presence of the ribosome complex (with GTP and initiation factors) does react with puromycin. Therefore, under these conditions, F-met-tRNA$_F$ is bound to the peptidyl site (Fig. 1.4, 1.15) (Brechter and Marcker, 1966; Leder and Bursztyn, 1966; Ohta et al., 1967; Salas et al., 1967). The release of peptides by release factor, R$_1$ in peptide termination (Fig. 1.15) has been shown to be blocked by gougerotin and sparsomycin (Capecchi 1969).

Bretscher and Marcker (1966) used puromycin to study the specificity of tRNA$_F$. They found that F-met-tRNA$_F$, met-tRNA$_F$, and acetyl-met-tRNA$_F$ were sensitive to puromycin when the triplet, AUG, was used. However, initial peptide bond formation is five times as fast when the formylated species is present. F-met-tRNA$_F$ is a better substrate for peptidyltransferase than is met-tRNA$_F$. In addition, Grunberg-Manago et al.

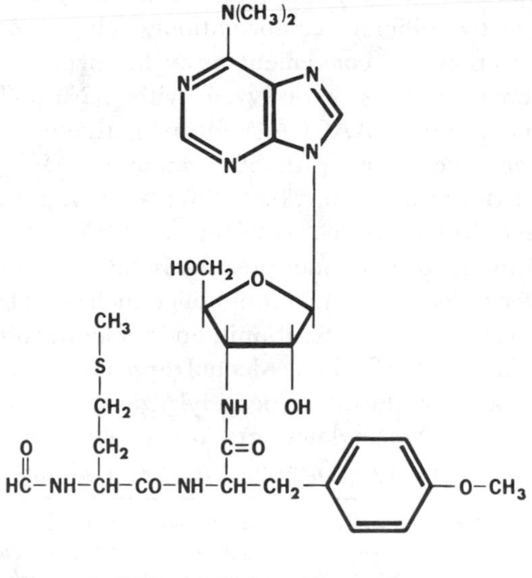

**N-Formyl-Methionyl-Puromycin**

FIGURE 1.14. *Structure for N-formylmethiony l-puromycin.* (*From Livingston and Leder*, 1969.)

(1969) have also shown that the 30-S subunit complex with F-met-tRNA$_F$ and AUG is much more stable than met-tRNA$_F$-AUG. The product of the reaction between F-met-tRNA$_F$ and puromycin is *N*-formyl-methionyl-puromycin (Fig. 1.14). F-met-tRNA$_M$, acetyl-met-tRNA$_M$, and met-tRNA$_M$ were insensitive to puromycin. These data indicate that F-met-tRNA$_F$ acts as an initiator of protein synthesis by its binding to the donor site (peptidyl site) on the ribosomal complex. Bretscher and Marcker (1966) concluded that the specificity lies in the tRNA$_F$. However, more recently, Rudland et al. (1969) presented evidence that both the ribonucleic acid moiety and the presence of the formyl group in F-met-tRNA$_F$ are recognized by bacterial initiation factors. From the general scheme in Figures 1.4, 1.15 for protein synthesis, it appears that the tRNA$_F$ species with its amino acid (methionyl, or an *N*-substituted methionine) is bound to the donor site. It then reacts with an aminoacyl-tRNA on the acceptor site. The cycle of peptide bond formation is repeated following the translocation of the peptidyl-tRNA from the acceptor to the donor site. The molar ratio of F-met-tRNA$_F$ to the AUG triplet is 1:1 (Bretscher, 1968a). When the puromycin reaction is carried out at 0°C (5m*M* $Mg^{2+}$), F-methionylpuromycin is removed from the ribosome, but the triplet (AUG) is not removed.

Miskin et al. (1968) used the fragment reaction (formyl-methionyl-hexanucleotide) and puromycin to demonstrate that peptidyl transferase (the enzyme responsible for peptide bond formation) on the ribosome from *E. coli* can exist in two different conformations, only one of which is active. The fragment reaction is a convenient assay for peptide bond formation. *N*-Formylmethionyl-tRNA is hydrolyzed with RNase $T_1$ to *N*-formyl-methionyl-hexanucleotide (CAACCA-*N*-formylmethionine). The 3′-terminal fragment reaction takes place with 50-S subunits, $Mg^{2+}$, $K^+$, and 30% ethanol at 0°C (Monro and Marcker, 1967; Monro, 1967; Monro et al., 1968; Maden and Monro, 1968; Hille et al., 1967; Monro et al., 1969; Monro, 1969). Ethanol or methanol plus the isolated 50-S subunits catalyze a peptidyl transfer of the formylmethionyl-oligonucleotide fragment reaction to puromycin. The template, 30-S subunit, and intact peptidyl-tRNA are not needed for the fragment reaction. Alcohol appears to have the effect of activating the normally dormant peptidyl transferase of 50-S subunits (Monro, 1969). Only *N*-acetylated or *N*-formylated transfer RNAs are active as peptidyl donors in the fragment reaction. The active form of ribo-

---

FIGURE 1.15*A*. *Schematic outline of steps in peptide chain initiation. Symbols:* $n-2$, $n-1$, $n$, $n+1$, $n+2$, *a series of adjacent codons in the mRNA segment shown; n, initiator codon; "bottomless bracket", tRNA*$^{fMet}$*;* $T_F$, *N*$^{10}$*-formyltetrahydrofolate-Met-tRNA*$_F$*-transformylase; fTHF, N*$^{10}$*-formyltetra-hydrofolate; THF, tetrahydrofolate; IF*$_s$*, initiation factors;* $E_{Met}$, *Met-tRNA synthetase. The oval shapes represent ribosomal subunits. A and P indicated in the* 50*S ribosomal subunit are hypothetical tRNA binding sites.*

*B*. *Schematic outline of steps in peptide chain elongation and termination. Symbols:* $S_1$, $S_2$, $S_3$,

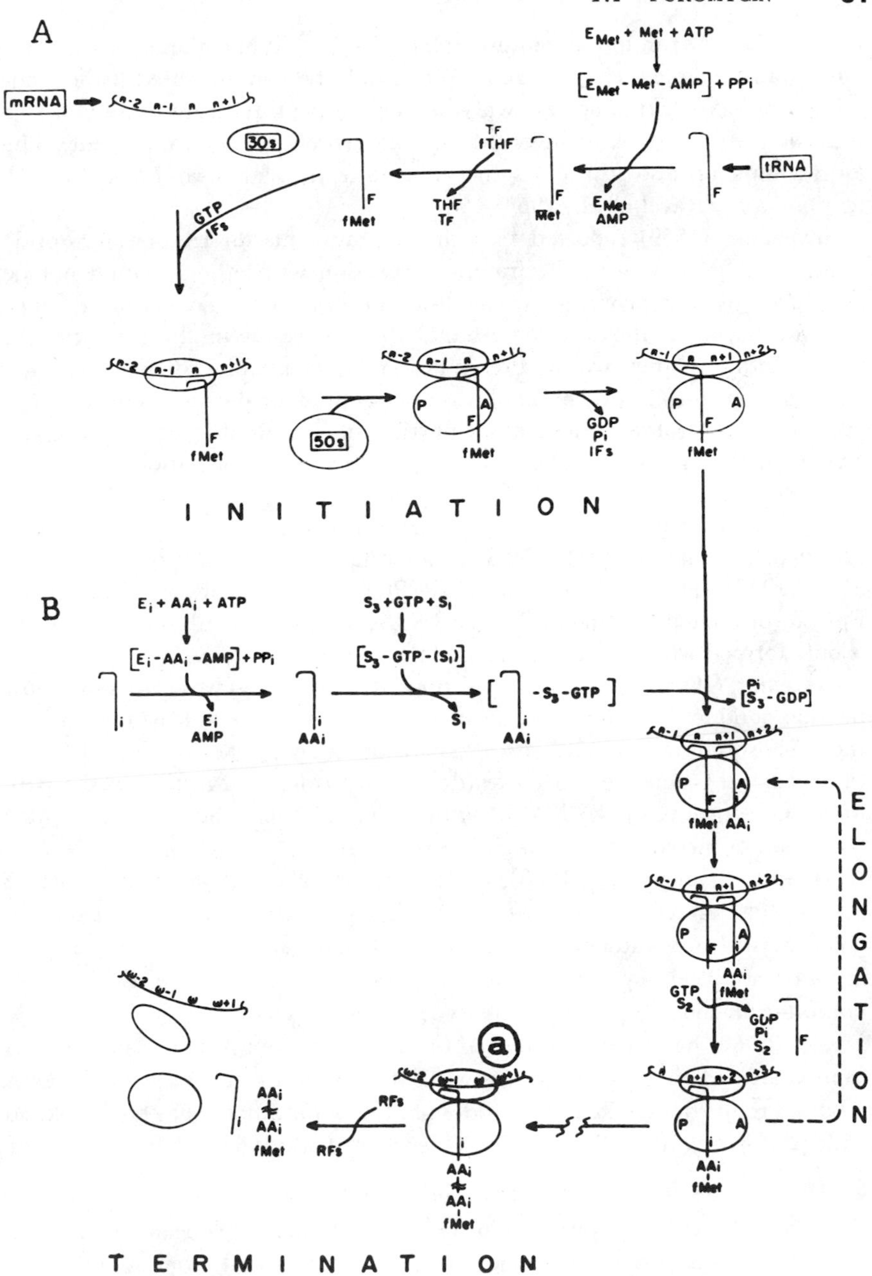


*elongation factors; "bottomless bracket", tRNA accepting $AA_i$; $\omega - 2$, $\omega - 1$, $\omega$, $\omega + 1$, a series of adjacent codons in the mRNA segment shown; $\omega$, a terminator codon; $RF_s$, release factors. In a (circled), a polypeptidyl residue is attached to the tRNA. (From Lengyel and Söll, 1969).*

somal-peptidyl transferase requires $NH_4^+$ or $K^+$. When these ions are removed and readded, enzyme activity can only be demonstrated if the ribosomes are heated. It is not known whether the peptidyl transferase alone or part of the ribosome is involved in this structural rearrangement. The ammonium ion also stimulates the enzymatic transfer of aminoacyl-tRNA to ribosomes (Ravel et al., 1968).

Silverstein (1969) reported that the requirements for ribosomal peptide bond-forming activity in the fragment reaction with ethanol could not be explained by the provision of medium of low dielectric constant, since substitution of 1,4-dioxane for ethanol did not result in the formation of formylmethionyl-puromycin. Precipitation or coprecipitation of ribosomes with aminoacyl-tRNA by ethanol was not essential for the fragment reaction. It may be that organic activators participate by binding at the catalytic center on the ribosome or enzyme and induce a conformation more conducive to catalysis.

The importance of GTP in the formation of F-met-puromycin has been reported by Ohta et al. (1967), Hershey and Thach (1967), and Anderson et al. (1967). They showed that GMP–PCP can substitute for GTP in the $F_1 + F_2$ promoted binding of F-met-$tRNA_F$, but that F-methionylpuromycin is only formed when GTP is added to the *in vitro* system.

Bretscher (1968b) has recently proposed a general scheme to show how peptide bond formation occurs. It appears that there are at least two sites on the 30-S subunit. One site is for the recognition of the aminoacyl-tRNA (acceptor site), and the other site (donor site) is for recognition of the peptidyl-tRNA or initiator-tRNA. Mukundan et al. (1968) showed that peptide formation occurred after adding 50-S subunits.

Bachmayer and Kreil (1968), Smith and Marcker (1968), and Schwartz et al. (1967) have shown that F-met-$tRNA_F$ is found in other bacteria, chloroplasts, and mitochondria from yeast and rat liver and may be the initiator of protein synthesis in all 70-S-type ribosomes. F-met-$tRNA_F$ was not found in the cytoplasm of eukaryotic cells.

Kim (1969) has recently reported on the isolation of *N*-formylseryl-tRNA from yeast and *E. coli*. It is not known if this *N*-substituted aminoacyl-tRNA is bound to the donor site of the ribosome. Lactylpuromycin has been isolated with protoplasts from *S. cereviseae* (Guenther *et al.*, 1968).

### 5. N-Formyl-methionyl-puromycin, a Substrate for Deformylase

*N*-Formyl-methionyl-puromycin (Fig. 1.14) has also been used as a substrate for the isolation of the enzyme (deformylase) from *E. coli*. The deformylase rapidly deformylates formyl-methionyl-puromycin (Fig. 1.16). Such activity is consistent with the redundancy of the formyl group in the subsequent steps of polypeptide elongation. This enzyme does not deformy-

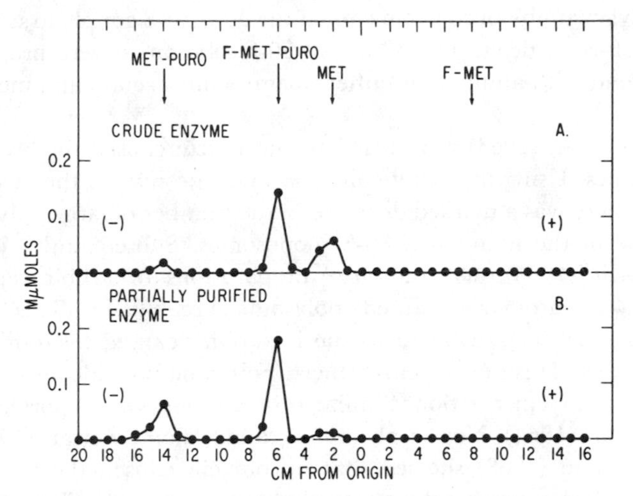


FIGURE 1.16. *Effect of crude and partially purified deformylase on the degradation of F-[$^{14}$C]Met-puromycin. (A) Crude enzyme; (B) partially purified enzyme. (From Livingston and Leder*, 1969.)

late formylmethionine, formylleucine, or formylmethionyl-tRNA. Deformylase rapidly removes the formyl group from formylleucylpuromycin. These findings are consistent with the notion that deformylation occurs only after transfer of formylmethionine to an incoming aminoacyl-tRNA and the formation of initial peptide bond during protein synthesis (Livingston and Leder, 1969).

### 6. Effect of Puromycin on Polyribosomes

Warner et al. (1966) utilized the two protein inhibitors cycloheximide and puromycin to study ribosomal RNA synthesis and ribosome maturation in HeLa cells. They reported that ribosomal RNA synthesis can be initiated and ribosomes can be completed in the presence of cycloheximide, which blocks 99% of protein synthesis. The results imply that a functional pool of ribosomal protein must exist. However, the effect with puromycin is entirely different. While cycloheximide blocks protein synthesis, but does not affect the maturation of ribosomes, puromycin inhibits protein synthesis and also effectively interrupts the completion of ribosomes. Judging from the results obtained with cycloheximide, some action other than the interruption of protein synthesis must be responsible for this particular action of puromycin. Kohler, Ron, and Davis (1968) showed that only 70-S particles were found as the products of polysome run-off by *E. coli* cells treated with puromycin. Treatment of the 70-S particle *in vitro* with puromycin did not result in the dissociation into subunits. Schlessinger et al. (1967) reported that release of the peptidyl chain by puromycin caused the 30-S 50-S couple to separate into 30-S and 50-S particles. The 30-S and 50-S ribosomes were prepared from *E. coli* ground with alumina in buffer, magnesium acetate and mercaptoethanol.

Hardesty et al. (1963) reported that puromycin caused the breakdown of polyribosomes. Using intact reticulocytes and puromycin, they were able to show that there was a marked decrease in the number of large polysomes and an increase in the number of 80-S monosomes. Subsequently, Williamson and Schweet (1965), with cell-free preparations of rabbit reticulocytes, showed that puromycin caused polysome breakdown. This polysomyl breakdown was not attributed to the instability caused by removal of the peptide chains. This puromycin-induced polysome breakdown is dissociable from the puromycin reaction. Similar observations were reported by Villa-Trevino et al. (1964), Noll et al. (1963), and Colombo et al. (1965). More recently, Bishop (1968) showed that puromycin caused the breakdown of polysomes when washed rabbit reticulocytes were used. There was a considerable increase in the amount of ribosomal monomers, but no increase in subribosomal particles. The incorporation of phenylalanine with either the ribosomal monomers of subribosomal particles was reduced. These results do not support the hypothesis that subribosomal particles are obligatory initiators of polypeptide chain synthesis. When the ribosomes and ribosomal monomers and the puromycin-treated cells were measured, the polysomes decreased 24% and the ribosomal monomers increased 44%. The amount of subribosomal particles was essentially unchanged. When puromycin was removed and the cells were incubated, polysomes formed rapidly, but the amount of free monomer decreased. Bishop concludes that the evidence accumulated to date argues against the hypothesis that subribosomal particles are responsible for chain initiation. He states that the evidence is not conclusive. There is the possibility that a limited pool of subribosomal particles is maintained and the monosomes pass rapidly through this pool when protein synthesis is resumed.

Yoshikawa-Fukada (1967) used puromycin to study the intermediate state of ribosome formation in animal cell in culture. The presome, the precursor of ribosomal RNA, accumulated in the nucleus when puromycin was present. When the antibiotic was removed by washing, presomes were converted to mature ribosomes. The results indicate that maturation of ribosome particles and the conversion of $q_1$RNA to ribosomal RNA, requires new protein synthesis. The conclusion is drawn that presomes represent a critical stage in the synthesis of ribosomes. Flessel (1968) studied the effect of puromycin on polysome breakdown in *M. lysodeikticus*. They concluded that polysome breakdown by puromycin results in the formation of 70-S ribosomes.

Since puromycin arrests the sedimentation profile of the high molecular weight RNA in cells, Tamaoki and Lane (1968) made an analytical comparison of the RNA of L cells incubated in the presence and absence of puromycin. They studied the methylation of sugar and bases in ribosomes and rapidly labeled ribonucleates. The RNA from puromycin-treated L cells displayed a proportionately greater labeling of adenosine than that of guanosine. However, the $O^{2'}$-methylribose content of ribonucleates from puromycin-treated cells was only slightly lower than that of ribonucleates from normal cells. Ribonucleates from puromycin-treated cells contained a larger proportion of fast-sedimenting ribonucleates than did ribonucleates from the untreated L cells.

Polyribosome disaggregation occurs by the selective omission of tryptophan from an otherwise complete medium at the time of hemoglobin synthesis (Hori et al., 1967; Freedman et al., 1968). Since it causes the premature release of the growing peptide chain in intact cells by competing with aminoacyl-tRNA for the peptidyl transferase, puromycin was used to determine whether, by substituting for a tryptophanyl-tRNA in tryptophan-deficient cells, ribosomal movement past the locus of tryptophan codons on mRNA could be promoted and normal polyribosome distribution thereby be restored. The addition of puromycin to rabbit reticulocyte ribosomes at the time of tryptophan deficiency prevented polyribosome disaggregation and also removed 80% of the nascent polypeptide chain. In addition, where polyribosome disaggregation had already occurred, puromycin brought about regeneration to the normal polyribosome pattern, even though the nascent chains were still being removed. It appears, therefore, that the action of puromycin on polyribosome may be specific for sites bearing peptidyl-tRNA and that under conditions of the experiments described, puromycin can substitute for tryptophanyl-tRNA to promote ribosomal movement. Laird and Studzinski (1968) concluded that protein synthesis is necessary for stable ribosome formation in the nucleolus.

Möller (1969) has recently proposed a structural model to explain the interaction among ribosomal components, tRNA, and mRNA during translation. The proposed outline in Figure 1.17*a* consists of protein and RNA organized in the 70-S ribosome. This model directs the tRNAs to a common peptide bond formation site. The model proposes that 40 substructures are present in the 70-S ribosomes and arranged cylindrically around a central axis. This axis is shared by two subparticles with 13 substructures in the 30-S ribosome and 26 in the 50-S ribosome. There are two predominantly double-helical RNA areas and a nonhelical (G rich region) area of 30 nucleotides. These interact preferentially with proteins which protect against nuclease. Another 10–20 nonhelical residues link the double-helical regions. Möller suggests that each substructure is composed of one protein and one RNA

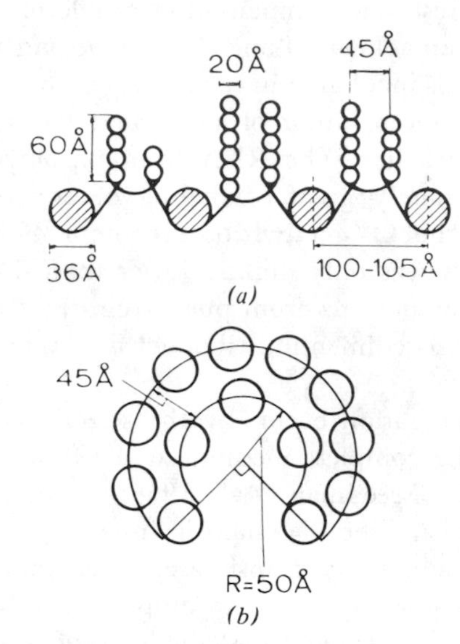


FIGURE 1.17. (*a*) *Proposed outline of part of the unfolded ribonucleoprotein strand of the ribosome from E. coli. The proteins are assumed to be globular in situ with an average diameter of* 36 A° *on the basis of a molecular weight, $MW$ =* 23,000, *and of a partial specific volume, $V$ =* 0.74. *Average length of the double-helical RNA regions is assumed to be* 60 A° *(corresponding with two turns of a double-stranded RNA helix); width of the double-helical RNA regions,* 20A°. *Dotted lines indicate proposed width of one substructure* (100–105 A°). (*b*) *View of the proposed ribosome model along the central channel of the particle. The inner arc contains five substructures and the outer eight. The gap defined by the angle subtended at the center is* 90°, *with the central hole having a diameter of about* 100 A°. *The growing polypeptide chain is assumed to be present in the central channel of the* 50*S particle and thus protected from protease.* (*From Möller*, 1969.)

chain of 120–130 nucleotide residues. About 60% of these are base-paired (Fig. 1.17*a*). The completed model is shown in Figure 1.17*b*. In view of Möller's structural model, the next logical step will be to determine the location of the sites that bind puromycin during polypeptide formation.

## II. Use of Puromycin for Studying Biochemical Reactions *in vivo* and *in vitro*

### 1. Phosphorylation of Precursor Protein

Puromycin has been used to determine the sequence for phosphorylation of casein. Singh et al. (1967) studied the effect of puromycin on the phosphorylation of precursor protein in lactating rat mammary-gland slices. When puromycin was added to mammary-gland slices, the incorporation of leucine into protein was inhibited 81%. However, the incorporation of inorganic-$^{32}$P was not inhibited in a 30-min period. It appears that a small pool of unphosphorylated or incompletely phosphorylated casein exists in the cell and that phosphorylation of this protein can occur once protein synthesis is inhibited by puromycin.

### 2. Antibody Production

Harris (1968) studied the effect of puromycin on antibody formation in spleen cell suspensions from rabbits immunized to sheep red blood cells. At high concentrations, puromycin suppressed antibody-producing cells when added to isolated cells. When puromycin was added after several days in culture in the presence of sheep red blood cells, the surviving suppressed antibody-producing cells were resistant. Suppressed antibody-producing cells that were stimulated by puromycin did not incorporate precursors for either protein synthesis or RNA synthesis. These results were taken as evidence that the stimulation of antibody production by suppressed antibody-producing cells *in vitro* occurs by interfering with represser mechanisms or by stimulating the completion of antibody molecules.

### 3. Interferon Inducers

Puromycin has been used to study the *in vitro* effect of interferon production with several naturally occurring and synthetic inducers of interferon in various cell types (Finkelstein et al., 1968). They reported that interferon can be induced by diverse agents in mammalian cells. This induction of interferon apparently takes place by two mechanisms. In one mechanism, the early appearance of interferon (2–10 hr) is relatively resistant to inhibition by puromycin. A second mechanism results in a late appearance of interferon. This second mechanism is more sensitive to inhibition by puromycin. Burke (1966) used puromycin and fluorophenylalanine to establish the existence of early protein intermediates that were essential for viral-induced interferon formation.

### 4. Effect of Puromycin on the Diphtheria Toxin–NAD Reaction

The puromycin reaction has recently been used by Maxwell and coworkers to elucidate the mechanism of action by which diphtheria toxin and nicotinamide adenine dinucleotide (NAD) inhibit protein synthesis. It has been established that diphtheria toxin inhibits mammalian protein synthesis in cell-free systems and in cultured cells. Collier and Pappenheimer (1964) and Collier (1967) reported that NAD was absolutely required for this inhibition. The toxin specifically inactivates the transferase II. Honjo et al. (1968) subsequently showed that the ADP-ribose portion of NAD was transferred to transferase II. The linkage of ADP-ribose to transferase II appeared to be covalent. Schneider et al. (1968) showed that diphtheria toxin and NAD prevented the puromycin reaction by crude soluble factors or partially purified transferase II. The peptidyl-puromycin reaction occurring following the addition of transferase II was not inhibited by either diphtheria toxin or NAD alone. Everse et al. (1970) have shown the formation of a ternary complex between diphtheria toxin, aminoacyltransferase II, and diphosphopyridine nucleotide.

### 5. Vaccinia Virus

Bablanian (1968) reported that puromycin prevented the early vaccinia virus-induced rounding of infected monkey kidney cells and prevented virus multiplication. At lower concentrations, puromycin did not prevent the cytopathic effects of vaccinia virus, even though virus multiplication was still inhibited. After 4 days, cell division was not altered. Upon removal of puromycin from the infected cultures, virus-induced cell damage began almost immediately. After 1 hr all cells were rounded.

### 6. Carbohydrate Metabolism

The action of puromycin on carbohydrate metabolism in fat cells differed depending on whether glucose or fructose was the substrate. When glucose was the substrate, basal uptake was greatly reduced with puromycin, but the amount of additional glucose utilized in the presence of insulin was not significantly different from that of controls without puromycin. The opposite situation was observed with fructose (Kuo, 1968). With puromycin, fructose utilization in isolated adipose cells was inhibited. The results obtained indicate that glucose and fructose uptake may not be effected by identical mechanisms. Puromycin also blocked glucose uptake by rat hemidiaphragms, but did not affect the glycogen content (Gershbein, 1966).

### 7. Amino Acid Transport

To determine the effect of puromycin on amino acid transport, Elsas and Rosenberg (1967) added puromycin to rat kidney cortex slices. They observed that $\alpha$-amino-isobutyric acid and glycine uptake was inhibited when kidney cortex slices were preincubated with puromycin. With no preincubation, there was no interference of amino acid uptake. The authors concluded that puromycin probably interferes with the synthesis of those peptides that catalyze amino acid transport. More recently, Elsas et al. (1968) showed that puromycin inhibited the uptake of amino acids by intact diaphragm muscle of male rats. The antibiotic was added 120 and 180 min before the addition of the amino acids. Puromycin inhibited amino acid transport, but did not change the tissue water spaces, oxygen consumption, the concentration of $\alpha$-amino nitrogen in the incubation medium, or the diffusion constant for $\alpha$-amino-isobutyric acid. Puromycin also reduced the stimulation of amino acid transport by insulin. The studies also showed that puromycin inhibited the synthesis of a rapidly turning over protein or proteins that are involved in the transport of amino-isobutyric acid (Elsas et al., 1968).

### 8. *Biosynthesis of Hyaluronic Acid*

Stoolmiller and Dorfman (1969) studied the effect of puromycin on the biosynthesis of hyaluronic acid with washed, suspended group A streptococci. Puromycin (500 μg/ml) completely blocked protein synthesis, but did not affect the biosynthesis of hyaluronic acid as measured by the incorporation of acetate-$^3$H. Thus, hyaluronic acid synthesis in streptococci does not appear to be dependent on concomitant protein synthesis. These findings are in contrast to the earlier report showing that puromycin-treated tissue minces inhibited both protein and polysaccharide synthesis. Chondroitin sulfate and hyaluronic acid synthesis are normal. Hurler fibroblasts grown in tissue culture are also inhibited by puromycin (Matalon and Dorfman, 1968). The lack of inhibition of hyaluronic acid synthesis by puromycin in streptococci indicates that this system is different with respect to a requirement of concomitant protein synthesis.

### 9. *Lipid Synthesis and Acid Mucopolysaccharides*

The inhibition of lipid synthesis in chick aortic cells by puromycin was reported by Murata et al. (1967). When puromycin (10–100 μg/ml) was added to chick aortic cells, the incorporation of acetate-1-$^{14}$C into lipids was inhibited. The authors stated that the mechanism for this inhibition was not clear. Puromycin also inhibited the biosynthesis of the acid mucopolysaccharides. De la Haba and Holtzer (1965) also reported that puromycin blocked the synthesis of the acid mucopolysaccharides when added to cartilagenous vertebral column from 10 day old chick embryos.

### 10. *Inhibition of 3′,5′-Cyclic AMP Phosphodiesterase*

Appleman and Kemp (1966) studied the effect of puromycin on cyclic-3′, 5′-adenylic acid (cyclic AMP) in rat hemidiaphragm *in vitro*. Puromycin was shown to increase the levels of cyclic-3′,5′-adenylic acid. This effect was attributed to the competitive inhibition of 3′,5′-cyclic AMP phosphodiesterase.

### 11. *Mitosis*

HeLa cells have been shown to enter and complete mitosis in the presence of puromycin (Buck et al., 1967). The daughter cells formed in the presence of puromycin contained complete nuclear membranes. The results obtained show that all cellular constituents necessary for initiation and completion of mitosis are produced prior to the onset of mitosis.

### 12. *Glycogen Synthesis*

Sövik (1967) studied the effect of puromycin on glycogen biosynthesis. Puromycin suppressed incorporation of glucose-$^{14}$C into glycogen. Glycogen synthetase and phosphorylase were measured in muscle tissue. The results showed that puromycin suppressed the activity of glycogen synthetase I. Phosphorylase activity was not affected.

### 13. Protein Synthesis in Plants

Allende and Bravo (1966) studied the effect of puromycin at various stages of germination. Cell-free systems obtained from ungerminated wheat embryos incorporated amino acids into polypeptides using either free amino acids or aminoacyl-tRNA. Puromycin inhibited the incorporation of phenylalanine.

More recently, Kulaeva and Klyachko (1967) studied the effect of puromycin on protein synthesis in *Nicotiana tabacum* leaf discs. Puromycin inhibited protein synthesis in chloroplasts, mitochondria, ribosomes, and the supernatant fraction of cells.

### 14. Effect of Puromycin on Memory

Several laboratories have been concerned with the inhibitory effect of puromycin on memory. Flexner, Flexner, and Stellar (1963) showed that a bitemporal injection of puromycin caused loss of memory in mice. The loss of memory is temporary, as shown by small intracerebral injections of saline up to two months following the experiment (Flexner and Flexner, 1967a). It appears that injection of puromycin interferes with retrieval, but does not alter the process that maintains the basic memory trace in mice. In two subsequent reports Flexner and his co-workers (Flexner and Flexner, 1968; Gambetti et al., 1968a) reported on the presence of peptidyl-puromycin in brain tissue and on the effect of puromycin on the alteration of neuronal mitochondrial membranes. They were able to show that puromycin-$^3$H injected into the brain of mice was retained for 16–18 days and peptidyl-puromycin-$^3$H could be isolated. These observations support their working hypothesis that puromycin reacts to form peptidyl-puromycin which interacts with neuronal membranes, particularly synaptic membranes. This conclusion is consistent with the observation that memory disappears for a period of 10–20 hr when peptidyl-puromycin is at a maximum. Puromycin does not affect memory in the presence of heximide, a known inhibitor of the formation of peptidyl-puromycin. Finally, peptidyl-puromycin alters neuronal mitochondrial membranes and they may well affect other neuronal membranes. Barondes and Cohen (1967) reported that intracerebral injection of puromycin into mice blocked cerebral protein synthesis and caused a marked impairment of memory 3 hr after training. This effect was not based on inhibition of protein synthesis required for memory storage. The authors concluded that protein synthesis in consolidation of memory in mice apparently does not play an important role. Peptidyl-puromycin complexes also cause neuronal mitochondrial swelling (Gambetti et al., 1968b). This swelling is related to a specific action of puromycin on ribosomal protein synthesis.

It appears that the type cation used with puromycin is important in interfering with the puromycin reaction. For example, puromycin dihydrochloride neutralized with KOH, LiOH, $Ca(OH)_2$ or $Mg(OH)_2$ fails to block the expression of memory of mice (Flexner and Flexner, 1969). This observation is in contrast to earlier studies in which memory of maze learning in mice was blocked when NaOH was used to neutralize puromycin dihydrochloride (Flexner et al., 1963). This failure may be due to cationic binding at anionic membrane sites. This assumed binding may result in exclusion of enough peptidyl-puromycin to reduce the effectiveness in blocking memory.

Agranoff and his co-workers have reported extensive studies on the effect of puromycin on memory fixation in goldfish (Agranoff and Klinger, 1964; Davis et al., 1965; Agranoff et al., 1965; Agranoff et al., 1966). The effect of puromycin on memory was studied by measuring the memory retention of shock avoidance in goldfish. Puromycin produced partial memory deficit when injected into the goldfish immediately after 20 trials, but had no effect when given 2 hr later (Davis et al., 1965). When the dose of puromycin injected intracranially varied between 0 and 20 μg, it was observed that puromycin could obliterate memory of the response on day 4 (Fig. 1.18)

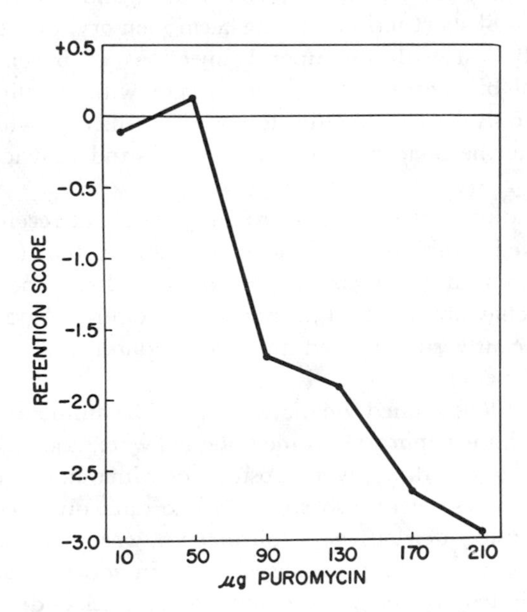


FIGURE 1.18. *The effect of memory measured on day* 4 *of different amounts of puromycin administered immediately following the trials on day* 1. *The retention scores obtained with* 90 *μg or more of puromycin represent significant memory deficits;* 170 *and* 210 *μg produced complete memory deficits.* (*From Agranoff et al.*, 1965.)

(Agranoff et al., 1965). When 170 μg of puromycin was injected immediately after trial 20 on day-1, there was a complete memory deficit on day 4. A 60-min delay in injecting puromycin after trial 20 did not produce a detectable effect on memory (Agranoff et al., 1966). Puromycin aminonucleoside or *O*-methyl-tyrosine were not effective. When puromycin is given shortly before the learning process, there is the usual increase in correct responses, but memory on day 4 is deficient. Thus, puromycin does not seem to affect short-term memory. It appears to act specifically on a process by which long-term memory is fixed. When long-term memory is fixed in a goldfish, it does not appear to turn over and appears to be stored in a metabolically inert form (Agranoff et al., 1966). These findings are in contrast to the reports by Flexner and his co-workers following the bilateral temporal injection of puromycin into mice (Flexner et al., 1965; Flexner et al., 1963). They showed that short-term memory can be destroyed by the injection of puromycin 3–6 days after a learning experience. Older memory in mice could be destroyed 6 weeks after learning by injection of puromycin into bilateral frontal and temporal ventricular sites.

The effects of different sites of injection of puromycin into mice on short-term and long-term memory are shown in Table 1.6 (Flexner et al., 1967b). Retention tests show that temporal, ventricular, and frontal injections resulted in the loss of short-term and long-term memory. Short-term memory was consistently lost with bitemporal injections of puromycin (90 μg). Bilateral frontal or ventricular injections were without effect. Long-term memory could only be consistently destroyed in mice in which puromycin was injected into the bilateral temporal, frontal, and ventricular portion of the brain.

Analysis of radioactivity in brain protein of fish that received puromycin showed a marked inhibition of $^{3}$H-leucine incorporation (Agranoff et al., 1966). The authors did not claim a simple correlation between the biochemical and behavioral effect of puromycin in goldfish. Agranoff and Davis (1968) have recently summarized the use of puromycin in studying the development of memory.

Wilson et al. (1960) studied the metabolism of the aminonucleoside moiety of puromycin and the nephrosyndrome obtained when injected into rats. The phosphorylation of the dimethyladenosine derivative and the inhibition of yeast adenosine kinase were also studied. The data obtained showed that the aminonucleoside derivative of puromycin was phosphorylated by a dialyzed yeast extract in the presence of ATP. In addition, yeast adenosine kinase was inhibited by the dimethyladenosine nucleoside. Several possibilities were advanced to explain why the aminonucleoside produced a nephrotic syndrome in rats, but not in guinea pigs or rabbits. It was concluded that further experiments were needed to demonstrate an alteration

**Table 1.6**

Effects of Different Sites of Injection of Puromycin on Short and Longer-Term Memory[a]

| Puromycin injections | | | No. of mice in which memory was | | |
|---|---|---|---|---|---|
| Site | Days | Dose, mg | L | I | R |
| | | *Short-term memory* | | | |
| T + V + F | 1 | 0.03–.06 | 7 | 0 | 0 |
| T | 1 | 0.09 | 10 | 0 | 0 |
| V | 1 | 0.09 | 0 | 0 | 5 |
| F | 1 | 0.09 | 0 | 0 | 5 |
| V + F | 1 | 0.09 | 0 | 1 | 2 |
| | | *Longer-term memory* | | | |
| T + V + F | 11–60 | 0.03 | 17 | 2 | 0 |
| T | 11–35 | 0.06–.09 | 0 | 0 | 7 |
| V | 12–38 | 0.06–.09 | 0 | 0 | 3 |
| F | 16–27 | 0.06–.09 | 0 | 0 | 3 |
| V + F | 28 | 0.06–.09 | 0 | 2 | 2 |
| V + T | 28–43 | 0.09 | 1 | 1 | 2 |
| T + F | 28 | 0.09 | 0 | 0 | 3 |

From Flexner et al., 1967b.

[a] L, lost; I, impaired; R, retained; Days, days after learning. T, V, and F refer, respectively, to temporal, ventricular, and frontal injections, all given bilaterally. For the mice with loss of memory, the means and standard deviations for percentages of savings of trials and of errors were respectively 1 ± 3 and 2 ± 6; for those with impaired memory, 26 ± 29 and 39 ± 12; for those with retention of memory, 90 ± 14 and 90 ± 9.

in the nucleotide profile of the rat kidney. Aminonucleoside-8-$^{14}C$ metabolism in guinea pigs and rats is markedly different. Guinea pigs, but not rats, excrete part of the aminonucleoside as $^{14}CO_2$. Demethylation of the aminonucleoside occurs in both animals (Nagasawa et al., 1967).

## III. Metabolism of the Aminonucleoside of Puromycin

### *1. Aminopurine Nucleoside*

The aminonucleoside isolated from puromycin is 6-dimethylamino-9-(3′-amino-3′-deoxy-β-D-ribofuranosyl)purine (**5**). It is 3–4 times as active against *Trypanosoma equiperdum* as is puromycin. In general, the aminonucleoside has no antibacterial activity that is characteristic of puromycin. Studzinski and Ellem (1968) reported that the aminopurine nucleoside of puromycin inhibited rRNA and the RNA content of cells, but did not affect the rate of cell division for several generations. The inhibition of cell division and RNA synthesis by aminopurine nucleoside in heteroploid and diploid cells was prevented by the simultaneous addition of inosine. Inosine did not act by preventing the entry of the nucleoside into the cells. Taylor and Stanners (1968) reported that the aminonucleoside of puromycin inhibited RNA synthesis in cultured hamster embryo cells, but had little effect on protein synthesis for several hours. On other systems, the aminonucleoside inhibition of RNA synthesis was interpreted as a specific inhibition of rRNA synthesis. No evidence was found in the hamster embryo cells for selective inhibition of the synthesis of rRNA, heavy nuclear RNA, or polysomal mRNA. These three RNAs were inhibited equally by the aminonucleosides. Nagasawa et al. (1967) reported that hepatic drug-metabolizing enzyme inhibitors did not stop the inhibition of nephrosis induced in rats by the aminonucleoside. Allopurinol, an inhibitor of xanthine oxidase, did not reverse the nephrotoxicity induced by the puromycin aminonucleoside.

When HeLa cells were exposed to the aminonucleoside (100 $\mu$g/ml), nucleolar changes were observed (Lewin and Moscarello, 1968). During a 96-hr exposure of HeLa cells to the aminonucleoside, RNA synthesis was completely inhibited, while protein synthesis continued. The nucleoli of aminonucleoside-treated cells first showed segmentation and clumping of fibrillar components. This was followed by extrusion of the fibrillar component. Actinomycin D caused similar nucleolar changes. The conclusion was drawn that these changes may indicate the site of defective 45-S RNA synthesis.

## Summary

Puromycin acts as a codon-independent functional analog of aminoacyl-tRNA by catalyzing the release of incomplete peptide chains from the peptidyl-tRNA–messenger–ribosome complex. Puromycin blocks protein synthesis by replacing an aminoacyl-tRNA and reacting with the nascent polypeptide on the peptidyl-tRNA site of the ribosome. The reaction terminates with the formation of peptidyl-puromycin. The puromycin reaction also takes place with *N*-formylmethionyl-tRNA$_F$ or methionyl-tRNA$_F$ in the initiation process. This species of tRNA is bound at the peptidyl site of the ribosome complex. Puromycin appears to enter the ribosomal structure at or near the aminoacyl-tRNA site. The translocation of peptidyl-tRNA by transferase II (or factor G) and GTP from the acceptor site to the donor site converts the peptide-synthesizing complex from a puromycin-insensitive state to a puromycin-sensitive state. Puromycin acts more like an amino acid antimetabolite rather than an analog of adenosine in that adenosine does not reverse puromycin inhibition. Although puromycin has been a valuable biochemical tool for studying peptide synthesis, the new experimental evidence cited here indicates that puromycin can not be used as the model nucleoside for the elucidation of the substeps involved in peptide bond formation. In addition to these studies, in which puromycin has been used to elucidate polypeptide elongation, it has also been a most useful biochemical tool to study a variety of processes from the simplest to the most complex organism.

The primary site of action of a number of antibiotics that inhibit the highly organized sequence of reactions involved in protein synthesis on polyribonucleotide templates has been clarified by the design of experiments utilizing puromycin. For example, puromycin has helped establish that erythromycin and fusidic acid inhibit translocation; lincomycin, chloramphenicol and sparsomycin inhibit peptide bond formation; and streptomycin inhibits translocation and binding of aminoacyl-tRNA.

## References

Agranoff, B. W., and R. E. Davis, in F. D. Carlson, Ed., 1968, *Physiological and Biochemical Aspects of Nervous Integration*, Prentice-Hall, New Jersey, p. 309.

Agranoff, B. W., and P. D. Klinger, *Science*, **146**, 952 (1964).

Agranoff, B. W., R. E. Davis, and J. J. Brink, *Proc. Natl. Acad. Sci. (U.S.)*, **54**, 788 (1965).

Agranoff, B. W., R. E. Davis, and J. J. Brink, *Brain Res.*, **1**, 303 (1966).

Albrecht, U., K. Prenzel and D. Richter, *Biochemistry* **9**, 361 (1970).

Allen, D. W., and P. C. Zamecnik, *Biochim. Biophys. Acta*, **55**, 865 (1962).

Allende, J. E., and M. Bravo, *J. Biol. Chem.*, **241**, 5813 (1966).

Anderson, J. S., M. S. Bretscher, B. F. C. Clark, and K. A. Marcker, *Nature*, **215**, 490 (1967).

Appleman, M. M., and R. G. Kemp, *Biochem. Biophys. Res. Commun.*, **24**, 564 (1966).

Arlinghaus, R., J. Shaffer, and R. Schweet, *Proc. Natl. Acad. Sci. (U.S.)*, **51**, 1291 (1964).

Bablanian, R., *J. Gen. Virol* , **3**, 51 (1968).

Bachmayer, H., and G. Kreil, *Biochim. Biophys. Acta*, **169**, 95 (1968).

Baker, B. R., and R. E. Schaub, *J. Org. Chem.*, **19**, 646 (1954).

Baker, B. R., J. P. Joseph, R. E. Schaub, and J. H. Williams, *J. Org. Chem.*, **19**, 1780 (1954).

Baker, B. R., and J. P. Joseph, *J. Amer. Chem. Soc.*, **77**, 15 (1955).

Baker, B. R., R. E. Schaub, and J. H. Williams, *J. Amer. Chem. Soc.*, **77**, 7 (1955a).

Baker, B. R., J. P. Joseph, and J. H. Williams, *J. Amer. Chem. Soc.*, **77**, 1 (1955b).

Baker, B. R., R. E. Schaub, J. P. Joseph, and J. H. Williams, *J. Amer. Chem. Soc.*, **77**, 12 (1955c).

Barondes, S. H., and H. D. Cohen, *Brain Res.*, **4**, 44 (1967).

Bishop, J. O., *Arch. Biochem. Biophys.*, **125**, 449 (1968).

Bodley, J. W., *Federation Proc.*, **29**, 538 (1970).

Bretscher, M. S., *Nature*, **220**, 1233 (1968a).

Bretscher, M. S., *Nature*, **218**, 675 (1968b).
Bretscher, M. S., and K. A. Marcker, *Nature*, **211**, 380 (1966).
Brot, N., R. Ertel, and H. Weissbach, *Biochem. Biophys. Res. Commun.*, **31**, 563 (1968).
Brot, N., C. Spears, and H. Weissbach, *Biochem. Biophys. Res. Commun.*, **34**, 843 (1969).
Buck, C. A., G. A. Granger, and J. J. Holland, *Current Mod. Biol.*, **1**, 9 (1967).
Burke, D. C., in *Interferons*, N. B. Finter, Ed., Saunders, Philadelphia, 1966, p. 55.
Cannon, M., *Europ. J. Biochem.*, **7**, 137 (1968).
Capecchi, M. R., 158th National Meetings, Am. Chem. Soc., 1969, New York, N.Y., Abstracts, Biol. 037.
Casjens, S. R., and A. J. Morris, *Biochim. Biophys. Acta*, **108**, 677 (1965).
Caskey, C. T., A. Beaudet, and M. Nirenberg, *J. Mol. Biol.*, **37**, 99 (1968).
Cathey, G. M., and S. J. Klebanoff, *Biochim. Biophys. Acta*, **145**, 806 (1967).
Clark, B. F. C., and K. A. Marcker, *J. Mol. Biol.*, **17**, 394 (1966).
Clark, Jr., J. M., and A. Y. Chang, *J. Biol. Chem.*, **240**, 4734 (1965).
Cohen, L. B., A. E. Herner, and I. H. Goldberg, *Biochemistry*, **8**, 1312 (1969a).
Cohen, L. B., I. H. Goldberg, and A. E. Herner, *Biochemistry*, **8**, 1327 (1969b).
Collier, R. J., *J. Mol. Biol.*, **25**, 83 (1967).
Collier, R. J., and A. M. Pappenheimer, Jr., *J. Exptl. Med.*, **120**, 1019 (1964).
Colombo, B., L. Felicetti, and C. Baglioni, *Biochem. Biophys. Res. Commun.*, **18**, 389 (1965).
Coutsogeorgopoulos, C., *Biochim. Biophys. Acta*, **129**, 214 (1966).
Coutsogeorgopoulos, C., *Biochem. Biophys. Res. Commun.*, **27**, 46 (1967a).
Coutsogeorgopoulos, C., *Biochemistry*, **6**, 1704 (1967b).
Coutsogeorgopoulos, C., *Fifth International Congress Chemotherapy*, Vienna, **IV**, 371 (1967c).
Coutsogeorgopoulos, C., *Federation Proc.*, **28**, 844 (1969).
Coutsogeorgopoulos, C., *Proc. 6th Intern. Congr. Chemotherapy Tokyo*, (1970).
Cundliffe, E., and K. McQuillen, *J. Mol. Biol.*, **30**, 137 (1967).
Davis, R. E., P. J. Bright, and B. W. Agranoff, *J. Comp. Physiol. Psychol.*, **60**,162 (1965).
de la Haba, G., and H. Holtzer, *Science*, **149**, 1263 (1965).
Eggers, S. H., S. I. Biedron, and A. O. Hawtrey, *Tetrahedron Letters* (1966) 3271.
Elsas, L. J., and L. E. Rosenberg, *Proc. Natl. Acad. Sci.* (*U.S.*), **57**, 371 (1967).
Elsas, L. J., I. Albrecht, and L. E. Rosenberg, *J. Biol. Chem.*, **243**, 1846 (1968).
Everse, J., D. A. Gardner, N. O. Kaplan, W. Galasinski, and K. Moldave, *J. Biol. Chem.*, **245**, 899 (1970).
Felicetti, L., G. P. Tocchini-Valentini, G. F. DiMatteo, *Biochemistry*, **8**, 3428 (1969).
Finkelstein, M. S., G. H. Bausek, and T. C. Merigan, *Science*, **161**, 465 (1968).
Fisher, L. V., W. W. Lee, and L. Goodman, *J. Med. Chem.*, **13**, 775 (1970).
Flessel, C. P., *Biochem. Biophys. Res. Commun.*, **32**, 438 (1968).
Flexner, J. B., and L. B. Flexner, *Proc. Natl. Acad. Sci.* (*U.S.*), **57**, 1651 (1967a).
Flexner, L. B., J. B. Flexner, and R. B. Roberts, *Science*, **155**, 1377 (1967b).
Flexner, L. B., and J. B. Flexner, *Proc. Natl. Acad. Sci.* (*U.S.*), **60**, 923 (1968).
Flexner, J. B., L. B. Flexner, and E. Stellar, *Science*, **141**, 57 (1963).
Flexner, J. B., and L. B. Flexner, *Science*, **165**, 1143 (1969).
Flexner, L. B., J. B. Flexner, G. de la Haba, and R. B. Roberts, *J. Neurochem.*, **12**, 535 (1965).
Freedman, M. L., J. M. Fisher, and M. Rabinovitz, *J. Mol. Biol.*, **33**, 315 (1968).

Galasinski, W., and K. Moldave, *J. Biol. Chem.*, **244**, 6567 (1969).
Gambetti, P., N. K. Gonatas, and L. B. Flexner, *J. Cell. Biol.*, **36**, 379 (1968a).
Gambetti, P., N. K. Gonatas, and L. B. Flexner, *Science*, **161**, 900 (1968b).
Gardner, R. S., A. J. Wahba, C. Basilio, R. S. Miller, P. Lengyel, and J. F. Speyer, *Proc. Natl. Acad. Sci. (U.S.)*, **48**, 2087 (1962).
Gershbein, L., *J. Pharm. Sci.*, **55**, 1303 (1966).
Gilbert, W., *J. Mol. Biol.*, **6**, 389 (1963).
Goldberg, I. H., and K. Mitsugi, *Biochemistry*, **6**, 383 (1967).
Gottesman, M. E., *J. Biol. Chem.*, **242**, 5564 (1967).
Grunberg-Manago, M., B. F. C. Clark, M. Revel, P. S. Rudland, and J. Dondon, *J. Mol. Biol.*, **40**, 33 (1969).
Guenther, K., H. Bachmayer, K. A. Davis, G. Polz, *FEBS Letters*, **1**, 97 (1968).
Haenni, A., and J. L. Lucas-Lenard, *Proc. Natl. Acad. Sci. (U.S.)*, **61**, 1363 (1968).
Hardesty, B., R. Miller, and R. Schweet, *Proc. Natl. Acad. Sci. (U.S.)*, **50**, 924 (1963).
Harris, G., *J. Exptl. Med.*, **127**, 675 (1968).
Heintz, R. L., H. McAllister, R. Arlinghaus, and R. Schweet, *Cold Spring Harbor Symp.* **31**, 633, (1966).
Heintz, R. L., M. L. Salas, and R. S. Schweet, *Arch. Biochem. Biophys.*, **125**, 488 (1968).
Herner, A. E., I. H. Goldberg, and L. B. Cohen, *Biochemistry*, **8**, 1335 (1969).
Hershey, J. W. B., and R. E. Thach, *Proc. Natl., Acad. Sci. (U.S.)*, **57**, 759 (1967).
Hewitt, R. I., A. R. Gumble, W. S. Wallace, and J. H. Williams, *Antibiot. Chemotherapy*, **5**, 139 (1955).
Hewitt, R. I., A. R. Gumble, W. S. Wallace, and J. H. Williams, *Antibiot. Chemotherapy*, **4**, 1222 (1954).
Hille, M. B., M. J. Miller, K. Iwasaki, and A. J. Wahba, *Proc. Natl. Acad. Sci. (U.S.)*, **58**, 1652 (1967).
Honjo, T., Y. Nishizuka, and O. Hayaishi, *J. Biol. Chem.*, **243**, 3553 (1968).
Hori, M., J. M. Fisher, and M. Rabinovitz, *Science*, **155**, 83 (1967).
Ibuki, F. and K. Moldave, *J. Biol. Chem.*, **243**, 791 (1968).
Igarashi, K., H. Ishitsuka, and A. Kaji, *Biochem. Biophys. Res. Commun.*, **37**, 499 (1969).
Igarashi, K., and A. Kaji, *Proc. Natl. Acad. Sci. (U.S.)*, **62**, 498 (1969).
Jayaraman, J. and I. H. Goldberg, *Biochemistry*, **7**, 418 (1968).
Kaji, H., *Biochemistry*, **7**, 3844 (1968).
Kim, W. S., *Science*, **163**, 947 (1969).
Kinoshita, T., G. Kawano, and N. Tanaka, *Biochem. Biophys. Res. Commun.*, **33**, 769 (1968).
Kohler, R. E., E. Z. Ron, and B. D. Davis, *J. Mol. Biol.*, **36**, 71 (1968).
Kulaeva, O. N., and N. L. Klyachko, *Dokl. Akad. Nauk. SSSR*, **175**, 958 (1967).
Kuo, J. F., *Biochim. Biophys. Acta*, **165**, 208 (1968).
Laird, G., and G. P. Studzinski, *Exptl. Cell Res.*, **52**, 408 (1968).
Leder, P. and H. Bursztyn, *Biochem, Biophys. Res. Commun.*, **25**, 233 (1966).
Lengyel, P. and D. Söll, *Bacterio. Rev.*, **33**, 264 (1969).
Lewin, P. K., and M. A. Moscarello, *Lab. Invest.*, **19**, 265 (1968).
Lin, S. Y., W. L. McKeehan, W. Culp and B. Hardesty, *J. Biol. Chem.*, **244**, 4340 (1969).
Lin, Y. C. and N. Tanaka, *J. Biochem. (Tokyo)*, **63**, 1 (1968).
Lin, Y. C., T. Kinoshita, and N. Tanaka, *J. Antibiotics (Tokyo)*, **21**, 471 (1968).

Lipmann, F., *Science*, **164**, 1024 (1969).
Livingston, D. M., and P. Leder, *Biochemistry*, **8**, 435 (1969).
Lucas-Lenard, J., and F. Lipmann, *Proc. Natl. Acad. Sci. (U.S.)*, **57**, 1050 (1967).
Lucas-Lenard, J., and F. Lipmann, *Proc. Natl. Acad. Sci. (U.S.)*, **55**, 1562 (1966).
Lucas-Lenard, J., and Haenni, A. L., *Proc. Natl. Acad. Sci. (U.S.)*, **59**, 554 (1968).
Lucas-Lenard, J., and Haenni, A. L., *Proc. Natl. Acad. Sci. (U.S.)*, **63**, 93 (1969).
Maden, B. E. H., and R. E. Monro, *European J. Biochem.*, **6**, 309 (1968).
Maden, B. E. H., R. R. Traut, and R. E. Monro, *J. Mol. Biol.*, **35**, 333 (1968).
Matalon, R., and A. Dorfman, *Proc. Natl. Acad. Sci. (U.S.)*, **60**, 179 (1968).
McKeehan, W. L. and B. Hardesty, *J. Biol. Chem.*, **244**, 4330 (1969a).
McKeehan, W. and B. Hardesty, *Biochem. Biophys. Res. Commun.*, **36**, 625 (1969b).
McKeehan, W., and B. Hardesty, *Federation Proc.*, **29**, 537 (1970).
Miskin, R., A. Zamir, and D. Elson, *Biochem, Biophys. Res. Commun.*, **33**, 551 (1968).
Möller, W., *Nature*, **222**, 979 (1969).
Monro, R. E., *Nature* **223**, 903 (1969).
Monro, R. E., *J. Mol. Biol.*, **26**, 147 (1967).
Monro, R. E., and K. A. Marcker, *J. Mol. Biol.*, **25**, 347 (1967).
Monro, R. E., and D. Vazquez, *J. Mol. Biol.*, **28**, 161 (1967).
Monro, R. E., J. Cerna, and K. A. Marcker, *Proc. Natl. Acad. Sci. (U.S.)*, **61**, 1042 (1968).
Monro, R. E., M. L. Celma, and D. Vazquez, *Nature*, **222**, 356 (1969).
Montgomery, J. A., and H. J. Thomas, *Advan. Carbohydrate Chem.*, **17**, 301 (1962).
Morris, A., R. Arlinghaus, S. Favelukes, and R. Schweet, *Biochemistry*, **2**, 1084 (1963).
Mukundan, M. A., J. W. B. Hershey, K. F. Dewey, and R. E. Thach, *Nature*, **217**, 1013 (1968).
Murata, K., J. J. Quilligan, and L. M. Morrison, *Biochim. Biophys. Acta*, **144**, 473 (1967).
Nagasawa, H. T., K. F. Swingle, and C. S. Alexander, *Biochem. Pharmacol.*, **16**, 2211 (1967).
Nathans, D., *Proc. Natl. Acad. Sci. (U.S.)*, **51**, 585 (1964).
Nathans, D., *Antibiotics*, **1**, 259 (1967).
Nathans, D., and F. Lipmann, *Proc. Natl. Acad. Sci. (U.S.)*, **47**, 491, (1961).
Nathans, D., and A. Neidle, *Nature*, **197**, 1076 (1963).
Nathans, D., J. E. Allende, T. W. Conway, G. I. Spyrides, and F. Lipmann, "Protein Synthesis from Aminoacyl-sRNA's," in *Symposium on Information Macromolecules*, H. J. Vogel, V. Bryson and J. O. Lampen, Eds., New York, Academic Press 1963.
Nirenberg, M. W., J. H. Matthae, and O. W. Jones, *Proc. Natl. Acad. Sci. (U.S.)*, **48**, 104 (1962).
Noll, H., T. Staehelin, and F. O. Wettstein, *Nature*, **198**, 632 (1963).
Ohta, T., S. Sarkar, and R. E. Thach, *Proc. Natl. Acad. Sci. (U.S.)*, **58**, 1638 (1967).
Oleinick, N. L., and J. W. Corcoran, *J. Biol. Chem.*, **244**, 727 (1969).
Ono, Y., A. Skoultchi, J. Waterson, and P. Lengyel, *Nature*, **222**, 645 (1969).
Pattabiraman, T. N., and B. M. Pogell, *Biochim. Biophys. Acta*, **182**, 245 (1969).
Pestka, S., *Biochem. Biophys. Res. Commun.*, **36**, 589 (1969a).
Pestka, S., *Proc. Nat'l. Acad. Sci.* (U.S.) , **61**, 726 (1968b).

Pestka, S., *Proc. Natl. Acad. Sci.* (*U.S.*), **64**, 709 (1969b).
Pestka, S., *J. Biol. Chem.*, **244**, 1533 (1969c).
Pestka, S., E. M. Scolnick, and B. H. Heck, *Anal. Biochem.*, **28**, 376 (1969d).
Pestka, S., *Arch. Biochem. Biophys.*, **136**, 80 (1970a).
Pestka, S., *Arch. Biochem. Biophys.*, **136**, 89 (1970b).
Pestka, S., *Cold Spring Harbor Symp.* **34**, 395 (1970c).
Pestka, S., *J. Biol. Chem.*, in press, (1970d).
Porter, J. N., R. I. Hewitt, C. W. Hesseltine, G. Krupka, J. A. Lowery, W. S. Wallace, N. Bohonos, and J. H. Williams, *Antibiot. Chemotherapy*, **2**, 409 (1952).
Porter, J. N., G. C. Krupka, and N. Bohonos, U.S. Pat. No. 2,763,642 (1956).
Rao, M. M., P. F. Rebello, and B. M. Pogell, *J. Biol. Chem.*, **244**, 112 (1969).
Ravel, J. M., R. L. Shorey, S. Froehner, and W. Shive, *Arch. Biochem. Biophys.*, **125**, 514 (1968).
Rebello, P. F., B. M. Pogell, and P. P. Mukherjee, *Biochim. Biophys. Acta*, **177**, 468 (1969).
Richter, D. and F. Klink, *Biochemistry*, **6**, 3569 (1967).
Rudland, P. S., W. A. Whybrow, K. A. Marcker, and B. F. C. Clark, *Nature*, **222**, 750 (1969).
Rychlík, I., *Biochim. Biophys. Acta*, **114**, 425 (1966).
Salas, M., M. B. Hille, J. A. Last, A. J. Wahba, and S. Ochoa, *Proc. Natl. Acad. Sci.* (*U. S.*), **57**, 387 (1967).
Sankar, S. and R. E. Thach, *Proc. Natl. Acad. Sci.* (*U. S.*), **60**, 1479 (1968).
Schlessinger, D., G. Mangiarotti, and D. Apirion, *Proc. Natl. Acad. Sci.* (*U.S.*), **58**, 1782 (1967).
Schneider, J. A., S. Raeburn, and E. S. Maxwell, *Biochem. Biophys. Res. Commun.*, **33**, 177 (1968).
Schwartz, J. H., R. Meyer, J. M. Eisenstadt, and G. Brawerman, *J. Mol. Biol.*, **25**, 571 (1967).
Shelton, K. R., and J. M. Clark, Jr., *Biochemistry*, **6**, 2735 (1967).
Sherman, J. F., D. J. Taylor, and H. W. Bond, *Antibiotics Ann.*, **2**, 757 (1954/1955).
Silverstein, E., *Biochim. Biophys. Acta.*, **186**, 402 (1969).
Singh, V. N., S. S. Dave, and T. A. Venkitasubramanian, *Biochem. J.*, **104**, 48c (1967).
Skogerson, L., and K. Moldave, *Arch. Biochem. Biophys.*, **125**, 497 (1968a).
Skogerson, L., and K. Moldave, *J. Biol. Chem.*, **243**, 5354 (1968b).
Skogerson, L., and K. Moldave, *J. Biol. Chem.*, **243**, 5361 (1968c).
Skoultchi, A., Y. Ono, H. M. Moon, and P. Lengyel, *Proc. Natl. Acad. Sci.*, **60**, 675 (1968).
Smith, A. E., and K. A. Marcker, *J. Mol. Biol.*, **38**, 241 (1968).
Smith, J. D., R. R. Traut, G. M. Blackburn, and R. E. Monro, *J. Mol. Biol.*, **13**, 617 (1965).
Sövik, O., *Biochim. Biophys. Acta*, **141**, 190 (1967).
Stoolmiller, A. C., and A. Dorfman, *J. Biol. Chem.*, **244**, 236 (1969).
Studzinski, G. P., and K. O. L. Ellem, *Cancer Res.*, **28**, 1773 (1968).
Symons, R. H., R. J. Harris, L. P. Clarke, J. F. Wheldrake, W. H. Elliott, *Biochim. Biophys. Acta*, **179**, 248 (1969).

Szumski, S. A., and J. J. Goodman, U. S. Pat. No. 2,797,187 (1957).
Tamaoki, T., and B. G. Lane, *Biochemistry*, **7**, 3431 (1968).
Tanaka, N., T. Kinoshita, and H. Masukawa, *Biochem. Biophys. Res. Commun.*, **30**, 278 (1968).
Tanaka, N., T. Kinoshita, and H. Masukawa, *J. Biochem.*, **65**, 459 (1969a).
Tanaka, N., T. Nishimura, T. Kinoshita, and H. Umezawa, *J. Antibiotics (Tokyo)*, **22**, 181 (1969b)
Taubman, S. B., A. G. So, F. E. Young, E. W. Davie, and J. W. Corcoran, *Antimicrobial Agents Chemotherapy*, **395** (1963).
Taylor, J. M., and C. P. Stanners, *Biochim. Biophys. Acta*, **155**, 424 (1968).
Traut, R. R., and R. E. Monro, *J. Mol. Biol.*, **10**, 63 (1964).
Vazquez, D., and R. E. Monro, *Agrochimica*, **12**, 485 (1968).
Villa-Trevino, S., E. Farber, T. Staehelin, F. O. Wettstein, and H. Noll, *J. Biol. Chem.*, **239**, 3826 (1964).
Waller, C. W., P. W. Fryth, B. L. Hutchings, and J. H. Williams, *J. Amer. Chem. Soc.*, **75**, 2025 (1953).
Warner, J. R., M. Girard, H. Latham, and J. E. Darnell, *J. Mol. Biol.*, **19**, 373 (1966).
Weber, M. J., and J. A. DeMoss, *J. Bacteriol.*, **97**, 1099 (1969).
Weissbach, H., D. L. Miller, and J. Hachmann, *Arch. Biochem. Biophys.*, **137**, 262 (1970).
White, J. R., and H. L. White, *Science*, **146**, 772 (1964).
Wiley, P. F. and F. A. MacKeller, *J. Am. Chem. Soc.*, **92**, 417 (1970).
Williamson, A. R., and R. Schweet, *J. Mol. Biol.*, **11**, 358 (1965).
Yarmolinsky, M. B., and G. L. de la Haba, *Proc. Natl. Acad. Sci. (U.S.)*, **45**, 1721 (1959).
Yoshikawa-Fukada, M., *Biochim. Biophys. Acta*, **145**, 651 (1967).
Yukioka, M. and S. Morisawa, *J. Biochem.*, **66**, 225 (1969a).
Yukioka, M. and S. Morisawa, *J. Biochem.*, **66**, 233 (1969b).
Yukioka, M. and S. Morisawa, *J. Biochem.*, **66**, 241 (1969c).

## 1.2. CORDYCEPIN (3′-DEOXYADENOSINE)

### INTRODUCTION

Cordycepin (3′-deoxyadenosine) (Fig. 1.19) was the first nucleoside antibiotic isolated. It has been isolated from the culture filtrates of *Cordyceps militaris* and *Aspergillus nidulans*. The structure and chemical syntheses have been completed. Cordycepin is a cytotoxic agent and a structural analog of adenosine. It is inhibitory to *Bacillus subtilis*, avian tubercle bacillus, and Ehrlich ascites tumor cells and is cytotoxic to KB cell cultures and H. Ep. #2 cells. Studies related to the effect of cordycepin phosphate on RNA and DNA synthesis will be described along with its effect on purine synthesis and the interaction at the catalytic and regulatory sites of enzymes. The biosynthesis of cordycepin and the metabolic fate in the producing organism will also be described.

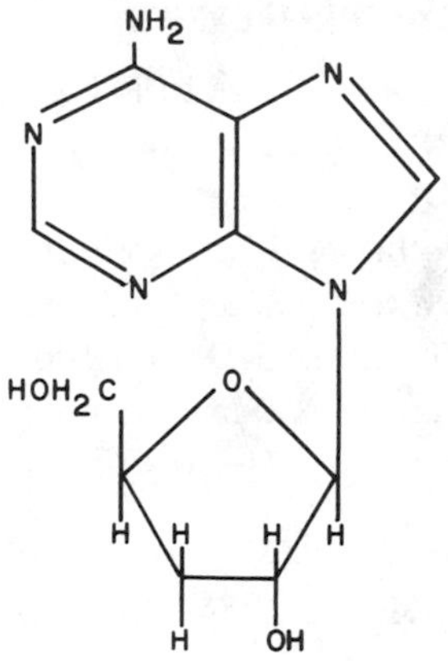

FIGURE 1.19. *Structure of cordycepin (3′-deoxyadenosine).*

Several recent reviews related to cordycepin have been written (Fox et al., 1966; Guarino, 1967; Suhadolnik, 1967).

## DISCOVERY, PRODUCTION, AND ISOLATION

Cunningham et al. (1951) were the first to report the isolation of the nucleoside antibiotic cordycepin from the culture filtrates of *C. militaris* (Linn) Link. More recently, cordycepin has been isolated from *A. nidulans* (Kaczka et al., 1964a) and from *C. militaris* (L. ex Fr.) Link by Frederiksen et al. (1965). The medium used for the growth and production of cordycepin is as follows: 0.50% casein hydrolyzate (enzymatically prepared) and 1% glucose (Kredich and Guarino, 1960, 1961). Inoculation is made by adding the conidia from ten agar slants of *C. militaris* (2–4 weeks old) per 13 liters of medium. The culture is allowed to grow without shaking. Cordycepin appears in the culture filtrate about 20 days after inoculation. The production of cordycepin from *A. nidulans* is carried out in shake flasks (Kaczka et al., 1964a). *Bacillus subtilis* is the test organism used to follow production (Cunningham et al., 1951). The yield of crystalline compound (per liter of medium) is 25–100 mg.

The isolation of cordycepin has been described by Cunningham et al. (1951), Kredich and Guarino (1960), Kaczka et al. (1964a), Frederiksen et al. (1965), and Chassy and Suhadolnik (1969). The procedure of Chassy and Suhadolnik is as follows: The culture filtrate is evaporated to dryness *in vacuo* and the residue is extracted with ethanol. The ethanol is evaporated *in vacuo* and the residue is dissolved in 15% aqueous methanol and is added to a Dowex-1-$OH^-$ column. Cordycepin is isolated as a pure compound by elution from the resin with 25% aqueous methanol. Cordycepin is crystallized from water.

## PHYSICAL AND CHEMICAL PROPERTIES

The molecular formula for cordycepin is $C_{10}H_{13}N_5O_3$ (mol. wt. 251); mp 230–231°C; $[\alpha]_D^{25}$ −35° ($c$ = 9.425 in water); $\lambda_{max}^{pH\ 4}$ 259 mμ ($\epsilon$ = 13,100); $\lambda_{max}^{pH\ 11}$ 260 mμ ($\epsilon$ = 13,700) (Kaczka et al., 1964a). Cordycepin is soluble in water, hot ethanol, and methanol. It is insoluble in benzene, ethyl ether, and chloroform. It is hydrolyzed to adenine and 3-deoxyribose by boiling in water with Dowex-50-$H^+$ for 90 min. 3-Deoxyribose has been oxidized to 3-deoxyribonic acid with bromine water and converted to 3-deoxyribonic acid phenylhydrazide (mp 151°C; $[\alpha]_D$ + 26°) (Bentley et al., 1951; Suhadolnik and Cory, 1964).

## STRUCTURAL ELUCIDATION

On the basis of degradative and chemical studies, Bentley et al. (1951) proposed that cordycepin was a nucleoside analog of adenine with the branched carbon chain sugar cordycepose (Fig. 1.20). Bentley et al. (1951) reported that cordycepin was hydrolyzed in acid. Adeline hydrochloride was isolated from the acid hydrolyzate. In addition, they showed that cordycepin could be deaminated with nitrous acid, and hypoxanthine could be isolated following acid hydrolysis. Proof that cordycepin was an adenine-glycoside was further demonstrated when a carbohydrate moiety, cordycepose, was isolated following acid hydrolysis. Spectral data showed that the glycosylic group was on $N_9$ and not $N_7$. The absorption spectrum of cordycepin was similar to that of 9-methyladenine and adenosine, but substantially different from that of 7-methyladenine. Additional studies were then performed to determine the structure of the carbohydrate moiety. Metaperiodate did not oxidize cordycepin. When the sugar moiety was treated with bromine water, cordyceponolactone (3-deoxyribonolactone) was formed which was then converted to cordyceponic acid phenylhydrazide. The physical properties of the phenylhydrazide were then compared with

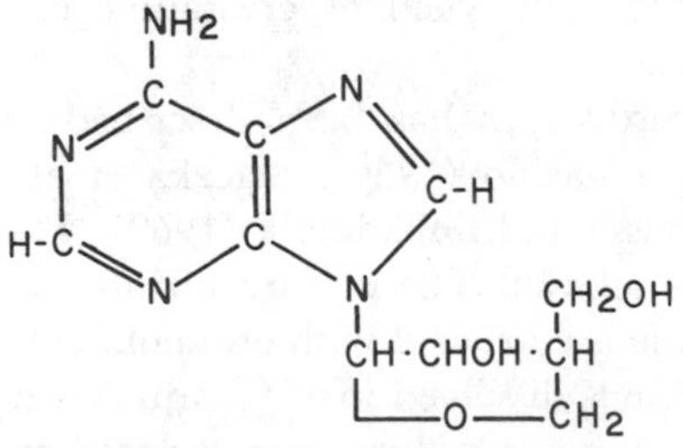

FIGURE 1.20. *Structure of cordycepin as originally proposed by Bentley et al.* (1951).

those of the known four authentic phenylhydrazide derivatives of the $\alpha,\gamma,\delta$-trihydroxyvaleric acids. Although cordyceponic acid phenylhydrazide and D-*erythro*-$\alpha,\gamma,\delta$-trihydroxyvaleric acid phenylhydrazide had essentially the same melting point, the optical rotation as reported by Bentley et al. (1951) for the phenylhydrazide of cordycepose was $+26°$, while the optical rotation of D-*erythro*-$\alpha,\gamma,\delta$-trihydroxyvaleric acid derivatives was $\pm 9°$.

Since cordycepose formed crystalline osazones with *p*-nitrophenylhydrazine,2,4-dinitrophenylhydrazine and *p*-bromophenylhydrazine, the presence of a 2-hydroxyl group was established. Bentley et al. (1951) concluded that cordycepose could not be a straight-chain 3-deoxyaldopentose.

The results as presented above led to the conclusion that the sugar moiety in cordycepin was a branched-chain 3-deoxypentose as shown in Figure 1.21. The structure as proposed in Figure 1.21 was closely related to the previously reported branched-chain pentose D-apiose isolated from the glycosides of *Apium petroselinum* (Hudson, 1949). Additional evidence substantiating cordycepose as a branched-chain 3-deoxypentose was presented by Raphael and Roxburgh (1955) in their report on the total chemical synthesis of racemic cordycepose and its conversion to the crystalline *p*-nitrophenylosazone. The osazone of the synthetic branched-chain sugar had the same melting point as that obtained from the cordycepose isolated from cordycepin. A mixture of the two compounds showed no depression in melting point. These findings were considered as additional proof that the structure assigned to the carbohydrate moiety of cordycepin (Figure 1.21) was indeed a branched-chain sugar. Subsequent studies show that the carbohydrate moiety is 3-deoxyribose and not a branched-chain sugar.

In retrospect, the first indication that the sugar moiety of cordycepin was not a branched-chain sugar was the report by Klenow (1963a) in which he showed that cordycepin 5′-monophosphate was dephosphorylated at about the same rate as that of an equimolar amount of 2′-deoxyAMP. Assuming that the structure for cordycepin as proposed by Bentley et al. (1951) did

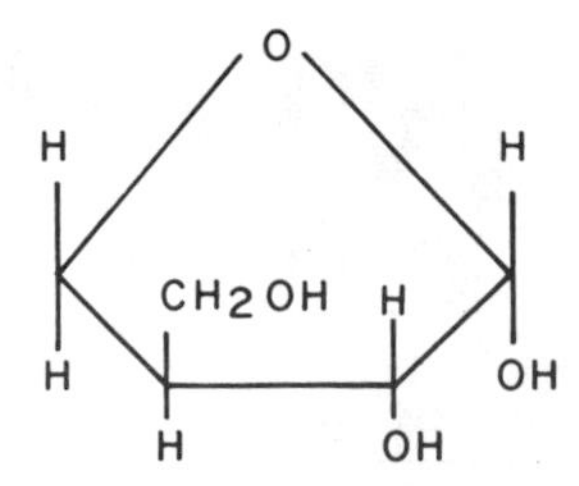


FIGURE 1.21. *Structure of cordycepose* (*From Bentley et al.*, 1951).

not contain a 5′-hydroxyl group, Klenow's results indicated that the specificity of snake venom 5′-nucleotidase included the hydrolysis of phosphate of branched-chain sugars with hydroxymethyl groups. Now that the structure of the sugar moiety of cordycepin is known to be 3-deoxyribose and not a branched-chain sugar, the findings reported by Klenow (1963a) would be expected.

Subsequent physical and chemical studies by Kaczka et al. (1964b), Suhadolnik and Cory (1964), Hanessian et al. (1966) and Guarino (private communication have shown that the structure of the carbohydrate moiety of cordycepin as proposed by Bentley et al. (1951) and Raphael and Roxburgh (1955) (Figs. 1.20 and 1.21) was incorrect and that cordycepin is a 3′-deoxynucleoside as shown in Figure 1.19. The nmr data of cordycepin as shown in Figure 1.22 show a two-proton multiplet with a center of gravity of 160 cps relative to tetramethylsilane as external reference (60 mc). By its position, this precluded the unknown compound from having two protons on C-4′ or adjacent ($\alpha$) to the ring oxygen atom. The nmr data showed that the two protons in question must be located on C-3′ or $\beta$ to the ring oxygen atom. Additional proof of the 3′-deoxyadenosine structure of cordycepin was supplied by the chemical studies reported by Suhadolnik and Cory (1964) and the biosynthetic studies of Suhadolnik et al. (1964).

Guarino (private communication) has demonstrated the formation of malonalddehyde following periodate oxidation of the sugar isolated from cordycepin. The ultimate consumption was 5 moles of periodate per mole of sugar oxidized. These data support the notion that cordycepin and 3′-deoxyadenosine are identical. Finally, the mass spectrometric evidence presented by Hanessian et al. (1966) established the straight-chain 3-deoxy-

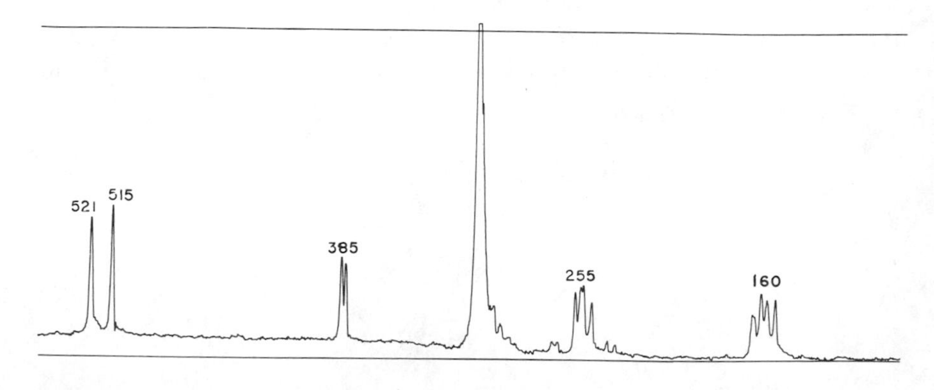


FIGURE 1.22. *The nmr spectrum of 3′-deoxyadenosine isolated from A. nidulans in $D_2O$ (60 Mc). (From Kaczka et al., 1964b).*

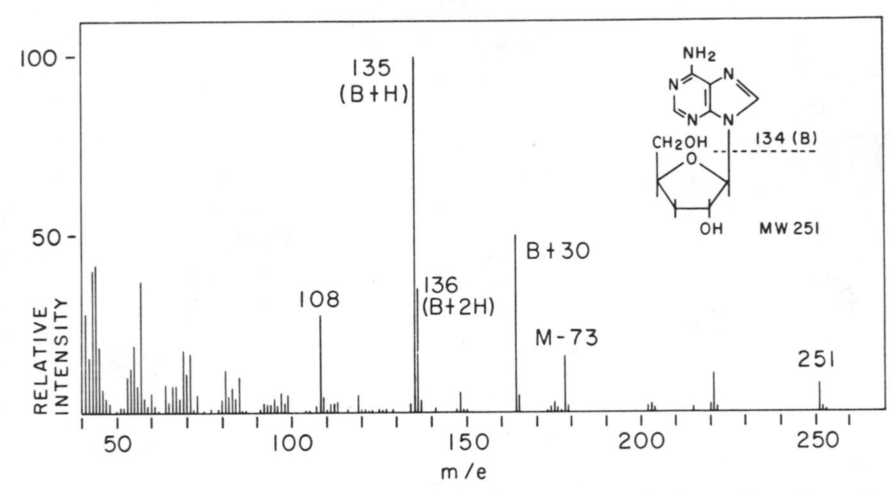


FIGURE 1.23. *Mass spectrum of cordycepin* (3′-*deoxyadenosine*). (*From Hanessian et al.*, 1966.)

pentose structure of the sugar component of cordycepin. The mass spectrum of 3′-deoxyadenosine is shown in Figure 1.23 and is identical to that of chemically synthesized cordycepin. The interpretation of the mass spectrum confirmed the structure of cordycepin. The molecular ion, *m/e* 251, eliminates the elements of formaldehyde from the C-5′ hydroxymethyl group to yield the fragment, *m/e* 221. Continued fragmentation yields *m/e* 178 (M-73). The peak at *m/e* 164 is similar to that of the mass spectra of adenosine and 2′-deoxyadenosine.

Since the branched-chain sugar synthesized by Raphael and Roxburgh (1955) was given the name cordycepose, Fox et al. (1966) suggested that further confusion in the chemical literature be avoided by abandoning the trivial name "cordycepose." They recommended the term "3-deoxyapiose" for the branched-chain sugar synthesized by Raphael and Roxburgh. This nomenclature would permit continued use of the trivial name "cordycepin" for the nucleoside antibiotic elaborated by *C. militaris* or *A. nidulans* and 3-deoxy-D-ribose for the carbohydrate moiety.

## CHEMICAL SYNTHESIS OF CORDYCEPIN (3′-DEOXYADENOSINE) AND 3′-THIOADENOSINE

The first total chemical synthesis of 3′-deoxyadenosine was reported by Todd and Ulbricht (1960).

Lee et al. (1961) and Walton et al. (1964) also described the direct synthesis of 3′-deoxyadenosine. The synthesis of cordycepin by Walton et al.

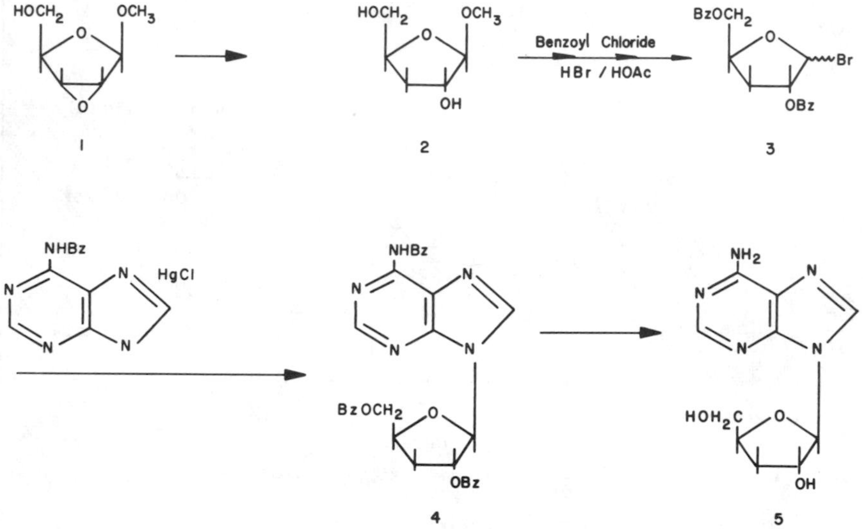

FIGURE 1.24. *Chemical synthesis of cordycepin* (*From Walton et al.*, 1964.)

(1964) is outlined in Figure 1.24. In this synthesis, methyl-2,3-anhydro-β-D-ribofuranoside (**1**), on treatment with Raney nickel, caused a stereospecific reduction of the epoxide function in **1** to give methyl-3-deoxy-β-D-ribofuranoside (**2**). The hydroxyl groups on carbons 2 and 5 were benzoylated. Subsequent treatment with HBr and acetic acid resulted in the formation of the 1-bromo compound (**3**). The bromo sugar (**3**) was coupled with monochloromercuri-6-benzamidopurine to yield the blocked nucleoside (**4**), which on removal of the protecting groups, yielded 3′-deoxyadenosine (**5**). Ikehara and Tada (1967) and Ikehara et al. (1968) have utilized the technique of converting their purine cyclonucleosides to 2′-deoxy- and 3′-deoxyadenosine (cordycepin) (Fig. 1.25). Adenosine (**6**) is converted to 5′-*O*-acetyl-8-bromoadenosine (**7**) through a series of reactions. Compound (**7**) was tosylated and deacetylated to the 3′-tosyl derivative (**8**), which on treatment with thiourea was subjected to cyclization to give 8,3′-anhydro-8-mercapto-9-β-D-xylofuranosyladenine (**9**). Desulfurization of **9** with Raney nickel afforded 3′-deoxyadenosine (**5**). Murray and Prokop (1968) described the synthesis of cordycepin by use of the titanium chloride technique. This procedure was used earlier by Baker et al. (1955) in the total chemical synthesis of purine nucleosides. The titanium chloride eliminates the need for an *O*-acylglycosyl halide.

Acton et al. (1967) reported the use of the neighboring group participation reaction of *trans*-*O*-thionbenzoate to synthesize 3′-thioadenosine.

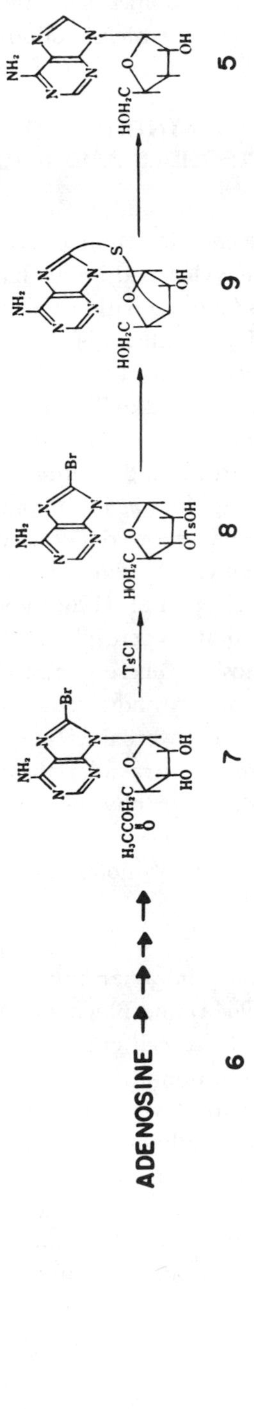

FIGURE 1.25. *Chemical synthesis of cordycepin.* (*From Ikehara and Tada*, 1967.)

Leonard and Rasmussen (1968) have described the synthesis of 3-β-(3′-deoxy-D-ribofuranosyl)adenine, an isomer of cordycepin. The synthesis makes use of 7-pivaloyloxymethyladenine which is obtained from sodium adenide and chloromethyl pivalate.

## CHEMICAL AND ENZYMATIC SYNTHESES OF 3′-DEOXYPURINE AND 3′-DEOXYPYRIMIDINE NUCLEOSIDES AND NUCLEOTIDES

The syntheses of the 5′-mono-, di-, and triphosphates of 3′-deoxyadenosine have been accomplished by chemical and enzymatic procedures. Cyanoethylphosphate and dicyclohexylcarbodiimide have been used in the 5′-phosphorylation of 3′-deoxyadenosine without protection of the hydroxyl on C-2′ of 3′-deoxyadenosine. The synthesis of the 5′ phosphates of 3′-deoxyadenosine as described by Rottman et al. (1963) and Suhadolnik et al. (1968a) was performed by the method of Tener (1961). The 9-(3-deoxy-β-D-ribofuranosides) of 2,6-diaminopurine, purine, purine-6-thiol, 6-methylaminopurine, 6-ethylaminopurine, and 6-dimethylaminopurine have also been synthesized, and their properties as inhibitors of the growth of KB and chick fibroblast cells have been reported (Walton et al., 1965). The synthesis of the pyrimidine 3′-deoxynucleosides and 3′-deoxy-2-fluoroadenosine has also been described by Walton et al. (1966) and Dickinson et al. (1967).

Klenow (1963a) reported that hyperdiploid Ehrlich ascites tumor cells phosphorylated 3′-deoxyadenosine. These nucleotides, resistant to periodate oxidation, were analyzed chemically and enzymatically. The results strongly suggested that the three compounds were the 5′-mono-, di-, and triphosphates of 3′-deoxyadenosine. Three major ultraviolet-absorbing peaks were obtained following ion-exchange chromatography of the acid-soluble fraction of tumor cells. Rottman and Guarino (1964c) showed that *B. subtilis* was able to phosphorylate cordycepin to the 5′-monophosphate.

## INHIBITION OF GROWTH

Cordycepin inhibits *B. subtilis* and avian tubercule bacillus (Cunningham et al., 1951). Kaczka et al. (1964a) and Bloch and Nichol (1964) reported on the inhibition of the growth of KB cell cultures by cordycepin, and Jagger et al. (1961) described the inhibition of Ehrlich ascites tumor cells.

The 9-(3-deoxy-β-D-ribofuranosides) of 2,6-diaminopurine, purine, purine-6-thiol, 6-methylaminopurine, 6-ethylaminopurine, and 6-dimethylaminopurine derivatives have been tested as inhibitors of KB and chick fibroblast cell cultures (Walton et al., 1965; Gitterman et al., 1965). 3′-Deoxyadenosine *N*′-oxide markedly inhibited the tumor growth and increased the survival time of mice (Frederiksen and Rasmussen, 1967).

## BIOSYNTHESIS OF 3′-DEOXYADENOSINE

The biosynthesis of cordycepin was first studied by Kredich and Guarino (1961). They reported that $^{14}C$-labeled adenine was incorporated into the adenine of cordycepin. Glucose-6-$^{14}C$ was incorporated more efficiently into the 3-deoxyribose moiety than was glucose-1-$^{14}C$. The incorporation of ribose-1-$^{14}C$ into 3-deoxyribose was also studied. The radioactive ribose was taken up from the culture medium very rapidly, but was not incorporated into cordycepin. Acetate-1-$^{14}C$ and isovalerate-1-$^{14}C$ were also used. These latter studies were based on the assumption that the sugar moiety of cordycepin was a branched-chain pentose. Neither acetate nor isovalerate was incorporated into the sugar moiety. In 1964 Suhadolnik et al. reported that adenosine-U-$^{14}C$ was incorporated into 3′-deoxyadenosine. The radioactive adenosine was added to cultures of *C. militaris* at the time of cordycepin biosynthesis. The per cent distribution of carbon-14 in the adenine/ribose of the adenosine added to the culture medium was 40:60, respectively. The distribution of carbon-14 in the adenine and 3-deoxyribose of the cordycepin isolated from these experiments was the same. The results obtained are consistent with the notion that adenosine is directly converted to 3′-deoxyadenosine without cleavage of the carbon–nitrogen riboside bond. Additional proof that the adenine–ribose bond of adenosine was not hydrolyzed during the biosynthesis of cordycepin was provided by the simultaneous addition of adenosine-U-$^{14}C$ and unlabeled D-ribose to the culture medium. The distribution of radioactivity in the adenine/3-deoxyribose of cordycepin from these experiments was the same as the labeling pattern when only adenosine-U-$^{14}C$ was added to the culture medium. Additional proof that adenosine is a direct precursor for 3′-deoxyadenosine was supplied by degradation of the 3-deoxyribose moiety of 3′-deoxyadenosine from the adenosine-U-$^{14}C$ experiments (Suhadolnik et al., 1964).

The reduction of carbon-3′ of preformed adenosine or a 5′-phosphorylated derivative of adenosine by *C. militaris* may occur in a manner analogous to the reduction of nucleoside triphosphates by ribonucleotide reductase from *Lactobacillus leichammanii* (Blakley and Barker, 1964) or the reduction of nucleoside diphosphates by the ribonucleotide reductase from *E. coli* (Reichard, 1962) and a rat tumor (Moore, 1967). However, studies with cell-free extracts of *C. militaris* failed to show the formation of cordycepin.

Chassy and Suhadolnik (1969) have studied the metabolic fate of adenosine-(U)-$^{14}C$ by *C. militaris* at the time of 3-deoxyadenosine biosynthesis. There is a very active transport system for uptake of adenosine by *C. militaris*. Examination of the acid-soluble pool of *C. militaris* showed very little cordycepin 5′-monophosphate. Fifteen per cent of the carbon-14 from adenosine-(U)-$^{14}C$ was incorporated into RNA 24 hr after the addition of the labeled adenosine. Cordycepin was not phosphorylated. Although adenosine is converted directly to cordycepin, the enzymatic mechanism by which the reduction of the 3′-hydroxyl occurs is not known.

One of the difficulties in extracting enzymes from *C. militaris* is the harsh mechanical methods needed to disrupt the cell. In order to find a more advantageous method to disrupt cells of *C. militaris* for *in vitro* studies, Marks, Keller and Guarino (1969) have studied the chemical components in the cell wall fractions from three separate cell cultures following disruption of frozen cells in a Sorvall Omnimixer at one-half maximum speed. The cell wall fraction from the three preparations was essentially reproducible in terms of its chemical components. Chitinase-treated cells were as difficult to break as the intact cells, even though the wall had apparently disappeared. The composition of the cell wall fraction of *C. militaris* is similar to other Ascomycetes that have been studied. Glucose is the major carbohydrate component. No trace of 3′-deoxypentose, the sugar moiety of cordycepin, was found. Marks, Keller and Guarino concluded that cordycepin is not used by *C. militaris* to any significant extent for cell wall structure even though this nucleoside antibiotic is produced by this organism in large quantities.

## BIOCHEMICAL PROPERTIES

### I. Effect of Cordycepin on Bacterial Systems

The first report of bacteria inhibited by cordycepin was by Cunningham et al. (1951). They showed that 43 of 45 strains of *B. subtilis* were inhibited by 3′-deoxyadenosine. The antibiotic did not inhibit the growth of *Staphylococcus aureus oxf. H*, *Sarcina lutea*, *E. coli*, *B. proteus*, *Streptococcus hemolyticus*, *S. flexneri*, *S. faecalis*, or *Pasteurella septica*. A most complete study on the effect of 3′-deoxyadenosine on a strain of *B. subtilis* (ATCC 10783) has been reported by Guarino and his co-workers (Guarino, 1967). Their studies showed that this nucleoside antibiotic inhibits purine biosynthesis *de novo* (Rottman and Guarino, 1964a). The hypoxanthine analog of cordycepin inhibited the growth of *B. subtilis*. However, much higher concentrations were required to produce the same degree of inhibition as with 3′-deoxyadenosine. In view of the reports that the desamino derivatives of tubercidin and formycin can be aminated in the cells (Acs et al., 1964; Bloch et al., 1969; Sawa et al., 1968), it may be that the toxicity of the hypoxanthine analog of cordycepin can also be attributed to the reamination of cordycepin. However, this has not been shown experimentally. Adenine and guanine were most effective in reversing the inhibition by this nucleoside antibiotic. The specific site inhibited by cordycepin appeared to involve an enzymatic reaction prior to the formation of glycinamide ribonucleotide. These conclusions are based on the uptake and incorporation of formate-$^{14}C$ and glycine-$^{14}C$ into the nucleic acid of *B. subtilis* in the presence or absence of 3′-deoxyadenosine (Rottman and Guarino, 1964b). Cordycepin inhibited the incorporation of formate-$^{14}C$ into the purines of RNA and DNA. It was also observed that the incorporation of 5-amino-4-imidazolecarboxamide-$^{14}C$ (AICA) into the purine ring was not markedly inhibited by cordycepin. These results are consistent with the idea that cordycepin inhibition occurs prior to the formation of the AICA-ribonucleotide. Since no specific compounds accumulated when either glycine-$^{14}C$ or formate-$^{14}C$ were added to *B. subtilis* in cultures in the presence of cordycepin, it appears that the metabolic block must occur in purine biosynthesis at a step prior to the incorporation of glycine. When bacteria are inhibited by 6-diazo-5-oxonorleucine(DON), formylglycinamide ribonucleotide (FGAR) is known to accumulate. Inhibition of any reaction prior to FGAR formation results in a decrease in the amount of glycine-$^{14}C$ incorporated into FGAR. When cordycepin was added to freshly harvested *B. subtilis* cells along with glycine-1-$^{14}C$ and DON, the FGAR was only one-tenth as radioactive as the control cells. These results indicate that cordycepin blocked an earlier stage in purine biosynthesis. Since glycinamide ribonucleotide (GAR) is the intermediate prior to the formation of FGAR, the accumulation of this radioactive intermediate was studied. It was assumed that cordycepin exerted its growth-inhibiting effect on either ribosephosphate pyrophosphokinase or phosphoribosyl pyrophosphate amidotransferase.

In subsequent studies using phosphoribosyl pyrophosphate amidotransferase from either pigeon liver or *B. subtilis*, Rottman and Guarino (1964c) were able to demonstrate that this enzyme was competitively inhibited by cordycepin 5′-monophosphate (Fig. 1.20). *B. subtilis* was able to phosphorylate cordycepin to the monophosphate.

The extent of inhibition of phosphoribosyl pyrophosphate amidotransferase by 3′-deoxyadenosine 5′-monophosphate and of adenosine 5′-monophosphate was essentially the same. The free nucleoside antibiotic was without effect. On the basis of these studies, Guarino and his co-workers have clearly demonstrated that the growth-inhibiting effect of cordycepin 5′-monophosphate in *B. subtilis* might be explained by the inhibition of phosphoribosyl pyrophosphate amidotransferase.

### II. Effect of Cordycepin 5′-Triphosphate (3′-Deoxy-5′-ATP) on DNA and RNA Polymerase

One additional bacterial enzyme that has been studied with 3′-deoxy-5′-ATP is the partially purified DNA polymerase from *E. coli*. Suhadolnik and Cory (1965) reported that 3′-deoxy-5′-ATP affected DNA synthesis by inhibiting the incorporation of either 2′-deoxy-ATP or 2′-deoxy-CTP into an acid-insoluble product. Evidence was also presented that 3′-deoxy-5′-

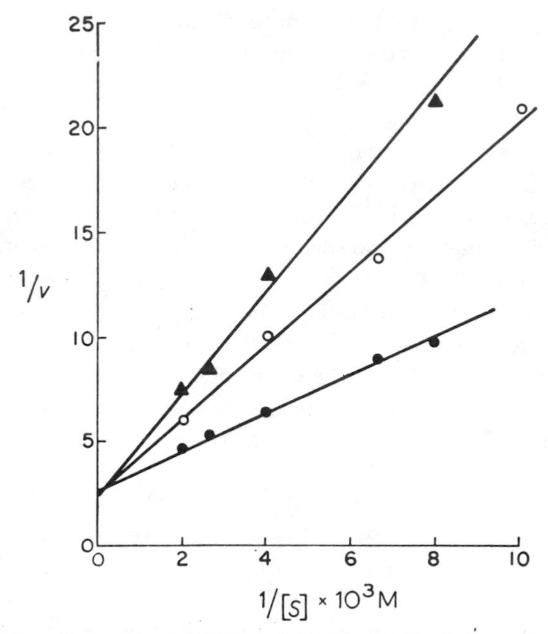


FIGURE 1.26. *Competitive inhibition of* 5-*phosphoribosyl*-1-*pyrophosphate binding to phosphoribosyl-pyrophosphate amidotransferase by cordycepin* 5′-*monophosphate.* (*From Rottman and Guarino*, 1964*c*.)

ATP was incorporated into a polymer by the *E. coli* DNA polymerase. The 3′-deoxy-5′-ATP used was synthesized by Ehrlich ascites tumor cells. Subsequent studies (Suhadolnik, unpublished results) with chemically synthesized tritium-labeled 3′-deoxy-5′-ATP showed that no radioactivity was incorporated into the acid-insoluble product. The results of earlier studies on the incorporation of the 3′-deoxy-5′-ATP (synthesized by Ehrlich ascites tumor cells) into DNA may be attributed to the deamination of 3′-deoxy-5′-ATP to form 3′-deoxy-5′-ITP.

Shigeura and Boxer (1964) reported that 3′-deoxy-5′-ATP-$^{14}$C was incorporated into RNA by partially purified RNA polymerase from *Micrococcus lysodeikticus*. Alkaline hydrolysis of the RNA isolated following the incorporation of 3′-deoxy-5′-ATP-$^{14}$C showed that all the radioactivity resided in the nucleoside fraction and essentially no radioactivity was in the nucleotide fraction. These results strongly suggested that 3′-deoxyadenosine was incorporated into the terminal position of the polynucleotide chain and prevented further elongation of the polymer. Shigeura and Gordon (1965) investigated the effect of 3′-deoxy-5′-ATP on the DNA-dependent RNA polymerase from *M. lysodeikticus*. They found that 3′-deoxy-5′-ATP inhibited RNA synthesis and poly A formation but did not inhibit the formation of polyuridylate (Table 1.7). The mono- and diphosphates of 3′-deoxyadenosine were not inhibitory. These data indicate that 3′-deoxy-5′-ATP competed with ATP in various polymerization reactions (Shigeura and Gordon, 1965).

**Table 1.7.**

Effects of 3′-Deoxyadenosine 5′-Triphosphate on Other Polymerization Reactions Catalyzed by RNA Polymerase from *Micrococcus lysodeikticus*

| Tube | $^{12}C$-Nucleotide triphosphate | $^{14}C$-Nucleotide triphos-phate | Polymer added | 3′-dATP, *mM* | $^{14}C$-Nucleotide triphosphate incor-porated, *mμmoles* | Inhibi-tion, % |
|---|---|---|---|---|---|---|
| 1 | ATP, CTP, UTP | GTP | DNA | | 27.30 | |
| 2 | ATP, CTP, UTP | GTP | | | 0.02 | |
| 3 | | ATP | DNA | | 47.60 | |
| 4 | | ATP | DNA | 0.12 | 1.10 | 98 |
| 5 | | ATP | | | 0.06 | |
| 6 | | UTP | DNA | | 2.40 | |
| 7 | | UTP | DNA | 0.12 | 2.50 | 0 |
| 8 | | UTP | | | 0.05 | |
| 9 | | ATP | Polyuridylic acid | | 0.64 | |
| 10 | | ATP | Polyuridylic acid | 0.08 | 0.06 | 94 |
| 11 | | ATP | | | 0.07 | |
| 12 | | UTP | Polyadenylic acid | | 3.70 | |
| 13 | | UTP | Polyadenylic acid | 0.08 | 3.76 | 0 |
| 14 | | UTP | | | 0.04 | |
| 15 | ATP, CTP, UTP | GTP | Yeast RNA | | 1.25 | |
| 16 | ATP, CTP, UTP | GTP | Yeast RNA | 0.12 | 1.10 | 12 |
| 17 | ATP, CTP, UTP | GTP | | | 0.02 | |
| 18 | ATP, CTP, UTP | GTP | *E. coli* RNA | | 1.58 | |
| 19 | ATP, CTP, UTP | GTP | *E. coli* RNA | 0.12 | 1.32 | 17 |

From Shigeura and Gordon, 1965.

More recently, Suhadolnik et al. (1968b), in their studies on the incorporation of the 5′-triphosphate of the pyrrolopyrimidine nucleoside antibiotic sangivamycin, reported that 3′-deoxy-5′-ATP inhibited the incorporation of tritium-labeled sangivamycin 5′-triphosphate into RNA with the partially purified DNA-dependent RNA polymerase from *M. lysodeikticus*.

Sentenac, Ruet, and Framageot (1968) also used cordycepin triphosphate to study the effect on initiation of RNA chains with partially purified RNA polymerase. When cordycepin triphosphate was present, the polymerase synthesized very short RNA chains. The first nucleotide of these acid-soluble chains was GTP. Filtration or Sephadex G-75 indicated that the ogligonucleotides were associated with a DNA-enzyme complex. They were not sensitive to RNase or DNase. When phage $T_7$ DNA was the primer, the number of chains initiated by adenine and guanine was about 50. This was observed at the point where RNA synthesis, as a function of RNA polymerase concentration, departed from proportionality.

## III. Binding of Cordycepin 5′-Diphosphate and 5′-Triphosphate at the Catalytic and Allosteric Sites of Ribonucleotide Reductases

3′-Deoxyadenosine 5′-mono, di-, and triphosphates have been used as biological tools to study the interaction between the catalytic and regulatory sites of several bacterial enzymes. Suhadolnik et al. (1968a) and Chassy and Suhadolnik (1968) have reported that 3′-deoxy-5′-ATP was not a substrate for the ribonucleotide reductase isolated from *L. leichmannii* and *E. coli*. However, 3′-deoxy-5′-ATP was 50% as good a prime effector as 2′-dATP for stimulating the reduction of CTP (Table 1.8).

**Table 1.8.**

Allosteric Stimulation of CTP Reduction by Nucleotide Antibiotics

| Addition | Deoxycytidine formed, *mμmoles* | Stimulation (ratio of activity) |
|---|---|---|
| None | 0.93 | 1.0 |
| 2′-dATP | 12.70 | 13.5 |
| 3′-Deoxy-5′-ATP | 6.75 | 7.2 |
| TuTP | 6.02 | 6.5 |
| ToTP | 3.79 | 4.1 |
| FTP | 1.19 | 1.3 |

From Suhadolnik et al., 1968a.

To determine if the stimulation of CTP reduction by 3′-deoxy-5′-ATP was kinetically similar to that observed for 2′-dATP, saturation curves in the presence of 2′-ATP or 3′-deoxy-5′-ATP were determined. Figure 1.27*a* (with $Mg^{2+}$) showed that 2′-dATP or 3′-deoxy-5′-ATP lowered the apparent half-saturation concentration of CTP to about $1 \times 10^{-3}$ *M*; 3′-deoxy 5′-ATP was about 60% as stimulatory as 2′-dATP. There was no $Mg^{2+}$ in the assays shown in Figure 1.27*b*. 3′-Deoxy-5′-ATP interacted at the allosteric site and modified the ribonucleotide reductase from *E. coli* such that the

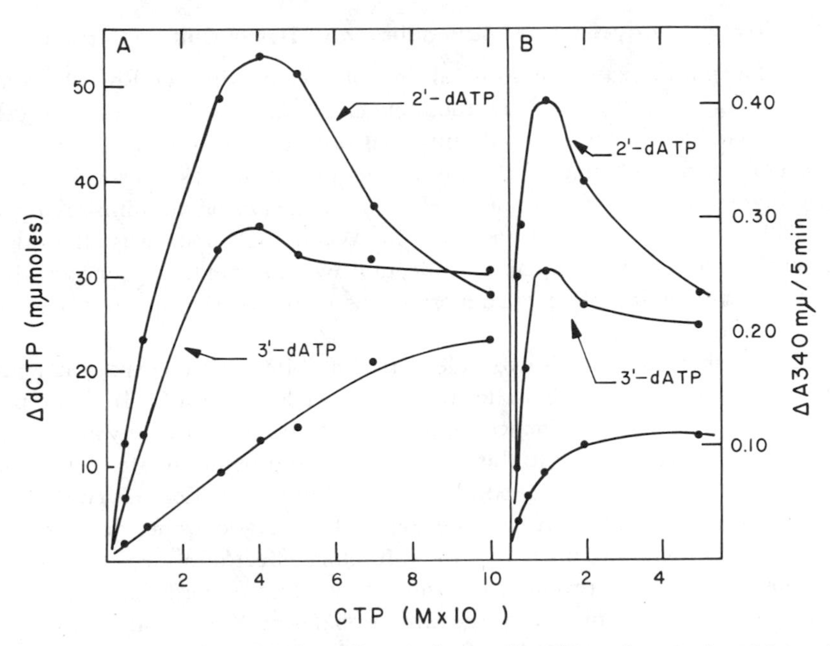


FIGURE 1.27. *Effect of 2′-dATP and 3′-deoxy-5′-ATP on the reduction of increasing concentrations of CTP. (From Suhadolnik et al.,* 1968*a*.)

reduction of ADP and GDP were inhibited. 3′-Deoxy-5′-ATP was not as good an allosteric modifier as was 2′-dATP. These data show that a hydroxyl on position 3′ of the ribonucleoside 5′-di- or 5′-triphosphate does not appear to be necessary for interaction at the allosteric site of the ribonucleotide reductases from *E. coli* or *L. leichmannii*. The data obtained from *E. coli* ribonucleotide reductase show that the structural requirements for interaction at the regulatory site are more specific than for the *L. leichmannii* ribonucleotide reductase.

It was concluded that the reactivity of 3′-deoxy-5′-ATP with ribonucleotide reductase may be an important aspect of the molecular biology of nucleoside antibiotics. However, the reductases are only one of a number of enzymes that have been reported to be affected by the nucleoside antibiotics.

Additional studies on the effect of nucleotide antibiotics on quaternary structure on alteration of AMP are reported by Nakazawa et al. (1967) and Rabinowitz et al., (1968). They reported that alterations at C-1′, C-2′, C-3′, and C-5′ of the ribosyl moiety of AMP abolished allosteric activity for threonine dehydrase. Those analogs differing in the structure of the adenine moiety had essentially the same allosteric activity as AMP.

## IV. Effect of Cordycepin on Mammalian and Tissue Culture Systems

Cordycepin increased the survival time of mice bearing the Ehrlich ascites tumor (Jagger et al., 1961). In those experiments in which the tumor cells and 3′-deoxyadenosine were administered simultaneously followed by daily injections of the antibiotic for 7 days, a significant increase in survival time was observed when the amount of 3′-deoxyadenosine administered was 15–200 mg per kilogram of body weight. When the tumor was allowed to develop before 3′-deoxyadenosine injections were started, the survival time of animals receiving 3′-deoxyadenosine was increased significantly over the control animals.

Studies showed that 3′-deoxyadenosine inhibited $^{32}P$ incorporation into DNA and RNA of Ehrlich ascites tumor cells (Klenow, 1961). In this report Klenow also presented evidence that Ehrlich ascites tumor cells were able to phosphorylate cordycepin to the 5′-mono-, di-, and triphosphates. Deamination of 3′-deoxyadenosine abolished the inhibitory action of cordycepin against RNA and DNA. Adenosine, but not deoxyadenosine, deoxyguanosine, or deoxycytidine, prevented the inhibitory effect of 3′-deoxyadenosine. Adenosine not only prevented the inhibitory effect of cordycepin, but also prevented the accumulation of the triphosphates of 3′-deoxyadenosine. The suggestion was made that the effect of 3′-deoxyadenosine on nucleic acid synthesis in these cells may be due to the formation of the corresponding triphosphate or a closely related compound. Klenow (1963a) subsequently showed that suspensions of Ehrlich ascites tumor cells, incubated in the presence of 3′-deoxyadenosine, were able to convert cordycepin to the 5′-mono-, di-, and triphosphates. Klenow was also able to show that 3′-deoxy-5′-ADP would replace ADP as a phosphate acceptor in the pyruvate kinase reaction when phosphoenolpyruvate was the phosphate donor. 3′-Deoxy-5′-ATP could also replace ATP as an energy source for the phosphorylation of D-fructose by purified rat liver fructokinase (Adelman et al., 1967).

Klenow (1963a) and Frederiksen (1963), using cordycepin-$N^1$-oxide, studied the nucleotide pool, RNA synthesis, and DNA synthesis with Ehrlich ascites tumor cells. Klenow reported that 80% of the cordycepin added to these tumor cells was converted to the 5′-mono-, di-, and triphosphates in 30 min. A considerable amount of deamination occurred. The addition of equimolar quantities of adenosine blocked the phosphorylation of cordycepin. Low concentrations of cordycepin (0.5–1.0 μmole/ml) resulted in the accumulation of the 5′-triphosphate derivative. At concentrations of 1.6 and 2.0 μmoles of cordycepin per milliliters of cells, the 5′-mono-, di-, and triphosphates were formed. At higher concentrations of cordycepin, there was a marked decrease in the acid-soluble pool of ribonucleotides and a pronounced inhibition of incorporation of $^{32}P$ into the RNA and DNA. In those experiments in which only 3′-deoxy-5′-ATP accumulated in the cell, there was no inhibition of the incorporation of $^{32}P$ into DNA (Klenow, 1963b). These results indicate that cordycepin 5′-monophosphate and cordycepin 5′-diphosphate, rather than the triphosphate derivative, were the more immediate inhibiting substances. Frederiksen's (1963) data with cordycepin-$N^1$-oxide were in agreement with Klenow's findings. RNA synthesis was blocked 1.5 hr after incubation and DNA synthesis was blocked 3 hr after incubation with cordycepin-$N^1$-oxide. Frederiksen concluded that cordycepin 5′-diphosphate probably causes the inhibition of DNA synthesis, while cordycepin 5′-di- or triphosphate or both inhibited RNA synthesis.

In a subsequent report, Klenow and Overgaard-Hansen (1964) showed that under experimental conditions in which cordycepin 5′-triphosphate accumulated, there was no significant inhibition in the incorporation of $^{32}P$ into the DNA of the tumor cells. However, the incorporation of adenine-$^{14}C$ into DNA was greatly inhibited. They concluded that 3′-deoxy-5′-ATP was inhibiting either reaction 1 or 2:

$$\text{ATP} + \text{ribose-5-P} \rightarrow \text{PRPP} + \text{AMP} \quad (1)$$

$$\text{Adenine} + \text{PRPP} \rightarrow \text{AMP} + \text{PP}_i \quad (2)$$

Reaction 1 is catalyzed by ribosephosphate pyrophosphokinase, and reaction 2 is catalyzed by adenine phosphoribosyl transferase. Overgaard-Hansen (1964) subsequently showed that 3′-deoxy-5′-ATP, with the high-speed supernatant from Ehrlich ascites tumor cells, inhibited reaction 1 (the formation of PRPP from ribose-5-phosphate and ATP). Since PRPP holds a key position for growth and is also an obligatory reactant in the *de novo* synthesis of purines, pyrimidines, and pyrimidine nucleotides, it is reasonable to assume that the inhibition by cordycepin in Ehrlich ascites tumor cells may be partially attributed to the inhibition that 3′-deoxy-5′-ATP exerts on the synthesis of PRPP.

Klenow and Frederiksen (1964) subsequently studied the effect of 3′-deoxy-5′-ATP on the enzyme purified from cell nuclei of Ehrlich ascites tumor cells that catalyzed the DNA-dependent synthesis of RNA. When 3′-deoxy-5′-ATP was added to the enzyme assay, the incorporation of AMP-$^{32}P$ into the RNA was completely prevented. No $^{32}P$ from 3′-deoxy-5′-ATP experiments could be found in the RNA. Frederiksen and Klenow (1964) subsequently showed that under conditions in which 3′-deoxy-5′-ATP accumulated in Ehrlich ascites tumor cells (1.4 $\mu$moles per gram of cells), the incorporation of $^{32}P$ into RNA was inhibited 50% after 60 min. When the RNA was fractionated, it was observed that the inhibitory effect of cordycepin on RNA labeling was not uniform. For example, the inhibition of incorporation of $^{32}P$ was found to occur in the cytoplasmic RNA (c-RNA)

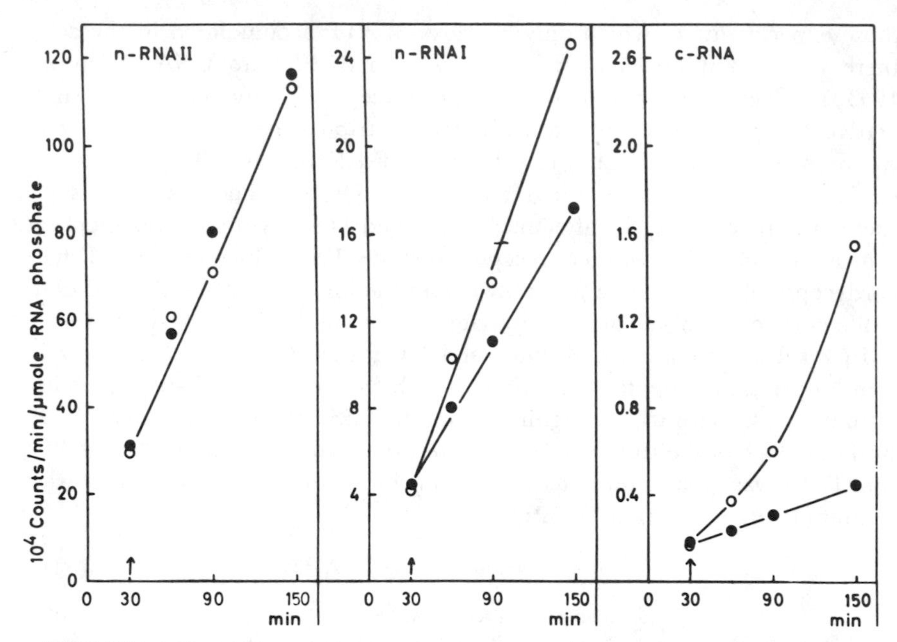


FIGURE 1.28. *Effect of* 3′-*deoxyadenosine on the incorporation of* $^{32}P_i$ *into different RNA fractions of Ehrlich ascites tumor cells in vitro. Each vessel contained per milliliter: Ehrlich ascites tumor cells,* 85 *mg (wet weight); ascites fluid,* 415 *µl; Robinson's medium containing glucose* (5.6 × $10^{-3}$ *M*), 500 *µl; folic acid,* 20 *mµmoles, and* $^{32}P_i$, 20 *µC. Additions:* (○) *none;* (●) 3′-*deoxyadenosine* (0 .35 *µmole*). 3-*Deoxyadenosine was added to experimental flask after incubation with shaking for* 30 *min at* 37° *C. Samples were withdrawn at time intervals and the cells were washed with* 0.15 *M NaCl. Suspensions of washed cells were treated with* 1 *vol. phenol at* 0°*C for about* 10 *min. After centrifugation, the aqueous phase containing cytoplasmic RNA was made* 10 *mM with respect to* $MgCl_2$ *and treated with* 2 *vol.* 96% *ethanol. The RNA precipitate was dissolved and reprecipitated. The interphase was washed several times with phenol–phosphate buffer mixtures and finally treated with sodium lauryl sulfate and phenol at* 70°*C. After centrifugation of the cooled mixture, n-RNA I was obtained from the aqueous phase by precipitation with* $MgCl_2$ *and* 2 *vol. ethanol. The n-RNA II fraction was obtained together with protein as a precipitate from the remaining interphase after addition of about* 10 *vol.* 66% *ethanol. (From Frederiksen and Klenow,* 1964.)

and one of the nuclear RNA fractions (n-RNA-I). The second nuclear fraction of RNA (n-RNA-II) did not show any inhibition of incorporation of $^{32}P$ (Fig. 1.28).

Truman and Frederiksen (1969) have studied the effect of cordycepin and 3′-amino-3′-deoxyadenosine on the differential labeling of RNA subspecies in Ehrlich ascites tumor cells. The cytoplasmic RNA and RNA of the nucleochromosomal apparatus were separated by the use of phenol and sodium dodecyl sulfate. Differential labeling of RNA was observed when uridine-$^3$H or adenine-$^3$H was used in place of $^{32}P$. The labeling of cytoplasmic RNA is inhibited about 75%, nuclear RNA-I 50%, and nuclear RNA-II only 10–20%. Subfractionation of the cytoplasmic RNA by sucrose-gradient centrifugation is shown in Figure 1.29. After a 4-hr incubation with 3′-deoxyadenosine, the 28-S ribosomal peak is inhibited 80%, the 18-S peak is inhibited 60%, and the 4–5-S peak only 40%. In contrast, the 50–60-S nuclear RNA is only slightly inhibited. The nuclear RNA-I fraction has a broad heterogeneous base with peaks at 28-S and 18-S. Cordycepin tends to flatten the 28-S and 18-S peaks, but does not inhibit the high molecular weight peaks. This RNA is very rapidly labeled and has a high specific activity and a short half-life. The elution of RNA-I from methylated albumin-kieselguhr requires higher NaCl concentrations than cytoplasmic RNA. The slowest eluting region of RNA-I is not inhibited by cordycepin. Nuclear RNA-II resembles the 50–60-S nuclear RNA-I. Three hypothetical explanations were offered to answer the question of how labeling of the 50–60-S nuclear RNA-I and nuclear RNA-II can occur and not be inhibited by the 3′-deoxy analogs, while cytoplasmic RNA is profoundly inhibited. First, the RNA species studied could be synthesized by RNA polymerases that differ in their specificity towards cordycepin and 3′-amino-3′-deoxyadenosine. Second, the nucleoside triphosphates may be located in compartments around the DNA that serve as templates for the various RNAs. The third

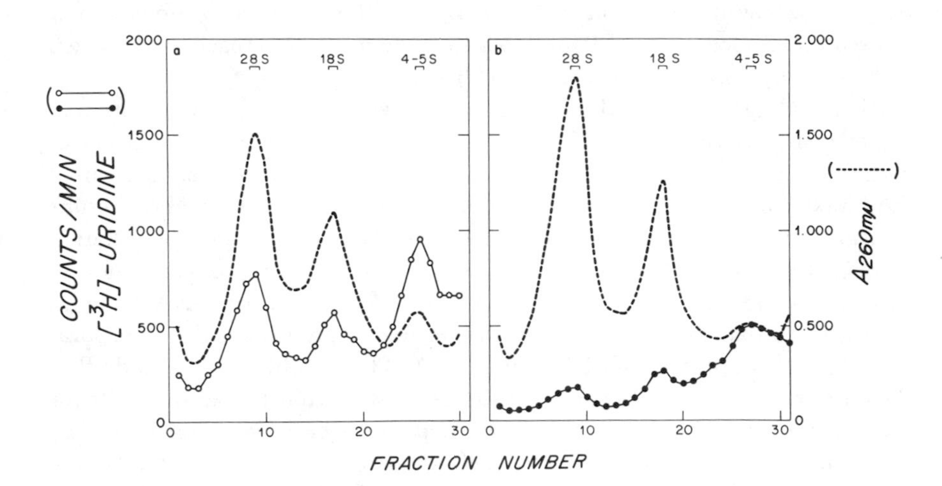


FIGURE 1.29. *Effect of* 3′-*deoxyadenosine on the labeling patterns of cytoplasmic RNA after sucrose-gradient centrifugation;* 33.3μC [³H]*uridine* (*specific activity,* 33.3 μC/μmole) *and continuous labeling for* 4 *hr. Separation of cytoplasmic RNA was done by the use of phenol and sodium docecyl sulfate. Centrifugation performed for* 16 *hr at* 21,000 *rpm in a Spinco SW*-25 *rotor* (*av.* 50,500 × *g*). (*a*) *Control;* (*b*) 3′-*deoxyadenosine,* 0.3 μ*mole/ml.* (*From Truman and Frederiksen,* 1969.)

suggestion requires an RNA polymerase that would add nucleotides to the 5′ end of a polyribotide chain. This type polymerase would not add 3′-deoxy analogs, and RNA synthesis would continue (Truman and Frederiksen, 1969). The inhibition of 28-S and 18-S ribosomal peaks of the cytoplasmic RNA by the 3′-deoxy analogs may be analogous to the selective inhibition of rRNA synthesis in mouse fibroblasts by toyocamycin as reported by Tavitian et al., (1968). They reported that this pyrrolopyrimidine nucleoside had a more pronounced inhibitory effect on RNA synthesis than on DNA synthesis. This finding would be similar to the inhibition of RNA synthesis in Erhlich ascites tumor cells by 3′-amino-3′-deoxyadenosine as reported by Truman and Klenow (1968). The sucrose gradient profile of RNA extracted from intact cells showed that low concentration of toyocomycin completely inhibited the synthesis of 28-S and 18-S RNA, while the syntheses of the 4-S and 4-5-S RNA were not inhibited (see Chap. 8, Fig. 8.16).

When the incorporation of formate, glycine, hypoxanthine, adenine, and guanine into the RNA of Ehrlich ascites tumor cells was studied at various concentrations of 3′-deoxyadenosine, Shigeura and Gordon (1965) showed that the incorporation of these five labeled compounds into purine precursors of the RNA was inhibited more than 80% by cordycepin. Since the incorporation of a *de novo* purine precursor or intact purine (hypoxanthine) into RNA was inhibited by 3′-deoxyadenosine, it appeared that the RNA polymerase might be inhibited. Shigeura and Gordon (1965) subsequently showed that 3′-deoxy-5′-ATP inhibited the polymerization reaction catalyzed by RNA polymerase isolated from *M. lysodeikticus*.

Rich et al. (1965) studied the effect of cordycepin on the growth of human tumor cells (H. Ep. #1 cells) grown in culture. The growth inhibition of cordycepin is cytostatic and not cytocidal. Adenosine competitively prevented inhibition, but did not reverse inhibition once it had occurred. These results indicate that adenosine competes with cordycepin for phosphorylation through the nucleotide. Exposure of cells to an inhibitory concentration of cordycepin did not markedly affect the protein, RNA, or DNA content. However, cordycepin caused a two- to four-fold depression of the incorporation of adenosine-8-$^{14}$C into RNA and DNA. The conclusion was made that these results were not in accord with the suggestion that the sensitive site for growth inhibition by cordycepin is limited to an early reaction in purine biosynthesis. In a subsequent study, Cory et al. (1965) reported on the incorporation of cordycepin into RNA and DNA of H. Ep. #1 cells.

When these tumor cells were incubated with cordycepin-G-$^{3}$H, the acid-soluble pool, the RNA and the DNA were radioactive. The incorporation of radioactivity into the RNA fraction was 0.023%, while the incorporation into the DNA fraction was 0.04%. To determine if cordycepin was incorporated into RNA at the terminal position or as an internal nucleotide, the

RNA was hydrolyzed with alkali. The neutralized RNA hydrolyzate was further treated by paper chromatography. The nucleotides and oligonucleotides remained at the origin, while the nucleosides, bases, and sugars migrated. Elution of the radioactivity remaining at the origin of the chromatogram and hydrolysis with alkaline phosphatase followed by rechromatography in the same solvent resulted in a radioactive spot having the same $R_f$ as cordycepin or 2′-deoxyadenosine. A total of $3.3 \times 10^3$ cpm were eluted from the paper chromatogram. Rechromatography in ammonia–water (pH 10) resulted in the isolation of $2 \times 10^3$ cpm in an area having a $R_f$ equivalent to that of authentic cordycepin. Authentic carrier cordycepin was added to the radioactive material eluted from this chromatogram. The mixture was dissolved in 95% ethanol and allowed to crystallize overnight at 2°C. The crystalline cordycepin was subsequently recrystallized from water. The specific activities of the cordycepin isolated from crystallization from these two solvents were the same. A control experiment in which tritium-labeled 2′-deoxyadenosine was added to cordycepin, followed by crystallization from the same two solvents showed that the tritium-labeled 2′-deoxyadenosine did not coprecipitate with cordycepin. Additional proof that the nucleoside isolated was cordycepin was obtained by hydrolysis with Dowex-50-$H^+$. Chromatography of 3-deoxyribose showed that the 3-deoxyribose was radioactive. Radioactive cordycepin was also isolated from the 3′-terminal position of RNA. If indeed cordycepin is incorporated into RNA, an abnormal 2′,5′-phosphodiester linkage must occur. It is difficult to explain why the 2′,5′-phosphodiester linkage in the RNA would be cleaved by alkali and enzymatic hydrolysis. This type bond should be chemically analogous to the 3′,5′-phosphodiester bond of DNA which is resistant to alkali. These results are in contrast to the reports of Klenow and Frederiksen (1964), Truman and Klenow (1968), Shigeura and Gordon (1965), Shigeura (1965), and Shigeura et al. (1966), in which they reported that cordycepin and 3′-amino-3′-deoxyadenosine were readily phosphorylated to the 5′-triphosphates and appear to inhibit RNA synthesis by incorporation into the terminal position of RNA and thereby block further growth of the polynucleotide chain. The *in vivo* studies of Cory et al. (1965) showed that a very small amount of cordycepin 5′-phosphate was actually incorporated into an internucleotide link.

Shigeura and Sampson (1967a) studied the demethylation and metabolism of 6-methylamino-9-(3′-deoxy-β-D-ribofuranosyl)-purine by KB cells. This nucleoside is converted to 3′-deoxyadenosine-5′-triphosphate, which is subsequently incorporated into the 3′-hydroxyl end of RNA. Another metabolic product is cordycepin. Although radioactivity from the cordycepin formed is present in the RNA in whole KB cells as adenosine, Shigeura and Sampson suggest that the isolation of radioactive adenosine from the RNA of whole KB cells following the administration and demethylation of 6-methylamino-9-(3′-deoxy-β-D-ribofuranosyl)purine to cordycepin can be attributed to the formation of hypoxanthine or the subsequent synthesis of inosine and IMP.

Until recently it was not known if alkaline phosphodiesterase was capable of hydrolyzing 2′,5′-linked dinucleoside phosphates. Zeleznick (1969) has now shown that soluble extracts from animal tissues contain an alkaline phosphodiesterase capable of hydrolyzing 2′,5′-linked dinucleoside phosphates of ara-C (see Chap. 3, Fig. 3.16, for more detailed discussion). Such a hydrolysis could explain the hydrolysis of cordycepin if it existed as a 2′,5′-linked dinucleoside phosphate. The *in vitro* evidence shows quite conclusively that the incorporation of cordycepin or 3′-amino-3′-deoxyadenosine is limited to the terminal position. If indeed cordycepin can actually be incorporated into the polynucleotide chain, it would appear unlikely that those cells incorporating a significant amount of this abnormal nucleoside could function normally. Cordycepin may well be a chain terminator, much the same as Atkinson and Kornberg (private communication) have shown for the addition of 9-β-D-arabinofuranosyladenine (ara-A) and 1-β-D-arabinofuranosylcytosine (ara-C) to DNA (see p. 150 of Chapter 3 for more detailed discussion).

## V. Phosphorylation and Hydrolysis of Cordycepin

In order to determine the structural basis for phosphorylation, a number of adenosine analogs were studied and summarized by Shigeura et al. (1966) Shigeura and Sampson (1967a; 1967b). They reported that 6-methylamino-9-(3′-deoxy-β-D-ribofuranosyl)purine was converted to the 5′-monophosphate. This nucleoside was also slowly demethylated to 3′-deoxyadenosine, which was then metabolized further along two separate pathways. The nucleosides containing *N*-dimethyl- and *N*-ethyl groups on C-6 of the purine ring were not phosphorylated. Removal of the 6-amino group or a substitution of one or both hydrogens on the 6-amino group of 3′-deoxyadenosine by an alkyl group drastically interfered with phosphorylation in the 5′ position of the nucleoside. A monomethyl group at the 6-amino position permitted the formation of only a small amount of 5′-mononucleotide. In addition, the 5′-di- and triphosphate kinases were inhibited by the nucleoside or nucleoside monophosphate. 6-Methylamino-9-(3′-deoxy-β-D-ribofuranosyl) purine 5′-monophosphate could not serve as a substrate for crystalline rabbit muscle myokinase. In contrast, 3′-deoxyadenosine-5′-monophosphate, 2-fluoroadenosine-5′-monophosphate, and 3′-amino-3′-deoxyadenosine-5′-monophosphate were converted to the 5′-diphosphates by rabbit muscle myokinase. Many of the $N^6$-alkylated nucleosides were not phosphorylated to the triphosphates.

Purified adenosine kinase from H. Ep. #2 cells did not phosphorylate cordycepin very readily (Schnebli et al., 1967). With rabbit liver adenosine kinase, cordycepin was a good substrate (Lindberg et al., 1967).

3′-Deoxynucleosides are resistant to glycosyl splitting by nucleoside phosphorylase (Walton et al., 1966, reference 13; LePage and Junga, 1965; Bloch and Nichol, 1964). Several 3′-deoxyadenosine analogs were synthesized by Walton et al. (1965) since an earlier report that 3′-deoxyadenosine was rapidly deaminated by Ehrlich ascites tumor cells (Klenow, 1963a; Weinbaum et al., 1964).

### VI. Inhibitory Effect of Cordycepin on C. *militaris*

Yoshida et al. (1966) reported that actively growing cells of *S. antibioticus* are sensitive to actinomycin. The organism is insensitive to the antibiotic during the time of antibiotic production. The antibiotic may function as a normal repressor in cellular metabolism. Similar studies by Chassy and Suhadolnik (1969) on the effect of adding cordycepin to *C. militaris* at various stages after inoculation show that cordycepin did not inhibit the growth of *C. militaris*. In addition, cordycepin-G-$^3$H was not taken up by *C. militaris* in the early stages of growth or at the time of cordycepin production. Apparently, once cordycepin is excreted into the medium, it is not taken up by the cell again. Since Rottman and Guarino (1964c) showed that phosphoribosyl pyrophosphate amidotransferase from pigeon liver and *B. subtilis* was inhibited by cordycepin 5′-monophosphate and that ribosephosphate pyrophosphokinase was inhibited by cordycepin 5′-triphosphate (Overgaard-Hansen, 1964), studies with cell-free extracts of *C. militaris* showed that these same two enzymes were inhibited in a similar manner (Chassy and Suhadolnik, 1969). Apparently, neither cordycepin 5′-monophosphate nor 5′-triphosphate accumulates in the acid-soluble pool of *C. militaris* at a sufficiently high concentration to inhibit these enzyme reactions by the producing organism.

## Summary

Cordycepin has been isolated from *C. militaris* and *A. nidulans*. The structure and chemical synthesis have been reported. Cordycepin is a cytostatic nucleoside. It accumulates in tumor cells as the 5′-mono-, di-, and triphosphates. It does not inhibit DNA synthesis, but is a strong inhibitor of RNA synthesis. It acts as a negative feedback inhibitor of purine nucleotide biosynthesis. Cordycepin 5′-monophosphate inhibits phosphoribosyl pyrophosphate amidotransferase, and cordycepin 5′-triphosphate inhibits ribosephosphate pyrophosphokinase. Adenosine reverses cordycepin inhibition. Adenosine is reduced to 3′-deoxyadenosine by *C. militaris* without cleavage of its *N*-riboside bond. The differential labeling of cytoplasmic RNA and nucleochromosomal RNA by cordycepin has been studied. The 28-S and 18-S ribosomal peaks are inhibited more than the 4–5-S peak. The 50–60-S nuclear RNA is only slightly inhibited.

## References


Acs, G., E. Reich, and M. Mori, *Proc. Natl. Acad. Sci. (U.S.)*, **52**, 493 (1964).

Acton, E. M., K. J. Ryan, and L. Goodman, *J. Amer. Chem. Soc.*, **89**, 467 (1967).

Adelman, R. C., F. J. Ballard, and S. Weinhouse, *J. Biol. Chem.*, **242**, 3360 (1967).

Baker, B. R., Schaub, R. E., Joseph, J. P., and Williams, J. H., *J. Am. Chem. Soc.*, **77**, 12 (1955).

Bentley, H. R., K. G. Cunningham, and F. S. Spring, *J. Chem. Soc.*, 1951, 2301.

Blakley, R. L., and H. A. Barker, *Biochem. Biophys. Res. Commun.*, **16**, 391 (1964).

Bloch, A., and C. A. Nichol, *Antimicrobial Agents Chemotherapy*, 1964, 530.

Bloch, A., E. Mihich, R. J. Leonard, and C. A. Nichol, *Cancer Res.*, **29**, 110 (1969).

Chassy, B. M., and R. J. Suhadolnik, *J. Biol. Chem.*, **243**, 3538 (1968).

Chassy, B. M., and R. J. Suhadolnik, *Biochim. Biophys. Acta*, **182**, 307 (1969).

Cory, J. G., R. J. Suhadolnik, B. Resnick, and M. A. Rich, *Biochim. Biophys. Acta*, **103**, 646 (1965).

Cunningham, K. G., S. A. Hutchinson, W. Manson, and F. S. Spring, *J. Chem. Soc.*, 1951, 2299.

Dickinson, M. J., F. W. Holly, E. Walton, and M. Zimmerman, *J. Med. Chem.*, **10**, 1165 (1967).

Fox, J. J., K. A. Watanabe, and A. Bloch, in "Progress in Nucleic Acid Research and Molecular Biology," J. N. Davidson and W. E. Cohn (Editors), **5**, Academic Press, N. Y. 1966, p. 251.

Frederiksen, S., *Biochim. Biophys. Acta*, **76**, 366 (1963).

Frederiksen, S., and H. Klenow, *Biochem. Biophys. Res. Commun.*, **17**, 165 (1964).

Frederiksen, S., and A. H. Rasmussen, *Cancer Res.*, **27**, 385 (1967).

Frederiksen, S., H. Malling, and H. Klenow, *Biochim. Biophys. Acta*, **95**, 189 (1965).

Gitterman, C. O., R. W. Burg, G. E. Boxer, D. Meltz, and J. Hitt, *J. Med. Chem.*, **8**, 664 (1965).

Guarino, A. J., in Anitbiotics, D. Gottlieb and P. D. Shaw (Editors), Springer Verlag, New York, 1967, p. 468.

Hanessian, S., D. C. DeJongh, J. A. McCloskey, *Biochim. Biophys. Acta*, **117**, 480 (1966).

Hudson, C. S., *Advan. Carbohydrate Chem.*, **4**, 57 (1949).

Ikehara, M., and H. Tada, *Chem. Pharm. Bull.*, **15**, 94 (1967).

Ikehara, M., H. Tada, M. Kaneko, and K. Muneyama, 156th National Meeting, American Chemical Society, Atlantic City, N.J., September 1968, Abstracts Medi., 28.

Jagger, D. V., N. M. Kredich. and A. J. Guarino, *Cancer Res.*, **21**, 216 (1961).

Kaczka, E. A., E. L. Dulaney, C. O. Gitterman, H. B. Woodruff, and K. Folkers, *Biochem. Biophys. Res. Commun.*, **14**, 452 (1964a).

Kaczka, E. A., N. R. Trenner, B. Arison, R. W. Walker, and K. Folkers, *Biochem. Biophys. Res. Commun.*, **14**, 456 (1964b).

Klenow, H., *Biochim. Biophys. Acta*, **76**, 347 (1963a).

Klenow, H., *Biochim. Biophys. Acta*, **76**, 354 (1963b).

Klenow, H., *Biochem. Biophys. Res. Commun.*, **5**, 156 (1961).

Klenow, H., and S. Frederiksen, *Biochim. Biophys. Acta*, **87**, 495 (1964).

Klenow, H., and K. Overgaard-Hansen, *Biochim. Biophys. Acta*, **80**, 500 (1964).

Kredich, N. M., and A. J. Guarino, *Biochim. Biophys. Acta*, **41**, 363 (1960),

Kredich, N. M., and A. J. Guarino, *Biochim. Biophys. Acta*, **47**, 529 (1961).

Lee, W. W., A. Benitez, C. D. Anderson, L. Goodman, and B. R. Baker, *J. Amer. Chem. Soc.*, **83**, 1906 (1961).

Leonard, N. J. and M. Rasmussen, *J. Org. Chem.*, **33**, 2488 (1968).

LePage, G. A., and I. G. Junga, *Cancer Res.*, **25**, 46 (1965).

Lindberg, B., H. Klenow, and K. Hansen, *J. Biol. Chem.*, **242**, 350 (1967).

Marks, D. B., B. J. Keller, and A. J. Guarino, *Biochim. Biophys. Acta*, **183**, 58 (1969).

Moore, E. C., *Biochem. Biophys. Res. Commun.*, **29**, 264 (1967).

Murray, D. H., and Prokop, J. in W. W. Zorbach and R. S. Tipson (Eds.), *Synthetic Procedures in Nucleic Acid Chemistry, Vol.* 1, John Wiley and Sons, Inc., New York, 1968, p. 193.

Nakazawa, A., M. Tokushige, and O. Hayaishi, *J. Biol. Chem.*, **242**, 3868 (1967).

Overgaard-Hansen, K., *Biochim. Biophys. Acta*, **80**, 504 (1964).

Rabinowitz, K. W., J. D. Shada, and W. A. Wood, *J. Biol. Chem.*, **243**, 3214 (1968).

Raphael, R. A., and C. M. Roxburgh, *J. Chem. Soc.*, 1955, 3405.

Reichard, P., *J. Biol. Chem.*, **237**, 3513 (1962).

Rich, M. A., P. Meyers, G. Weinbaum, J. G. Cory, and R. J. Suhadolnik, *Biochim. Biophys. Acta*, **95**, 194 (1965).

Rottman, F., M. L. Ibershof, and A. J. Guarino, *Biochim. Biophys. Acta*, **76**, 181 (1963).

Rottman, F., and A. J. Guarino, *Biochim. Biophys. Acta*, **80**, 632 (1964a).

Rottman, F., and A. J. Guarino, *Biochim. Biophys. Acta*, **80**, 640 (1964b).

Rottman, F., and A. J. Guarino, *Biochim. Biophys. Acta*, **89**, 465 (1964c).

Sawa, T., Y. Fukagawa, I. Homma, T. Wakashiro, T. Takeuchi, and M. Hori, *J. Antibiotics (Tokyo)*, **21A**, 334 (1968).

Schnebli, H. P., D. L. Hill, and L. L. Bennett, Jr., *J. Biol. Chem.*, **242**, 1997 (1967).

Sentenac, A., A. Ruet, and P. Fromageot, *European J. Biochem.*, **5**, 385 (1968).

Shigeura, H. T., *Federation Proc.*, **24**, 668 (1965).

Shigeura, H. T., and G. E. Boxer, *Biochem. Biophys. Res. Commun.*, **17**, 758 (1964).

Shigeura, H. T., and C. N. Gordon, *J. Biol. Chem.*, **240**, 806 (1965).

Shigeura, H. T., and S. D. Sampson, *Biochim. Biophys. Acta*, **138**, 26 (1967a).

Shigeura, H. T., and S. D. Sampson, *Nature*, **215**, 419 (1967b).

Shigeura, H. T., G. E. Boxer, M. L. Meloni, and S. D. Sampson, *Biochemistry*, **5**, 994 (1966).

Suhadolnik, R. J., in Antibiotics, D. Gottlieb and P. D. Shaw (Editors), Springer-Verlag, New York, 1967, p. 400.

Suhadolnik, R. J., and J. G. Cory, *Biochim. Biophys. Acta*, **91**, 661 (1964).

Suhadolnik, R. J., and J. G. Cory, 150th National Meeting, American Chemical Society, Atlantic City, N. J., September 1965, Abstracts 86C.
Suhadolnik, R. J., G. Weinbaum, and H. P. Meloche, *J. Amer. Chem. Soc.*, **86**, 948 (1964).
Suhadolnik, R. J., S. I. Finkel, and B. M. Chassy, *J. Biol. Chem.*, **243**, 3532 (1968a).
Suhadolnik, R. J., T. Uematsu, and H. Uematsu, *J. Biol. Chem.*, **243**, 2761 (1968b).
Tavitian, A., S. C. Uretsky, and G. Acs, *Biochim. Biophys. Acta*, **157**, 33 (1968).
Tener, G. M., *J. Amer. Chem. Soc.*, **83**, 159 (1961).
Todd, A. R., and T. L. V. Ulbricht, *J. Chem. Soc.*, 1960, 3275.
Truman, J. T., and S. Frederiksen, *Biochim. Biophys. Acta*, **182**, 36 (1969).
Truman, J. T., and H. Klenow, *Mol. Pharmacol.*, **4**, 77 (1968).
Walton, E., R. F. Nutt, S. R. Jenkins, and F. W. Holly, *J. Amer. Chem. Soc.*, **86**, 2952 (1964).
Walton, E., F. W. Holly, G. E. Boxer, R. F. Nutt, and S. R. Jenkins, *J. Med. Chem.*, **8**, 659 (1965).
Walton, E., F. W. Holly, G. E. Boxer, and R. F. Nutt, *J. Org. Chem.*, **31**, 1163 (1966).
Weinbaum, G., J. C. Cory, R. J. Suhadolnik, P. Meyers, and M. A. Rich. The 6th International Congress of Biochemistry, New York, July 1964, p. 343, Abstract No. IV.
Yoshida, T., H. Weissbach, and E. Katz, *Arch. Biochem. Biophys.*, **114**, 252 (1966).
Zeleznick, L. D., *Biochem. Pharmacol.*, **18**, 855 (1969).

## 1.3. 3′-AMINO-3′-DEOXYADENOSINE

### INTRODUCTION

3′-Amino-3′-deoxyadenosine (Fig. 1.30) is found in the culture filtrates of *Helminthosporium*, *Cordyceps militaris*, and *Aspergillus nidulans*. It has antitumor, antimitotic activity and inhibits *Cryptococcus neoformans* 4806 and *Candida albicans*, but does not have antibacterial properties. 3′-Amino-3′-deoxy-5′-ATP is a strong inhibitor of RNA and DNA synthesis in Ehrlich ascites tumor cells. It is readily phosphorylated and deaminated, but is not cleaved enzymatically. The chemical synthesis, mass spectrum, and biosynthesis have been reported.

### DISCOVERY, PRODUCTION, AND ISOLATION

Ammann and Safferman (1958) were the first to report on the isolation of 3′-amino-3′-deoxyadenosine from the culture filtrates of *Helminthosporium species* no. 215. This compound has also been isolated from the culture filtrates of *C. militaris* ((Guarino and Kredich, 1963) and *A. nidulans* (Suhadolnik, unpublished results).

The production of 3′-amino-3′-deoxyadenosine is carried out in shake flasks containing 1% yeast extract, 1% glucose, and 1.25% calcium carbonate (Gerber and Lechevalier, 1962). Seven days are required for maximum

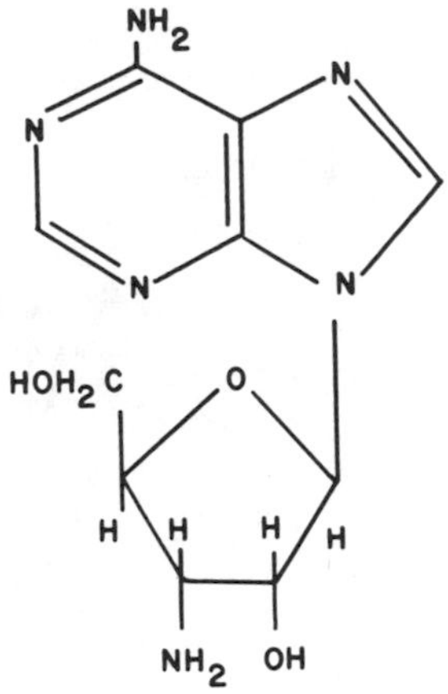

FIGURE 1.30. *Structure of* 3′-*amino*-3′-*deoxyadenosine.*

nucleoside production. The isolation and crystallization of this compound are accomplished by filtration and extraction of the mycelium. The combined filtrates are adjusted to pH 5.0 and added to a Dowex-50-$H^+$ column. The column is washed with water (pH 3.5) and the nucleoside eluted with ammonium hydroxide. The eluant containing the nucleoside antibiotic is concentrated to a small volume, made up to 15% with methanol, and added to a Dowex-1-$OH^-$ column. The nucleoside is eluted with 25% aqueous methanol and crystallized from water. The yield is 105 mg/liter of medium (Suhadolnik et al., 1969).

## PHYSICAL AND CHEMICAL PROPERTIES

The molecular formula for 3′-amino-3′-deoxyadenosine is $C_{10}H_{14}N_6O_3$; mp 271–273°C (decomp.) [chemically synthesized, mp 265–267°C (decomp.)] (Reist and Baker, 1958); $[\alpha]_D^{25}$ −37° (in 0.1*N* HCl); $\lambda_{max}^{H_2O}$ 260 mμ ($\epsilon$ = 17,300) (Gerber and Lechevalier, 1962). It is readily hydrolyzed in acid to adenine and 3-amino-3-deoxyribose hydrochloride (mp 154–158°C). The mass spectrum of 3′-amino-3′-deoxyadenosine is shown in Figure 1.31. The structural assignment is in part confirmed by the appearance of the molecular ion at *m/e* at 266. A strong fragment is observed at B + 2H(*m/e*−136). Evidence for the adenine ring is the loss of HCN (*m/e* −108). The loss of HCN from the purine and pyrrolopyrimidine rings had been reported earlier (Rice and Dudek, 1967; Smulson and Suhadolnik, 1967; Suhadolnik et al., 1969; Eggers et al., 1966). The mass spectra for 3′-amino-3′-deoxyadenosine (Fig. 1.31) can be compared with cordycepin (see p. 55, Fig. 1.23) and 3′-acetamido-3′-deoxyadenosine (see p. 88, Fig. 1.35).

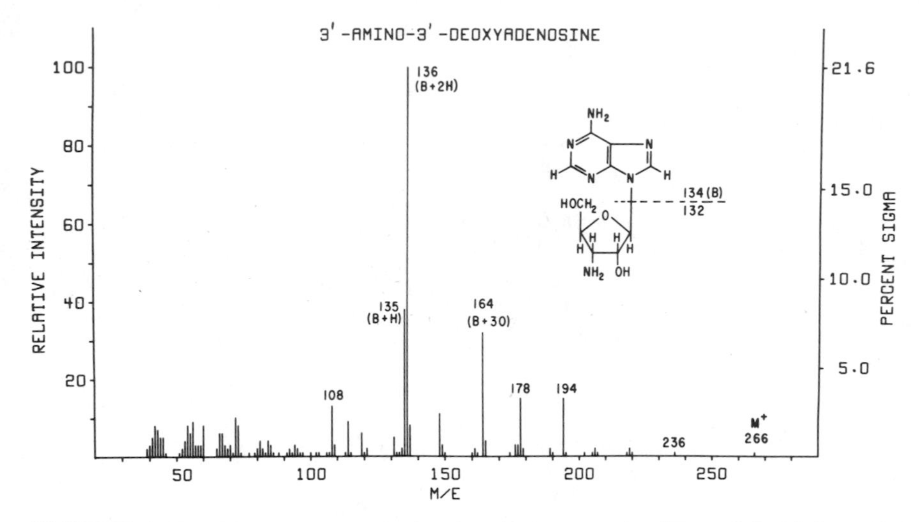


FIGURE 1.31. *Mass spectrum of* 3′-*amino*-3′-*deoxyadenosine* (*From Suhadolnik et al.*, 1969.)

## STRUCTURAL ELUCIDATION

Gerber and Lechevalier (1962) and Guarino and Kredich (1963) studied the structure of 3′-amino-3′-deoxyadenosine by hydrolyzing the nucleoside and isolating adenine and 3-amino-3-deoxy-D-ribose. The mass spectrum of 3′-amino-3′-deoxyadenosine is additional proof for the structure of this nucleoside antibiotic (Suhadolnik et al., 1969).

## CHEMICAL SYNTHESIS

Baker et al. (1955b) described the chemical synthesis of 3′-amino-3′-deoxyadenosine from adenine and D-xylose. A number of 3′-amino-3′-deoxypurine and pyrimidine nucleosides have been synthesized. Kissman and Weiss (1958) described the synthesis of 3′-amino-3′-deoxypyrimidines; Goldman and Marisco (1963) reported on the synthesis of 3′-amino-3′-deoxyribosides of 6-chloropurine; Gerber (1964) described the synthesis of 3′-amino-3′-deoxyadenosine 1-*N*-oxide, and Watanabe and Fox (1966) synthesized the 3′-amino-3′-deoxyhexopyranosyl nucleosides.

The synthesis of the carbohydrate moiety, 3-amino-3-deoxy-D-ribose, has been reported from several laboratories (Baker and Schaub, 1953, 1954; Baker et al., 1955a; Lemieux and Chu, 1958; Baer and Fischer, 1959; Sowa, 1968).

## INHIBITION OF GROWTH

Pugh et al. (1962) reported that 3′-amino-3′-deoxyadenosine exhibited toxicity to ascitic tumors of mice when administered intraperitoneally. Bloch and Nichol (1964) subsequently reported on the inhibition of adenocarcinoma S3A (ascitic), Ehrlich carcinoma (ascitic), Gardner lymphosarcoma (ascitic), sarcoma 180 (ascitic), and J-tumor.

Gerber and Lechevalier (1962) showed that *Crytococcus neoformans* 4806 and *Candida albicans* were inhibited at 500 μg/ml. The nucleoside was not inhibitory to bacteria. Only partial inhibition occurred with yeast protoplasts from *Saccharomyces cerevisiae* LK2G12 (Gerber and Lechevalier, 1962). Strong antimitotic activity in onion root tips was also observed (Ammann and Safferman, 1958).

## BIOSYNTHESIS

The only study on the biosynthesis of this nucleoside analog has been the report of Chassy and Suhadolnik (1969). They have shown that adenosine-U-$^{14}$C is the direct precursor for 3′-amino-3′-deoxyadenosine. This direct conversion of adenosine to the aminonucleoside antibiotic resembles the biosynthesis of cordycepin (Suhadolnik et al., 1964). 3′-Deoxyadenosine-(U)-$^{14}$C was not a precursor for the 3′-aminonucleoside. Since *Helminthosporium* also synthesizes the nucleoside analog 3′-acetamido-3′-deoxyadenosine along with 3′-amino-3′-deoxyadenosine, experiments were performed to determine if there was an interconversion of these two nucleoside antibiotics (Chassy and Suhadolnik, 1969). The results obtained showed that there is a rapid interconversion of 3′-amino-3′-deoxyadenosine and 3′-acetamido-3′-deoxyadenosine. These findings are similar to the reports of Hoeksema et al. (1964) and Chassy et al. (1966) on the interconversion of psicofuranine and decoyinine by *S. hydroscopicus*. Since adenosine or a 5′-phosphorylated derivative appears to be the precursor for cordycepin, 3′-amino-3′-deoxyadenosine, and 3′-acetamido-3′-deoxyadenosine, it may be that adenosine is converted to an intermediate that is the common precursor for these three deoxynucleoside antibiotics.

## BIOCHEMICAL PROPERTIES

### I. Effect on Bacterial and Mammalian Cells

Shigeura and Sampson (1967), Shigeura et al. (1966a), and Truman and Klenow (1968) have supplied considerable information on the phosphorylation or lack of phosphorylation by intact Ehrlich ascites cells with various nucleosides structurally related to 3′-amino-3′-deoxyadenosine (Table 1.9). Structural alteration of 3′-amino-3′-deoxyadenosine by modification of the amino group markedly decreased the inhibitory activity. Similarly, 3′-amino-3′-deoxyadenosine was a more effective inhibitor of 5-phosphoribosylamine synthesis in whole ascites cells than the 6-dimethyl-3′-amino-3′-deoxyadenosine (Table 1.10). 3′Amino-3′-deoxyadenosine was phosphorylated to 5′-mono-, 5′-di-, and 5′-triphosphates. As a consequence of the phosphorylation of this nucleoside antibiotic, Shigeura et al. (1966a) were able to show that the synthesis of RNA was inhibited. In contrast, 6-methylamino-9-(3′-deoxy-β-D-ribofuranosyl)purine was only converted to the monophosphate level (Shigeura et al., 1966b). Similarly, 3′-amino-3′-deoxyadenosine-5′-monophosphate was 50% as good a substrate as AMP for crystalline rabbit muscle myokinase, while the 6-methylamino purine mononucleotides were not substrates. 3′-Amino-3′-deoxyadenosine-5′-monophosphate was not as good a substrate for myokinase as 3′-deoxyadenosine-5′-monophosphate. The results with rabbit muscle myokinase appear to be consistent with results of the phosphorylation of the nucleoside analogs with intact Ehrlich ascites cells. Lindberg et al. (1967) reported on the phosphorylation of 3′-amino-3′-deoxyadenosine by adenosine kinase. The $K_m$ for 3′-amino-3′-deoxyadenosine for partially purified adenosine kinase was $6.1 \times 10^{-4}$ *M*, while the $K_m$ for 3′-deoxyadenosine was $4.6 \times 10^{-4}$ *M* ($K_m$ for adenosine was $1.6 \times 10^{-6}$ *M*). Shigeura et al. (1966a) and Bloch and Nichol (1964) showed that 3′-amino-3′-deoxyadenosine was not cleaved during incubation with whole Ehrlich ascites cells or *Streptococcus faecalis*. Adenosine reverses the inhibition by this analog (Table 1.11). 3′-Amino-3′-deoxyadenosine, like 3′-deoxyadenosine, is a powerful inhibitor of the incorporation of adenine-$^{14}$C and uridine-$^{3}$H into both RNA and DNA Ehrlich ascites tumor cells (Figs. 1.32*a*,*b*) (Shigeura et al., 1966a; Truman and Klenow, 1968). At $0.5 \times 10^{-3}$ *M*, 3′-amino-3′-deoxyadenosine inhibited the incorporation of hypoxanthine-8-$^{14}$C into nucleic acids of whole Ehrlich ascites cells by 57% (Table 1.9, Expt. 7). This compares with a 95% inhibition of incorporation of its structural analog, 3′-deoxyadenosine. 3′-Amino-3′-deoxy-5′-ATP is an effective inhibitor of RNA synthesis with the partially purified RNA polymerase from *Micrococcus lysodeikticus* and cell-free DNA-dependent RNA polymerase from Ehrlich ascites tumor cells (Shigeura et al., 1966a; Truman and Klenow, 1968). As with cordycepin, the kinetics of inhibition are the type expected from the irreversible incorporation of analogs lacking a 3′-hydroxyl group that are incorporated into the 3′ end of RNA growing in the 5′ to 3′ direction. No conclusions have been drawn concerning the inhibitory effect of 3′-amino-3′-deoxyadenosine on the incorporation of adenine-$^{14}$C into DNA of Ehrlich ascites tumor cells *in vitro*. 3′-Amino-3′-deoxy-5′-ATP did not inhibit the activity of DNA polymerase *in vitro*. These results suggest that the effect of this nucleoside antibiotic on DNA synthesis in the whole cell may be attributed to the inhibition of ribonucleotide reductase. Suhadolnik et al. (1968) and Chassy and Suhadolnik (1968) have reported that the nucleotide antibiotics are both substrates and allosteric effectors for the partially purified ribonucleotide reductase isolated from *L. leichmannii* and *E. coli*. Therefore, the notion that 3′-amino-3′-deoxy-5′-ATP could affect DNA synthesis by inhibiting ribonucleotide reductase is a possibility.

**Table 1.9.**

Effects of Purine 3′-Deoxyribonucleosides on the Incorporation of Hypoxanthine-8-$^{14}$C into Nucleic Acids in Whole Ehrlich Ascites Cells

| Expt no. | Compound | Conc., m*M* | % inhibition |
|---|---|---|---|
| 1 | 6-Methylamino-9-(3′-deoxy-β-D-ribofuranosyl)purine | 2.0 | 30 |
| 2 | 6-Dimethylamino-9-(3′-deoxy-β-D-ribofuranosyl)purine | 2.0 | 25 |
| 3 | 6-Ethylamino-9-(3′-deoxy-β-D-ribofuranosyl)purine | 2.0 | 15 |
| 4 | Purine 3′-deoxyribonucleoside | 2.0 | 20 |
| 5 | 2,6-Diamino-9-(3′-deoxy-β-D-ribofuranosyl)purine | 2.0 | 8 |
| 6 | 3′-Deoxyadenosine | 0.5 | 95 |
| 7 | 3′-Amino-3′-deoxyadenosine | 0.5 | 57 |
| 8 | Puromycin aminonucleoside | 1.5 | 18 |

From Shigeura et al., 1966a.

**Table 1.10.**

Effects of Nucleosides on the Synthesis of 5-Phosphoribosylamine in Whole Ascites Cells[a]

| Expt. no. | Compound | % inhibition |
|---|---|---|
| 1 | Adenosine | 70 |
| 2 | 3′-Deoxyadenosine | 92 |
| 3 | 2-Fluoroadenosine | 94 |
| 4 | 6-Methylamino-9-(β-D-ribofuranosyl)purine | 62 |
| 5 | 6-Methylamino-9-(3′-deoxy-β-D-ribofuranosyl)purine | 52 |
| 6 | 6-Dimethylamino-9-(3′-deoxy-β-D-ribofuranosyl)purine | 9 |
| 7 | 6-Ethylamino-9-(3′-deoxy-β-D-ribofuranosyl)purine | 7 |
| 8 | 2,6-Diamino-9-(3′-deoxy-β-D-ribofuranosyl)purine | 14 |
| 9 | 3′-Amino-3′-deoxyadenosine | 74 |
| 10 | 6-Dimethyl-3′-amino-3′-deoxyadenosine | 24 |
| 11 | Uridine | 14 |
| 12 | Xanthosine | 8 |

From Shigeura et al., 1966a.

[a] The concentration of nucleosides was 1.0 m*M*.

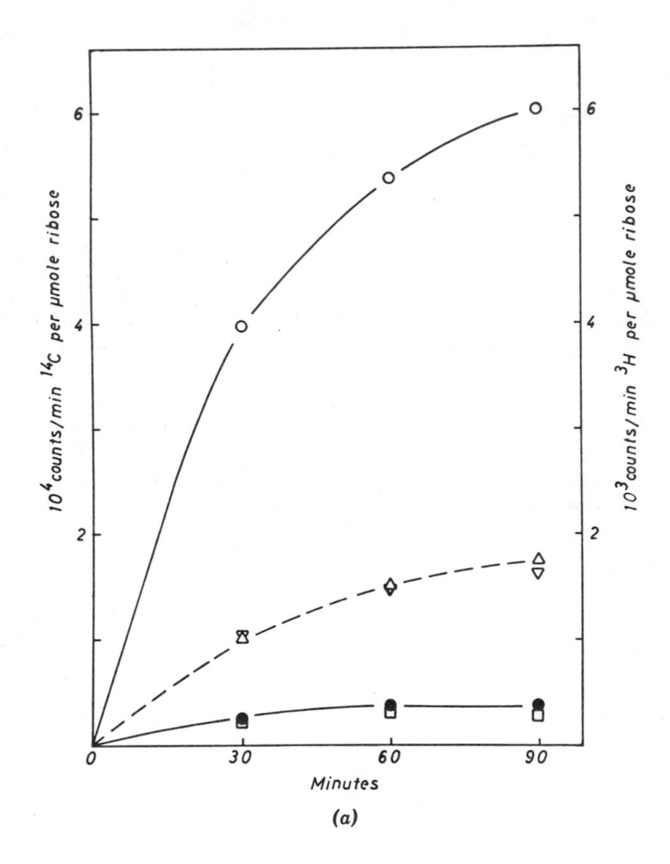


FIGURE 1.32. (*a*) *Effect of* 3′-*amino*-3′-*deoxyadenosine on the incorporation of adenine*-$^{14}$*C and uridine*-$^{3}$*H into RNA of Ehrlich ascites tumor cells in vitro. Adenine*-$^{14}$*C incorporation with* 3′-*amino*-3′-*deoxyadenosine added* (*per ml*)*:* (○) *none;* (△) 0.3 *mole;* (●) 2.2 *moles. Uridine*-$^{3}$*H incorporation with* 3′-*amino*-3′-*deoxyadenosine added* (*per ml*)*:* (○) *none;* (▽) 0.3 *mole;* (□) 2.2 *moles.* (*From Truman and Klenow*, 1968.) (*b*) *Effect of* 3′-

One striking difference in the effect of 3′-amino-3′-deoxyadenosine in contrast to 3′-deoxyadenosine is that the former adenosine analog does not cause any pronounced decrease in the specific activity of the acid-soluble ribonucleotide pool labeled with adenine-$^{14}$C. Truman and Klenow (1968) suggested that the effect of these two nucleotides on PRPP formation may be markedly different. Shigeura et al. (1966a) showed that the addition of 3′-amino-3′-deoxyadenosine to whole Ehrlich ascites tumor cells inhibited the synthesis of 5-phosphoribosylamine by 74% (Table 1.10, Expt. 9). Although the inhibition of ribosephosphate pyrophosphokinase or phos-

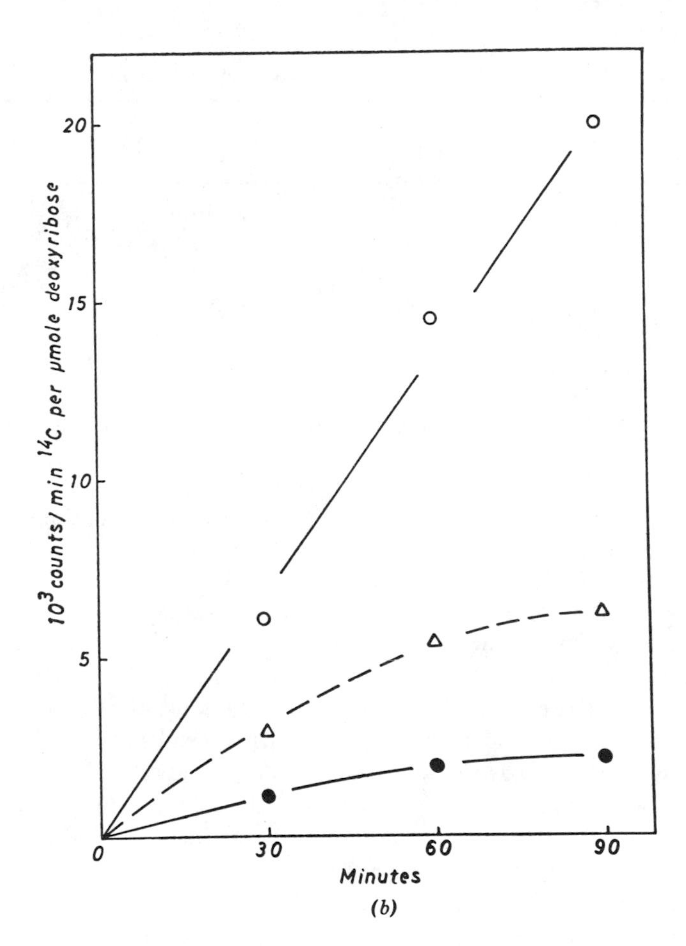


*amino-3′-deoxyadenosine on the incorporation of adenine-$^{14}$C into DNA of Ehrlich ascites tumor cells in vitro. Conditions of incubation were the same as in a.* (*From Truman and Klenow*, 1968.)

phoribosylpyrophosphate amidotransferase was not determined [as has been reported for the 5′-mono- and 5′-triphosphates of 3′-deoxyadenosine (Overgaard-Hansen, 1964; Rottman and Guarino, 1964)], the data presented by Shigeura et al. (1966a) suggest that 3′-amino-3′-deoxyadenosine may inhibit these same enzymes.

Truman and Frederiksen (1969) have recently reported on the inhibitory effect of 3′-amino-3′deoxyadenosine on RNA labeling in Ehrlich ascites tumor cells. This 3′-deoxy analog was shown to exert a differential inhibitory effect on the labeling of nucleochromosomal and cytoplasmic RNA. The 28-S and 18-S cytoplasmic RNA are very strongly inhibited, and the cyto-

**Table 1.11.**
Prevention of Adenosine-Antibiotic Inhibition by Adenosine

| Inhibitor | Inhibition index[a] |
|---|---|
| Tubercidin | 0.3 |
| Decoyinine | 0.5 |
| Psicofuranine | 0.5 |
| 3′-Amino-3′-deoxyadenosine | 10 |
| 3′-Deoxyadenosine | 15 |
| Puromycin | No reversal |

From Bloch and Nichol, 1964.
[a] Inhibition index = [I]/[S] for 50% growth inhibition. Range tested: $10^{-3}$ to $10^{-6}$ *M*.

plasmic 4–5-S RNA is not as inhibited by 3′-amino-3′-deoxadenosine. The 50–60-S nuclear RNA is only slightly inhibited. Similar findings were reported for 3′-deoxyadenosine (cordycepin) (see Sect. 1.2, Fig. 1.29) and toyocamycin (see Chapter 8, Figs. 8.16 and 8.17).

Studies on the structural requirements of nucleoside binding to adenosine aminohydrolase have utilized 3′-amino-3′-deoxyadenosine (Cory and Suhadolnik, 1965; Coddington, 1965; Frederiksen, 1966). The results published from these three laboratories show that substitution of an amino group for the hydroxyl on carbon 3′ of adenosine results in a nucleoside that is a good substrate for adenosine aminohydrolase. The relative initial velocity was 0.46 (adenosine = 1.0) (Cory and Suhadolnik, 1965). The $K_m$ for purified adenosine aminohydrolase from calf intestine for 3′-amino-3′-deoxyadenosine was reported to be $1.67 \times 10^{-4}$ (adenosine = $3.3 \times 10^{-5}$ *M*) (Frederiksen, 1966).

## Summary

3′-Amino-3′-deoxyadenosine is an analog of adenosine and is a substrate for adenosine kinase, but not for nucleoside phosphorylase. It is converted to the 5-mono-, di-, and triphosphates by Ehrlich ascites tumor cells in culture and is a potent inhibitor of RNA and DNA synthesis in Ehrlich ascites tumor cells *in vivo*. 3′-Amino-3′-deoxyadenosine inhibited partially purified RNA polymerase, which appears to explain the inhibition of this nucleoside analog. DNA synthesis *in vitro* was not inhibited. The suggestion was made that 3′-amino-3′-deoxy-5′-ATP inhibits DNA synthesis by inhibiting ribonucleotide reductase.

The mass spectrum of 3′-amino-3′-deoxyadenosine is similar to the mass spectra of adenosine, puromycin, and 3′-acetamido-3′-deoxyadenosine.

The biosynthesis of 3′-amino-3′-deoxyadenosine is similar to that of cordycepin in that the 3′ carbon of adenosine or a phosphorylated derivative is reduced without cleavage of the *N*-riboside bond.

## References

Ammann, C. A., and R. S. Safferman, *Antibiot. Chemotherapy*, **8**, 1 (1958).
Baer, H. H., and H. O. L. Fischer, *J. Amer. Chem. Soc.*, **81**, 5184 (1959).
Baker, B. R., and R. E. Schaub, *J. Amer. Chem. Soc.*, **75**, 3864 (1953).
Baker, B. R., and R. E. Schaub, *J. Org. Chem.*, **19**, 646 (1954).
Baker, B. R., R. E. Schaub, and J. H. Williams, *J. Amer. Chem. Soc.*, **77**, 7 (1955a).
Baker, B. R., R. E. Schaub, and H. M. Kissman, *J. Amer. Chem. Soc.*, **77**, 5911 (1955b).
Bloch, A., and C. A. Nichol, *Antimicrobial Agents Chemotherapy*, 1964, 530.
Chassy, B. M., and R. J. Suhadolnik, *J. Biol. Chem.*, **243**, 3538 (1968).
Chassy, B. M., and R. J. Suhadolnik, *Biochim. Biophys. Acta*, **182**, 315 (1969).
Chassy, B. M., T. Sugimori, and R. J. Suhadolnik, *Biochim. Biophys. Acta*, **130**, 12 (1966).
Coddington, A., *Biochim. Biophys. Acta*, **99**, 442 (1965).
Cory, J. G., and R. J. Suhadolnik, *Biochemistry*, **4**, 1729 (1965).
Eggers, S. H., S. I. Biedron, and A. O. Hawtrey, *Tetrahedron Letters*, 1966, 3271.
Frederiksen, S., *Arch. Biochem. Biophys.*, **113**, 383 (1966).
Gerber, N. N., *J. Med. Chem.*, **7**, 204 (1964).
Gerber, N. N., and H. A. Lechevalier, *J. Org. Chem.*, **27**, 1731 (1962).
Goldman, L., and J. W. Marsico, *J. Med. Chem.*, **6**, 413 (1963).
Guarino, A. J., and N. M. Kredich, *Biochim. Biophys. Acta*, **68**, 317 (1963).
Hoeksema, H., G. Slomp, and E. E. van Tamelen, *Tetrahedron Letters*, 1964, 1787.
Kissman, H. M., and M. J. Weiss, *J. Amer. Chem. Soc.*, **80**, 2575 (1958).
Lemieux, R. U., and P. Chu, *J. Amer. Chem. Soc.*, **80**, 4745 (1958).
Lindberg, B., H. Klenow, and K. Hansen, *J. Biol. Chem.* **242**, 350 (1967).
Overgaard-Hansen, K., *Biochim. Biophys. Acta*, **80**, 504 (1964).
Pugh, L. H., H. A. Lechevalier, and M. Solotorovsky, *Antibiot. Chemotherapy*, **12**, 310 (1962).
Reist, E. J. and B. R. Baker, *J. Org. Chem.*, **23**, 1083 (1958).
Rice, J. M., and G. O. Dudek, *J. Amer. Chem. Soc.*, **89**, 2719 (1967).
Rottman, F., and A. J. Guarino, *Biochim. Biophys. Acta*, **89**, 465 (1964).
Shigeura, H. T., and S. D. Sampson, *Nature*, **215**, 419 (1967).
Shigeura, H. T., G. E. Boxer, M. L. Meloni, and S. D. Sampson, *Biochemistry*, **5**, 994 (1966a).
Shigeura, H. T., D. S. Sampson, and M. L. Meloni, *Arch. Biochem. Biophys.*, **115**, 462 (1966b).
Smulson, M. E., and R. J. Suhadolnik, *J. Biol. Chem.*, **242**, 2872 (1967).
Suhadolnik, R. J., G. Weinbaum, and H. P. Meloche, *J. Amer. Chem. Soc.*, **86**, 948 (1964).
Suhadolnik, R. J., S. I. Finkel, and B. M. Chassy, *J. Biol. Chem.*, **243**, 3532 (1968).

Suhadolnik, R. J., B. M. Chassy, and G. R. Waller, *Biochim. Biophys. Acta*, **179**, 258 (1969).
Sowa, W., *Can. J. Chem.*, **46**, 1586 (1968).
Truman, J. T., and H. Klenow, *Mol. Pharmacol.*, **4**, 77 (1968).
Truman, J. T. and S. Frederiksen, *Biochim. Biophys. Acta*, **182**, 36 (1969).
Watanabe, K. A., and J. J. Fox, *J. Org. Chem.*, **31**, 211 (1966).

## 1.4. 3′-ACETAMIDO-3′-DEOXYADENOSINE

### INTRODUCTION

3′-Acetamido-3′-deoxyadenosine (Fig. 1.33) is found in the culture filtrates of *Helminthosporium* together with 3′-amino-3′-deoxyadenosine. The proof of structure, synthesis, biosynthesis, and inhibition of growth are discussed in this section. Although this nucleoside analog is not an inhibitor of growth, it is included in this material due to its simultaneous occurrence with 3′-amino-3′-deoxyadenosine in the culture filtrates of *Helminthosporium*.

### DISCOVERY, PRODUCTION, AND ISOLATION

3′-Acetamido-3′-deoxyadenosine has recently been reported as a naturally occurring adenosine analog found in the culture filtrates of *Helminthosporium sp.* 215 by Suhadolnik and his co-workers (1969). This was the same culture from which Gerber and Lechevalier (1962) isolated 3′-amino-3′-deoxyadenosine. The medium and growth conditions are the same as described in the chapter for 3′-amino-3′-deoxyadenosine. During nucleoside production,

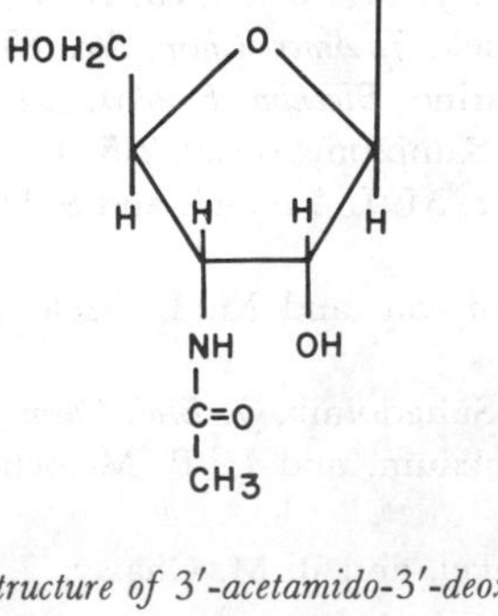

FIGURE 1.33. *Structure of 3′-acetamido-3′-deoxyadenosine.*

3′-acetamido-3′-deoxyadenosine appears in the culture filtrate about 40 hr after inoculation. 3′-Amino-3′-deoxyadenosine is found a short time later, but in significantly smaller amounts. About 60 hr after inoculation, the molar concentrations of the two nucleoside antibiotics are the same. With increasing time, 3′-amino-3′-deoxyadenosine becomes a predominant nucleoside analog in the culture medium. (Fig. 1.34). Adenosine is also isolated from the culture filtrates. The isolation of adenosine, 3′-amino-3′-deoxyadenosine, and 3′-acetamido-3′-deoxyadenosine from the culture filtrates of *Helminthosporium* is accomplished by evaporation of the culture filtrate to dryness, extraction with hot ethanol, evaporation to dryness, and addition of 15% aqueous methanol. This solution is then added to a Dowex-1-$OH^-$ column. 3′-Amino-3′-deoxyadenosine is eluted with 25% aqueous methanol; 3′-acetamido-3′-deoxyadenosine is eluted with 35% aqueous methanol, and adenosine is eluted with 60% aqueous methanol. The addition of ammonium chloride to the medium completely blocks the biosynthesis of 3′-acetamido-3′-deoxyadenosine, but does not affect the production of 3′-amino-3′-deoxyadenosine or adenosine.

## PHYSICAL AND CHEMICAL PROPERTIES

The molecular formula for 3′-acetamido-3′-deoxyadenosine is $C_{12}H_{16}N_6O_4$; mp 263–265°C; $\lambda_{max}$ 260 m$\mu$ ($\epsilon$ = 15,600) (in water); $[\alpha]_D^{25}$ = 8 ± 2° ($c$ =

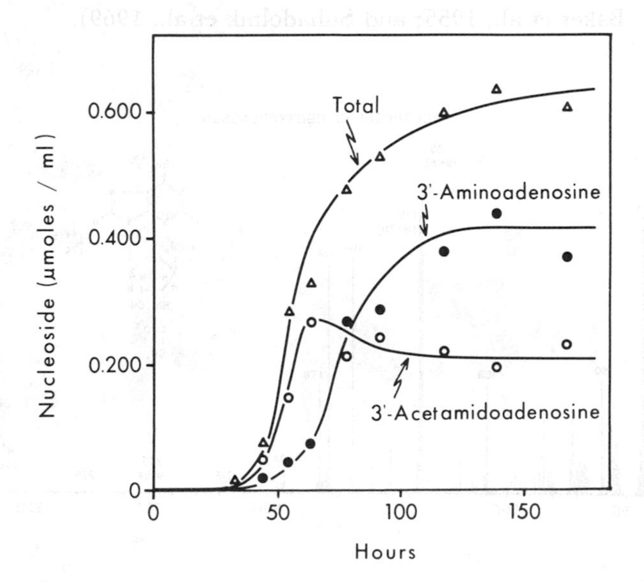


FIGURE 1.34. *Production of 3′-amino-3′-deoxyadenosine and 3-′acetamido-3′-deoxyadenosine by Helminthosporium sp.* 215. (*From Suhadolnik et al.*, 1969.)

0.1 *N* HCl); nmr spectrum showed a singlet absorption at $\tau$ 8.1 (3H), characteristic of an acetyl methyl group (Suhadolnik et al., 1969). The mass spectrum of 3′-acetamido-3′-deoxyadenosine is shown in Figure 1.35 and the the fragmentation pattern is shown in Figure 1.36. The loss of the acetamido group (M-59) is shown by *m/e* 249. The B+2H ion (*m/e* 136) intensity for 3′-acetamido-3′-deoxyadenosine is less than that observed for 3′-amino-3′-deoxyadenosine and cordycepin (Suhadolnik et al., 1969; Hanessian et al., 1966). Additional fragments include the loss of: (*1*) a hydroxyl (M-17, *m/e* 291); (*2*) formaldehyde (M-30, *m/e* 278); (*3*) HCN [(B + H) −29, *m/e* 108]; and (*4*) the direct loss of B (M-134, *m/e* 174). A peak was observed for 3-acetamidoribofuranose (*m/e* 184). The infrared spectrum of chemically synthesized nucleoside and the nucleoside isolated from *Helminthosporium* were identical (Suhadolnik et al., 1969).

## STRUCTURAL ELUCIDATION

The structure of 3′-acetamido-3′-deoxyadenosine was determined by hydrolysis in acid to adenine and 3-amino ribose, in alkali to 3′-amino-3′-deoxyadenosine, and in part by mass spectrometry (Suhadolnik et al., 1969).

## CHEMICAL SYNTHESIS

3′-Acetamido-3′-deoxyadenosine was synthesized from 3′-amino-3′-deoxyadenosine (Baker et al., 1955; and Suhadolnik et al., 1969).

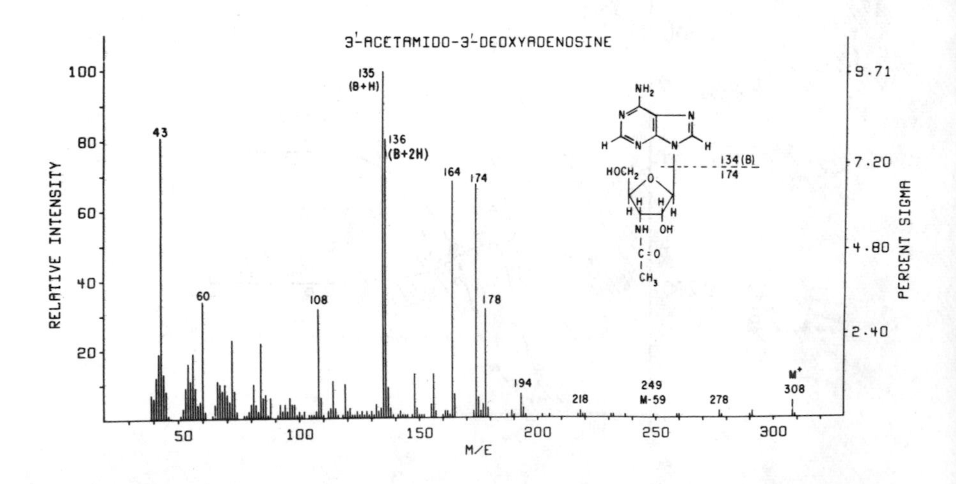


FIGURE 1.35. *Mass spectrum for 3′-acetamido-3′-deoxyadenosine.* (*From Suhadolnik et al.,* 1969.)

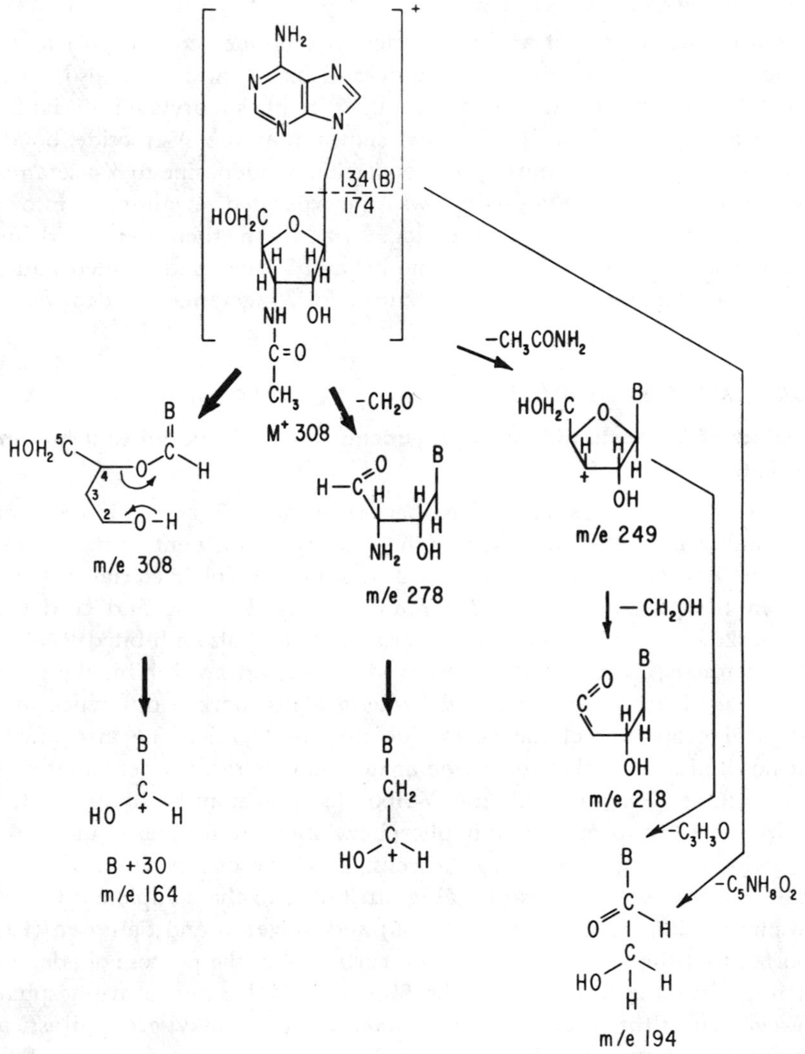

FIGURE 1.36. *Fragmentation pattern for 3′-acetamido-3′-deoxyadenosine.* (*From Suhadolnik et al.*, 1969.)

## INHIBITION OF GROWTH

3′-Acetamido-3′-deoxyadenosine did not inhibit the growth of the ascitic form of Ehrlich-Lettre carcinoma, nor did it inhibit the growth of bacteria (Suhadolnik et al., 1969).

## BIOSYNTHESIS

As with the structural analogs, 3′-deoxyadenosine (cordycepin) and 3′-amino-3′-deoxyadenosine, 3′-acetamido-3′-deoxyadenosine is also formed directly from preformed adenosine or its 5′-phosphorylated derivatives. Chassy and Suhadolnik (1969) have shown that the *N*-riboside bond of adenosine is not cleaved during the conversion of adenosine to 3′-acetamido-3′-deoxyadenosine. Acetate-1-$^{14}C$ was incorporated exclusively into the acetyl moiety of this adenosine analog. In addition, there is a rapid interconversion of 3′-amino-3′-deoxyadenosine and 3′-acetamido-3′-deoxyadenosine. 3′-Deoxyadenosine is not a precursor for 3′-acetamido-3′-deoxyadenosine.

## BIOCHEMICAL PROPERTIES

### I. Effect of 3′-Acetamido-3′-deoxyadenosine on Bacterial and Mammalian Cells

Growth inhibition studies were performed with 3′-acetamido-3′-deoxyadenosine to determine the effect of the *N*-acetyl substituent on the 3′-amino position. Ehrlich ascites tumor cells and a number of bacterial cells were used in these studies. While 3′-amino-3′-deoxyadenosine and cordycepin had marked antimitotic and antitumor activity and also inhibited RNA and DNA synthesis, 3′-acetamido-3′-deoxyadenosine did not inhibit the growth of Ehrlich ascites tumor cells and bacteria. This marked difference in the biological properties of the acetamido nucleoside might be attributed to four possibilities: (*1*) this nucleoside analog may be rapidly deaminated to a hypoxanthine derivative; (*2*) the *N*-riboside bond may be hydrolyzed; (*3*) the *N*-acetyl group may inhibit phosphorylation by a kinase, and (*4*) the nucleoside is not permeable to the cells. The lack of inhibition of 3′-acetamido-3′-deoxyadenosine is probably attributed to the complete absence of phosphorylation. Shigeura et al. (1966) and Shigeura and Sampson (1967) reported that the functional groups on carbon 3′ of the pentose of adenosine did not affect the formation of the 5′-mono-, 5′-di-, and 5′-triphosphates. However, substitution of the 6-amino moiety with an alkyl group drastically interfered with phosphorylation in the 5′ position of the sugar. It may follow that an *N*-acetyl group on the 3′-amino group of the nucleoside analog may be sufficiently different from 3′-amino-3′-deoxyadenosine that 5′-phosphorylation does not occur. Klenow (private communication) showed that adenosine kinase from liver could not phosphorylate 3′-acetamido-3′-deoxyadenosine. There is apparently no significant enzymatic cleavage of the *N*-acetyl group of 3′-acetamido-3′-deoxyadenosine. If this hydrolysis did take place, the nucleoside formed would be 3′-amino-3′-deoxyadenosine, which is a known inhibitor in these cells (Truman and Klenow, 1968).

It is interesting to note that resistance of *E. coli* carrying the R factor to streptomycin, chloramphenicol, and kanamycin has been shown to be attributed to the formation of either phosphorylating or acetylating enzymes (Umezawa et al., 1967a; Umezawa et al., 1967b). Once the antibiotic is acetylated, it is no longer toxic. Similarly, the data presented by Suhadolnik et al. (1969) show that the *N*-acetyl nucleoside antibiotic 3′-acetamido-3′-deoxyadenosine is not toxic to bacteria, yeast, or tumor cells.

## Summary

3′-Acetamido-3′-deoxyadenosine, a naturally occurring nucleoside, is elaborated by *Helminthosporium sp.* 215. This nucleoside does not inhibit Ehrlich ascites tumor cells or bacteria. The biosynthetic data indicate that adenosine (or a phosphorylated derivative) is the immediate precursor. The mass spectrum has been reported. This nucleoside is not a substrate for adenosine kinase.

## References

Baker, B. R., R. E. Schaub, and H. M. Kissman, *J. Amer. Chem. Soc.*, **77**, 5911 (1955).
Chassy, B. M., and R. J. Suhadolnik, *Biochem. Biophys. Acta*, **182**, 315 (1969).
Gerber, N. N., and H. A. Lechevalier, *J. Org. Chem.*, **27**, 1731 (1962).
Hanessian, S., D. C. DeJongh, and J. A. McCloskey, *Biochim. Biophys. Acta*, **117**, 480 (1966).
Shigeura, H. T., G. E. Boxer, M. L. Meloni, and S. D. Sampson, *Biochemistry*, **5**, 994 (1966).
Shigeura, H. T., and S. D. Sampson, *Nature*, **215**, 419 (1967).
Suhadolnik, R. J., B. M. Chassy, and G. R. Waller, *Biochim. Biophys. Acta*, **179**, 258 (1969).
Truman, J. T., and H. Klenow, *Mol. Pharmacol.*, **4**, 77 (1968).
Umezawa, H., M. Okanishi, S. Kondo, K. Hamana, R. Utahara, K. Maeda, and S. Mitsuhashi, *Science*, **157**, 1559 (1967a).
Umezawa, H., M. Okanishi, R. Utahara, K. Maeda, and S. Kondo, *J. Antibiotics* (*Tokyo*), **20A**, 136 (1967b).

# 1.5. HOMOCITRULLYLAMINOADENOSINE

## INTRODUCTION

Homocitrullylaminoadenosine (Fig. 1.37*A*) is one of four adenine nucleoside analogs found in the culture filtrate or mycelia of *Cordyceps militaris*. The other three nucleosides that have been isolated are cordycepin, 3′-amino-3′-deoxyadenosine, and lysylamino adenosine.

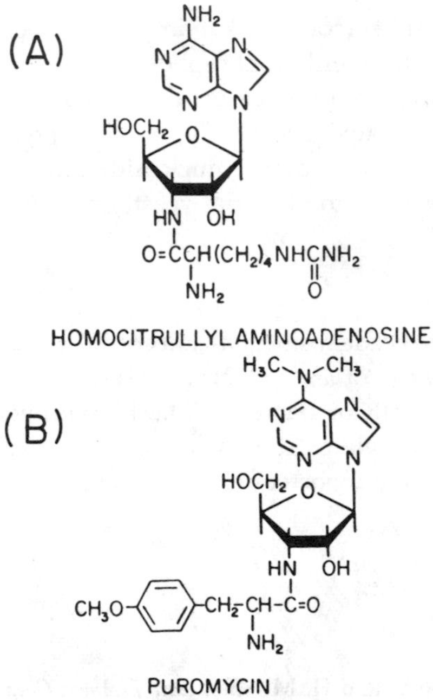

(A)

HOMOCITRULLYLAMINOADENOSINE

(B)

PUROMYCIN

FIGURE 1.37. *Structures of (A) homocitrullylaminoadenosine and (B) puromycin.*

## DISCOVERY, PRODUCTION AND ISOLATION

Kredich and Guarino (1961) isolated homocitrullylaminoadenosine from the mycelia of *C. militaris* grown for 30 days in static culture. The medium used was 1% glucose and 0.5% enzymatically prepared casein hydrolyzate. The mycelia was homogenized in a Waring Blendor and filtered, and the filtrate was evaporated to dryness. The residue was dissolved in water, adjusted to pH 3.5, and passed through a Dowex-50-$NH_4^+$ column, which was washed with water and then with 0.1 *N* $NH_4OH$. The eluant, containing homocitrullylaminoadenosine, was taken to dryness and a small volume of water was added along with carbon to remove impurities and the solution was then filtered. The nucleoside crystalized from water.

## PHYSICAL AND CHEMICAL PROPERTIES AND STRUCTURAL ELUCIDATION

The molecular formula of homocitrullylaminoadenosine is $C_{17}H_{27}N_9O_5$. It is hydrolyzed (6 *N* HCl, 100°C, 1 hr) to form adenine, 3-amino-3-deoxyribose, and L-homocitrulline. Treatment with methanolic sodium hydroxide (1.0 *N*, 17 hr) resulted in the isolation of 3′-amino-3′-deoxyadenosine and homocitrulline. On the basis of the physical and chemical data obtained, Kredich and Guarino (1961) reported the structure of this nucleoside as homocitrullylaminoadenosine.

## BIOCHEMICAL PROPERTIES

Guarino et al. (1963) reported that homocitrullylaminoadenosine inhibited the incorporation of amino acids into protein in cell-free extracts of *E. coli* and rat liver. The level of inhibitor necessary for half-maximal inhibition for homocitrullylaminoadenosine was $1.1 \times 10^{-4}$ *M*; for puromycin, $2.2 \times 10^{-6}$ *M*; and for chloromycetin, $1.8 \times 10^{-5}$ *M*. Homocitrulline and 3′-amino-3′-deoxyadenosine were only mild inhibitors.

To determine if homocitrullylaminoadenosine exerts its inhibition on protein synthesis in a manner similar to that of puromycin, experiments were designed to establish the site of inhibition. The inhibition did not appear to involve the activation of amino acids nor their transfer to sRNA, but did inhibit the overall conversion of amino acids from sRNA charged with leucine into ribosomal protein (Table 1.12).

**Table 1.12.**
Inhibition of Amino Acid Incorporation from sRNA Charged with Leucine-$^{14}$C into Protein

| Expt. no. | Additions | Total counts incorporated |
|---|---|---|
| 1 | Complete | 296 |
| 2 | Complete + amino acids | 240 |
| 3 | Complete + HCAA[a] | 92 |
| 4 | Complete + puromycin | 12 |

From Guarino et al. (1963).
[a] Homocitrullylaminoadenosine.

Homocitrullylaminoadenosine resembles puromycin more than chloromycetin, since the former two nucleosides inhibit protein synthesis in both bacterial and mammalian systems. Chloromycetin inhibits the bacterial, but not the mammalian, system. The structural similarities of homocitrullyaminoadenosine and puromycin are shown in Fig. 1.37A and B.

Since 3′-amino-3′-deoxyadenosine is not an effective inhibitor of protein synthesis, it appears that the amino acid on the 3′-amino group is an important structural requirement. For example, the leucine analog of puromycin is a less effective inhibitor of protein synthesis than puromycin (Rabinovitz and Fisher, 1962).

The inhibition of protein synthesis by homocitrullylaminoadenosine is consistent with the notion that 3′-substituted amino-3′-deoxyadenosine compounds act as analogs of aminoacyl-sRNA. For example, the addition of an acetyl group to 3′-amino-3′-deoxyadenosine results in a nucleoside that is not toxic to mammalian cells or bacteria (Suhadolnik et al., 1969).

## References

Guarino, A. J., M. L. Ibershof, and R. Swain, *Biochim. Biophys. Acta*, **72**, 62 (1963).
Kredich, N. M. and A. J. Guarino, *J. Biol. Chem.*, **236**, 3300 (1961).
Rabinovitz, M., and J. M. Fisher, *J. Biol. Chem.*, **237**, 477 (1962).
Suhadolnik, R. J., Bruce M. Chassy, and G. R. Waller, *Biochim. Biophys. Acta*, **179**, 258 (1969).

# 1.6. LYSYLAMINO ADENOSINE

This nucleoside antibiotic was also isolated from the mycelia of *Cordyceps militaris* grown in static culture for 30 days (Guarino and Kredich, 1964).

Lysylamino adenosine (Fig. 1.38) was isolated following homogenization of the mycelia, adsorption into Dowex-50-$NH_4^+$, and elution with 3 *N* $NH_4OH$. The nucleoside did consume periodate. It was ninhydrin positive and had an ultraviolet absorption spectrum similar to adenosine. Hydrolysis in alkali produced 3′-amino-3′-deoxyadenosine and lysine.

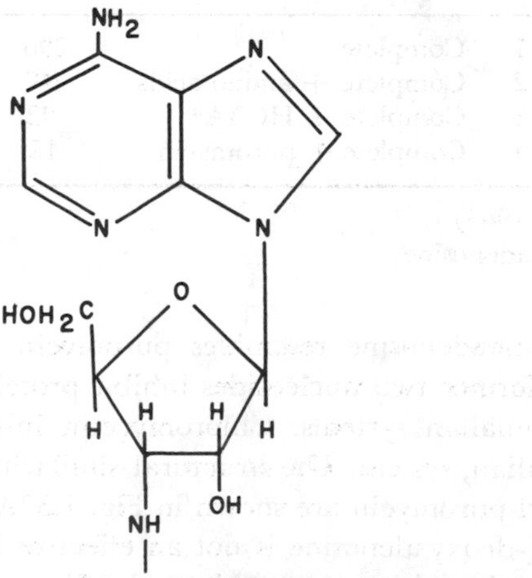

FIGURE 1.38. *Structure of lysylamino adenosine.*

A bis-dinitrophenyl lysine derivative was synthesized, indicating that the carboxyl end of lysine was attached to the 3′-amino group of 3′-amino-3′-deoxyadenosine. Since the lysine moiety in the nucleoside was stable to acid and the original compound did not react with periodate, Guarino and Kredich suggested that the lysine forms an amide linkage with the 3′-amino group.

## Reference

Guarino, A. J., and N. M. Kredich, *Federation Proc.*, **23**, 371 (1964).

CHAPTER 2

# Ketohexose Nucleosides



## INTRODUCTION

Psicofuranine (angustmycin C) and decoyinine (angustmycin A) make up the two adenine-ketose nucleoside antibiotics elaborated by the *Streptomyces*. Angustmycin C and psicofuranine have been shown to be the same nucleosides and angustmycin A and decoyinine have been shown to be the same. These trivial names will be used interchangeably in this chapter. The structure assigned to psicofuranine is 6-amino-9-(β-D-psicofuranosyl)purine (Fig. 2.1*a*) and the structure assigned to decoyinine is 9-β-D-(5,6-psicofuranoseenyl)-6-aminopurine (Fig. 2.1*b*). Psicofuranine and decoyinine are structural analogs of adenosine with modifications in the glycoside moiety. They are antibacterial and antitumor nucleoside antibiotics. Angustmycin C was isolated from *Streptomyces sp.* 6A-704 from a soil sample in Yayiocho, Bunkyo-ku, Tokyo by Yüntsen et al. (1954). In 1959, Schroeder and Hoek-

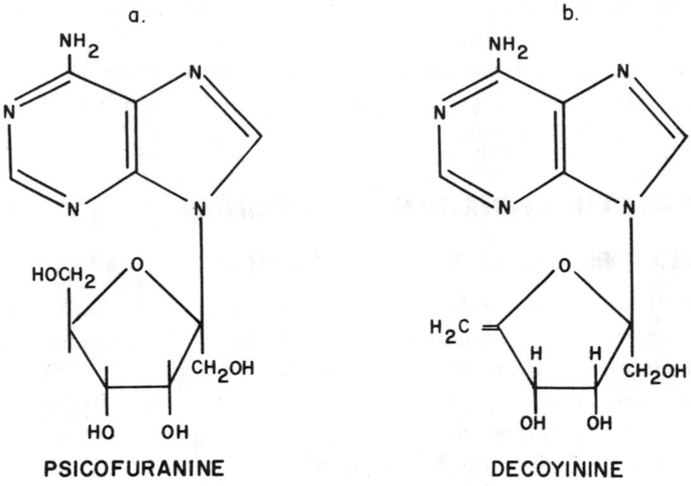

FIGURE 2.1. *Structure of psicofuranine and decoyinine.*

sema isolated psicofuranine from a culture of *S. hygroscopicus* var. *decoyicus*. Angustmycin A was isolated from the same *Streptomyces* that produced angustmycin C (Yüntsen et al., 1956). Adenine (angustmycin B) was also isolated and identified. As a result of degradation studies, Yüntsen (1958a) proposed the structure of angustmycin A to be 6-amino-9-(L-1,2-fucopyranoseenyl)purine. This same nucleoside antibiotic was isolated from the *S. hygroscopicus* culture by Hoeksema et al. (1964). They showed that angustmycin A and decoyinine were the same compound. On the basis of nuclear magnetic resonance studies, derivatives, hydrolysis, and fermentation data they concluded that the structure for decoyinine as reported by Yünsten (1958a) was in need of revision. The revised structure for decoyinine is shown in Figure 2.1*b* (Hoeksema et al., 1964). The *Streptomyces*-producing angustmycin A and angustmycin C were shown to be *S. hygroscopicus* var. *angustmyceticus* (Yüntsen, 1958a). Based on degradation studies and physical properties, Yüntsen proposed the structure of angustmycin C to be 6-amino-9-($\beta$-D-psicofuranosyl)purine. Sakai et al. (1954) subsequently identified the strain *Streptomyces sp.* 6-A704 to be *S. hygroscopicus*. Angustmycin A and angustmycin C were reported to be inhibitors of *Mycobacterium* 607 (Yüntsen et al., 1956). It was subsequently reported to protect mice from experimental infections with *Streptococcus haemolyticus*, *Staphylococcus aureus*, and *Escherichia coli* when administered orally or subcutaneously. Subsequent studies have shown that psicofuranine and decoyinine both inhibit xanthosine 5′-phosphate aminase. The inhibition takes place at the level of the nucleoside. In addition, both these nucleoside antibiotics act as feedback inhibitors of purine biosynthesis.

The chemistry, mechanism of action, and biosynthesis of psicofuranine and decoyinine have been recently reviewed (Fox et al., 1966; Hanka, 1967; Guarino, 1967; Suhadolnik, 1967).

## 2.1. PSICOFURANINE (ANGUSTMYCIN C)

### DISCOVERY, PRODUCTION, AND ISOLATION

The culture medium of Sakai et al. (1954) or that of Vavra et al. (1959) can be used for the production of psicofuranine. The procedure described here involves the medium as described by Vavra et al. (1959). Stock cultures of *S. hygroscopicus* var. *decoyicus* were maintained as spore preparations on sterile soil. A vegetative seed was prepared by incubating spores from the stock culture for 48 hr with shaking in a seed medium of the following composition: 2.5% glucose, 2.5% cotton seed flour; pH 7.2. Inoculum was 0.2–1.0 ml per 100 ml production medium. The production medium was as follows: 3% soy flour, 0.5% ammonium sulfate, 4% glycerol, and 0.4% calcium carbonate. Fermentations were carried out in baffled flasks on an incubator shaker at 28°C. Psicofuranine was isolated 48 hr after inoculation. The yield was about 1 g/liter.

De Boer et al. (1965, 1967) described the maintenance of *S. hygroscopicus* var. *decoyicus* (NRRL 2666) and the production and isolation of psicofuranine. Spores were used to inoculate 100 ml medium containing (per liter): glucose, 25g; soy peptone, 10g; corn steep liquor, 3g; yeast extract, 3g; N-Z amine A, 2g; $(NH_4)_2SO_4$, 3g; $MgSO_4$, 0.2g; NaCl, 0.1g; hydrated $FeSO_4$, 0.02g; hydrated $MnSO_4$, 0.003g; hydrated $ZnSO_4$, 0.004g; $KH_2PO_4$, 1.1g; pH adjusted to 7.2 and water to 1 liter. Incubation was for 72 hr at 28°C on a rotary shaker at 250 rpm. The culture was used to inoculate medium containing Kay-soy, 30g; $(NH_4)_2SO_4$, 5g; glycerol, 40g; cerelose, 20g; $CaCO_3$, 4g; pH adjusted to 7.2 and water to 1 liter. Incubation was in 100-ml volumes for 5 days at 30°C on a rotary shaker at 250 rpm.

The isolation of psicofuranine as described by Eble et al. (1959) was as follows: The culture medium was adjusted to pH 2.0 and filtered. The filtrate was adjusted to pH 9 and psicofuranine and decoyinine were adsorbed on charcoal. Psicofuranine was eluted with 80% acetone (pH 9.7–10.0). The acetone eluant was adjusted to about pH 7.5. The acetone was removed *in vacuo* and the aqueous solution held at 2°C until psicofuranine crystallized. More recently, Chassy et al. (1966) reported on the isolation of psicofuranine and decoyinine from the culture filtrates of *S. hygroscopicus* by use of the Dowex-1-$OH^-$–aqueous methanol method described by Dekker (1965).

## PHYSICAL AND CHEMICAL PROPERTIES

The molecular formula for psicofuranine is $C_{11}H_{15}N_5O_5$; mp 212–214°C (decomp.); $[\alpha]_D^{25} = -53.7°$ ($c$ = 1% in dimethyl sulfoxide); $\lambda_{max}$ 259 mμ ($E_{1\,cm}^{1\%}$ = 508) in 0.01 $N$ $H_2SO_4$ and 261 mμ ($E_{1\,cm}^{1\%}$ = 530) in 0.01 $N$ NaOH (Schroeder and Hoeksema, 1959). Psicofuranine is very easily hydrolyzed in acid (Eble et al., 1959).

## STRUCTURAL ELUCIDATION

The structural elucidation of angustmycin C (psicofuranine) was first reported by Yüntsen (1958b). The degradation of psicofuranine and identification of the aglycone and sugar moieties are shown in Figure 2.2. Acid hydrolysis of psicofuranine (**1**a) resulted in the isolation of adenine (**2**) and the ketohexose D-psicose (**3**). Treatment of this ketohexose with phenylhydrazine gave rise to a crystalline osazone (**4**) that was identical with that synthesized from D-psicose. Angustmycin C consumed 1 mole of periodate per mole of nucleoside. This was in agreement with a furanosyl structure. When D-psicose was reduced with sodium borohydride, D-talitol (**5**) and allitol (**6**) were isolated and identified. Although the anomeric configuration of psicofuranine was not known, Yüntsen proposed the structure of angustmycin C as 6-amino-9-(β-D-psicofuranosyl)purine (Fig. 2,1a).

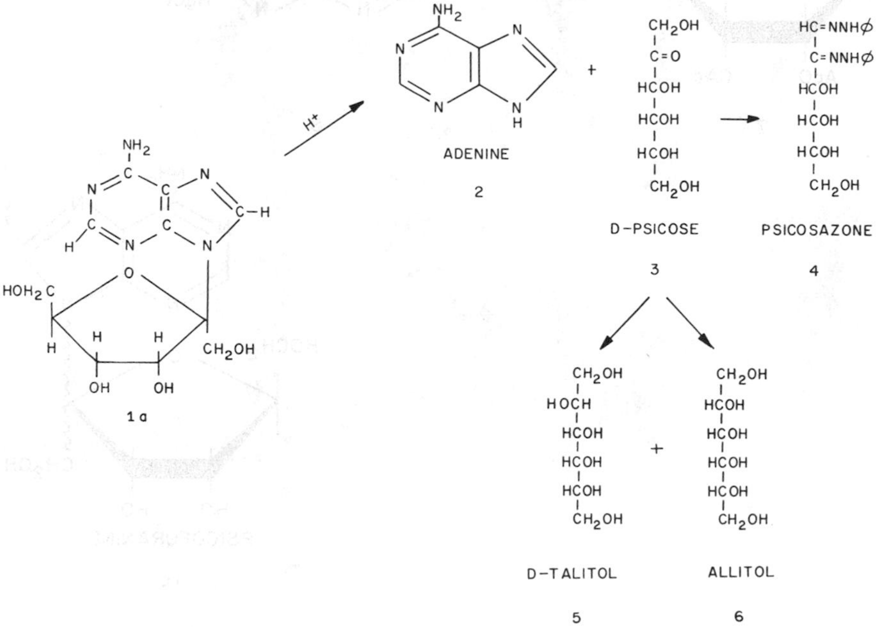

FIGURE 2.2. *Hydrolysis of psicofuranine.*

## CHEMICAL SYNTHESIS OF PSICOFURANINE (ANGUSTMYCIN C)

Schroeder and Hoeksema (1959) and Schroeder (1961) confirmed the structure of psicofuranine by describing a chemical synthesis of this nucleoside antibiotic (Fig. 2.3). When D-psicosyl chloride tetraacetate (**7**) was condensed with chloromercuri-6-acetamidopurine (**8**) followed by deacylation, psicofuranine (**1**a) was isolaled. The chemical and physical properties of the synthetic nucleoside antibiotic were identical with those of angustmycin C. This report constituted the first example of the chemical synthesis of a naturally occurring ketose nucleoside. More recently, Farkaš and Šorm (1962, 1963) reported their procedure for the chemical synthesis of psicofuranine. D-Psicose was synthesized according to the procedure of Wolfrom et al. (1945). The *O*-isopropylidine derivative of D-psicose (D-allulose) has been synthesized by Cree and Perlin (1968). The first unequivocal proof of the anomeric configuration of psicofuranine was established by McCarthy,

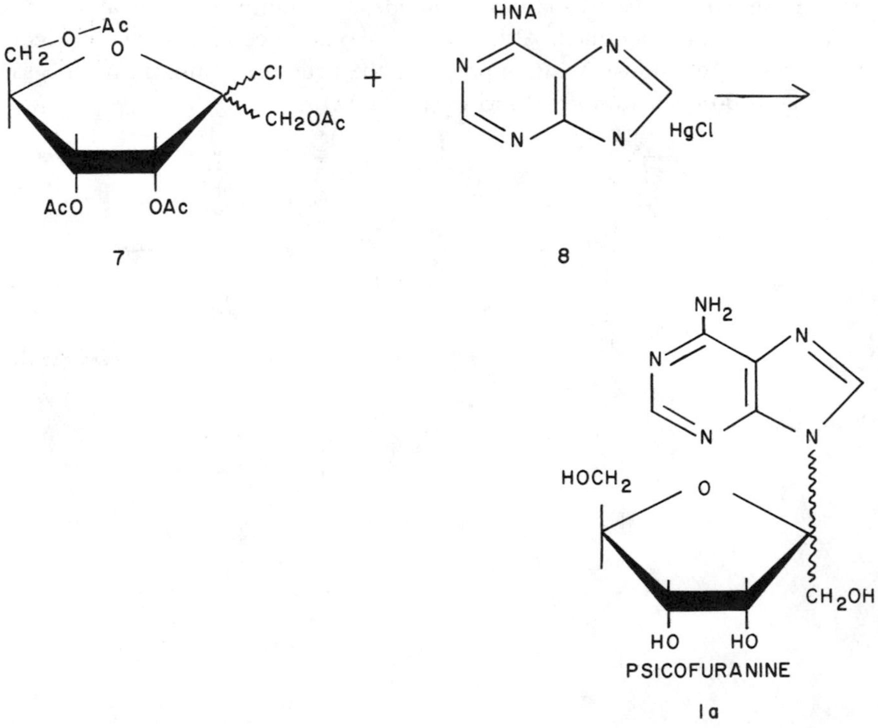

CH2—O—Ac O Cl CH2OAc AcO OAc 7 + HNA N N N N HgCl 8 → NH2 N N N N HOCH2 O CH2OH HO HO PSICOFURANINE 1a

FIGURE 2.3. *Chemical synthesis of psicofuranine; A, acetamido-* (*From Schroeder and Hoeksema*, 1959.)

Robins, and Robins (1968). Farkaš and Šorm (1967) also described the chemical synthesis of 6-amino-9-(1-deoxy-β-D-psicofuranosyl)purine.

## INHIBITION OF GROWTH

The study of the antibacterial spectrum of angustmycin C was performed on the agar discs (Yüntsen et al., 1954) (Table 2.1). *Mycobacterium tuberculosis* 607 was the only organism inhibited. Lewis et al. (1959) reported that psicofuranine protected mice from experimental infections with *Streptococcus hemolyticus*, *Staphylococcus aureus*, and *E. coli* when administered orally or subcutaneously. Psicofuranine was inactive in experimental virus and nematode infections. Forist et al. (1959) described a method to determine psicofuranine in blood plasma and serum. Psicofuranine did not inhibit KB cells in tissue culture. Wallach and Thomas (1959) reported that most of the psicofuranine administered orally was excreted through the kidney. While free psicofuranine was absorbed in animals, it was not absorbed in man unless it was converted to the tetraacetate. Hanka et al. (1959) reported on the assay techniques and special media to demonstrate *in vitro* activity of psicofuranine, and Evans and Gray (1959) reported that the oral administration of psicofuranine increased the number of survivors and caused tumor regression in rats bearing Walker adenocarcinoma, Murphy-Sturm lymphosarcoma, and Jensen sarcoma. Magee and Eberts (1961) reported that psicofuranine caused a marked regression and sharp drop in the uptake of phosphate in rats with subcutaneously transplanted Walker 256 adenocarcinoma.

**Table 2.1.**
Antimicrobial Spectrum of Angustmycin

| Test organisms | Inhibitory diameters, mm | Test organisms | Inhibitory diameters, mm |
|---|---|---|---|
| *Myc. tuberculosis* 607 | 15.0 | Grisein-fast *E. coli* | 0 |
| *Myc. phlei* | 13.5 | *Asp. oryzae* | 0 |
| *B. subtilis* | 0 | *Asp. niger* | 0 |
| *Staph. aureus* 209 P | 0 | *Rhizopus japanicus* | 0 |
| *B. agri* | 0 | *Pen. chrysogenum* Q-176 | 0 |
| *Pseud. fluorescence* | 0 | *Sacch. cerevisiae* | 0 |
| *E. coli* | 0 | *Torula utilis* | 0 |
| Streptomycin-fast *E. coli* | 0 | *Zyg. sulsus* | 0 |
| Streptothricin-fast *E. coli* | 0 | *Trichophyton intergitales* | 0 |
| Neomycin-fast *E. Coli* | 0 | *Candida albicans* | 0 |

From Yüntsen et al., 1954.

The $LD_{50}$ dose for mice receiving psicofuranine subcutaneously is 6.1–41 mg/kg, orally 13–69 mg/kg; and intraperitoneally 1695 mg/kg. The $LD_{50}$ for psicofuranine given orally to rats is 10,000 mg/kg.

## BIOSYNTHESIS

Since adenosine was shown to be the direct precursor in the biosynthesis of cordycepin (Suhadolnik et al., 1964), one pathway involved a $C_1$ unit added to carbon 1′ of the ribose moiety in adenosine (or a phosphorylated adenosine derivative) to form psicofuranine. The second pathway involved the condensation of adenine with a phosphorylated derivative of D-psicose in a manner analogous to the condensation of adenine with PRPP. The product formed would be psicofuranine-6-phosphate. Hydrolysis would result in the formation of psicofuranine. The third pathway proposed involved adenine condensation with fructose-2-phosphate to give 6-amino-9-β-D-fructofuranosylpurine. This nucleoside would then undergo a 3-epimerization to form psicofuranine. The fourth scheme postulated that adenine would condense with psicose-2-phosphate to give psicofuranine. The fifth scheme postulated that adenine condensed with psicose-6-phosphate to give psicofuranylic acid. The sixth scheme involved the condensation of adenine with a nucleoside diphosphate-hexose (glucose, fructose, or psicose) to give psicofuranine. Finally, all the above reactions might also take place if hypoxanthine replaced adenine. This notion would require amination of the nucleoside to give psicofuranine.

When adenosine-U-$^{14}C$ was added to the culture filtrates of *S. hygroscopicus*, only the adenine moiety of the psicofuranine was radioactive (Sugimori and Suhadolnik, 1965). Formate-$^{14}C$ was incorporated into the adenine ring, but not into the sugar moiety of psicofuranine. These results strongly suggested that the first pathway (adenosine + $C_1$) was not operating in the biosynthesis of psicofuranine and that cordycepin and psicofuranine biosynthesis proceeded by different pathways. When glucose-1-$^{14}C$, glucose-6-$^{14}C$, fructose-U-$^{14}C$, allose-1-$^{14}C$, or psicose-$^{3}H$ was added to cultures of *S. hygroscopicus*, there was a considerable amount of radioactivity in the psicofuranine. All the radioactivity resided in the ketohexose moiety. Hydrolysis of the psicofuranine from the glucose-1-$^{14}C$ and glucose-6-$^{14}C$ experiments, formation of the crystalline, radioactive psicosazone, and degradation by periodate showed that glucose was a direct precursor for the psicose moiety of psicofuranine (Table 2.2).

The data in Table 2.2 show that 81% of the carbon-14 from the D-glucose-1-$^{14}C$ experiment resided in carbons 1, 2, and 3 of D-psicose. Presumably, all the radioactivity in these 3 carbons was located in carbon 1. Only 10% of the carbon-14 was found in carbon 6. Likewise, none of the radioactivity resided in carbons 1, 2, and 3 from the glucose-6-$^{14}C$ experiment, and 88% of the

**Table 2.2.**
Distribution of Carbon-14 in D-Psicose from Glucose-1-$^{14}$C and Glucose-6-$^{14}$C

| Derivative | Carbon atoms | Glucose-1-$^{14}$C cpm/mmole | % $^{14}$C | Glucose-6-$^{14}$C cpm/mmole | % $^{14}$C |
|---|---|---|---|---|---|
| Psicosazone | 1,2,3,4,5,6 | 52,200 | 100 | 4,580 | 100 |
| Mesoxaldehyde-1,2-bisphenylhydrazone | 1,2,3 | 42,000 | 81 | 0 | 0 |
| Formaldimedone | 6 | 5,000 | 10 | 4,020 | 88 |

From Sugimori and Suhadolnik, 1965.

radioactivity resided in carbon 6. These data suggested that glucose underwent an isomerization from the aldohexose and a 3-epimerization to form D-psicose. This biological transformation could be conceived as occurring by any of the following pathways (X = purine or pyrimidine):

Glucose→glucose-6-P→fructose-6-P→psicose-6-P

Hexose→Hexose-1-P→XDP-hexose→XDP-allose→psicose-1-P

Fructose→fructose-2-P→XDP-fructose→XDP-psicose→psicose

To determine which pathway for the biosynthesis of psicofuranine was operating, cell-free extracts of *S. hygroscopicus* were made (Suhadolnik and Sugimori, 1966). The cells were ground with alumina and centrifuged. Incubations of the supernatant were carried out with glucose-1-$^{14}$C, ATP, $Mg^{2+}$, and phosphate buffer (pH 7.2) for 90 min. A water-soluble, alcohol-insoluble barium phosphate compound formed that represented 40% of the radioactivity. Hydrolysis of the sugar phosphates with alkaline phosphatase, paper chromatography, elution of radioactivity in the psicose area, and treatment with phenylhydrazine resulted in the isolation of radioactive, crystalline psicosazone. No radioactive psicose was found in control experiments when ATP was omitted from the incubation mixture. Neither GTP nor UTP increased the conversion of D-glucose to D-psicose. To determine if fructose-2-phosphate was involved in the biosynthesis of psicofuranine, fructose-2-phosphate was synthesized according to the method of Pontis and Fischer (1963). The addition of fructose-2-phosphate to the cell-free preparation of *S. hygroscopicus* along with adenine, ATP, $Mg^{2+}$, and phosphate buffer (pH 7.2) did not result in the isolation of psicofuranine. The data reported here might be similar to the report of Gibbins and Simpson (1964) and Simpson and Gibbins (1966) in which they suggested that cell-free extracts of *Aerobacter aerogenes* converted fructose-6-phosphate to D-psicose-6-phosphate by means of a 3-epimerase.

Although formate-$^{14}C$ was not incorporated into the psicose moiety of psicofuranine, Kemp and Quayle (1967) reported that formate condensed with carbon 1 of ribose-5-phosphate to give psicose-6-phosphate. This reaction took place in *Pseudomonas methanica*. Hough and Stacey (1966) also reported on the biosynthesis and metabolism of allitol and D-psicose in *Itea* plants. The distribution of the carbon-14 in D-psicose isolated from *Itea* plants following the incorporation of radioactive carbon dioxide was not reported. Strecker et al. (1965) reported on the isolation and identification of D-psicose from human urine. Passeron and Recondo (1965) discovered that DCC in methanol rearranges fructose to glucose, mannose and psicose. These reports show that D-psicose has been isolated from several different biological sources. Undoubtedly, D-psicose will be found in other bacterial and animal tissues in the future. Although neither GTP nor UTP increased the conversion of glucose to psicose, it should be emphasized that UDP-fructose is a naturally occurring compound. It has been isolated from the tubers of Jerusalem artichoke and germinating pea seeds (Umemura et al., 1967; Brown and Mangat, 1967). GDP-fructose was also isolated from *Eremothecium ashbii* by Pontis et al. (1960) and Gonzales and Pontis (1963). Although the cell-free studies reported here did not implicate a nucleoside diphosphate-ketohexose as a precursor in psicofuranine biosynthesis, there is precedence for the importance of the phosphorylation of the keto group of fructose in biological systems. It must also be emphasized that the conversion of glucose or glucose-6-phosphate to psicose-phosphate need not be related to psicofuranine biosynthesis.

Studies on the interconversion of psicofuranine and decoyinine were first reported by Hoeksema et al. (1964). These studies were performed with tritium-labeled psicofuranine and decoyinine. More recently, Chassy et al. (1966) reported that psicofuranine-U-$^{14}C$ was converted to decoyinine-U-$^{14}C$ without cleavage of the *N*-glycoside bond. These studies were performed with psicofuranine labeled in the adenine moiety and in the sugar moiety at carbon 6′. The carbon-14 labeled decoyinine isolated following the addition of psicofuranine to the culture filtrates of *S. hygroscopicus* had the same distribution of radioactivity in the adenine and sugar moiety as the psicofuranine used in these experiments. The mechanism by which the C-6′ hydroxymethyl group (—$CH_2OH$) of carbon C-6′ of psicofuranine is converted to a C=C structure is not known.

## BIOCHEMICAL PROPERTIES OF PSICOFURANINE

The first study related to the mechanism of action of psicofuranine was that of Hanka (1960), in which he used the technique of quantitative reversal of antimicrobial action against *Staphylococcus aureus*. The antimicrobial activity of psicofuranine was reversed by guanine and guanosine by using the semi-synthetic medium of Hanka and Burch (1960). The reversal was noncompetitive. Psicofuranine at concentrations of 1 mg/ml did not cause inhibition once the concentration of guanine in the sugar medium was 5 μg/ml (Hanka, 1960) (Table 2.3).

**TABLE 2.3.**

Reversal of Inhibition by Guanine or Guanosine Added to Growth Medium

| Compound tested | Conc., μg/ml | Concentration of psicofuranine, μg/ml | | | | |
|---|---|---|---|---|---|---|
| | | 10 | 20 | 40 | 80 | 160 |
| Control | — | 15.5[a] | 18 | 21 | 24 | 27 |
| Guanine | 2.5 | 2 | 0 | Trace | 21 | 24.5 |
| | 5.0 | 0 | 0 | 0 | 0 | 0 |
| Guanosine | 2.5 | Trace | 16 | 18 | 21 | 24.5 |
| | 5.0 | 0 | 0 | 0 | 0 | 19 |

From Hanka, 1960.
[a]Zone of inhibition in mm. diameter.

When adenosine or inosine was added to the growth media in place of adenine, the reversal of inhibition of *S. aureus* by psicofuranine was not as good. Xanthosine had little effect and the pyrimidines did not reverse the inhibition of psicofuranine. Hanka (1960) suggested that these data indicated that the mechanism of action of psicofuranine was not directly related to the biosynthesis of adenosine or inosine. Hanka concluded that the main effect of psicofuranine in *S. aureus* was related to the interference of the biosynthesis of GMP from XMP. In subsequent studies on the inhibition of growth of *E. coli* B, Slechta (1960a, 1960b) reported that psicofuranine was bacteriostatic and that guanine and guanosine prevented inhibition by psicofuranine. These findings with *E. coli* were in agreement with those reported by Hanka (1960) with *S. aureus*. When Slechta (1960b) examined the culture filtrates of the *E. coli* grown in the presence of psicofuranine, xanthosine was found to accumulate. In addition, psicofuranine markedly changed the incorporation of carbon-14 labeled glycine into the purine bases. Slechta concluded that the mechanism of action in *E. coli* was related to the interference in the conversion of XMP to GMP. In a subsequent study, Slechta (1960b) used cell-free extracts of *E. coli* and reported that xanthosine 5′-phosphate aminase activity was inhibited by psicofuranine (Fig. 2.4). Magee and Eberts (1961), also reported that psicofuranine blocked the incorporation of glycine-2-$^{14}C$ into the purine bases of animal tumors.

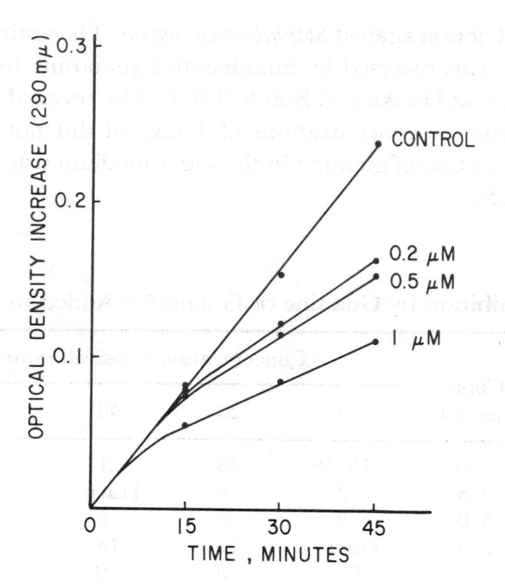


FIGURE 2.4. *Inhibition of XMP aminase by psicofuranine. The amounts of psicofuranine are given in the figure* (*From Slechta,* 1960*b*.)

Moyed and his co-workers studied the mechanism of inhibition of psicofuranine in *E. coli* B and in psicofuranine-resistant mutants of *E. coli* B to determine if psicofuranine resulted in a change of the XMP aminase following resistance to this nucleoside analog. These studies have supplied most interesting data on the manner in which psicofuranine binds to XMP aminase. Moyed (1961) showed that the production of IMP dehydrogenase and XMP aminase was controlled by a feedback control mechanism. Moyed suggested that the inhibition of XMP aminase by psicofuranine in *E. coli* decreased the supply of GMP. With this decrease in GMP, XMP aminase increased rapidly. In a subsequent study, Udaka and Moyed (1963) reported that the inhibition of XMP aminase by psicofuranine was not a simple, common type of inhibition. The authors reported that the level of XMP aminase in a psicofuranine-resistant mutant of *E. coli* B (type B-24) was about six times as great as that in the wild-type cell. A resistant mutant (type B-35) with a reduced XMP aminase activity was also found. Inhibition of XMP aminase by psicofuranine was found to be dependent on XMP as well as inorganic pyrophosphate. Udaka and Moyed concluded that the inhibition by psicofuranine is a two-step process. The first is a freely reversible, pyrophosphate-dependent reaction that occurs between psicofuranine and the enzyme.

$$\text{XMP aminase} + \text{psicofuranine} + \text{PPi} \rightarrow (\text{inhibited enzyme})_1 \qquad (1)$$

The second reaction is the irreversible step. It requires the addition of XMP.

$$(\text{inhibited enzyme})_1 \xrightarrow{\text{XMP}} (\text{inhibited enzyme})_2 \quad (2)$$

The aminase from the psicofuranine-resistant mutant appears to have lost the ability to undergo reaction 2, i.e., the irreversible inhibition step. A study of the kinetic constants of the enzymatic reaction with this mutant showed that psicofuranine is a noncompetitive inhibitor and is apparently bound at a site separate from the binding or regulatory site on the enzyme.

In a subsequent report, Fukuyama and Moyed (1964) reported that XMP and inorganic pyrophosphate stimulated the binding of psicofuranine to XMP aminase. ATP and ammonia reduced the inhibitory action of psicofuranine. Since there was no competition for binding sites on the enzyme between psicofuranine and XMP, Fukuyama and Moyed concluded that psicofuranine was bound at a specific site. The binding of psicofuranine to a site of XMP aminase separate from the catalytic site by the use of specific protein-modifying reagents was reported by Kuramitsu and Moyed (1964, 1966). They showed that chemical modifications of XMP aminase by urea, mercaptoethanol, and ethylene glycol permitted desensitization of XMP aminase. The enzyme from the mutant was more readily desensitized than was the enzyme from the parent strain. This desensitization and enzyme inactivation were reversible. Urea caused a threefold reduction in the capacity of the aminase to bind psicofuranine. 2-Mercaptoethanol reduced the sensitivity of the aminase to inhibition by psicofuranine. This reduction did not affect the activity of the aminase nor its ability to bind psicofuranine. Methylene blue desensitized the enzyme by selectively reducing the ability of XMP aminase to bind psicofuranine. The substrate-binding capacities of the aminase were not affected by photooxidation. The addition of psicofuranine to XMP aminase increased the availability of sulfhydryl groups. The authors felt that the binding of psicofuranine at a separate binding site probably acts by distorting the tertiary structure at the active center. Therefore, this noncompetitive inhibition of XMP aminase by psicofuranine meets the two important requirements for allosteric inhibition of regulatory proteins. They are: (*1*) the psicofuranine is bound at a site distinct from the active site of XMP aminase, as shown by kinetic studies and selective desensitization and (*2*) the inhibition by psicofuranine is accompanied by a change in the conformation of the aminase.

In order to determine how psicofuranine reduced the velocity of XMP aminase, Fukuyama (1966) studied the formation and inhibition of an adenylxanthosine monophosphate intermediate by XMP aminase.

By analogy with studies on the formation of GMP by mammalian XMP aminase, Fukuyama proposed that GMP formation in *E. coli* proceeds in two steps:

$$\text{ATP} + \text{XMP} \rightarrow \text{adenyl-XMP} + \text{PPi} \quad (3)$$

$$\text{Adenyl-XMP} + NH_3 \rightarrow \text{AMP} + \text{GMP} \quad (4)$$

In the absence of $NH_3$, incubation of the aminase with ATP-8-$^{14}$C and XMP-8-$^{14}$C resulted in the formation of AMP without formation of GMP. Data were provided from electrophoretic studies that showed that an intermediate was formed that contained equal amounts of radioactivity from ATP-8-$^{14}$C and XMP-8-$^{14}$C. When $NH_3$ was added, AMP and GMP were isolated. In the absence of $NH_3$, the enzyme complex gave rise to AMP and XMP. Fukuyama found that psicofuranine did not inhibit the hydrolytic cleavage of the intermediate (adenyl-XMP) to form AMP and XMP. However, the psicofuranine-inhibited enzyme could not catalyze the aminolysis of this preformed intermediate to AMP and GMP (Table 2.4). Although the psicofuranine-inhibited aminase can bind both ATP and XMP, it cannot join them together to form the adenyl-XMP intermediate. Despite the inability of the psicofuranine-inhibited aminase to condense ATP and XMP to form the adenyl-XMP intermediate, both XMP and ATP are bound to the enzyme. More recently, Fukuyama (personal communication) reported that the nucleoside antibiotics decoyinine, tubercidin, and cordycepin

**Table 2.4.**

Aminolysis of Adenyl-XMP Intermediate[a]

| Additions | Intermediate | XMP, Δ*cpm* | GMP |
|---|---|---|---|
| Aminase | −1801 | +1637 | +353 |
| Aminase + $NH_3$[b] | −2101 | +713 | +1466 |
| Aminase + psicofuranine + XMP + PPi + $NH_3$ | −2714 | +2357 | +164 |

From Fukuyama, 1966.

[a] The intermediate was formed in a reaction mixture (0.78 ml) containing 100 μmoles of Tris buffer (pH 7.1), 20 μmoles of $MgCl_2$, 1.0 μmole of NaF, 100 mμmoles of ATP, 35 mμmoles of XMP-8-$^{14}$C, ($3.59 \times 10^5$ dpm), and 9.8 units of aminase. After 45 sec of incubation at room temperature, 0.20 mg of crystalline hexokinase and 0.25 μmole of glucose were added and the reaction mixture was incubated for an additional 30 sec. The reaction was terminated by chilling and the addition of 0.05 ml of cold 50% trichloracetic acid. Precipitated protein was removed by centrifugation, and the excess trichloracetic acid was removed by extraction with ether. Samples (0.1 ml) of the preparation were incubated in reaction mixtures (0.34 ml) containing 100 μmoles of Tris buffer (pH 7.1), 2.0 μmoles of $MgCl_2$, 0.45 unit of aminase, and, as indicated, 20 mμmoles of XMP, 400 mμmoles of ATP, 400 mμmoles of psicofuranine, and 4.0 μmoles of $(NH_4)_2SO_4$. After 6.0 min of incubation, these reaction mixtures were deproteinized as usual and subjected to paper electrophoresis in 0.04 *M* sodium citrate buffer, pH 2.3, at 3000 V for $1\frac{3}{4}$ hr. The data are expressed as changes in radioactive counts in comparison to a control mixture without fresh aminase.

[b] Aminolysis of the intermediate was not observed with $NH_3$ alone.

inhibit XMP-aminase. The inhibition of the enzyme with decoyinine is enhanced when XMP and pyrophosphate are added. Treatment of the aminase with mercaptoethanol desensitized the aminase to the inhibitory action of these nucleoside antibiotics.

Moyed (private communication) and Zyk, Citri, and Moyed (1969; 1970) now believe that there is not a separate site for the binding of adenine glycosides as psicofuranine and decoyinine. In view of more recent data on the nature of conformational changes that occur with XMP aminase and certain of its substrates, the original conclusion that psicofuranine is a non-competitive inhibitor of XMP aminase has now been reevaluated by Moyed and his co-workers. Their principal finding is that XMP induces a conformative response in XMP aminase (Fig. 2.5) (Zyk, Citri, and Moyed, 1969). This conformative response has been expressed in terms of the conformative response constants or $K_{cr}$, which is equal to the concentration of XMP causing a half maximal response. This value is $10^{-4}$ *M* XMP. This conformative response is the reversible change in conformation of XMP aminase that occurs following binding of XMP, pyrophosphate, and $Mg^{2+}$. The conformative response to XMP is reflected in resistance to inactivation by heat or hydrolysis by proteolytic enzymes. The $K_{cr}$ values for XMP obtained by heat treatment (Fig. 2.5*a*), digestion by trypsin (Fig. 2.5*b*), and digestion by Pronase (Fig. 2.5c) are in excellent agreement. The excellent agreement by these three criteria led Moyed to assume that each method yields valid information on the conformative response on the enzyme to XMP. Although

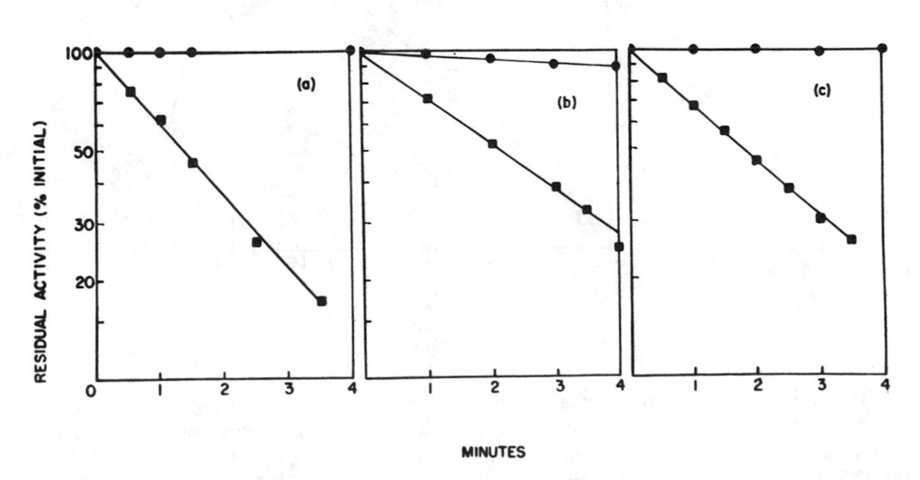


FIGURE 2.5. *The effect of XMP on the rate of inactivation of XMP aminase. XMP aminase* (6.5 *μg of protein*) *was incubated* (●) *with and* (■) *without* 1.0 *mM XMP. The methods of inactivation are:* (*a*) *heat treatment;* (*b*) *digestion by trypsin;* (*c*) *and digestion with Pronase.* (*From Zyk, Citri, and Moyed,* 1969.)

ATP does not induce a detectable conformational change, it enhances or modifies the conformative response to XMP by a factor of 2000, that is, the $K_{cr}$ for XMP decreases by $10^{-4}$ to $5 \times 10^{-8}$ in the presence of ATP. Adenosine and pyrophosphate at equimolar concentrations produce identical effects. Adenosine alone is without effect; pyrophosphate causes a 33-fold decrease in $K_{cr}$ (Fig. 2.6). These data clearly show that the adenosine–pyrophosphate combination causes an increase in XMP binding similar to that of ATP and, thereby, an enhancement of the conformative response to XMP under conditions which the catalytic intermediate could not form.

According to this argument, the adenosine–pyrophosphate–XMP combination serves as a collective analog of the catalytic intermediate (Zyk, Citri, and Moyed, 1969; Moyed, private communication). The validity

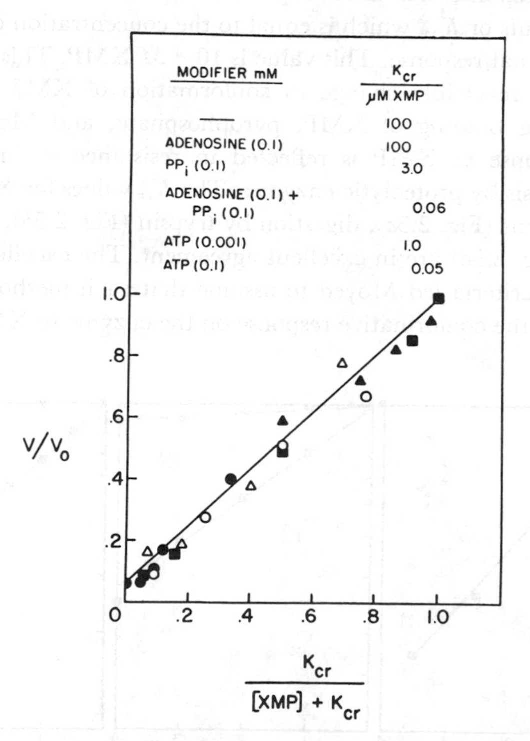


FIGURE 2.6. *Replacement of ATP with structurally related modifiers of the conformative response. XMP aminase* (6.5 *μg protein*) *was exposed to proteolytic inactivation by trypsin with varying amounts of XMP along with these additional variations:* (●) 0.1 *mM ATP;* (○) 0.001 *mM ATP;* (▲) 0.1 *mM* $PP_i$*;* (△) 0.1 *mM adenosine and* 0.1 *mM* $PP_i$*; and* (■) *no additions or adenosine alone. Rates of inactivation in the V, presence, and* $V_0$*, absence, of XMP and the* $F_{cr}$ *values were determined. The* $K_{cr}$ *values are listed in the insert.* (*From Zyk, Citri, and Moyed*, 1969.)

of this argument is demonstrated as follows: Ammonia indeed reduces modification of the conformative response by ATP and XMP to that of XMP alone since ammonia causes aminolysis of the catalytic intermediate. On the other hand, the adenosine–pyrophosphate–XMP combination is not subject to aminolysis and, therefore, ammonia cannot eliminate the modified response to this combination.

Since Fukuyama and Moyed (1964) observed that the adenine glycoside antibiotic psicofuranine and pyrophosphate also cause considerable enhancement of XMP binding, Zyk, Citri, and Moyed (1970) have examined the role of conformative response in reversible and irreversible binding of adenine glycosides. By using *E. coli* mutants, with reduced susceptibility to inhibition to psicofuranine and decoyinine, they have been able to show that the conformative response of the mutant to XMP alone is identical with that of the parental aminase. However, ATP or adenosine and pyrophosphate have greatly diminished abilities to enhance the conformative response to XMP: in the parental aminase, ATP or adenosine and pyrophosphate decrease the $K_{cr}$ for XMP from $10^{-4}$ to $5 \times 10^{-8}$ $M$, but in the mutants the reduction is only to $10^{-4}$ $M$ in both cases. The most drastic change observed in the mutants is that pyrophosphate alone is incapable of modifying the conformative response to XMP: in the parental aminase, pyrophosphate decreases the $K_{cr}$ value for XMP from $10^{-4}$ $M$ to $10^{-6}$ $M$, but in the mutants the conformative response constant remains at $10^{-4}$ $M$ with or without inorganic pyrophosphate.

The inhibition patterns of enzyme preparations from psicofuranine-resistant mutants are shown in Figure 2.7 (Zyk, Citri, and Moyed, 1970). XMP aminases from psicofuranine-resistant strains (B-96-7, B-96-17, B-96-24) derived from *E. coli* B-96 have altered susceptibility to inhibition by the adenine glycosides. The mutant aminases (B-96-7, B-96-24) are completely resistant to irreversible inhibition by piscofuranine and decoyinine. The B-96-17 mutant retains sensitivity to irreversible inhibition by psicofuranine and decoyinine; however, its rate of inactivation is lower than that of the parental aminase. The evidence presented shows that the genetic alteration of XMP aminase for the psicofuranine-resistant *E. coli* resulted in a loss or diminution of sensitivity to irreversible inhibition by psicofuranine and decoyinine, as well as decreased sensitivity to reversible inhibition by adenosine. This effect was shown to be attributed to the loss of responsiveness of pyrophosphate. The loss of responsiveness to pyrophosphate due to genetic changes confers resistance to the adenine glycosides. This effect appears to be accomplished primarily by disruption of the pyrophosphate-induced conformation either chemically (urea or guanidine–HCl) or by mutational events, presumably appropriate replacements of one of several amino acid residues, which impair the flexibility of the XMP aminase.

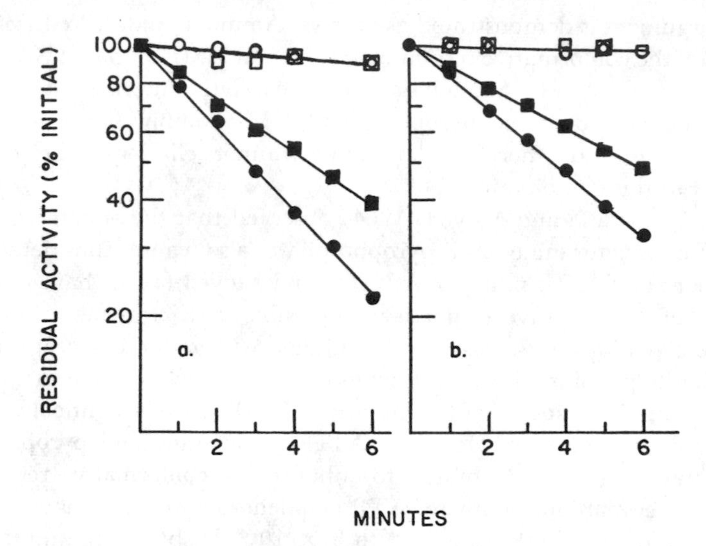


FIGURE 2.7. *Inactivation of parental and mutant XMP aminases by psicofuranine and decoyinine. The rate of inhibition was determined without added ATP and with either* 0.30 *mμ mole of psicofuranine* (*a*) *or* 0.30 *mμ mole of decoyinine* (*b*). *Purified aminases were obtained from parental strain* (●) *B* 96, (■) *B* 96–17, (○) *B* 96–7, *or* (□) *B* 96–24. (*From Zyk, Citri, and Moyed,* 1970.)

Chemical treatment of the aminase also eliminates the ability of adenine glycosides to modify the conformative response. Neither mutation of the aminase nor the chemical treatment reduces catalytic activity, but both kinds of alteration prevent the modification of the conformative response to XMP. Therefore, modification of this response is not related to catalytic activity. Mutants in which the aminase cannot undergo modification of its conformative response produce fewer molecules of XMP aminase. Moyed suggests that this modification of the conformative response has a physiological function in controlling the rate of aminase synthesis and is independent of the repression system controlling the guanine operon. Finally, this putative control mechanism must be involved in a late stage of the translation, since the enzyme must be very nearly completely synthesized before it can be involved in conformational changes.

In recent studies by Beppu et al. (1968a) angustmycin C increased the specific activity of inosine 5′-phosphate dehydrogenase of *E. coli* six-fold. Xanthosine was excreted into the medium. They reported that angustmycin C preferentially inhibited the biosynthesis of RNA and DNA (Fig. 2.8). The inhibition of protein synthesis was not as severe.

In a subsequent paper, Beppu et al. (1968b) reported that a small, but significant, amount of tritium-labeled angustmycin C was incorporated into

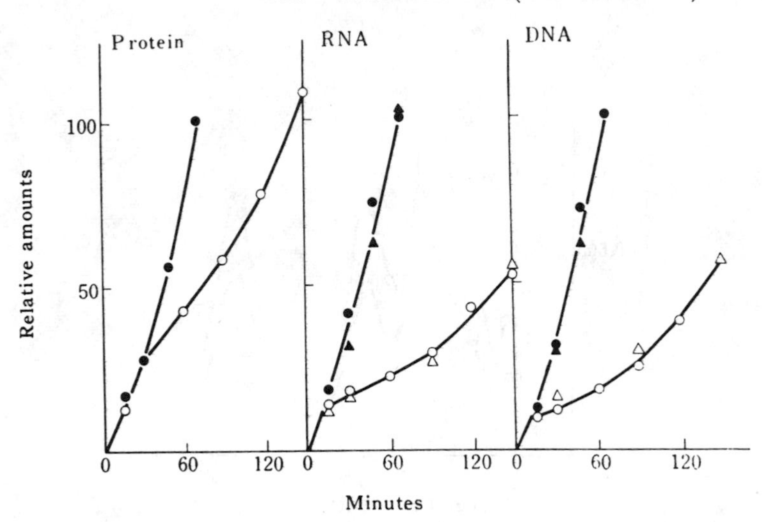


FIGURE 2.8. *Effect of angustmycin C on the synthesis of protein, RNA, and DNA. Angustmycin was added at time* 0 *at the concentration of* 50 *μg/ml.* $^{32}$*P-Phosphate was added simultaneously.* (○) *Chemical amounts in the presence of the antibiotic;* (●) *chemical amounts in the absence of the antibiotic;* (△) *radioactivities in the presence of the antibiotic;* (▲) *radioactivities in the absence of the antibiotic.* (*From Beppu et al.*, 1968*a*.)

the RNA of *E. coli*. The amount of psicofuranine taken up by the cells was 2.3%, and the largest amount of radioactivity in the cells was found in the RNA fraction. The distribution of radioactivity in the RNA following phenol treatment of broken cells shows that the $^3$H coincide with the 3 peaks (Fig. 2.9). The bands are rRNA and one is sRNA. Angustmycin C was isolated when the RNA was hydrolyzed by an enzyme mixture of snake venom, pancreatic ribonuclease I, and alkaline phosphatase. The authors concluded that these data provided evidence that angustmycin C was incorporated into the polyribonucleotide chain of RNA. Magee and Eberts (1961) also reported on the isolation of psicofuranine phosphate from the tumor tissue of animals that received tritium-labeled psicofuranine. Beppu et al. (1968b) stated that the incorporation of angustmycin C into RNA by *E. coli* suggests that some activation enzymes forming angustmycin C phosphate must be present in *E. coli* cells. They showed that β-galactosidase synthesis in *E. coli* cells preincubated with angustmycin C was inhibited. The addition of guanosine reversed this inhibition. The authors concluded that it was reasonable to assume that such an abnormal nucleotide formed in the cells of nucleic acids may be present for a considerable period of time after the removal of angustmycin C from the medium. This would cause a residual inhibitory effect on β-galactosidase synthesis.

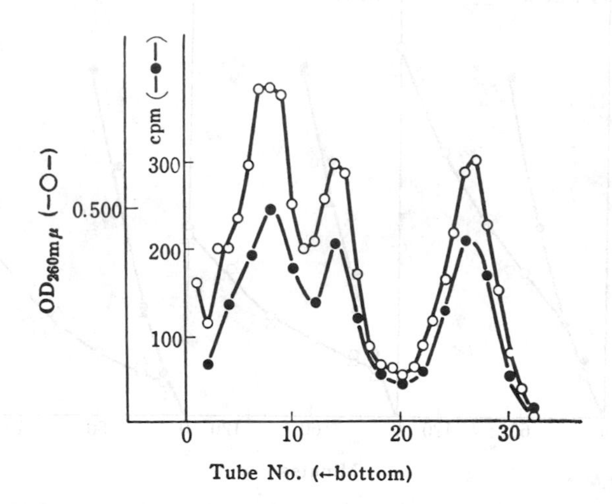


FIGURE 2.9. *Zonal centrifugation of labeled RNA.* (*From Beppu et al.*, 1968*b*.)

The presence of the hydroxymethyl group on C-1 of adenosine results in a nucleoside (psicofuranine) that is not a substrate for adenosine kinase from Ehrlich ascites tumor cells or rat liver (Shigurea and Sampson, 1967; Lindberg et al., 1967) and intestinal adenosine deaminase (Cory and Suhadolnik, 1965; Bloch and Nichol, 1964a).

The recent report on the synthesis of 6′-*O*-*p*-toluenesulfonyl-1′, 3′, 4′-*O*-orthoformyl-psicofuranine (see Fig. 2.10) compound **10**, Sect. 2.2) by McCarthy, Robins, and Robins (1968) provides an excellent compound for the chemical synthesis of the psicofuranine 6′-mono-, di-, and triphosphate. Because of the two hydroxymethyl groups in the ketose moiety of psicofuranine, the specific synthesis of the C-6′ phosphates has been very difficult.

Henderson (1963) studied the feedback inhibition of purine biosynthesis in Ehrlich ascites tumor cells by purine analogs since psicofuranine is structurally related to adenosine. Of the 37 purine analogs studied for their ability to inhibit purine biosynthesis *de novo* by a feedback mechanism, only 8 analogs were active feedback inhibitors. Psicofuranine was approximately as active as a feedback inhibitor as the less active natural purines. There did not appear to be a structural relationship with feedback inhibition activity. Of the purine inhibitors that were shown to be active feedback inhibitors, only five were phosphorylated to the nucleotides by these tumor cells.

In a subsequent report, Henderson and Khoo (1965) stated that psicofuranine was not a feedback inhibitor of purine biosynthesis *de novo* in Ehrlich ascites tumor cells. Psicofuranine did not inhibit the activity of phos-

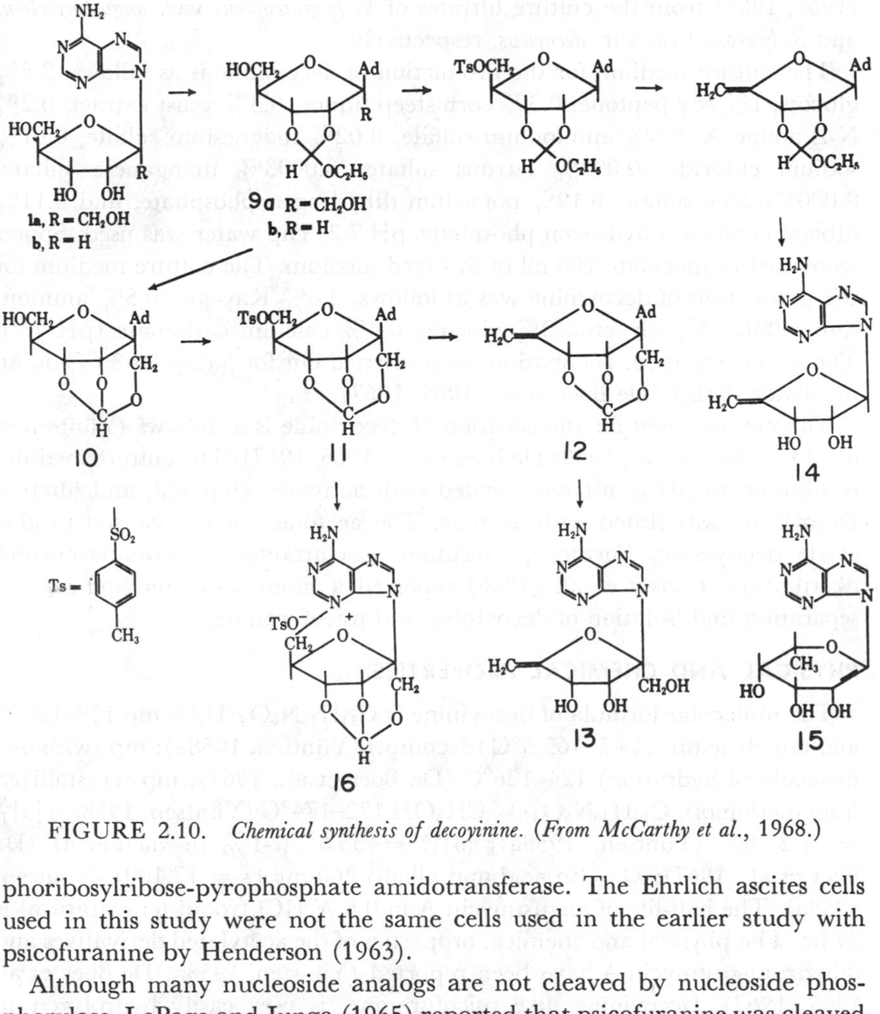

FIGURE 2.10. *Chemical synthesis of decoyinine.* (*From McCarthy et al.*, 1968.)

phoribosylribose-pyrophosphate amidotransferase. The Ehrlich ascites cells used in this study were not the same cells used in the earlier study with psicofuranine by Henderson (1963).

Although many nucleoside analogs are not cleaved by nucleoside phosphorylase, LePage and Junga (1965) reported that psicofuranine was cleaved by nucleoside phosphorylase from mouse tumors. Komuro et al. (1969) have recently shown that the addition of psicofuranine to cultures of *B. ammoniagenes* resulted in the accumulation of XMP in the medium. Guanine and xanthine derivatives completely suppressed the accumulation of XMP.

## 2.2. DECOYININE (ANGUSTMYCIN A)

### DISCOVERY, PRODUCTION, AND ISOLATION

Decoyinine (angustmycin A) was isolated by Yünsten et al. (1956), Hoeksema et al. (1964; and reference 5 of this paper), and De Boer et al.

(1965, 1967) from the culture filtrates of *S. hygroscopicus* var. *angustmyceticus* and *S. hygroscopicus* var. *decoyicus*, respectively.

The culture medium for the production of decoyinine is as follows: 2.5% glucose, 1% soy peptone, 0.3% corn steep liquor, 0.3% yeast extract, 0.2% N-Z amine A, 0.3% ammonium sulfate, 0.02% magnesium sulfate, 0.01% sodium chloride, 0.002% ferrous sulfate, 0.0003% manganese sulfate, 0.0004% zinc sulfate, 0.19% potassium dihydrogen phosphate, and 0.11% dibasic potassium hydrogen phosphate, pH 7.2. Tap water was used. Spores were used to inoculate 100 ml of this seed medium. The culture medium for the production of decoyinine was as follows: 3.0% Kay-soy; 0.5% ammonium sulfate, 4% glycerol, 2% glucose, 0.4% calcium carbonate (pH 7.2). Tap water was used. Incubations were carried out for 5 days at 30°C on an incubator shaker (De Boer et al., 1965, 1967).

The method used for the isolation of decoyinine is as follows (Yüntsen et al., 1956; Eble et al., 1959; De Boer et al., 1965, 1967): The culture medium is adjusted to pH 2, filtered, treated with activated charcoal, and filtered. Decoyinine was eluted with acetone. The acetone was evaporated to give crude decoyinine. Further purification was attained by counter-current distribution. Chassy et al. (1966) reported a more rapid method for the separation and isolation of decoyinine and psicofuranine.

## PHYSICAL AND CHEMICAL PROPERTIES

The molecular formula of decoyinine is $C_{11}H_{13}N_5O_4 \cdot H_2O$; mp 128–130°C and remelt temp. 164.5–165.5°C (decomp.) (Yüntsen, 1958a); mp (with one molecule of hydration) 124–126°C (De Boer et al., 1967); mp (crystallized from methanol), $C_{11}H_{13}N_5O_4 \cdot \frac{1}{2} CH_3OH$ 172–174°C (Yüntsen, 1958a); $[\alpha]_D^{25} = +27.02°$ (Yüntsen, 1958a); $[\alpha]_D^{25} = 35.6°$ ($c$-1% in methaeol) (De Boer et al., 1967); $\lambda_{max}$ (in acid and alkali) 260 m$\mu$ ($\epsilon = 17{,}100$) (Yüntsen, 1958a). The half-life of angustmycin A in 0.5 *N* HCl (room temperature) is 24 hr. The physical and chemical properties of the acetylated derivatives and dihydroangustmycin A have been reported (Yünsten, 1958a; De Boer et al. 1965, 1967). Decoyinine, like psicofuranine, is very easily hydrolyzed in dilute acid.

## STRUCTURAL ELUCIDATION

Based on the degradation studies of angustmycin A (decoyinine), Yüntsen originally identified this nucleoside antibiotic as 6-amino-9-(L-1,2-fucopyranoseenyl)purine. Hoeksema et al. (1964) suggested that angustmycin A and decoyinine were the same compound. By using nuclear magnetic resonance, derivatives, periodate oxidation, hydrolysis, and interconversion between psicofuranine and decoyinine, they concluded that the structure for decoyinine is 6-amino-9-(6-deoxy-$\beta$-D-*erythro*-hex-5-enofuran-2-ulosyl) purine (Fig. 2.1*b*).

## CHEMICAL SYNTHESIS OF DECOYININE

Robins and his co-workers have recently reported on the total chemical synthesis of decoyinine (McCarthy et al., 1968). The chemical synthesis involved the conversion of psicofuranine to decoyinine. This synthetic approach is extremely interesting since the biological interconversion of psicofuranine and decoyinine by *S. hygroscopicus* had been reported earlier by Hoeksema et al. (1964) and Chassy et al. (1966). The introduction of a 2′–3′ double bond into adenosine was reported (McCarthy et al., 1966). With this synthetic approach as a background, the possibility of utilizing an E2 elimination of *p*-toluenesulfonate from carbon 6′ was attempted. The synthesis of decoyinine is shown in Figure 2.10.

Treatment of psicofuranine (**1a**) with triethyl orthoformate resulted in the isolation of 3′,4′-*O*-ethoxymethylidinepsicofuranine (**9a**). The orthoformate proton resonance (pmr) appeared as a sharp singlet at 6.07 $\delta$ in DMSO-$d_6$. This indicated the selective formation of only one of the two possible diasterioisomers in this equilibrium reaction (McCarthy et al., 1968). Compound **9**a was treated with boron trifluoride etherate to give **10**. The formation of this latter compound provided the first unequivocal proof of the anomeric configuration of psicofuranine. These data show that only the $\beta$ anomer with the 1′-hydroxymethyl group below the plane of the sugar could form this tridentate orthoformate. When compound **10** was treated with *p*-toluenesulfonyl chloride, the product was 6′-*O*-*p*-toluenesulfonyl-1′,3′,4′-*O*-orthoformylpsicofuranine (compound **11**). When compound **11** was treated with potassium *t*-butoxide, *t*-butanol-pyridine, 6-amino-9-(1,3,4-*O*-orthoformyl-6-deoxy-$\beta$-D-*erythro*-hex-5-enofuran-2-ulosyl)purine (compound **12**) was isolated. Compound **12** was converted to 6-amino-9-(6-deoxy-$\beta$-D-*erythro*-hex-5-enofuran-2-ulosyl)purine (compound **13**, angustmycin A or decoyinine). Analytically pure deocyinine (angustmycin A) monohydrate melted at 130–133°C (softens at 125°C), resolidified by 150°C, and decomposed at 156–159°C; $\lambda_{max}^{H_2O}$ 259 m$\mu$ ($\epsilon = 15{,}500$); $\lambda_{max}^{pH\ 11}$ 259 m$\mu$ ($\epsilon = 15{,}500$); synthetic decoyinine, $[\alpha]_D^{26}$ +43.5° (*c* 1% in water); naturally occurring decoyinine, $[\alpha]_D^{26}$ +43.8° (*c* 1% in water). The infrared and nmr spectra of the synthetic and naturally occurring nucleosides were identical. The original structure for decoyinine as proposed by Yüntsen (1958a) is shown as compound **15**.

6-Amino-9-(5-deoxy-$\beta$-D-*erythro*-pent-4-enofuranosyl)purine (compound **14**) was also synthesized from adenosine (**1b**). A new synthesis of 5′-deoxyadenosine from adenosine was also reported. Compound **14** was reported to exhibit about equal antibacterial potency against *S. faecalis* as did angustmycin A. The conversion of **11** to **13** supplies additional evidence for the anomeric configuration of psicofuranine. Verheyden and Moffatt (1966) also described the synthesis of the 4′,5′-unsaturated pyrimidine nucleoside 1-(5-deoxy-$\beta$-D-*erythro*-pent-4-enofuranosyl)uracil.

## INHIBITION OF GROWTH

Yüntsen et al. (1956) reported that the antimicrobial activity of angustmycin A (decoyinine) was exclusively against *Mycobacteria*. Angustmycin A completely inhibited the growth of *Mycobacterium* 607 at 20 $\mu$g/ml. Tanaka et al. (1959) reported that gram-positive organisms were more sensitive to angustmycin A than were the gram-negative organisms. Tanaka et al. (1959) compared the antibiotic spectra of angustmycin A and angustmycin C. They showed that angustmycins A and C were distinctly inhibitory against gram-positive microorganisms and the *Mycobacterium*. Both angustmycins were equally inhibitory against a strain of *Bacillus subtilus*. Angustmycin A was more inhibitory against *Mycobacterium* 607 than was angustmycin C. The inhibition of both nucleoside antibiotics was reversed by adenosine, guanosine, and guanine. Angustmycin C was more inhibitory against *Staphylococcal* and *Streptococcal* infections in mice than angustmycin A. Angustmycin A was effective against Walker adenocarcinoma 256 in rats (Tanaka et al., 1961c).

The oral and parenteral $LD_{50}$ for decoyinine was 2.5 g/kg (De Boer et al., 1967).

## BIOSYNTHESIS OF PSICOFURANINE AND DECOYININE

Since psicofuranine and decoyinine are adeninehexoside nucleosides, several pathways were considered for their biosynthesis (see section on Psicofuranine, Biosynthesis). The structural similarity of psicofuranine and decoyinine and their occurrence in the culture filtrates strongly suggests the interconversion of these two nucleosides by *S. hygroscopicus*. Hoeksema et al. (1964) described the interconversion of tritium-labeled psicofuranine and decoyinine. Chassy et al. (1966) subsequently reported on the conversion of psicofuranine-U-$^{14}$C to decoyinine-U-$^{14}$C. A method was described for the isolation of carbon 6′ of decoyinine as the iodoform derivative. It is not known if the conversion of psicofuranine to decoyinine occurs as the nucleoside or nucleotide.

## BIOCHEMICAL PROPERTIES OF DECOYININE

Yüntsen et al. (1956) reported that angustmycin A (decoyinine) inhibited *Mycobacteria tuberculosis*. This inhibition by decoyinine was reversed in organic media. Tanaka et al. (1959) reported that guanine or guanosine reversed the inhibition of decoyinine against *B. subtilis* and *Mycobacterium* 607. Tanaka et al. (1961a, 1961b, 1961c) reported on the activity of angustmycin A with a number of organisms and compared its activity on antibiotic-resistant, coagulase-positive *Staphylococci*.

The inhibition of *B. subtilis* by angustmycin A was subsequently shown to excrete xanthosine into the culture medium (Tanaka, 1963). Tanaka concluded that angustmycin A inhibits XMP aminase in a manner similar to that reported for angustmycin C (psicofuranine). In view of the close similarity of structures of these two nucleoside antibiotics, it is likely that they act as noncompetitive inhibitors of XMP aminase. Fukuyama (personal communication) stated that XMP aminase from *E. coli* was inhibited by decoyinine. The inhibitory action of decoyinine was enhanced when it was preincubated with XMP aminase along with XMP and pyrophosphate. When the aminase was treated with mercaptoethanol, it was desensitized to the inhibitory action of decoyinine. Fukuyama stated that decoyinine acts much like psicofuranine as an allosteric inhibitor of XMP aminase. These adenosine analogs may be exerting their inhibition by inducing the same kind of conformational change that is caused by the catalytic intermediate.

Additional evidence that decoyinine acts by blocking XMP aminase was provided by Bloch and Nichol (1964b). Decoyinine decreased the growth of *S. faecalis* by 50% at $5 \times 10^{-6}$ *M* in a medium free of purines or pyrimidines. Guanine and guanosine competitively reversed this inhibition. These findings are in agreement with the work cited earlier by Tanaka et al. (1959).

Bloch and Nichol (1964c) studied the inhibition of ribosephosphate pyrophosphokinase by decoyinine. When cell-free extracts of *S. faecalis* were incubated with ribose 5′-phosphate, ATP, and guanine, the formation of GMP, GDP, and GTP was inhibited. The addition of PRPP to the decoyinine-inhibited incubation resulted in the formation of GMP. The authors concluded that decoyinine interfered with ribose phosphate pyrophosphokinase activity. Nucleoside kinase was not detected in these preparations. Decoyinine was not converted to its nucleotides. Decoyinine was not a substrate for adenosine phosphorylase nor did it inhibit this enzyme. These data taken together make it unlikely that the conversion of guanine to its nucleotides occurred by either a nucleoside phosphorylase or nucleoside kinase pathway. Bloch and Nichol also reported that decoyinine inhibited the conversion of XMP to GMP. This reaction and the formation of PRPP both require ATP. The authors suggested that decoyinine acts by occupying the ATP site in reactions involving pyrophosphate cleavage from ATP.

## Summary

Psicofuranine and decoyinine are ketohexose nucleosides elaborated by the *Streptomyces* that are structural analogs of adenosine. They inhibit bacterial cells and Walker adenocarcinoma 256 in rats. The gram-positive organisms are more sensitive to decoyinine than the gram-negative organisms. Both nucleoside antibiotics inhibit *B. subtilus* and *M. tuberculosis*.

A site of action of psicofuranine and decoyinine in the cell appears to involve inhibition of XMP aminase. This inhibition can be reversed by guanine-containing compounds. The binding of psicofuranine to XMP aminase is noncompetitive and reversible. XMP and pyrophosphate stimulate the binding of psicofuranine to the aminase. XMP aminase can be desensitized to psicofuranine. The binding of psicofuranine to the aminase blocks ammonolysis of the adenyl-XMP intermediate, but does not inhibit the binding of ATP or XMP to the aminase. Recent evidence indicates that the adenine glycosides that inhibit XMP aminase do so by interacting at a site near the catalytic site and not at an allosteric site. This binding causes a conformative response in the enzyme. Psicofuranine is not a substrate for adenosine deaminase, but is phosphorylated to a nucleotide. Finally, psicofuranine has been isolated from the RNA of bacterial and mammalian tissue.

Whereas cordycepin biosynthesis has been shown to arise from adenosine or a 5′-phosphorylated derivative (without cleavage of the *N*-riboside bond), psicofuranine and decoyinine biosynthesis takes place by the enzymatic condensation of adenine and/or an aldohexose or a ketohexose. Psicofuranine and decoyinine are interconverted by *S. hygroscopicus*.

## References


Beppu, T., M. Nose, and K. Arima, *Agr. Biol. Chem.* (*Tokyo*), **32**, 197 (1968a).

Beppu, T., M. Nose, and K. Arima, *Agr. Biol. Chem.* (*Tokyo*), **32**, 203 (1968b).

Bloch, A., and C. A. Nichol, *Antimicrobial Agents Chemotherapy*, (1964a,) 530.

Bloch, A., and C. A. Nichol, *Federation Proc.*, **23**, 324 (1964b).

Bloch, A., and C. A. Nichol, *Biochem. Biophys. Res. Commun.*, **16**, 400 (1964c).

Brown, E. G., and B. S. Mangat, *Biochim. Biophys. Acta*, **148**, 350 (1967).

Chassy, B. M., T. Sugimori, and R. J. Suhadolnik, *Biochim. Biophys. Acta*, **130**, 12 (1966).

Cory, J. G., and R. J. Suhadolnik, *Biochemistry*, **4**, 1729 (1965).

Cree, G. M., and A. S. Perlin, *Can. J. Biochem.*, **46**, 765 (1968).

De Boer, C., A. Dietz, L. E. Johnson, T. E. Eble, and H. Hoeksema, U. S. Pat. No. 3,207,750 (1965).

De Boer, C., A. Dietz, L. E. Johnson, T. E. Eble, and H. Hoeksema, French Pat. No. 1,465,395 (1967).

Dekker, C. A., *J. Amer. Chem. Soc.*, **87**, 4027 (1965).

Eble, T. E., H. Hoeksema, G. A. Boyack, and G. M. Savage, *Antibiot. Chemotherapy*, **9**, 419 (1959).

Evans, J. S., and J. E. Gray, *Antibio. Chemotherapy*, **9**, 675 (1959).

Farkaš, J., and F. Šorm, *Tetrahedron Letters*, 1962, 813.

Farkaš, J., and F. Šorm, *Collect. Czech. Chem. Commun.*, **28**, 882 (1963).

Farkaš, J., and F. Šorm, *Collect. Czech. Chem. Commun.*, **32**, 2663 (1967).

Forist, A. A., S. Theal, and H. Hoeksema, *Antibiot. Chemotherapy*, **9**, 685 (1959).

Fox, J. J., K. A. Watanabe, and A. Bloch, in Davidson, J. N. and Cohn W. E.,

(Editors), *Progress in Nucleic Acid Research and Molecular Biology, Vol. V*, Academic Press, New York, 1966, p. 251.
Fukuyama, T. T., *J. Biol. Chem.*, **241**, 4745 (1966).
Fukuyama, T. T., and H. S. Moyed, *Biochemistry*, **3**, 1488 (1964).
Gibbins, L. N., and F. J. Simpson, *Can. J. Microbiol.*, **10**, 829 (1964).
Gonzalez, N. S., and H. G. Pontis, *Biochim. Biophys. Acta*, **69**, 179 (1963).
Guarino, A. J., in Gottleib, D., and Shaw, P. D. (Editors), *Antibiotics, I, Vol. I*, Springer-Verlag, New York, 1967, p. 464 (1967).
Hanka, L. J., *J. Bacteriol.*, **80**, 30 (1960).
Hanka, L. J., in Gottleib, D., and Shaw, P. D. (Editors), *Antibiotics, Vol. I*, Springer-Verlag, New York, 1967, p. 457 (1967).
Hanka, L. J., and M. R. Burch, *Antibiot. Chemotherapy*, **10**, 484 (1960).
Hanka, L. J., M. R. Burch, and W. T. Sokolski, *Antibiot. Chemotherapy*, **9**, 432 (1959).
Henderson, J. F., *Biochem. Pharmacol.*, **12**, 551 (1963).
Henderson, J. F., and M. K. Y. Khoo, *J. Biol. Chem.*, **240**, 3104 (1965).
Hoeksema, H., G. Slomp, and E. E. van Tamelen, *Tetrahedron Letters*, 1964, 1787.
Hough, L., and B. E. Stacey, *Phytochemistry*, **5**, 215 (1966).
Kemp, M. B., and J. R. Quayle, *Biochem. J.*, **102**, 94 (1967).
Komuro, T., T. Nara, M. Misawa, S. Kinoshita, *Agr. Biol. Chem. (Tokyo)*, **33**, 230 (1969).
Kuramitsu, H., and H. S. Moyed, *Biochim. Biophys. Acta*, **85**, 504 (1964).
Kuramitsu, H., and H. S. Moyed, *J. Biol. Chem.*, **241**, 1596 (1966).
Le Page, G. A., and Junga, I. G., *Cancer Res.*, **25**, 46 (1965).
Lewis, C., Reames, H. R., and L. E. Rhuland, *Antibiot. Chemother.*, **9**, 421 (1959).
Lindberg, B., Klenow, H., and Hansen, K., *J. Biol. Chem.*, **242**, 350 (1967).
Magee, W. E., and F. S. Eberts, Jr., *Cancer Res.*, **21**, 611 (1961).
McCarthy, Jr., J. R., R. K. Robins, and M. J. Robins, *J. Amer. Chem. Soc.*, **90**, 4993 (1968).
McCarthy, Jr., J. R., M. J. Robins, L. B. Townsend, and R. K. Robins, *J. Amer. Chem. Soc.*, **88**, 1549 (1966).
Moyed, H. S., *Cold Spring Harbor Symp. Quant. Biol.*, **26**, 323 (1961).
Passeron, S., and E. Recondo, *J. Chem. Soc.*, 813 (1965).
Pontis, H. G., and C. L. Fischer, *Biochem. J.*, **89**, 452 (1963).
Pontis, H. G., A. L. James, and J. Baddiley, *Biochem. J.*, **75**, 428 (1960).
Sakai, H., Yüntsen, H., and F. Ishikawa, *J. Antibiot. (Tokyo)*, **7A**, 116 (1954).
Schroeder, W., U. S. Pat. No. 2,993,039 (1961).
Schroeder, W., and H. Hoeksema, *J. Amer. Chem. Soc.*, **81**, 1767 (1959).
Shigeura, H. T., and S. D. Sampson, *Nature*, **215**, 419 (1967).
Simpson, F. J., and L. N. Gibbins, in S. P. Colowick, and N. O. Kaplan (Eds.), *Methods in Enzymology, Vol. IX*, Academic Press, New York, 1966, p. 412.
Slechta, L., *Biochem. Pharmacol.*, **5**, 96 (1960a).
Slechta, L., *Biochem. Biophys. Res. Commun.*, **3**, 596 (1960b).
Strecker, G., B Goubet, and J. Montreuil, *Comptes Rendus*, **260**, 999 (1965).
Sugimori, T., and R. J. Suhadolnik, *J. Amer. Chem. Soc.*, **87**, 1136 (1965).

Suhadolnik, R. J., and T. Sugimori, *Federation Proc.*, **25**, 525 (1966).
Suhadolnik, R. J., in D. Gottlieb, and P. D. Shaw (Eds.),*Antibiotics*, Vol. II, Springer-Verlag, 1967, p. 400.
Suhadolnik, R. J., G. Weinbaum, and H. P. Meloche, *J. Amer. Chem. Soc.*, **86**, 948 (1964).
Tanaka, N., *J. Antibiotics (Tokyo)*, **16A**, 163 (1963).
Tanaka, N., N. Miyairi, and H. Umezawa, *J. Antibiotics (Tokyo)*, **13A**, 265 (1959).
Tanaka, N., N. Miyairi, T. Nishimura, and H. Umezawa, *J. Antibiotics (Tokyo)*, **14A**, 18 (1961a).
Tanaka, N., N. Miyairi, and H. Umezawa, *J. Antibiotics (Tokyo)*, **14A**, 23 (1961b).
Tanaka, N., T. Nishimura, H. Yamaguchi, and H. Umezawa, *J. Antibiotics (Tokyo)*, **14A**, 98 (1961c).
Udaka, S., and H. S. Moyed, *J. Biol. Chem.*, **238**, 2797 (1963).
Umemura, Y., M. Nakamura, and S. Funahashi, *Arch. Biochem. Biophys.*, **119**, 240 (1967).
Vavra, J. J., A. Dietz, B. W. Churchill, P. Siminoff, and H. J. Koepsell, *Antibiot. Chemotherapy*, **9**, 427 (1959).
Verheyden, J., and J. G. Moffatt, *J. Amer. Chem. Soc.*, **88**, 5684 (1966).
Wallach, D. P., and R. C. Thomas, *Antibiot. Chemotherapy*, **9**, 722 (1959).
Wolfrom, M. L., A. Thompson, and E. F. Evans, *J. Amer. Chem. Soc.*, **67**, 1793 (1945).
Yüntsen, H., *J. Antibiotics (Tokyo)*, **11A**, 233 (1958a).
Yüntsen, H., *J. Antibiotics (Tokyo)*, **11A**, 244 (1958b).
Yüntsen, H., H. Yonehara, and H. Ui, *J. Antibiotics (Tokyo)*, **7A**, 113 (1954).
Yüntsen, H., K. Ohkuma, Y. Ishii, and H. Yonehara, *J. Antibiotics (Tokyo)*, **9A**, 195 (1956).
Zyk, N., N. Citri, and H. S. Moyed, *Biochemistry*, **8**, 2787 (1969).
Zyk, N., N. Citri and H. S. Moyed, *Biochemistry*, **9**, 677 (1970).

CHAPTER 3

# Spongosine and Arabinosyl Nucleosides

---

## ABBREVIATIONS

9-β-D-Ribofuranosyl-2-methoxyadenine, spongosine; 9-β-D-arabinofuranosyladenine, ara-A; 1-β-D-arabinofuranosylthymine, ara-T or spongothymidine; 1-β-D-arabinofuranosylcytosine, ara-C; 1-β-D-arabinofuranosyluracil, ara-U or spongouridine; 9-β-D-arabinofuranosylhypoxanthine, ara-Hx; 9-β-D-arabinofuranosyladenine-5′-mono-, di-, and triphosphates, ara-AMP, ara-ADP and ara-ATP; 1-β-D-arabinofuranosylcytosine-5′-mono-, di-, and triphosphates, ara-CMP, ara-CDP, ara-CTP; 1′-β-D-arabinofuranosylcytosine-3′-monophosphate, 3′-ara-CMP; deoxythymidine 5′-triphosphate, TTP; deoxycytidine 5′-diphosphate, dCDP; deoxycytidine 5′-triphosphate, dCTP; deoxyadenosine 5′-diphosphate, dADP; polyadenylic acid, poly-A; polycytidylic acid, poly C.

Five nucleosides will be reviewed in this section: ara-A, ara-C, ara-T, ara-U, and spongosine. Of these, ara-A and ara-T are the only naturally occurring nucleoside antibiotics. Although ara-C is not a naturally occurring nucleoside antibiotic, it is reviewed here since many biochemical studies have involved both ara-A and ara-C. Ara-U and spongosine are not nucleoside antibiotics, but are included since they are naturally occurring nucleosides. Three purine and pyrimidine nucleosides, spongosine, ara-T, and ara-U, were isolated from the Caribbean sponge, *Cryptotethia crypta*. The purine nucleoside spongosine was assigned the structure 9-β-D-ribofuranosyl-2-methoxyadenine. Spongothymidine, (1-β-D-arabinofuranosylthymine) (ara-T) and spongouridine (1-β-D-arabinofuranosyluracil) (ara-U) are pyrimidine nucleosides. Most recently, the research group at the Parke, Davis Laboratories reported activity against herpes simplex virus in an antibiotic concentrate obtained from a microbial fermentation. The active component has been identified as 9-β-D-arabinofuranosyladenine (ara-A) (Fig. 3.1) (Parke, Davis and Company, Belgium Pat. No. 671,557). This nucleoside is elaborated by *Streptomyces antibioticus*.

Ara-A and ara-C are cytocidal agents and have been extensively studied as inhibitors of mammalian cells in culture, tumor cells, bacterial cells, and viruses. Ara-C and ara-A are DNA inhibitors. Both ara-ATP and ara-CTP are incorporated as terminal residue into DNA *in vitro*. Ara-A is incorporated into the 3′-terminal end of tRNA by the partially purified RNA polymerase from *E. coli*. The glycosyl linkage of ara-U is cleaved by uridine phosphorylase of *E. coli*.

Several recent reviews have been written on these nucleosides. The reader is referred to the excellent reviews and publications by Cohen (1966), Smith (1968), and Schabel (1967,1968) on the arabinosyl nucleosides.

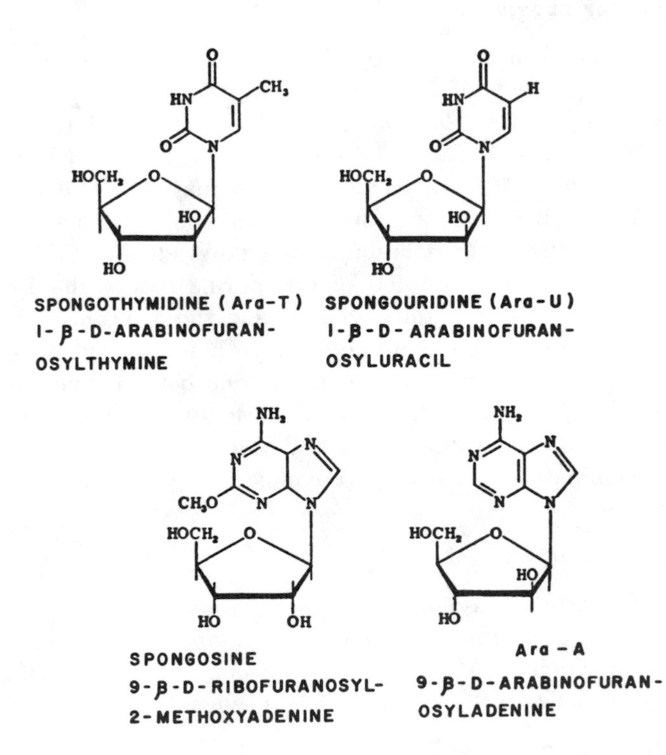


FIGURE 3.1. *Structures for spongosine, ara-A, ara-T, and ara-U.*

## 3.1. 9-β-D-RIBOFURANOSYL-2-METHOXYADENINE (SPONGOSINE)

### DISCOVERY AND ISOLATION

The first isolation of spongosine (9-β-D-ribofuranosyl-2-methoxyadenine) was reported by Bergmann and Feeney (1950; 1951) from the sponges of the species *C. crypta*. The tissue was vacuum dried, ground, and extracted with acetone. Spongothymidine and spongosine were extracted into the acetone, from which spongothymidine crystallized during the extraction process. A second crop was isolated on concentration of the original acetone extract. Further evaporation of the acetone extract resulted in the crystallization of spongosine. The spongosine was recrystallized from water and ethanol. Bergmann and Burke (1955) subsequently isolated spongosine, spongothymidine, and spongouridine by means of ion-exchange chromatography.

### CHEMICAL PROPERTIES

The molecular formula for spongosine is $C_{11}H_{15}N_5O_5$; mp 192–193°C (decomp.); $[\alpha]_D^{25} = -42.5°$ ($c$ = 24.6 mg in 3.09 ml of 8% NaOH); $\lambda_{max}$ 274 mμ ($\epsilon = 11.9 \times 10^3$), 249 mμ ($\epsilon = 8.35 \times 10^3$) at pH 1; $\lambda_{max}$ 267 mμ ($\epsilon = 12.5 \times 10^3$) at pH 7; $\lambda_{max}$ 268 mμ ($\epsilon = 12.7 \times 10^3$) at pH 13 (Bergmann and Stempien, 1957; Schaeffer and Thomas, 1958). Bergmann and Feeney (1951) and Bergmann and Burke (1956) reported that acid hydrolysis of spongosine resulted in the isolation of 2-methoxyadenine, $C_6H_7N_5O$. They also reported on the deamination of this purine base with nitrous acid. Chemically synthesized 2-methoxyadenine and the aglycone of spongosine were identical (Bergmann and Burke, 1956). These data led the authors to conclude that 2-methoxyadenine was the purine base of spongosine. Bergmann and Burke (1956) and Bergmann and Stempien (1957) compared the spectra of crotonoside and spongosine. Both nucleosides were deaminated by adenosine deaminase. The carbohydrate moiety of spongosine was shown to be D-ribose.

### CHEMICAL SYNTHESIS

Final proof for the absolute structure of spongosine was obtained by chemical synthesis of this nucleoside (Bergmann and Stempien, 1957; Schaeffer and Thomas, 1958). The chemical synthesis of spongosine as reported by Schaeffer and Thomas (1958) is shown in Figure 3.2.

### BIOCHEMICAL PROPERTIES

Spongosine is not an inhibitor of mammalian cells or bacterial cells as is the corresponding analog spongoadenosine (ara-A). Rockwell and Maguire

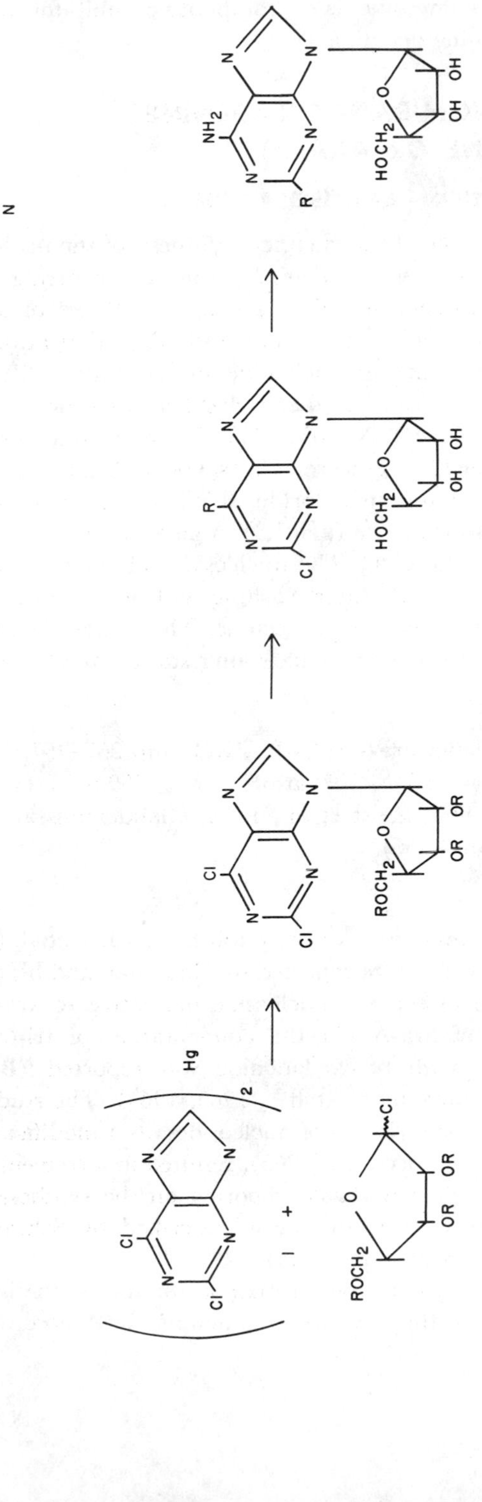

2, R = $C_6H_5CO-$

3, R = $C_6H_5CO-$

4, R = $NH_2-$

5, R = $CH_3O-$

FIGURE 3.2. *Chemical synthesis of 6-amino-2-methoxy-9-β-D-ribofuranosylpurine (spongosine). (From Schaeffer and Thomas, 1958.)*

(1966) reported that spongosine is a competitive inhibitor of partially purified ox heart adenosine deaminase.

## 3.2. 9-β-D-ARABINOFURANOSYLADENINE (SPONGOADENOSINE OR ARA-A)

### DISCOVERY, PRODUCTION, AND ISOLATION

Ara-A is another example of the chemical synthesis of the nucleoside preceding the discovery and isolation as a naturally occurring antibiotic. Goodman and his co-workers described the first syntheses of ara-A (Lee et al., 1960; Reist et al., 1962). It was not until 1967 that ara-A was discovered as a naturally occurring nucleoside in the culture filtrates of *S. antibioticus* by the research group at the Parke, Davis Laboratories (Parke, Davis and Company, 1967, Pat. No. 671,557). The nucleoside is synthesized in a medium containing 2% glucose, 1% soybean meal, 0.5% peptone (Wilsons 159), 0.2% ammonium chloride, 0.5% sodium chloride, 0.25% calcium carbonate, and tap water (pH 7.5) in flasks which are maintained in a shaker incubator at 25–27°C. The nucleoside is isolated by adsorption from the filtrate onto charcoal. After washing with water, 9-(β-D-arabinofuranosyl)adenine is eluted with 50% acetone. The acetone is removed by evaporation and the nucleoside crystallizes after storing at 5°C for 48 hr.

### CHEMICAL PROPERTIES

The molecular formula for ara-A is $C_{10}H_{13}N_5O_4$; mp 257–257.5°C; $[\alpha]_D^{27} = -5°$; $\lambda_{max}$ 257.5 mμ ($\epsilon = 12.7 \times 10^3$) at pH 1; $\lambda_{max}$ 259 mμ ($\epsilon = 14 \times 10^3$) at pH 13 (Lee et al., 1960; Reist et al., 1962; Glaudemans and Fletcher, 1963, 1964; Lepine et al., 1966).

### CHEMICAL SYNTHESIS

Early reports of syntheses of ara-A were published by Lee et al. (1960) and Reist et al. (1962). The procedure reported by Goodman and his co-workers involved the conversion of the *xylo*-nucleoside derivative to ara-A. Subsequently, the synthesis of ara-A via the condensation of tri-*O*-benzyl-D-arabinofuranosyl halide with benzoyladenine was reported (Barker and Fletcher, 1961; Glaudemans and Fletcher, 1963, 1964). The study of Reist et al. (1964) on the *O*-debenzylation of nucleosides, as a modification of the Glaudemans and Fletcher procedure (1963), resulted in a convenient, large-scale synthesis of ara-A. Cohen (1966) reported on the synthesis of ara-A using the polyphosphate ester technique as described by Schramm et al. (1962) and Pollman and Schramm (1964).

Lepine et al. (1966) also described a method for the synthesis of ara-A. This procedure involved the heating of adenine, arabinose, and ethyl polyphosphate in HCl with *N*,*N*-dimethylformamide. Reist et al. (1968) reported on the synthesis of 8-bromo-9-(2,3,5-tri-*O*-acetyl-β-D-arabinofuranosyl) adenine. A number of 8-substituted purine nucleosides were synthesized from this 8-bromopurine nucleoside.

## BIOSYNTHESIS OF SPONGONUCLEOSIDES BY *CRYPTOTETHIA CRYPTA*

Cohen (1966) reported that freshly isolated cells of *C. crypta* containing sea water supplemented with radioactive glucose, uracil, or D-arabinose did not incorporate these components to thymine or the arabinonucleosides. Incubation of the cells in the original supernatant with the radioactive compounds did not result in the biosynthesis of the spongonucleosides. Cohen concluded that these cells appeared to lack biosynthetic mechanisms. In addition, they did not consume oxygen or produce radioactive carbon dioxide in the presence of carbon-14 labeled glucose or D-arabinose. A small amount of RNA could be found in these cells, but essentially no DNA was present. However, the detection of nucleic acids in the cells and the sponge residues is difficult due to the false carbohydrate reactions from unknown components. Extracts of the washed cells did contain a small amount of pyrimidine nucleoside phosphorylase. Cohen speculated that if this enzyme was indeed from the sponge cells, ara-U and ara-T could be derived by the enzymatic condensation of D-arabinose-1-phosphate and the corresponding pyrimidines.

No free pentoses were detected on paper chromatograms and only one single ultraviolet-absorbing component (containing phosphorus) was detected in the eluent from a Dowex-1-formate column. This compound had an absorption maximum at 305 mμ. There were no ultraviolet-absorbing compounds containing phosphorus in the supernatant fluid. Cytosine nucleosides were not detected. Of 16 different species of sponge examined, only *C. crypta* was found to contain arabinonucleosides (Bergmann et al., 1957). Hydrolysis of the nucleic acids from whole dried sponges resulted in the isolation of only normal ribo- and deoxyribo-components. Hubert-Habart and Cohen (1962) reported that 7 g of uracil, 6 g of thymine, and 2 g each of ara-U and ara-T were isolated from the alcoholic and acetone extracts of fresh samples of *C. crypta*. Yields were recorded per kilogram of dried sponge. The sponge nucleosides and pyrimidine bases were isolated following removal of the protective ectosomal layer. The cells were squeezed in a potato ricer between layers of cheese cloth. Brown, motile cells (3–4 μ in diameter) were found in the turbid fluid. Very few bacteria were found. Cold 3% perchloric acid extracts were prepared from these cells following sedimentation. The supernatant fluid following centrifugation possessed an osmolarity comparable to that of sea water. Essentially all the arabinonucleosides in the sponge were found in these sponge cells.

## INHIBITION OF MAMMALIAN CELLS IN CULTURE, BACTERIA, AND VIRUSES BY ARA-A

Ara-A has been reported to inhibit mammalian cell cultures, purine-requiring *E. coli* mutants, virus-infected cells, and tumors. These studies have been very thoroughly summarized and clearly presented by Cohen (1966) and Schabel (1968).

Schabel (1968) reported that ara-A has broad-spectrum activity against DNA viruses in cell culture (Table 3.1), but only limited activity against RNA viruses in cell culture (Table 3.2). In addition, ara-A has significant therapeutic activity against herpes keratitis in hamsters. It has similar activity in mice inoculated intracerebrally with herpes simplex virus (HSV) and vaccinia virus. Peroral administration of ara-A to mice previously inoculated intracerebrally with herpes simplex virus resulted in survivor rates as high as 60%. Ara-A is also effective against herpes encephalitis when the nucleoside is administered subcutaneously or percutaneously.

**Table 3.1**

Activity of Ara-A against DNA Viruses in Cell Culture

| Virus | Cell | Activity[a] |
|---|---|---|
| Herpes simplex | HEp 2 | + |
| Herpes marmoset | HEp 2 | + |
| Herpes simiae B | Primary MK | + |
| Varicella | HEL WI-38 | + |
| Cytomegalo | HEL WI-38 | + |
| Vaccinia | HEp 2 | + |
| Polyoma | $C_3H$ | − |
| Adeno 3 | HEp 2 | ± |

From Schabel, 1968.

[a] + = Antiviral activity in cell culture with one or more *in vitro* test procedures (plaque reduction, inhibition of viral cytopathogenicity, reduction of titratable virus, or hemagglutination). − = No antiviral activity in cell culture seen with one or more *in vitro* test procedures.

The collaborative investigations of the Parke, Davis Laboratories and the Southern Research Institute have led to unequivocal evidence that ara-A has marked antiviral activity against DNA viruses in cell culture and in animals. This nucleoside appears to be a potentially useful antiviral agent (Miller et al., 1968; Sidwell et al., 1968a; Schardein and Sidwell, 1968; Sloan et al., 1968; Dixon et al., 1968; Kurtz et al., 1968). The studies described by Miller et al. (1968) showed that ara-A has marked *in vitro* activity against a broad spectrum of DNA viruses implicated in human and veterinary

**Table 3.2**

Activity of Ara-A against RNA Viruses in Cell Culture

| Virus | Cell | Activity[a] |
|---|---|---|
| Parainfluenza 3 (HA-1) | HEp 2 | − |
| Respiratory syncytial | HEp 2 | − |
| Rhino 1B | Primary MK | − |
| ECHO 9 | Primary MK | − |
| Measles | HEp 2 | − |
| Newcastle disease | HEp 2 | − |
| Rous sarcoma | Chick embryo | + |
| Poliomyelitis II | Primary MK | − |
| Coxsackie B 1 | Primary MK | − |

From Schabel, 1968.

[a] + = Antiviral activity in cell culture with one or more *in vitro* test procedures (plaque reduction, inhibition of viral cytopathogenicity, and/or reduction of titratable virus). − = No antiviral activity in cell culture seen with one or more *in vitro* test procedures.

diseases. Rous sarcoma virus, an RNA virus that requires cell DNA synthesis, was also inhibited by ara-A. The above investigations suggested six characteristics for a compound to be acceptable as an antiviral agent: "(i) it should have activity against a broad spectrum of viruses; (ii) these viruses should be of current public health importance; (iii) the antiviral activity demonstrated should be dose-responsive and significant at nontoxic levels; (iv) the activity should be readily reproducible, preferably in more than one test system and in more than one laboratory; (v) the antiviral activity should be apparent in both *in vivo* and *in vitro* systems; and (vi) the compound should be readily available" (Miller et al., 1968). Ara-A met a number of those requirements and in addition, herpes simplex did not seem to acquire resistance to ara-A following three passages in H. Ep. #2 cells. The findings by Miller et al. (1968) on the inhibitory activity of ara-A against HSV *in vitro*, prompted Sidwell et al. (1968a) to study this drug as an antiviral agent against herpes simplex keratitis infections in hamsters when the agent is applied at a concentration of 5% in ophthalmic ointment. Under these test conditions, ara-A has a therapeutic index of greater than 60 (a therapeutic index of 4 is considered significant for such *in vivo* antiviral chemotherapy experiments). 5-Iodo-2′-deoxyuridine and ara-C have a therapeutic index of 4 for the same hamster herpes keratitis test (Sidwell et al., 1966). Ara-A appears to inhibit multiplication of the herpes virus and thereby prevents major damage to the eye. In addition, no virus could be isolated from the brains of ara-A-treated HSV-infected hamsters. Untreated controls had titers as high as $10^4$ mouse infectious units per milligram of brain. Ara-A was not toxic to the eyes. This is in contrast to ara-C which is also an antiviral agent, but is limited by the toxic effects exhibited when applied topically to the eye (Kaufman, et al., 1964).

Similarly, Schardein and Sidwell (1968) showed that ara-A results in a marked reduction of encephalitis in hamsters and a lessened incidence and severity of hamsters infected with HSV. Ara-A is the most active drug against herpes simplex keratitis and appears worthy of additional consideration for clinical studies.

To determine if ara-A could block intracerebral virus multiplication and prevent subsequent death in HSV-infected mice, Sloan et al. (1968) studied the suppression of virus multiplication in the mouse brain following intracerebral inoculation with HSV. Ara-A was effective against HSV infections in mice if the drug was administered intraperitoneally, subcutaneously, perorally, or percutaneously. The therapeutic index of ara-A is about 4.

Ara-A is also active against intracerebral vaccinia virus infections in Swiss mice when administered intraperitoneally, perorally, or percutaneously (Dixon et al., 1968) and it is more active against vaccinia virus than *N*-methyl β-thiosemicarbazone. The evidence indicates that ara-A crosses the blood–brain barrier, since peroral or intraperitoneal administration markedly inhibits vaccinia virus in mice infected intracerebrally. In addition, a high rate of survival was observed in mice inoculated intracerebrally with lethal doses of vaccinia virus and treated percutaneously with ara-A dissolved in DMSO. Apparently ara-A can inactivate the virus that is already replicating in the central nervous system (Dixon et al., 1968). A similar observation was made by Sloan et al. (1968). These results and the report by Kurtz et al. (1968) that ara-A has little or no toxicity when administered percutaneously or parenterally in animals strongly suggest that ara-A may be extremely effective against diseases caused by DNA viruses. The $LD_{50}$ for an intraperitoneal injection of ara-A is 4677 mg/kg (Kurtz et al., 1968; Schabel, 1968).

Since ara-A is rapidly deaminated to ara-Hx in the mouse (Brink and LePage, 1964b), Schabel (1968) compared the activity of ara-A and ara-Hx. The *in vivo* activity of ara-Hx appears to be quantitatively similar to that of ara-A. An added significant strength of the therapeutic effectiveness of ara-A is its lack of toxicity in mice.

## 3.3. 1-β-D-ARABINOFURANOSYLURACIL (SPONGOURIDINE OR ARA-U)

### DISCOVERY AND ISOLATION

Spongouridine was isolated by Burke (1955) and Bergmann and Burke (1955) from *C. crypta*. Spongouridine, the third nucleoside in the crude extracts of *C. crypta*, was detected by paper chromatography. The three nucleosides were subsequently separated on a large scale by means of ion-exchange chromatography (Bergmann and Burke, 1955). Small amounts of spongothymidine were removed from the spongouridine by rechromatography on a larger Dowex-1-$OH^-$ column.

## CHEMICAL PROPERTIES

The molecular formula for spongouridine is $C_9H_{12}N_2O_6$, mp 226–228°C; $[\alpha]_D = +97°$ ($c = 0.6$ in 8% NaOH); $[\alpha]_D = +126°$ ($c = 1\%$ in water); $pK_a$ 9.3; $\lambda_{max}$ 262.5 ($\epsilon = 10{,}500$), pH 7.0 (Bergmann and Burke, 1955; Brown et al., 1956). Cushley, Watanabe, and Fox (1967) studied the nmr spectra of 36 acetylated derivatives of 3′-aminohexosyl and pentosyl nucleosides. The data showed that chemical shifts of acetyl signals were not reliable for determining the configuration of the sugar moiety. With acetylated pentofuranosyl nucleosides, a *cis*-C-1′–C-2′ relationship caused a paramagnetic (downfield) shift in the C-2′-acetoxy resonance signal upon the reduction of the double bond in the pyrimidine ring. Similar treatment of the transnucleosides led to a small diamagnetic shift. Included in these studies was the acetylated pentofuranosyl nucleoside ara-U. These authors showed that the pentofuranosyl nucleosides have a conformational arrangement similar to that of the pyranosyl nucleosides. Apparently the pyrimidine nucleosides exist predominantly in a preferred conformation in which the 5,6 double bond of the aglycon is "*endo*" to the 5-membered sugar ring. Figure 3.3 shows the predominant conformation for the acetylated derivative of 1-(β-D-*aldo*-arabinofuranosyl)uracil. It can be seen from the figure that the *cis*-acetoxy group (arabino R = OAc, R′ = H) is located in the cone of positive shielding of the 5,6 double bond. The significance of the "*endo*" conformation of the arabinopyrimidine nucleoside became apparent following removal of the anisotropy of the 5,6 double bond by hydrogenation of the uracil ring. The C-2′-acetoxy resonance was shifted downfield by 0.05 ppm. With the *ribo*- and *xylo*- derivatives, where a *trans* relationship exists between the C-1′–C-2′ substituents, either an upfield shift or no appreciable effect is observed in the C-2′-acetoxy resonance. These nmr data were interpreted in terms of a preferred *anti* conformation for the pyrimidine nucleosides.

Emerson et al. (1967) studied the factors that affect the sign and magnitude of the Cotton effect to determine the conformation of pyrimidine nucleosides in solution. They reported the ORD of 50 pyrimidine nucleosides and related compounds. The data obtained showed that the sign of the Cotton effect is as expected on the basis of the anomeric configuration at C-1′. To rationalize the ORD results, the authors assumed that the variation in both the sign and magnitude of the Cotton effect was due to the changing

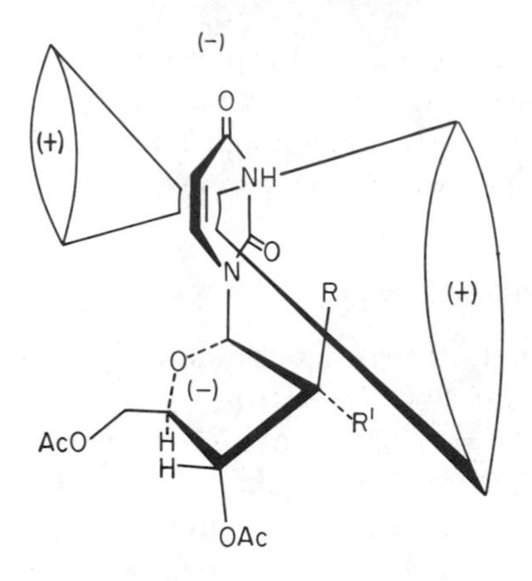


*arabino*, R=OAc, R'=H

*ribo*, R=H, R'=OAc

FIGURE 3.3. *Diagram showing the preferred* "endo" *conformation for the triacetyl derivatives of arabinofuranosyl- and ribofuranosyluracil. Note that the* cis-*acetoxy group of the* arabino *nucleoside is found in the cone of positive shielding of the 5,6 double bond or uracil. (From Cushley, Watanabe, and Fox, 1967.)*

conformation of the planar pyrimidine ring with reference to the sugar ring, i.e., the position of the chromophore relative to the asymmetric centers in the sugar. The ORD curves for ara-U and ara-C were compared with those for uridine and cytidine. The relative stereochemistry of the pyrimidine nucleosides studied are shown more clearly in Figure 3.4. If $O^6$-5′-cyclo-6-hydroxyuridine (compound **9**) is taken as the reference plane, then the angle that the pyrimidine ring makes with this reference plane in compounds **6–9** varies from 0 to 180°. As the angle of the pyrimidine ring varies, the positive Cotton effect is reduced in amplitude and finally becomes negative. The authors concluded that normal pyrimidine β-nucleosides have the *anti* conformation in aqueous solution. Miles et al. (1969a, 1969b) have also reported studies on the conformation of nucleosides in solution as predicted from their circular dichroism (CD) curves. Structural changes in the uracil, thymine, or cytosine bases have little or no effect on the absorption spectra; however, changes in the sign and intensities of the Cotton effects were observed. The isopropylidene derivatives of the nucleosides have little effect on the magnitude of the CD bands (Emerson et al., 1967; Miles et al., 1969a). A maximum of four CD bands are present in the CD spectra of uridine and uridine and cytidine derivatives. Only two bands are observed in the absorption spectrum. Since there is a red shift of all CD bands upon

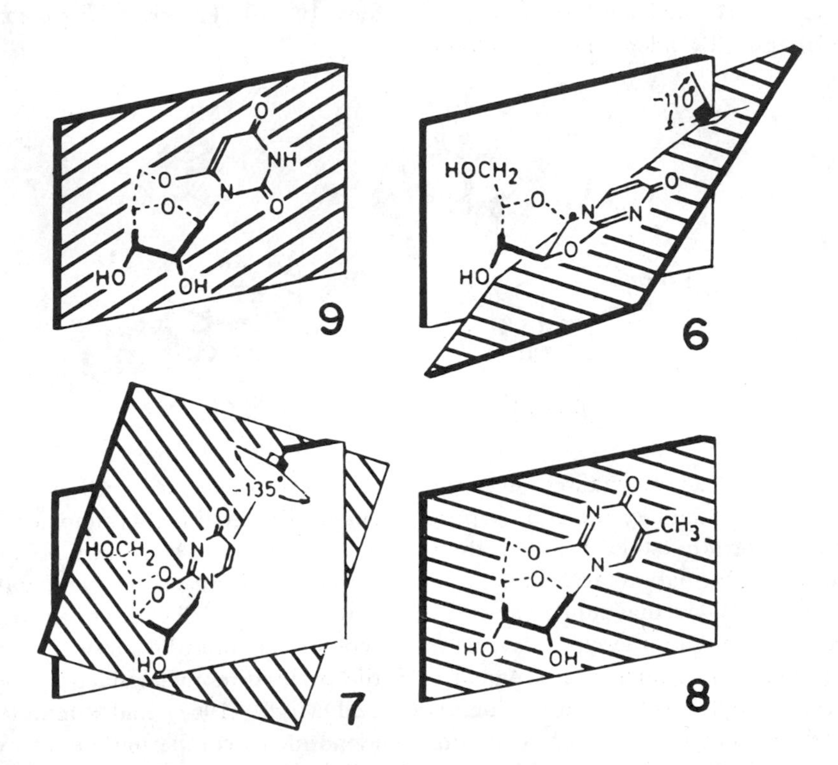


FIGURE 3.4. *Stereochemical drawings of the cyclonucleosides, 6–9. The plane of the chromophore is shaded and the reference plane is unshaded. Thick lines are in front of the planes, broken lines behind them, and plain lines are in the reference plane.* (*From Emerson et al. 1967.*)

conjugative substitution on the uracil ring, Miles et al. (1969a) stated that this was further evidence that all CD bands in uridine arise from $\pi$–$\pi^*$ transitions. Ulbricht's rule (Emerson et al., 1967) does not appear to be in agreement with the data obtained with the cyclouridine analogs studied by Miles et al. (1969b). In addition, the 6-methyluracil and 6-methylcytosine nucleosides do not follow Ulbricht's rule. Miles et al. (1969b) concluded that the application of ORD and CD to configurational analysis of nucleosides and their derivatives should be made with caution. The exceptions to Ulbricht's rule *as stated* are sufficient to preclude its usefulness (Miles et al., 1969b). Hart and Davis (1969a) recently reported that the preferred conformations for adenosine and guanosine were *syn* and *anti*, respectively. However, with the pyrimidine nucleosides, it appears that the conformation of the glycosidic bond must include the *syn*, intermediate-range conformations and the accepted *anti* form (Hart and Davis, 1969b). The *anti* and *syn* conformations for adenosine are as follows:

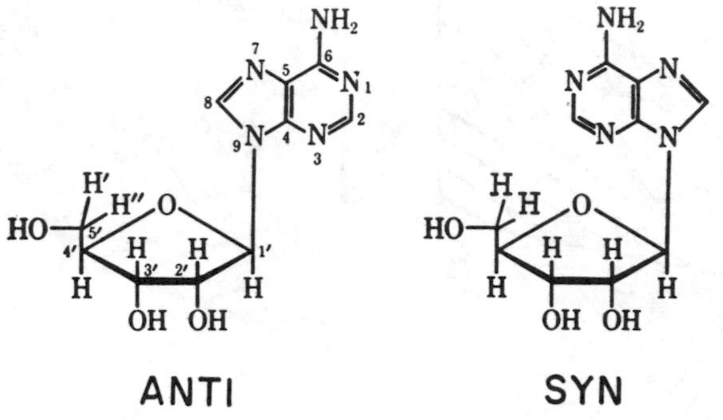

ANTI SYN

The nuclear Overhauser effect experiments were performed with uridine, cytidine, thymidine, and 2′,3′-isopropylidene uridine in dimethyl sulfoxide-$d_6$ and deuterium oxide. By using these solvents, Hart and Davis were able to study conformation as a function of nonbonded interactions. The data strongly suggest that subtle changes in the sugar moiety affect the conformer equilibrium and, therefore, several major conformers must be considered for the pyrimidine nucleosides. Apparently the pyrimidine nucleosides are not limited to the *anti* conformation. Klee and Mudd (1967) had attempted earlier to determine the conformation of adenosine by comparing the rates of acid hydrolysis of 5′-methylthioadenosine and adenosine. The data did not permit an unambiguous interpretation of *syn* versus *anti* preference of β-purine ribosides in solution. Ward and Reich (1968) and Koyama et al. (1966) reported that the pyrazolopyrimidine nucleoside antibiotic formycin existed in the *syn* conformation.

Ts'o et al. (1969) most recently reported that the shielding effect of the ring current on the neighboring bases and the specific deshielding effect of the 5′-phosphate on the H-6 proton of the pyrimidine nucleotide determine the nmr spectra of nucleosides and nucleotides. From these studies they constructed a general conformational model for all nucleosidyl units of purine and pyrimidine dimers. All nucleosidyl units have the *anti* conformation with respect to the sugar base torsion angle. The turn of the (3′–5′) screw axis of the stack is right handed.

## CHEMICAL SYNTHESIS OF ARA-U

The chemical synthesis of 1-β-D-arabinofuranosyluracil was reported by Brown et al. (1956). Doerr et al. (1965) reported on the synthesis of the 1-β-D-arabinofuranosyluracil from derivatives of lyxofuranosyluracil. Since the $O^2$,2′-anhydroarabinosyluracil is a necessary intermediate in the chemical synthesis of arabinosyluracil, Fox and Wempen (1965) described a new method for the synthesis of this important intermediate. Fox and Wempen (1965) and Doerr et al. (1967) used the thionocarbonate procedure to convert the 1-(5-*O*-trityl-β-D-ribofuranosyl)pyrimidine to the 1-β-D-arabinofuranosyl pyrimidines. The synthesis of 1-β-D-arabinofuranosyluracil is shown in Figure 3.5 (Fox and Wempen, 1965). Uridine (**10**; R = H) was converted directly into 2,2′-anhydro-1-(β-D-arabinofuranosyl)uracil (**12a**) in a 40% yield. From 5′-*O*-trityluridine (**10b**) an 85% yield of the anhydronucleoside (**12b**) was obtained which, after detritylation and anhydro ring opening, afforded ara-U (**13**, R′ = H).

Tono and Cohen (1962) reported on the enzymatic synthesis of ara-U and ara-T via the reaction of D-arabinose-1-phosphate and uracil or thymine. Naito et al. (1968a; 1968b) described the chemical synthesis of 3′-deoxyspongouridine by the catalytic reduction of 3′-deoxy-3′-bromospongouridine or 3′-deoxy-3′-ethylthiospongouridine. This nucleoside shows interesting antiviral and antitumor properties.

Otter, Falco, and Fox (1968) have described the transformations of the pyrimidine nucleosides in alkaline media. Treatment of 1-(β-D-arabino-

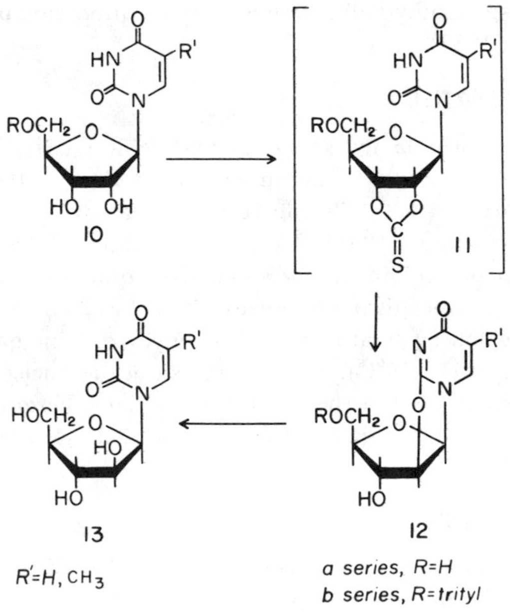

FIGURE 3.5. *Syntheses of 1-β-D-arabinofuranosyluracil and thymine. (From Fox and Wempen, 1965; Doerr, Codington, and Fox, 1967.)*

furanosyl)-5-bromouracil with warm aqueous sodium hydroxide resulted in the isolation of the imidazole nucleoside 1-(β-D-arabinofuranosyl)-2-oxo-4-imidazoline-4-carboxylic acid. The formation of polyarabino-U from polyribo-U without breaking the 3′–5′ internucleotide bond by the conversion of polyribo-U to polyara-U has been reported (Nagyvary, 1967a; Nagyvary and Provenzale, 1969; Provenzale and Nagyvary, 1970).

## 3.4. 1-β-D-ARABINOFURANOSYLTHYMINE (SPONGOTHYMIDINE OR ARA-T)

### DISCOVERY AND ISOLATION

The procedure for the isolation of spongothymidine (ara-T) has been described under the isolation of spongosine (Bergmann and Feeney, 1950). A second method for the isolation and purification of spongothymidine has been described by Bergmann and Burke (1955). The three spongonucleosides were separated by ion exchange with Dowex-1-$OH^-$. Spongosine was eluted with ammonium hydroxide–ammonium formate (pH 9.5). Further elution with the same buffer (pH 8.3) resulted in the separation and isolation of spongothymidine, thymine, uracil, and spongouridine. The proof of structure for spongothymidine as 1-β-D-arabinofuranosylthymine was determined by degradations of spongothymidine followed by identification of the hydrolysis products (Burke, 1955).

### CHEMICAL PROPERTIES

The molecular formula for spongothymidine is $C_{10}H_{14}N_2O_6$; mp 246–247°C; $[\alpha]_D^{25} = +80°$ ($c$ = 33.5 mg in 3.09 ml of 8% NaOH); $\lambda_{max}$ 268 mμ ($\epsilon = 9590$) at pH 7.0; $\lambda_{max}$ 270 mμ ($\epsilon = 7870$ at pH 13. (Bergmann and Feeney, 1951; Reist et al., 1961). Bergmann and Burke (1955) reported that reduction of spongothymidine by sodium in dry liquid ammonia and ethanol and passing the product through Dowex-50-$H^+$ resulted in the isolation of a yellow gum, which on treatment with phenylhydrazine gave the osazone for arabinose (mp 154–155°C). Acid hydrolysis of the nucleoside resulted in the isolation of thymine. On the basis of these data, Bergmann and Burke proposed the structure of spongothymidine as 1-β-D-arabinofuranosylthymine (Figure 3.1).

### CHEMICAL SYNTHESIS

Since the arabinofuranosyl nucleosides with the β configuration cannot be prepared directly by the standard procedure of acetohalofuranosyl sugar and mercury salt of the pyrimidine, Fox et al. (1957) described a method for the synthesis of 1-β-D-arabinofuranosylpyrimidines. This synthesis of arabinosylthymine from β-ribosylthymine is shown in Figure 3.6. The significance of this synthesis is the formation of the anhydro-nucleoside and epimerization at C-2′. Doerr et al. (1967) subsequently described their synthesis of 1-β-D-arabinofuranosylthymine using an improved method reported earlier by Fox and Wempen (1965) (Fig. 3.5). The important intermediate compound in this synthesis is the $O^2$:2′-anhydronucleoside (**14**) (Brown et al., 1956; Fox et al., 1956; 1957; Walwick et al., 1959). 1-β-D-Xylofuranosylthymine

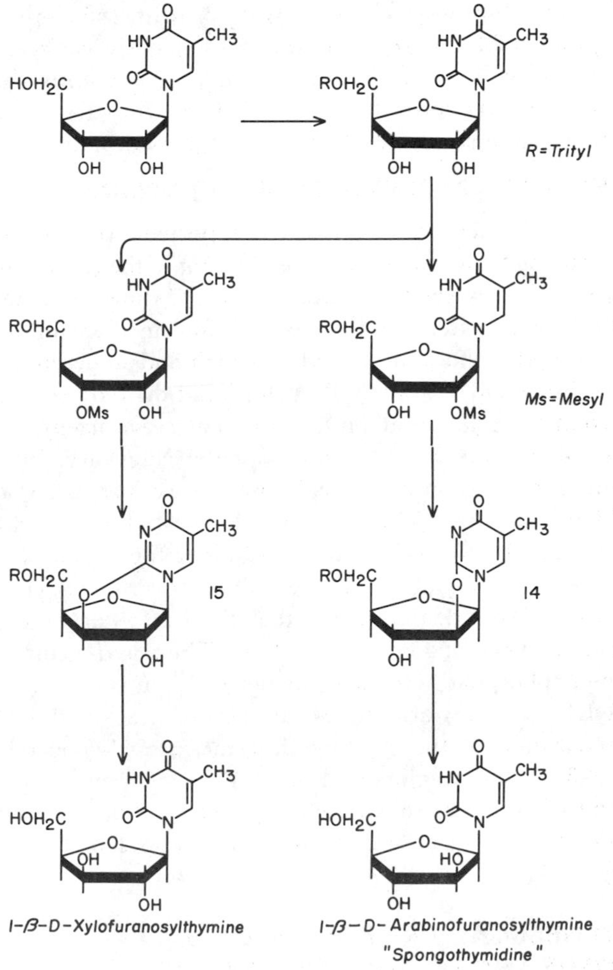

FIGURE 3.6. *The chemical synthesis of arabinofuranosylthymine and xylofuranosylthymine from β-ribofuranosylthymine. (From Fox et al., 1957.)*

was not isolated. Reist et al. (1961) reported an alternate improved method for the synthesis of ara-T from 1,2-diacetyl-5-methoxycarbonyl-3-(*p*-toluenesulfonyl)-D-xylofuranose. The yield was 11%. Nishimura and Shimizu (1965) also described the synthesis of 1-β-D-arabinofuranosyl-pyrimidines via the tri-*O*-benzyl-α-D-arabinofuranosyl bromide and bis-(trimethylsilyl)pyrimidine. The synthesis of ara-T from 1-β-D-xylofuranosyl-5-methyl-uracil has been reported [Upjohn Co., Pat. No. 1,396,003, (1965)].

Keller and Tyrrill (1966) described a convenient synthesis of 1-β-D-arabinofuranosylthymine (ara-T), involving the reaction of 2,3,5-tri-*O*-benzyl-α-D-arabinofuranosyl chloride with 2,4-dimethoxy-5-methylpyrimidine. The product of this reaction was subsequently converted to ara-T. The overall yield was 38%. The product is predominantly the β anomer. No α anomer was isolated or detected. All these syntheses served unequivocally to establish the structure of ara-T as 1-β-D-arabinofuranosylthymine.

### 3′-ARA-CMP AND ARABINOURIDYL(3′→5′) URIDINE

Nagyvary and Tapiero (1969) recently reported two syntheses of 3′-ara-CMP. The starting materials were quite different in the two methods, 5′-*O*,$N^4$ diacetylcytidine 2′:3′-cyclic phosphate and $N^4$-dimethylaminomethylene-cytidine 2′(3′)-phosphate, but the reactive intermediate in both cases involved a 2′:3′-cyclic phosphoryl tosylate which had undergone a rearrangement and hydrolysis to 3′-ara-CMP. A third and the most simple synthesis of the same compound is based on the thermal rearrangement of fully trimethylsilylated cytidine 2′:3′-cyclic phosphate (Nagyvary, 1969).

A certain degree of prebiological significance has been attributed to polyarauridylic acid by Schramm and Ulmer-Schürnbrand (1967) and Nagyvary et al. (1968). While the first mentioned team performed the synthesis of polyarauridylate from the monomeric ribouridylate, Nagyvary and Provenzale (1969) demonstrated that the ribo → arabino conversion is equally possible at the oligonucleotide level. They used a suitably protected diuridine monophosphate (**16**) as a model which was eventually converted to arauridylyl-(3′ → 5′)-uridine(**19**, aU-rU). Three methods were given for the preparation of the reactive intermediate 5′-*O*-acetyluridylyl-[(2′:3′) → 5′]-(2′,3′-*O*-isopropylidene)uridine (**17**), which on heating rearranged to compound cU-rU (**18**) in a base-catalyzed reaction. Alkaline hydrolysis of the isourea linkage in cU-rU followed by acidic deblocking gave aU-rU (**19**) in 90% yield.

## 3.5. BIOCHEMICAL PROPERTIES OF 9-β-D-ARABINOFURANOSYLADENINE (ARA-A)

The reader is referred to the excellent review by Cohen (1966) on the biochemical effects of 9-β-D-arabinofuranosyladenine on mammalian cell

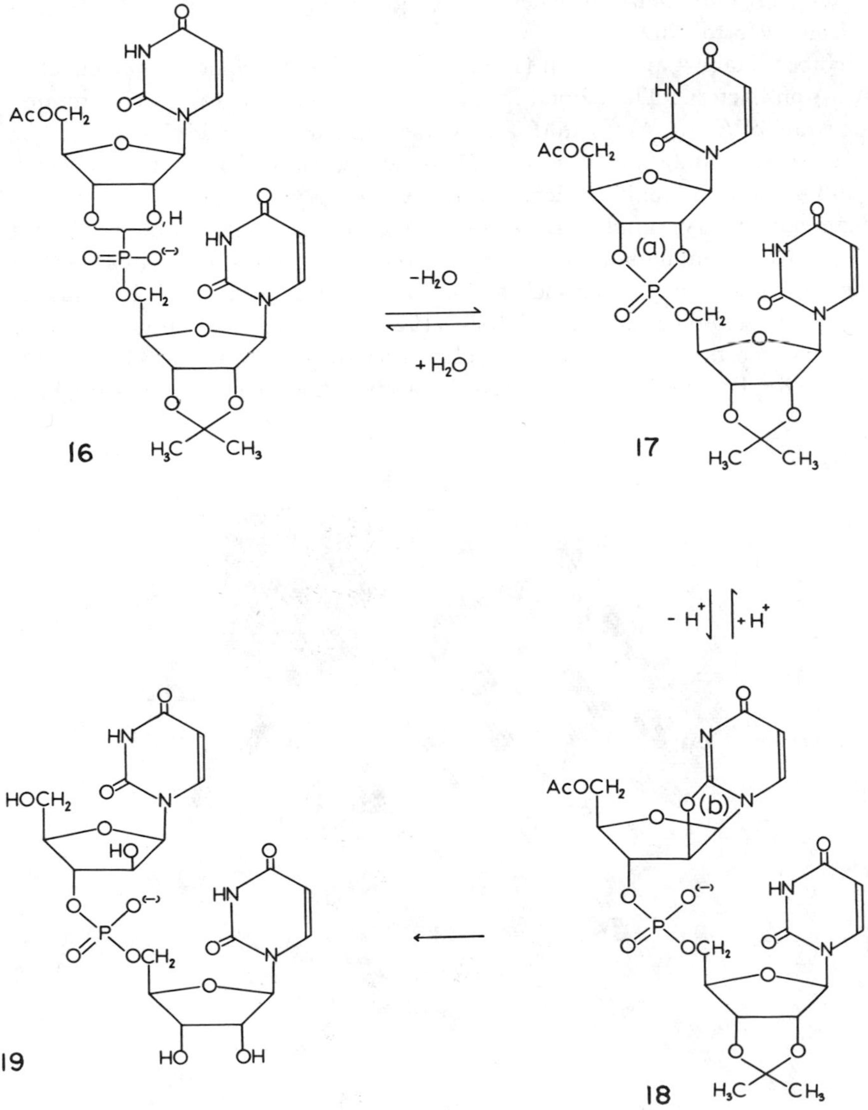

FIGURE 3.7. *The chemical synthesis of arabinouridyl(3′→5′)uridine (ᵃUpʳU). (From Nagyvary and Provenzale, 1969.)*

cultures, tumors, bacteria, and viruses. Since this review included the literature up through 1965, only a brief treatment will be given here for that material. A more detailed review will be given on the studies related to ara-A from 1965 to 1969.

Hubert-Habart and Cohen (1962) were the first to report on the effect of ara-A on bacteria. They found that ara-A was very toxic to a purine-requiring strain of *E. coli*. At 0.1 m*M*, ara-A was lethal and the addition of adenine reversed the inhibition of ara-A. There was a marked inhibition of DNA synthesis and a smaller effect on RNA and protein synthesis. Ara-A is deaminated very rapidly to 9-β-D-arabinofuranosylhypoxanthine. Several studies with mammalian cell cultures have shown that ara-A is a more potent inhibitor than compounds such as fluorodeoxyuridine or ara-C in bringing about chromosome breaks. Nichols (1964) reported that ara-A induced chromosome breaks in human peripheral leukocytes during RNA synthesis (Fig. 3.8). Rao and Natarajan (1965) studied the effect of ara-A on *Vicia*

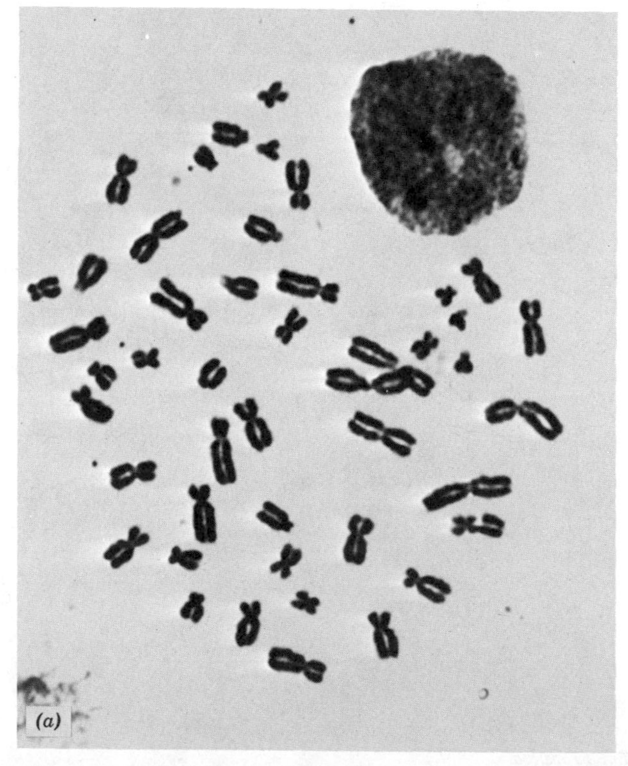


FIGURE 3.8.(a). *Chromosome preparation from human leukocyte treated with ara-A at a concentration of* $1 \times 10^{-4}$ M *exhibiting one chromatid gap. (From Nichols, 1964.)*

*faba* chromosomes. Their data indicate that ara-A breaks chromosomes following as short a treatment as 2 hr, and the chromosome breakage pattern resembles that produced by other inhibitors of DNA synthesis.

Kihlman and Odmark (1966) reported that ara-A produced chromosomal aberrations of the fragment type in *V. faba* root tip cells that were fixed immediately or only a few hours after treatment. A large portion of the exchanges were localized in the nucleolar constriction. Ara-A inhibited the incorporation of $^{32}P$ into DNA of excised *V. faba* root tips, but had no effect on the incorporation into RNA. In contrast to this study, 9-β-D-xylofuranosyladenine and 3′-deoxyadenosine inhibited the incorporation of $^{32}P$ into both DNA and RNA.

Ara-A was shown to be active against two strains of ascites tumors, but had little activity against leukemia L-1210 or subcutaneous solid tumor in mice (Brink and LePage 1964a, 1964b). Ara-A does not inhibit purine nucleoside phosphorylase (Brink and LePage, 1965). In addition, it does not

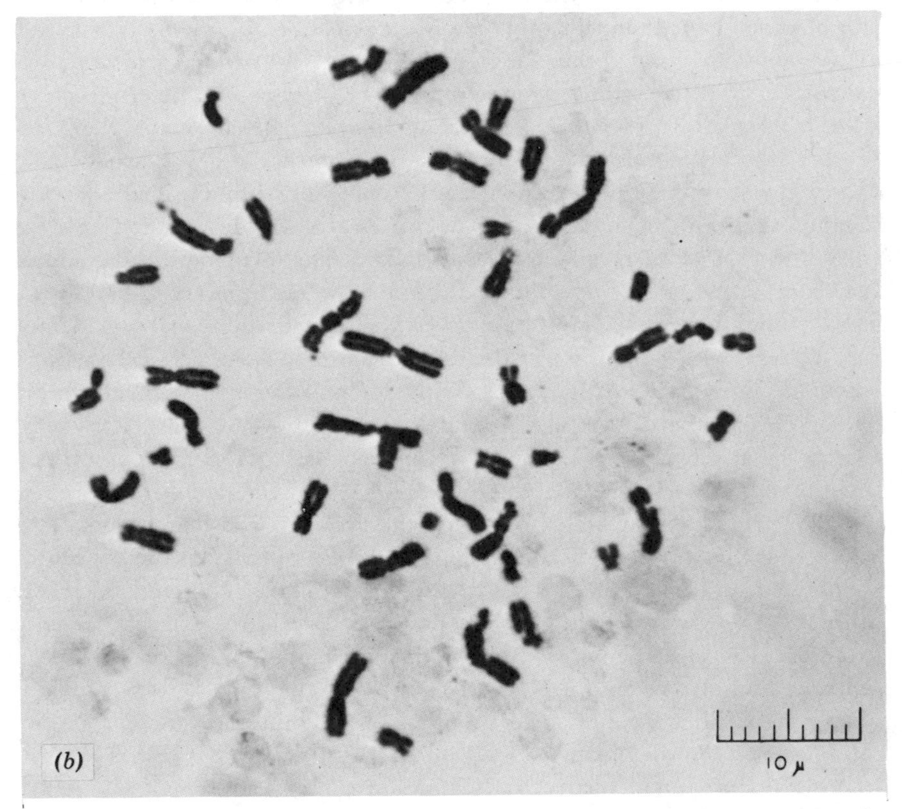


(*b*) *Chromosome preparation from human leukocyte treated with ara-A at a concentration of* $3 \times 10^{-4}$ M *exhibiting multiple breaks.* (*From Nichols, 1964.*)

undergo glycosidic cleavage by cell-free extracts of ascites cells or mouse tissue (Brink and LePage, 1964a; LePage and Junga, 1965). Bloch and Nichol (1964) reported that ara-A was a substrate for adenosine phosphorylase with cell-free extracts from *Streptococcus faecalis*, but the nucleoside did not inhibit adenosine kinase. The deamination of ara-A by adenosine deaminase from several biological sources has been studied in a number of laboratories. Hubert-Habart and Cohen (1962) reported that *E. coli* converted ara-A to ara-Hx. In a subsequent report, Cory and Suhadolnik (1965) reported studies on the structural requirements of nucleosides for binding by adenosine deaminase from calf intestinal mucosa. Ara-A was deaminated at a rate about 20% that of adenosine, while xylofuranosyladenine was more rapidly deaminated than ara-A. LePage and Junga (1965) reported that these two nucleoside analogs were deaminated at the same rate by transplanted tumors. York and LePage (1966b) subsequently reported that the configuration of the 2′-hydroxyl group is very important for binding of adenosine deaminase to the nucleoside. On the basis of $K_m$ and $V_{max}$ values, they stated that the 2′-hydroxyl and 3′-hydroxyl groups play an important role in binding of substrate to the enzyme and determining the rate of breakdown of enzyme–substrate complex. Bloch et al. (1967) and Frederiksen (1966) also reported on the deamination of ara-A. Cohen compared the deamination of ara-A with deoxyadenosine by purified adenosine deaminase from calf intestine and *E. coli*. Ara-A was deaminated at a rate one-fifth that of deoxyadenosine by calf intestinal adenosine deaminase, while *E. coli* adenosine deaminase did not show such marked differences in the deamination rates. The $K_m$ values for the deamination of ribosyladenine, ara-A, and xylosyladenine by adenosine deaminase from TA3 cells was not the same. This greater affinity for ribosyladenine was used by LePage and Junga (1965) to protect ara-A from deamination by TA3 cells.

LePage and his co-workers have studied extensively the effect of ara-A in mammalian systems. LePage and Junga (1963) and Brink and LePage (1964a) reported that ara-A inhibited the growth of ascites tumors, TA3, and 6C3HED, but not L1210 leukemia and solid tumors. At the concentrations used, ara-A was not cleaved by ascites tumors TA3 and 6C3HED, although there was a slight cleavage of ara-A by cell-free extracts prepared from L1210 cells. The inhibition of incorporation of carbon-14 labeled adenine and glycine into RNA and DNA by ara-A and ara-G is shown in Table 3.3. With ara-A there was a more profound inhibition of incorporation of adenine-8-$^{14}C$ and glycine-2-$^{14}C$ into DNA than into RNA. RNA synthesis remained unaffected or even stimulated. Ara-A was more inhibitory to DNA synthesis than was ara-G. Hubert-Habart and Cohen (1962) reported similar findings when ara-A was added to cultures of *E. coli*.

**Table 3.3**

Incorporation of Adenine-$^{14}$C and Glycine-$^{14}$C into Nucleic Acids of TA3 Ascites Cells Treated with Arabinosylguanine and Arabinosyladenine

| Treatment | $^{14}$C-precursor[a] | Group[b] | Counts per ml packed cells × $10^{-3}$ | |
|---|---|---|---|---|
| | | | RNA | DNA |
| Arabinosylguanine (0.5 mg/mouse) | Adenine-8-$^{14}$C | 1 | 41.7 | 7.1 |
| | | 2 | 46.5 | 5.6 |
| | | 3 | 41.7 | 3.9 |
| Arabinosylguanine (0.5 mg/mouse) | Glycine-2-$^{14}$C | 1 | 21.2 | 7.8 |
| | | 2 | 24.0 | 11.5 |
| | | 3 | 20.9 | 9.3 |
| Arabinosyladenine (0.5 mg/mouse) | Adenine-8-$^{14}$C | 1 | 39.7 | 9.5 |
| | | 2 | 52.0 | 3.0 |
| | | 3 | 38.5 | 2.5 |
| Arabinosyladenine (0.5 mg/mouse) | Glycine-2-$^{14}$C | 1 | 28.4 | 7.0 |
| | | 2 | 39.6 | 6.9 |
| | | 3 | 35.8 | 5.4 |

From Brink and LePage, 1964a.

[a] Groups of four mice (± 20-g females) were used in each experiment from which the pooled cells were assayed for radioactivity in the nucleic acids. Each mouse received the same dose of $^{14}$C-precursor.

[b] Group 1: saline and labeled substrate. Group 2: ara-A in saline and labeled substrate. Group 3: ara-A or ara-G in saline and then labeled substrate 15 min. later. Mice were killed 1 hr after injection of labeled compound.

Brink and LePage (1964a, 1964b, 1965) presented data showing that ara-A-8-$^{14}$C was readily phosphorylated to the triphosphate level and incorporated into the nucleotide fraction. Ara-A was rapidly deaminated by cell-free extracts of TA3 tumors, but was not cleaved by extracts of TA3 or 6C3HED ascites cells. Cohen (1966) was critical of these experiments since arabinose was not established as being associated with the labeled adenine in the nucleotide fraction studied. Although LePage and his workers reported that there was no cleavage of ara-A or ara-Hx, Cohen (1966) reported that ara-A can be deaminated to ara-Hx and cleaved to give small amounts of hypoxanthine. In addition, Tono and Cohen (1962) reported that the cleavage of ara-U occurred at high concentrations. Therefore, it may be that the radioactivity in the RNA from the ara-A-8-$^{14}$C could be explained by a small amount of carbon-14 labeled hypoxanthine resulting from cleavage of ara-Hx. LePage (1970) has now reported findings in which he clearly shows that ara-A does not enter the RNA of TA3 ascites tumor cells. Incorporation of radioactivity from this nucleoside occurred after cleavage to the base and subsequent reutilization of the labeled base. The metabolite, 9-β-D-arabinofuranosylhypoxanthine, appears to be reaminated to form ara-A. This reamination may explain the activity of the hypoxanthine analogue as an antiviral agent.

Ara-A is rapidly cleared from the blood of normal mice and excreted in the urine as ara-Hx (Brink and LePage, 1964b). In addition, no evidence for the cleavage of ara-A could be detected in the organs or blood of normal mice, although rapid deamination occurred. The authors also presented evidence that ara-A was incorporated into the acid-soluble nucleotides and the RNA from the liver. A single injection of ara-A inhibited DNA synthesis for about 3 hr in L1210, 8 hr in TA3, and more than 24 hr in 6C3HED ascites tumors. More recently, Cohen (1966) reported experiments with L-cells and tritium-labeled ara-A. There was no incorporation into the acid-soluble pool of the cells. When cultured cells were incubated with tritium-labeled ara-A for 5 hr, the total isotope incorporated into RNA and DNA was less than 0.02% of the ara-A added.

Ara-A did not inhibit the incorporation of lysine or glutamic acid into protein (Brink and LePage, 1964b), but the incorporation of glycine into protein was slightly decreased. Since ara-A was shown to be more specific with respect to the inhibition of DNA compared with RNA synthesis, LePage and his co-workers suggested that the effect of ara-A on DNA synthesis might be explained by three possibilities: (*1*) ara-A could interfere with the formation of DNA precursors, (*2*) it could have a direct effect on DNA polymerase, and (*3*) it could interfere with protein synthesis.

## EFFECT ON RIBONUCLEOTIDE REDUCTASE

Moore and Cohen (1967) studied the effects of arabinonucleotides on the reduction of ribonucleotides by the ribonucleotide reductase from Novikoff ascites tumor cells grown in rats. The 5′-triphosphate of ara-A inhibited the ribonucleotide reductase from Novikoff ascites tumor cells (Fig. 3.9). They showed that ara-A was a much stronger inhibitor than ara-C. Ara-A has a lower inhibition than dATP. Ara-C is not a potent inhibitor of ribonucleotide reductase. Furth and Cohen (1968) subsequently proposed that ara-ATP might produce a physiologically significant inhibition of the ribonucleotide reductase isolated from the tumor cell. York and LePage (1966a) reported earlier that ara-ATP inhibited the reduction of CTP.

## EFFECT ON BACTERIAL AND MAMMALIAN POLYMERASES

Cardeilhac and Cohen (1964) reported that ara-CTP was not an inhibitor of the bacterial system. These findings were subsequently confirmed by Furth and Cohen (1968).

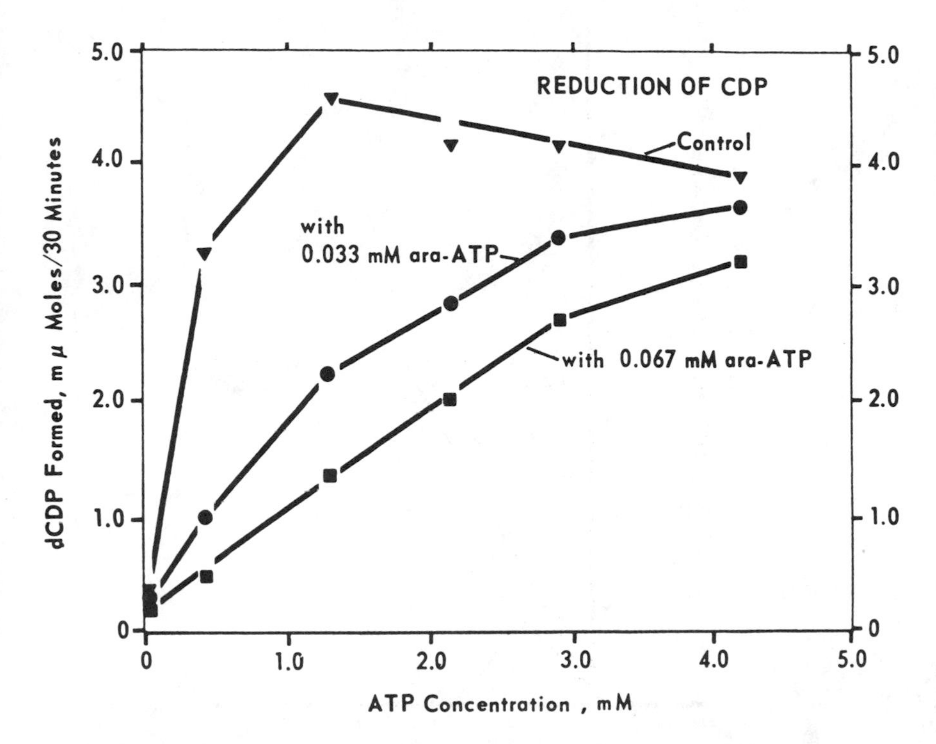


FIGURE 3.9. *Effect of ATP concentration on inhibition by ara-ATP of the reduction of CDP. Each incubation mixture, containing 0.17 mg of partially purified nucleotide reductase from rat tumor in a total volume of 0.12 ml, with 1 μmole of potassium phosphate, pH 7, 0.5 μmole of magnesium acetate, 0.005 μmole of ferric chloride, 0.75 μmole of dithioerythritol, 0.02 μmole of CDP labeled with* $^{32}P$ (1.2 × 10⁶ *cpm per μmole), and ara-ATP and ATP as indicated, was incubated for 30 min at 38°C.* (*From Moore and Cohen, 1967.*)

Cohen, LePage, Kornberg, and their co-workers have made significant contributions to elucidate the biochemical effect of ara-A and ara-C on the bacterial and mammalian polymerases. York and LePage (1966a) studied the effect of ara-ATP on the incorporation of thymidine 5′-monophosphate into DNA by DNA polymerase with cell-free extracts of TA3 ascites tumor cells. They reported that ara-ATP is a noncompetitive inhibitor of the incorporation of dATP into DNA by DNA polymerase ($K_i = 9 \times 10^{-5}$ *M*). The concentration of ara-ATP in the intact cell was shown to be $5.6 \times 10^{-4}$ *M*, which was four times as high as that required with the cell-free extracts to give a 40% inhibition of DNA polymerase activity. With a more highly purified calf thymus, DNA polymerase, ara-ATP was shown to be a competitive inhibitor of dATP (Fig. 3.10) (Furth and Cohen, 1968). The $K_i$ for ara-ATP is 1.3 $\mu M$. The $K_m$ for dATP is 2.5 $\mu M$. Ara-ATP-$^3$H did not appear to be incorporated into DNA. RNA polymerase from lympho-

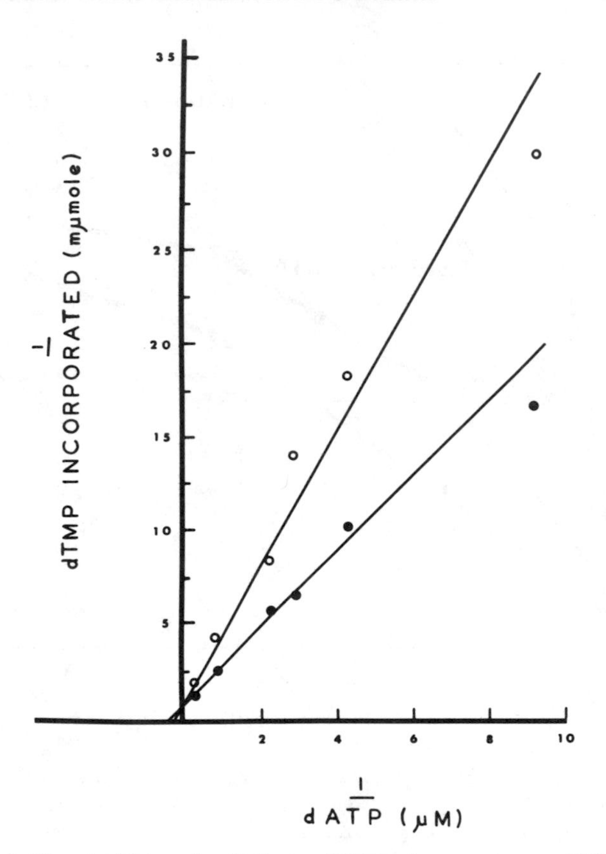


FIGURE 3.10. *Inhibition of calf thymus DNA polymerase by ara-ATP. The reaction mixture (0.5 ml) contained: 50* mM *phosphate buffer, (pH 7.0); 4* mM *$MgCl_2$, 2* mM *2-mercaptoethanol, 50* μM *each of dCTP and dGTP, 56* μM *dTTP-$^{14}$C* (1.35 × $10^6$ *cpm/μmole); and 54 μg of enzyme.* (●) *no ara-ATP;* (○) *plus 1.7* μM *ara-ATP.* (*From Furth and Cohen, 1968.*)

sarcoma, and *E. coli* was not inhibited by ara-ATP. The authors suggested that an incorporation of one molecule of ara-ATP per 100 nucleotides could not be detected.

The effects of ara-A on the growth, multiplication, and RNA and DNA syntheses of mouse fibroblasts in suspended cell culture were studied by Doering et al. (1966). They showed a considerable enlargement of the cells in the presence of ara-A and ara-C, but a marked inhibition of multiplication. Both nucleosides inhibited DNA synthesis and reduced the cloning capacity of the cultures. There was no measurable inhibition of RNA synthesis, although ara-A produced a slight inhibition of protein synthesis. These studies are shown in Figure 3.11. The inhibition of DNA synthesis and

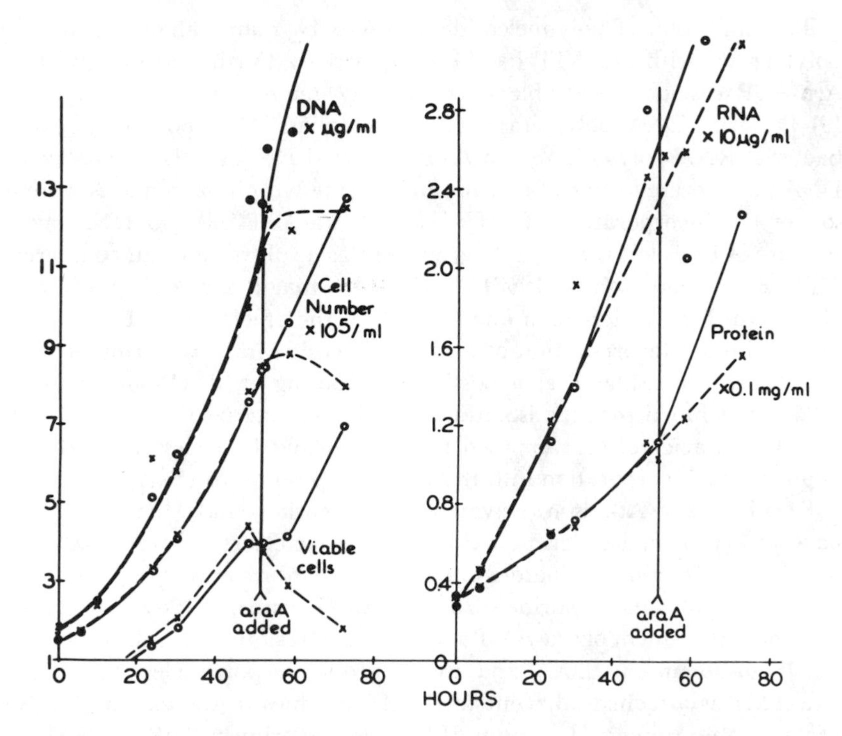


FIGURE 3.11. *Multiplication and synthesis of RNA, DNA, and protein in the cells.* (×) *Curves for paired cultures to which ara-A* ($2 \times 10^{-4}$M) *was added.* (○) *Comparative controls.* (*From Doering et al., 1966.*)

cell division was severe and the loss of cloning ability approached the effect observed with ara-C. Ara-A was rapidly deaminated to ara-Hx and the authors stated that the latter compound may have also contributed to the observed inhibitions. The data presented indicated that the inhibition of DNA synthesis and of cell multiplication by ara-A and ara-C cannot be attributed to irreversible lesions in the DNA that would result from the terminal addition of these ara-nucleotides, since removal of ara-A or ara-C from the medium resulted in an incorporation of thymidine-2-$^{14}$C into DNA of L cells without a time lag. It appears that the killing of the cell by ara-A or ara-C is attributed to causes other than the incorporation of ara-A into the ends of the DNA molecules. It was suggested by the authors and shown by more recent auto radiographic studies that there may be multiple initiation sites on the same chromosomes which could give the same kinetics even though some DNA synthesis had been stopped (Cohen, private communication).

The inhibition of polynucleotide synthesis by mammalian and bacterial polymerases with ara-ATP has been studied by Furth and Cohen (1967). Ara-ATP was not a detectable substrate for (*1*) mammalian DNA polymerase, (*2*) bacterial DNA polymerase, (*3*) mammalian RNA polymerase, or (*4*) bacterial RNA polymerase. Ara-ATP inhibited DNA synthesis catalyzed by DNA polymerase from calf thymus and bovine lymphosarcoma. A comparison of the incorporation of TTP-$^3$H and ara-ATP-$^3$H into DNA showed that ara-ATP is less than 1/100 as efficient as a substrate when compared to TTP. Furth and Cohen (1967) concluded, "that if ara-ATP is utilized at all, it is utilized at less than one-hundredth the efficiency of TTP." Therefore, a limited incorporation of an ara-nucleoside into the terminal residue in DNA was considered as a possibility. Doering et al. (1966) and Cohen (1966) concluded that the isolation of well-characterized 3′-ara-CMP from the nucleic acids of cells exposed to ara-C would be necessary evidence to establish the incorporation into the terminal position of DNA.

Kornberg and Atkinson (private communication) and Momparler (1969) have now provided experimental evidence that clearly characterizes ara-A and ara-C as chain terminators in DNA in *in vitro* systems. They have shown that DNA polymerase purified from *E. coli* (Kornberg, 1969) catalyzes the terminal attachment of ara-AMP and ara-CMP residues from the corresponding triphosphates to DNA and the appropriate polynucleotide template. Ara-AMP is attached adjacent to dTMP residues to the extent of 1 AMP per available primer when poly d(A-T) is the primer. Both ara-AMP and ara-CMP are incorporated into DNA that has been nicked with DNase; the extent of incorporation again indicates terminal attachment. Only one ara-AMP or ara-CMP residue is added to the terminal position of DNA. Xylo-ATP is not a substrate. Proof of the addition of the analogs to the terminal end of DNA was established by nearest neighbor frequency analysis with micrococcal nuclease and spleen diesterase. The nature of the incorporated nucleotide was confirmed by chromotography following hydrolysis to the 5′-phosphate with pancreatic DNase and snake venom diesterase. The rate of incorporation of ara-ATP or ara-CTP is about 1/1000 that of the normal rate.

Although the findings of Kornberg and Atkinson clearly demonstrate the incorporation of ara-CMP and ara-AMP residues into DNA, the problem of lethality as caused by arabinosyl compounds may still be obscure. When comparing the relationship of the terminal data and the lethality, one must consider that (*1*) the rate of addition of ara-A or ara-C is very low and (*2*) there has been no evidence that the same type of incorporation occurs *in vivo*. For example, Leung et al. (1966) showed that an *E. coli* TAUAd mutant, killed by ara-A in the presence or absence of thymine, is capable of extensive DNA synthesis. The distribution of DNA in the nucleus of ara-

A-treated cells appeared to be normal (Nass, Doering, and Cohen, unpublished data). Similarly, Doering, Keller, and Cohen (1966) reported that the inhibition of DNA synthesis in mouse fibroblasts by ara-A is not irreversible, even when cells have lost their ability to undergo cell division.

The data presented by York and LePage (1966a), Furth and Cohen (1967, 1968), and Kornberg and Atkinson (unpublished results) emphasized the importance of the stereochemistry of the hydroxyl group on the C-2′ position of nucleotides for those enzymes catalyzing polynucleotide synthesis. RNA polymerase requires the ribofuranosyl moiety, in which the hydroxyl groups on C-2′ and C-3′ are *cis*; however, this enzyme cannot utilize the arabinofuranosyl moiety in which the C-2′ and C-3′ hydroxyl groups are *trans*. On the other hand, bacterial DNA polymerase can utilize the ribofuranosyl nucleotides and can also add the arabinofuranosyl nucleotides to the 3′-terminal positions of DNA *in vitro*. The mammalian DNA polymerase is inhibited by the arabinofuranosyl nucleotides. The structural similarity of the arabinofuranosyl nucleotides and the 2′-deoxynucleotides have been demonstrated in two ways. First, ara-ATP is a better inhibitor of mammalian DNA polymerase than is ATP. Second, ara-ADP and 2′-dADP both inhibit polynucleotide phosphorylase (Lucas-Lenard and Cohen, 1966; Cardeilhac and Cohen, 1964).

In conclusion, it can be stated that ara-A and ara-C can act as chain terminators for DNA synthesis, although this observation may not explain the problem of lethality exerted on the cell by arabinosyl nucleotides.

A similar study on the incorporation of ara-AMP into RNA has been reported by Ilan and co-workers (private communication). Ilan, while studying the prevention of malaria by ara-A, reported that ara-AMP is incorporated into the 3′-terminal end of tRNA. These findings were based on double-labeling experiments (with tritum-labeled ara-A in the presence of $^{32}Pi$) and the subsequent incorporation into RNA. While only 1% of ara-A was phosphorylated by the liver of control mice, 3% was phosphorylated by the liver of the infected mice; however, 26% of the ara-A was phosphorylated by *Plasmodium berghei*. A closer study of the incorporation of ara-AMP from the corresponding triphosphate into tRNA of *E. coli* indicated that CCA-pyrophosphorylase added CMP and ara-AMP to the 3′-terminal position of pyrophosphorylized tRNA. The tRNA containing ara-AMP in the 3′-terminal end (following periodate treatment) was able to accept activated-amino acids. It is not yet clear if tRNA, with ara-AMP in the 3′-terminal position, can act as a donor for protein synthesis. Ara-ATP is a potent inhibitor of leucyl-tRNA synthetase ($K_m$ for ATP of leucyl-tRNA synthetase is 0.4 $\mu M$; $K_i$ is 6 $\mu M$). Ara-ATP is very easily synthesized by the photophosphorylation system using spinach chloroplasts.

The evidence presented here indicates that the stereochemistry of the hydroxyl group on C-2′ of ribose does not appear to be important for the esterification of tRNA by activated amino acids. These data may well answer the long standing question, "Is the activated amino acid transferred to the 2′ hydroxyl carbon or the 3′ hydroxyl carbon of ribose when adenosine is located in the 3′-terminal end of tRNA?"

## PHOSPHORYLATION STUDIES OF ARA-A *IN VITRO*

Although Brink and LePage (1964a) reported that ara-A was phosphorylated to the 5′-mono-, di-, and triphosphates, ara-A is a poor substrate with nucleoside kinase isolated from ascites tumor cells (Pierre et al., 1967) or adenosine kinase from H. Ep. #2 cells or rabbit liver (Schnebli et al., 1967; Lindberg et al., 1967).

## DEAMINATION OF THE 1-*N*-OXIDE OF ARA-A

The deamination of ara-A has been reported by a number of laboratories (Hubert-Habart and Cohen, 1962; Brink and LePage, 1964b; Bloch and Nichol, 1964; Cory and Suhadolnik, 1965; LePage and Junga, 1965; York and LePage, 1966a, 1966b; Cohen, 1966, pp. 62, 64, 65; Doering et al., 1966; Frederiksen, 1966; Bloch et al., 1967). To prevent deamination of ara-A, the 1-*N*-oxide was synthesized (Reist et al., 1967). It had been reported earlier that the deamination of 3′-deoxyadenosine (cordycepin) was nearly eliminated by the use of cordycepin 1-*N*-oxide (Frederiksen, 1963). Cohen (1966, pp. 64, 65) also reported on the synthesis of the 1-*N*-oxide of ara-A by the method of Stevens et al. (1958). This compound inhibited growth, but was slowly lethal for an adenosine-requiring mutant of the *E. coli* strain 15 polyauxotroph requiring exogenous thymine, arginine, and uracil for growth and survival. The effect of ara-A, ara-Hx, adenosine 1-*N*-oxide, and the 1-*N*-oxide of ara-A for the polyauxotrophic *E. coli* strain TAUAd is shown in Figure 3.12.

## EFFECT OF ARA-A ON A POLYAUXOTROPHIC STRAIN OF *E. COLI*

Cohen (1966, p. 66) and Leung et al. (1966) compared the relation of lethality imposed by a thymine deficiency and that produced by ara-A in the purine-requiring *E. coli* strain 15 TAUAd. The simultaneous omission of adenosine from a thymine-deficient medium markedly inhibited thymineless death. Killing with ara-A under these conditions occurred in the presence or absence of thymine. Adenine reversed the lethality caused by ara-A.

Leung et al. (1966) reported the effect of ara-A with the purine requiring polyauxotrophic strain of *E. coli*. The deamination product, ara-Hx, is lethal to this organism. The absence of uracil reduced ara-A toxicity, while the lack of arginine almost eliminated lethality. DNA synthesis was com-

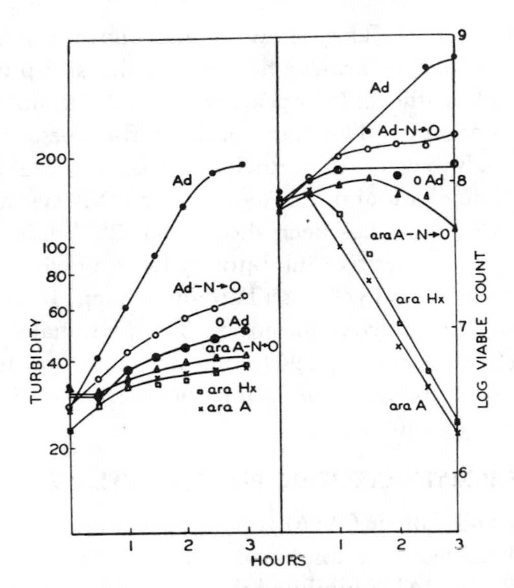


FIGURE 3.12. *The lethality of* D-*arabinosyladenine and other* D-*arabinosyl derivatives for the polyauxotrophic* E. coli *strain $T^-A^-U^-Ad$ when exposed to ara-A, ara-Hx, the 1-N-oxides, of ara-A, and adenosine. The symbol "O" designates growth in the complete medium except for a lack of the component named after the "O."* (*Ad, adenosine; Ad*-N→*O, adenosine 1*-N-*oxide; ara-A*-N→*O, 1*-N-*oxide of ara-A*). (*From Cohen, 1966, p. 64; Leung et al., 1966.*)

pletely inhibited when ara-A was added at lethal concentrations to a medium lacking exogenous purine, and there was only a slight decrease in protein synthesis, but no inhibition of RNA synthesis. The inhibition of DNA synthesis by ara-A is in agreement with the earlier report by Moore and Cohen (1967), in which they discussed the activity of ara-A triphosphate in inhibiting the reduction of the ribonucleoside diphosphates to their corresponding 2′-deoxy nucleotides by partially purified ribonucleotide reductase from animal tissue. Finally, the ability of cells (that have been killed by ara-A) to incorporate thymine into DNA strongly suggests that none of the enzymes involved in DNA synthesis is irreversibly damaged. The authors also concluded that lethality did not arise from irreversible damage to the ends of the DNA templates due to the incorporation of arabinosyl nucleotides.

In a subsequent paper, *E. coli* strain 15 TAUAd was used to study polymer synthesis (RNA, DNA, and protein) following killing with ara-A or 2′,3′-dideoxyadenosine (Doering-McGovern et al., 1966). The lethality of dideoxyadenosine was similar to that of ara-A, i.e., RNA synthesis and protein synthesis continued, while DNA synthesis was inhibited. The structure of the adenine nucleosides used in these studies by Doering-McGovern et al. (1966)

are shown in Figure 3.13. They showed that following inhibition of DNA synthesis and cell killing by ara-A, the DNA chains still provided templates for continued DNA synthesis. It appears that ara-A did not block DNA synthesis by terminal addition. The data obtained from the 2′,3′-dideoxyadenosine studies strongly suggest that this nucleoside added to the end of a polydeoxynucleotide chain and blocked further DNA synthesis. With DNA polymerase from *E. coli*, it has been shown that 2′,3′-dideoxyadenosine triphosphate is both a competitive inhibitor of DNA polymerase and a terminator of polydeoxynucleotide chains (Toji and Cohen, 1969). Recent studies have shown that 2′,3′- dideoxynucleoside 5′-triphosphate binds to the triphosphate site (Englund et al., 1969) and are substrates for addition to a DNA chain (Cohen, private communication; Deutscher and Kornberg, 1969; Atkinson et al., 1969).

## INHIBITION OF POLYNUCLEOTIDE PHOSPHORYLASE BY ARA-A

Lucas-Lenard and Cohen (1966) reported their studies on the kinetics of inhibition of polynucleotide phosphorylase by ara-ADP, dCDP, and dADP. Ara-ADP, dCDP, and dADP inhibited the rate of ADP and CDP polymeriza-

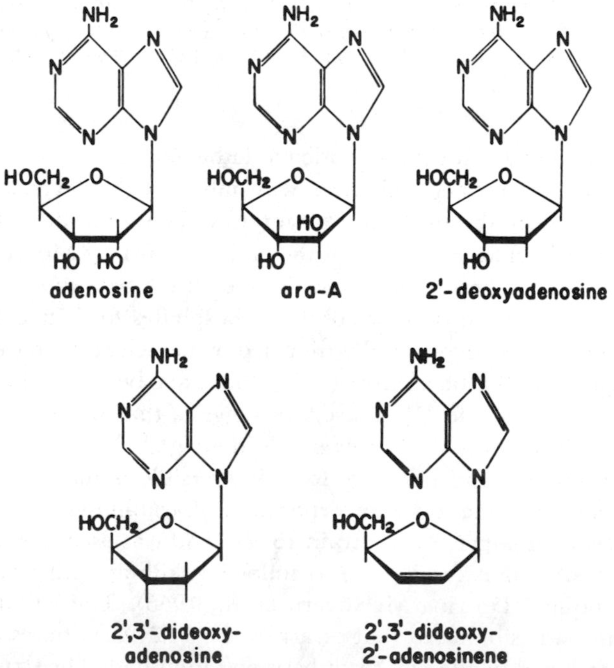

FIGURE 3.13. *Structures of adenosine analogs. (From Doering-McGovern et al., 1966.)*

tion by polynucleotide phosphorylase from *Ps. fluorescens* and *E. coli.* (Fig. 3.14). There is a lag in the polymerization of ADP and CDP in the presence of these analogs. With increasing time, the rate of reaction increases such that the final incorporation is the same as in the control reaction mixture that did not contain the inhibitor. When ara-ADP-$^3$H was added to the incubation mixture, there was no radioactivity in the washed acid-insoluble precipitate. These results are in agreement with the earlier findings by Cardeilhac and Cohen (1964) indicating that ara-CDP was not incorporated into either poly-C or poly-A. Lucas-Lenard and Cohen (1966) also showed that the lag produced by these substrate analogs could not be eliminated. To determine if the lag period was the result of a slow formation of initial oligonucleotides due to the presence of ara-ADP, experiments were designed to determine if there was a lag when ara-ADP was added after the start of the polymerization of ADP. The results obtained showed that a lag period was observed when ara-ADP was added 10 min after the onset of polymerization. Although ara-ADP and dADP inhibited the rate of ADP and CDP polymerization, these substrate analogs had no effect on the phosphorolysis of poly-A. Lucas-Lenard and Cohen suggested that ara-ADP may act in a manner similar to dADP by preventing polymerization of ribonucleoside diphosphates by polynucleotide phosphorylase.

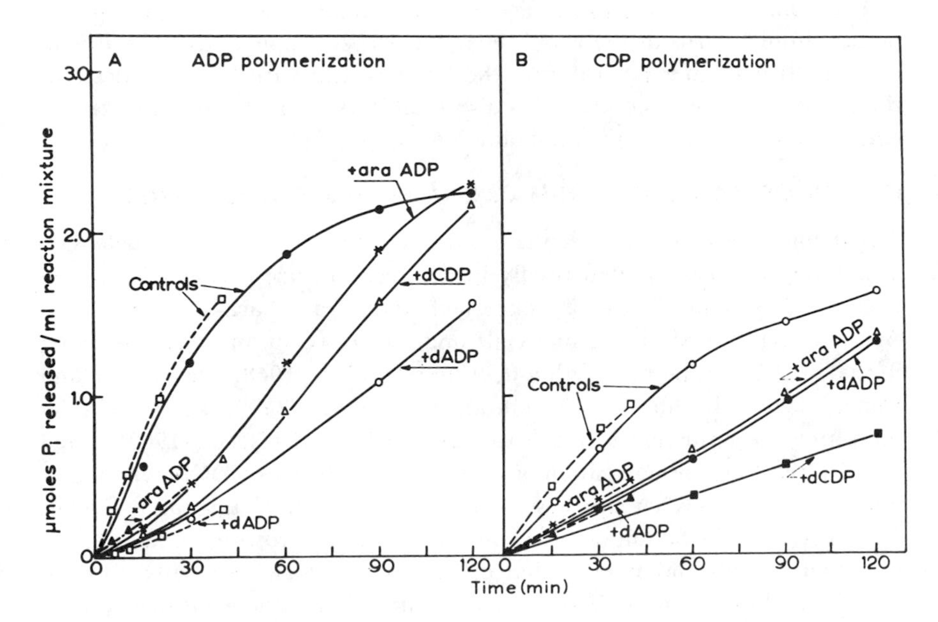


FIGURE 3.14. *Inhibition of ADP and CDP polymerization by dADP, dCDP, and ara-ADP. (Solid lines) polynucleotide phosphorylase from* Ps. fluorescens; *(broken lines),* E. coli *polynucleotide phosphorylase. (From Lucas-Lenard and Cohen, 1966.)*

## 3.6. BIOCHEMICAL PROPERTIES OF ARA-C AND ARA-U

The current knowledge concerning the biological activity of ara-C has been elegantly reviewed by Cohen (1966), Schabel (1967, 1968), and Smith (1968). The reader should consult this material for a complete coverage of the work related to these nucleoside analogs. This section will be primarily concerned with those publications related to ara-C and ara-U since 1966.

Ara-C, an analog of cytidine, has the C-2′ hydroxyl which is *cis* to the glycosyl linkage, while the C-2′ hydroxyl of the ribonucleoside is *trans*. Ara-C was first synthesized by Walwick et al. (1959), and Evans et al. (1961) were the first to show that ara-C was effective against animal neoplasms. Ara-C has been shown to be cytotoxic for mammalian cells in culture and inhibits numerous DNA viruses. Deoxycytidine reverses this inhibition. Ara-C is also an effective inhibitor of mice bearing L1210 cells (Schabel, 1967). Ara-C is similar to other purine and pyrimidine antagonists in that L1210 cells are only inhibited by ara-C if the cells are in exponential growth and are rapidly synthesizing DNA (Schabel, 1967; Schabel et al., 1965). The conversion of cytidine to deoxycytidine and the activity of certain pyrimidine kinases is inhibited by ara-C, which is also incorporated into the 3′-terminal position of DNA of bacteria and mammalian cells in culture. Cytidine deaminase converts ara-C to the less active ara-U.

Ara-C inhibits neoplasms in mice, rats, and humans and is active against herpes simplex virus in the rabbit eye and herpes simplex virus keratitis in humans. It compares favorably in the latter system with 5-iodo-2′-deoxyuridine, but is more toxic. Ara-C is also inhibitory to a strain of *Streptococcus faecalis* that is resistant to actinobolin.

### EFFECTS OF ARA-C ON ANIMAL CELL CULTURES AND VIRUSES

Although ara-C has not been isolated as a naturally occurring nucleoside antibiotic, it will be treated briefly in this section since it is closely related to those biochemical studies concerned with ara-A, ara-T, and ara-U. Ara-C is highly toxic to animal cells and viruses (Chu and Fischer, 1962; Buthala, 1964; Kim and Eidinoff, 1965; Schabel, 1968; and Furth and Cohen, 1968). It inhibits the reproduction of L5178Y leukemic cells by inhibiting the reduction of CMP to dCDP. Chu and Fischer (1962) stated that by blocking the formation of dCDP, the synthesis of DNA was subsequently blocked. RNA synthesis was not inhibited. Young and Fischer (1968) studied the action of ara-C on synchronously growing populations of mammalian cells and showed that cell division of logarithmically growing cells is inhibited; however, they increase in volume and continue to synthesize protein (Table 3.4). Although the lethal effect of ara-C could be prevented by deoxycytidine, cells which are making DNA are inhibited by this nucleoside and the addition of deoxycytidine does not reverse this inhibition. The data indicate that ara-C acts specifically and irreversibly on the S-phase of the cell cycle. The inability of deoxycytidine to reverse the cytotoxicity caused by ara-C in S-phase cells suggests that the biochemical lesion involves a process in DNA replication. Skipper et al. (1967) investigated the effect of ara-C and its S-phase specifically against leukemia cells in mice. They observed that the average doubling time of leukemia cells *in vivo* is 0.4–0.6 a day; the S-phase averaged about 90% of this time. A single dose of ara-C reduced the survival of leukemic cells by only 50–70%. Doses spaced over a 24-hr period reduce the survival of the hosts' leukemia cell population to $10^{-4}$. Multiple, shortly spaced doses of ara-C are capable of eradicating as much as $10^6$ leukemia cells in the leukemic animals. By using this *in vivo* procedure with L1210 leukemia cells, Skipper et al. (1967) found a high "cell cure" rate so long as the animals do not bear more than $10^6$ leukemia cells at the time of therapy.

**Table 3.4**

Protein Synthesis by Arabinofuranosylcytosine-Treated Cells

| Sample | Cell number $\times 10^6$ | Mean cell volume, cubic microns | Total protein per culture, mg | Protein per $10^6$ cells, $\mu g$ |
|---|---|---|---|---|
| Control | 6.47 | 1250 | 1.26 | 195 |
| G1 | 3.07 | 2536 | 1.22 | 397 |
| S | 3.12 | 2536 | 1.18 | 379 |

From Young and Fischer, 1968.

Doering et al. (1966) studied the effects of ara-C on the growth and multiplication of mouse fibroblasts in suspended cell culture. They reported that ara-C induced a considerable enlargement of the cells, although there was a marked inhibition of multiplication. Ara-C markedly inhibited DNA synthesis, but did not affect RNA or protein syntheses. Cultures treated with ara-C rapidly lose their cloning efficiency, an effect defined as cell death. Since ara-C markedly inhibited DNA synthesis in suspended cultures of mouse fibroblasts, Doering et al. (1966) studied the ability of these cells to synthesize DNA following the addition and removal of ara-C. The data obtained showed that a prior inhibition of DNA synthesis with ara-C did not prevent a rapid rate of DNA synthesis once the cells were washed free of this agent. The authors suggested that the ends of the DNA molecules of the inhibited cells did not contain ara-C, which might prevent subsequent addition of deoxyribonucleotides. Of extreme interest was the observation that this DNA synthesis proceeded at high rates in cultures that contained a large number of "dead" cells. This phenomenon was also observed in bacterial cultures that had suffered thymineless death (Barner and Cohen, 1956). It was concluded that the inhibition of DNA synthesis and cell multiplication by ara-C cannot be attributed to irreversible lesions in the DNA that would result from the addition of ara-C to the terminal ends of DNA polymers. However, the findings by Kornberg and Atkinson show conclusively that ara-AMP and ara-CMP are incorporated into the terminal positions of DNA *in vitro* (see p. 150 for more details of these experiments). It is not known if ara-C adds in a similar manner *in vivo*.

Kaplan et al. (1968) reported on the effect of ara-C on the nucleic acid metabolism of normal rabbit kidney (RK) cells. They reported that low concentrations of ara-C inhibited the incorporation of cytidine and thymidine into DNA. However, ara-C caused an increase in incorporation of deoxycytidine into the DNA of RK cells. The level of activity of deoxycytidine kinase increased. There was also a corresponding increase in the intracellular pool of phosphorylated deoxycytidine. Of interest was the observation that although ara-C did not affect the rate of reduction of CDP to its corresponding 2′-deoxyribonucleoside diphosphate in RK cells incubated in a medium free of deoxycytidine, the reduction was somewhat inhibited when deoxycytidine and ara-C were added. The inhibition of DNA synthesis by ara-C was reversed by deoxycytidine. The inhibition of phosphorylation of ara-C by deoxycytidine is shown in Figure 3.15. This decrease in phosphorylation of ara-C with increasing concentrations of deoxycytidine appears to explain the reversal of the drug by deoxycytidine. Apparently deoxycytidine can successfully compete for ara-C at the level of phosphorylation and can therefore stop the accumulation of the nucleotide(s) of ara-C that becomes an effective inhibitor of DNA synthesis (Kaplan et al., 1968). In a subsequent paper, Ben-Porat et al. (1968) studied the effect of ara-C on DNA synthesis in rabbit kidney cells infected with herpes viruses. They reported that ara-C inhibited DNA synthesis in noninfected rabbit kidney cells to a greater degree than in those cells infected with herpes simplex and pseudorabies viruses. The resistance of the viral-infected cells to ara-C was attributed to the low phosphorylation of this nucleoside analog. This decreased phosphorylation was related to a decrease following infection in the level of activity of deoxycytidine kinase. Although ara-C induced noninfected cells to increase the level of activity of deoxycytidine kinase, there was no increase in the level of activity of the kinase in the infected cells. The authors concluded that ara-C is not a specific antiviral agent. Smith (1968) has reviewed the antiviral studies with ara-C.

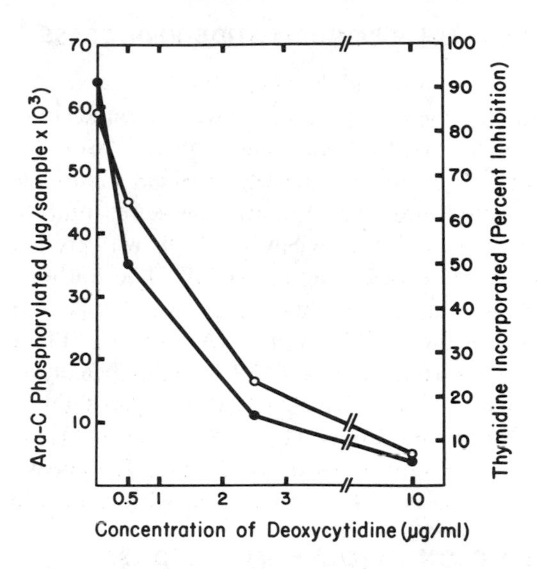


FIGURE 3.15. *Inhibition by deoxycytidine of the phosphorylation of ara-C.* (●) *ara-C phosphorylated;* (○) *thymine incorporated.* (*From Kaplan et al., 1968.*)

The biochemical and pharmacological effects of ara-C in man have been reported by Creasey et al. (1966). The administration of tritium-labeled ara-C to patients with advanced neoplasms showed that ara-C was rapidly deaminated. About 86–96% of the radioactivity excreted in the urine was accounted for as ara-U. Ara-C was reported to inhibit the incorporation of cytidine-$^3$H into DNA by suspensions of human leukemic leukocytes. They reported that there was a small, but significant, incorporation of ara-C into RNA and DNA.

## DEAMINATION OF ARA-C

The deamination of ara-C by cytosine nucleoside deaminase was reported by Pizer and Cohen (1960). Camiener and Smith (1965) have also shown in tissue that ara-C is deaminated by human liver and other animal preparations. Dollinger et al. (1967) and Camiener (1967) have reported on numerous ara-C derivatives that might be less easily deaminated and therefore possibly more effective. Many of the nucleoside analogs were not deaminated and some inhibited the deamination of ara-C, but did not potentiate the action of ara-C either in cell culture (where ara-C is not deaminated) or in the mouse. Smith et al. (1965) and Smith (1968) reported that ara-C was deaminated by PPLO and KB cells. High concentrations of ara-U appear to inhibit the deamination of ara-C.

## EFFECT OF ARA-C ON RIBONUCLEOTIDE REDUCTASE

The suggestion by Chu and Fischer (1962) that ara-C inhibited DNA synthesis by blocking the reduction of cytosine ribonucleotides to 2′-deoxyribonucleotides was studied by Moore and Cohen (1967). Ara-CDP and ara-CTP were weak inhibitors of partially purified ribonucleotide reductase from Novikoff ascites tumor cells, but ara-C was not inhibitory. There was a slight increase in the inhibition when ara-CDP was preincubated with the ribonucleotide reductase before adding CDP. The authors concluded that the inhibition of ribonucleotide reduction by ara-C nucleotides does not play an important role in the inhibition of DNA synthesis. The observation that ara-CDP inhibits the reduction of CDP by ribonucleotide reductase from *E. coli* has been confirmed by Reichard (1967). Ara-CDP is an inhibitor at $10^{-3}$ to $10^{-2}$ *M*. Moore and Cohen (1967) also reported that ara-CDP gave a 50% inhibition at concentrations below 2 m*M*. It appears that the *E. coli* system and the mammalian system are similar with respect to ara-CDP.

## EFFECT OF ARA-C ON CYTOCHROME OXIDASE

Alterations in cytochrome oxidase activity in tissues of normal and leukemic mice that were treated with ara-C were reported by Saslaw et al. (1968). When normal mice were maintained on a toxic level of ara-C, there were marked alternations in cytochrome oxidase activity of spleen and marrow. Splenic cytochrome oxidase activity increased initially and then decreased. Similar phenomena were observed in mice inoculated with leukemia L1210 during treatment with ara-C. Ara-C did not inhibit normal splenic cytochrome oxidase *in vitro*.

## EFFECT OF DINUCLEOSIDE PHOSPHATES OF ARA-C

A series of nucleotides and dinucleoside phosphates of ara-C were synthesized in an attempt to increase the inhibitory activity of this cytotoxic nucleoside (Wechter, 1967). Renis et al. (1967) and Smith et al. (1967) used a DNA virus and KB cells, respectively, for their studies. Smith et al. (1967) showed that simple nucleotides of ara-C were good inhibitors of KB carcinoma cells and mouse L5178Y cells in culture. The 3′ → 5′-dinucleoside phosphates of ara-C were more active than the corresponding 2′ → 5′ analogs. The 5′ → 5′ analogs of ara-C were also active. No conclusions were drawn concerning the penetration and/or the action of the intact dinucleoside phosphates. Deoxycytidine reversed the cytotoxicity of the nucleotides or dinucleoside phosphates of ara-C to KB or L5178Y cells. It is possible that the inhibition by these ara-C nucleotides is due to hydrolysis to ara-C. More recently, Zeleznick (1969) showed that soluble extracts of mouse liver, rabbit liver, and rabbit kidney contain an alkaline phos-

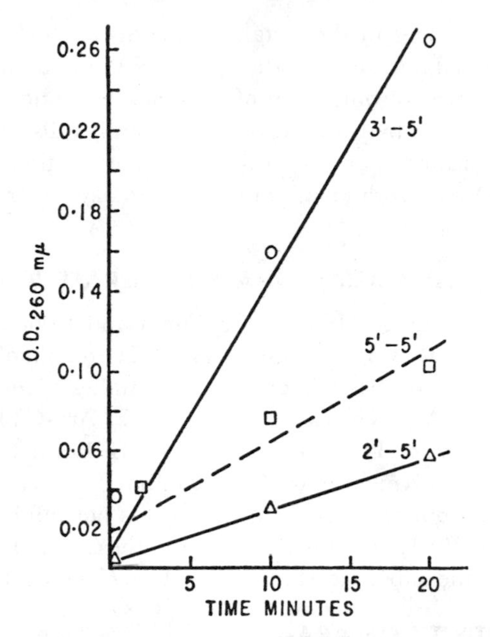


FIGURE 3.16. *Hydrolysis of ara-C containing dinucleoside phosphates, with a 3′-5′, 5′-5′, or 2′-5′ linkage, by rabbit kidney phosphodiesterase.* (*From Zeleznick, 1969.*)

phodiesterase capable of hydrolyzing 2′-5′-linked dinucleoside phosphates of ara-C (Fig. 3.16). The enzymatic hydrolysis of the 2′-5′-linked ara-C dinucleoside occurred at about one-third the rate of hydrolysis of the 3′-5′-ara-C dinucleoside. Nagyvary (private communication) has also shown that $O^2$, 2′-cyclocytidine 3′-phosphate is active against Rauscher virus and leukemia caused by this virus in mice. This activity is assumed to be attributed to the slow hydrolysis of $O^2$,2′-cyclocytidine 3′-phosphate in the blood to ara-C. Similarly, Montgomery et al. (1963) reported that the 5′ → 5′-dinucleoside phosphate bis(thioinosine)-5′,5″-phosphate inhibited mammalian cells in culture that were resistant to 6-mercaptopurine.

## EFFECT OF ARA-C ON HUMAN LEUKEMIC LEUKOCYTES

Creasey et al. (1968) reported their studies with ara-C in human leukemic leukocytes and normal bone marrow cells. Four of the six patients with leukemia, following treatment with ara-C, showed significant response in terms of a reduction of more than 90% peripheral leukemic cell count. The leukocytes of these cells showed a reduced DNA and elevated RNA and protein synthesis. The uptake of cytidine into DNA was depressed while the specific activity of RNA was increased. With normal bone marrow cells, there was an increase in the amount of DNA, RNA, and protein, with fewer consistent changes in the uptake of cytidine into DNA per cell. The unexpected increased uptake of cytidine into RNA after treatment with ara-C *in vivo* was attributed to some form of feedback inhibition. Examination of nucleic acids and their hydrolyzates from leukemic cells exposed to tritium-labeled ara-C indicated that only a limited incorporation of the nucleoside had occurred. Deoxycytidine inhibited the uptake of ara-C by leukemic leukocytes.

### INHIBITION OF MAMMALIAN DNA POLYMERASE BY ARA-CTP

Since York and LePage (1966a) and Furth and Cohen (1967) reported on the inhibition of DNA polymerase by ara-ATP in animal and tumor cells, Furth and Cohen (1968) have extended these studies to include the effect of ara-CTP on the DNA polymerase of animal cells. Ara-CTP did not replace dCTP as a substrate for DNA polymerase or CTP as a substrate for RNA polymerase. RNA polymerase was not inhibited by ara-CTP. Similarly, RNA polymerase from animal and bacteria was not inhibited by ara-ATP. Ara-CTP competitively inhibited DNA synthesis by DNA polymerase, when dCTP was the substrate (Fig. 3.17). The $K_i$ was 1.1 $\mu M$.

### ANTIVIRAL ACTIVITY OF ARA-C

Ara-C inhibits a broad spectrum of DNA viruses in cell culture (Renis and Johnson, 1962; Underwood, 1962; Feldman and Rapp, 1966; Renis et al., 1967). Ara-C inhibits the replication in cell cultures of herpes simplex, pseudorabies, herpes simiae B, vaccinia, swine pox, and fowl pox viruses (DNA viruses) (Buthala, 1964), but it is inactive against RNA viruses in cell culture. In cultures of mammalian cells, low concentrations of ara-C will inhibit the replication of cellular DNA to a much greater degree than that of viral DNA. However, ara-C has very little antiviral activity *in vivo*. It is not effective *in vivo* against vaccinia virus infections (Sidwell et al., 1968b). Kaufman and Maloney (1963) and Underwood (1962) reported that ara-C was active against herpes simplex virus *in vivo*, but Kaufman et al. (1964) reported that ara-C is an extremely toxic antiviral agent when applied topically to the eye and, therefore, has limited usefulness. Although ara-C is deaminated *in vivo*, antiviral activity is still demonstrable (Camiener and Smith, 1965; Creasey et al., 1966; Camiener, 1967). Ara-U has little or no biological activity (Camiener and Smith, 1965).

## 3.7. BIOCHEMICAL PROPERTIES OF ARA-U

Ara-U is inactive as a nucleoside analog against tumor cells (Evans et al., 1961). Detailed studies of ara-U in *E. coli* have been described by Tono and Cohen (1962) and reviewed by Cohen (1966). The metabolic activities of

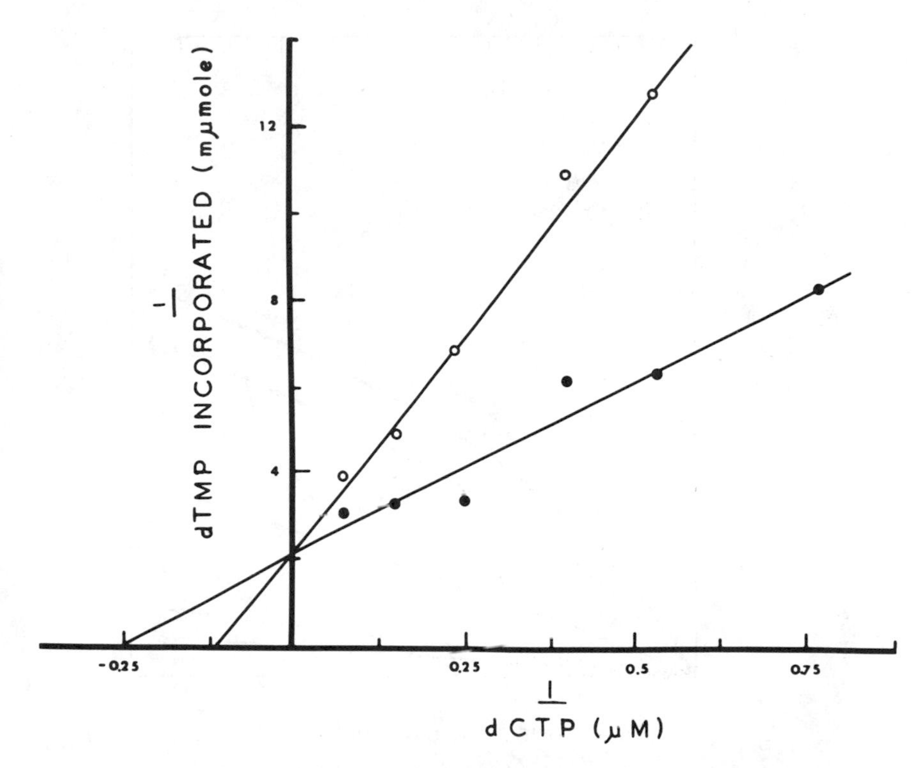


FIGURE 3.17. *Inhibition of DNA polymerase of calf thymus by ara-CTP. The reaction mixture (0.25 ml) contained 50*mM *phosphate buffer pH 7.0; 4* mM $MgCl_2$*; 2* mM *2-mercaptoethanol; 40* μM *each of dATP and dGTP; 36* μM *dTTP-*$^{14}C$ (1.10 × $10^6$ *cpm/μmole); heated calf thymus DNA; 24 mμmoles of deoxynucleotide; 1.25* μM *ara-CTP; and varying concentrations of dCTP. The reaction was initiated by the addition of enzyme and terminated after 60 min at 37°C.* (●) *No ara-CTP;* (○) *plus ara-CTP.* (*From Furth and Cohen 1968.*)

ara-U have been studied by Pizer and Cohen (1960), who found that uracil auxotrophs of two strains of *E. coli* could have their uracil requirements satisfied by either ara-U or ara-C. Ara-U was not hydrolyzed to free uracil by nucleoside phosphorylase. Phosphorylated ara-U was prepared enzymatically by the phosphotransferases present in wheat germ or carrot extracts. Ara-U is inactive as an inhibitor of growth of leukemic cells (Chu and Fischer, 1962). Ara-U is phosphorylated less rapidly than is ara-C (Schrecker and Urshel, 1967).

Nagyvary et al. (1968) used polyarabinouridylic acid to determine if this polyribonucleotide could serve as a primordial messenger or inhibitor in prebiological systems. They used the method of Nagyvary (1967a, 1967b) for the synthesis of oligonucleotides of arabinose and ribose. This novel reaction

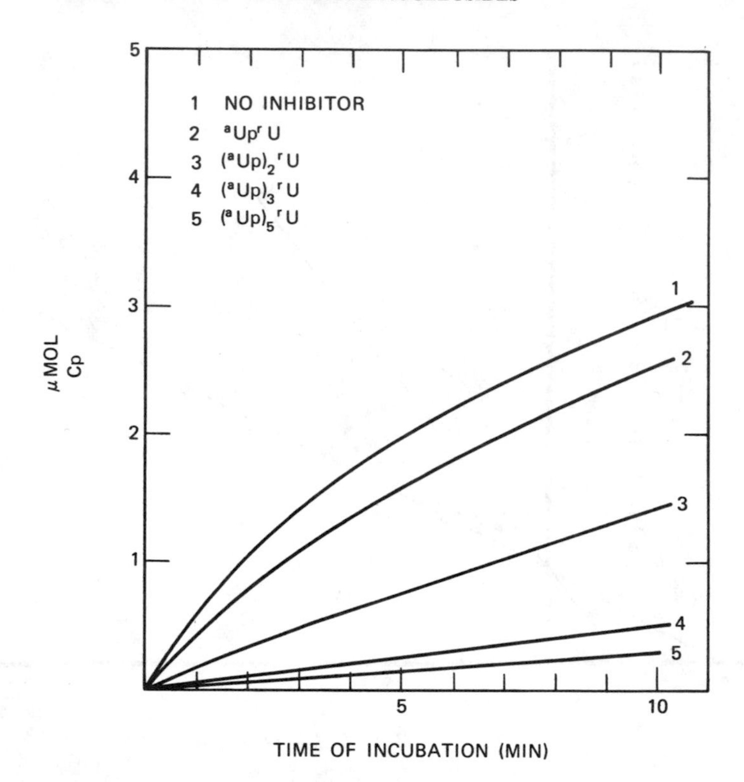


FIGURE 3.18. *Chain length dependent inhibition of pancreatic RNase. (Line 1) no inhibitor; (line 2) ($^{a}Up^{r}U$); (line 3) $(^{a}Up)_2{}^{r}U$; (line 4) $(^{a}Up)_3{}^{r}U$; (line 5) $(^{a}Up)_5{}^{r}U$. $^{a}U$ represents 1-β-D-arabinofuranosyluracil and $^{r}U$ represents uridine; Cp, 3′-cytidylic acid. (From Nagyvary et al., 1968.)*

is based on the formation of (2′:3′) → 5′-cyclic triesters from ribopolymers of the pyrimidine bases and subsequent treatment with a tertiary base which catalyzed the rearrangement of this system to $O^2$,2′-cyclouridine derivatives. Hydrolysis of this product yielded oligoarabino-nucleotides of small chain length. Some ribouridine units were at the 3′ end.

The effect of oligoarabino-(3′-ribo)-uridylates on the inhibition of pancreatic RNase is shown in Figure 3.18. This inhibition of RNase was shown to be competitive. While Schramm and Ulmer-Schürnbrand (1967) reported that their polyspongouridylates (polyarabinonucleotides) were bound to phenylalanyl-tRNA of washed *E. coli* ribosomes, Nagyvary et al. (1968) reported that their polyarabinouridylic acid did not function as a messenger under the conditions where polyribouridylic acid formed polyphenylalanine synthesis. Mixing of polyarabinouridylic acid with poly-A did not cause any hyperchromicity. These results excluded the formation of a double-stranded structure. Nagyvary et al. (1968) showed that parallel stacking of bases within polyarabinonucleotides is sterically impossible and concluded that the prebiotic evolutionary role of polyarabinonucleotides was that of an inhibitor rather than that of a primeval template. Schramm (unpublished results) supports this contention and indicates a wide range of inhibitory activities by poly ara-U.

## Summary

The isolation, chemical properties, chemical synthesis, and inhibition of the four following naturally occurring nucleoside analogs have been described: spongosine, ara-A, ara-U, and ara-T. One of these, ara-A, is a very active biological antagonist of mammalian cells in culture, tumor cells, bacterial cells, and viruses. Ara-C, the synthetic, abnormal nucleoside has antitumor and antiviral activity and is cytotoxic for mammalian cells in culture at concentrations of 0.1 $\mu$g/ml or greater. The inhibition of mammalian cells or of DNA viruses by ara-C is reversed by deoxycytidine. Ara-C is rapidly deaminated to ara-U.

Ara-A is cytocidal; this antibiotic inhibits DNA synthesis in animal cells in culture, bacteria, and viruses. Protein synthesis is not inhibited in animal cells, although cell division stops and chromosome breaks occur. Although ara-A inhibits DNA synthesis in animal and bacterial cells, this inhibition is not irreversible, even in cells incapable of undergoing cell division. Ara-ATP and ara-CTP are incorporated into DNA *in vitro* as terminal residues by partially purified DNA polymerase from *E. coli*. The incorporation of ara-AMP and ara-CMP into the terminal positions of DNA has not been demonstrated *in vivo*. Ara-A is incorporated into the 3′-terminal position of tRNA. Polynucleotide phosphorylase does not incorporate ara-ADP or ara-CDP into polynucleotides. Ara-ATP and ara-CTP inhibit the activity of animal DNA polymerase, but not the activity of the DNA polymerase present in *E. coli*. The inhibition of multiplication of *E. coli* by ara-ATP might be explained by the inhibition of ribonucleotide reductase. The mechanism of inhibition of DNA synthesis by ara-C and ara-A is still unknown. Dinucleoside phosphates of ara-C are hydrolyzed by an alkaline phosphodiesterase to ara-C.

Ara-A is a very effective antiviral agent for herpes simplex keratitis infections in hamsters and vaccinia virus infections in Swiss mice. Ara-A is active against lethal herpes encephalitis in mice following intracerebral inoculation of the virus. The drug can be administered by the oral, parenteral, or percutaneous routes. It also inactivates DNA viruses already replicating in the central nervous system of animals. These findings, plus the low toxicity of ara-A, suggest that this nucleoside may be extremely effective against diseases caused by DNA viruses.

Ara-U is not very toxic to tumor cells. The glycosyl linkage of ara-U is cleaved by uridine phosphorylase of *E. coli*. Uracil and arabinose-1-phosphate are the products of hydrolysis. Ara-U was not incorporated into the nucleic acids and very little phosphorylation of this nucleoside has been demonstrated.

## References

Atkinson, M. R., M. P. Deutscher, A. Kornberg, A. F. Russell and J. G. Moffatt, *Biochemistry*, **8**, 4897 (1969).

Barker, R., and H. G. Fletcher, *J. Org. Chem.*, **26**, 4605 (1961).

Barner, H. D., and S. S. Cohen, *J. Bacteriol.*, **72**, 115 (1956).

Ben-Porat, T., M. Brown, and A. S. Kaplan, *Mol. Pharmacol.*, **4**, 139 (1968).

Bergmann, W., and D. C. Burke, *J. Org. Chem.*, **20**, 1501 (1955).

Bergmann, W., and D. C. Burke, *J. Org. Chem.* **21**, 226 (1956).

Bergmann, W., and R. J. Feeney, *J. Amer. Chem. Soc.*, **72**, 2809 (1950).

Bergmann, W., and R. J. Feeney, *J. Org. Chem.*, **16**, 981 (1951).

Bergmann, W., and M. F. Stempien, Jr., *J. Org. Chem.*, **22**, 1575 (1957).

Bergmann, W., J. C. Watkins, and M. F. Stempien, *J. Org. Chem.*, **22**, 1308 (1957).

Bloch, A., and C. A. Nichol, *Antimicrobial Agents Chemotherapy*, 1964, 530.

Bloch, A., M. J. Robins, and J. R. McCarthy, Jr., *J. Med. Chem.*, **10**, 908 (1967).

Brink, J. J., and G. A. LePage, *Cancer Res.*, **24**, 312 (1964a).

Brink, J. J., and G. A. LePage, *Cancer Res.*, **24**, 1042 (1964b).

Brink, J. J., and G. A. LePage, *Can. J. Biochem.*, **43**, 1 (1965).

Brown, D. M., A. Todd, and S. Varadarajan, *J. Chem. Soc.*, 1956, 2388.

Buthala, D. A., *Proc. Soc. Exptl. Biol. Med.*, **115**, 69 (1964).

Burke, D. C., *J. Org. Chem.*, **20**, 643 (1955).

Camiener, G. W., *Biochem. Pharmacol.*, **16**, 1691 (1967).

Camiener, G. W., and C. G. Smith, *Biochem. Pharmacol.*, **14**, 1405 (1965).

Cardeilhac, P. T., and S. S. Cohen, *Cancer Res.*, **24**, 1595 (1964).

Chu, M. Y., and G. A. Fischer, *Biochem. Pharmacol.*, **11**, 423 (1962).

Cohen, S. S., in Progress in Nucleic Acid Research and Molecular Biology, **5**, J. N. Davidson and W. E. Cohn, Eds., New York, 1966, p. 1.

Cory, J. G., and R. J. Suhadolnik, *Biochemistry*, **4**, 1729 (1965).

Creasey, W. A., R. J. Papac, M. E. Markiw, P. Calabresi, and A. D. Welch, *Biochem. Pharmacol.*, **15**, 1417 (1966).

Creasey, W. A., R. C. DeConti, and S. R. Kaplan, *Cancer Res.*, **28**, 1074 (1968).

Cushley, R. J., K. A. Watanabe, and J. J. Fox, *J. Amer. Chem. Soc.*, **89**, 394 (1967).

Dixon, G. J., R. W. Sidwell, F. A. Miller, and B. J. Sloan, *Antimicrobial Agents Chemotherapy*, 1968, 172.

Deutscher, M. P., and A. Kornberg, *J. Biol. Chem.*, **244**, 3019, 3029 (1969).

Doering, A., J. Keller, and S. S. Cohen, *Cancer Res.*, **26**, 2444 (1966).

Doering-McGovern, A., M. Jansen, and S. S. Cohen, *J. Bacteriol.*, **92**, 565 (1966).
Doerr, I. L., J. F. Codington, and J. J. Fox, *J. Med. Chem.*, **10**, 247 (1967).
Doerr, I. L., J. F. Codington, and J. J. Fox, *J. Org. Chem.*, **30**, 467 (1965).
Dollinger, M. R., J. H. Burchenal, W. Kreis, and J. J. Fox, *Biochem. Pharmacol.*, **16**, 689 (1967).
Emerson, T. R., R. J. Swan, and T. L. V. Ulbricht, *Biochemistry*, **6**, 843 (1967).
Englund, P. T., J. A. Huberman, T. M. Jovin, and A. Kornberg, *J. Biol. Chem.*, **244**, 3038 (1969).
Evans, J. S., E. A. Musser, G. D. Mengel, K. R. Forsblad, and J. H. Hunter, *Proc. Soc. Exptl. Biol. Med.*, **106**, 350 (1961).
Feldman, L. A., and F. Rapp, *Proc. Soc. Exptl. Biol. Med.*, **122**, 243 (1966).
Fox, J. J., and I. Wempen, *Tetrahedron Letters*, 1965, 643.
Fox, J. J., N. Yung, J. Davoll, and G. B. Brown, *J. Amer. Chem. Soc.*, **78**, 2117 (1956).
Fox, J. J., N. Yung, and A. Bendich, *J. Amer. Chem. Soc.*, **79**, 2775 (1957).
Frederiksen, S., *Biochim. Biophys. Acta*, **76**, 366 (1963).
Frederiksen, S., *Arch. Biochem. Biophys.*, **113**, 383 (1966).
Furth, J. J., and S. S. Cohen, *Cancer Res.*, **27**, 1528 (1967).
Furth, J. J., and S. S. Cohen, *Cancer Res.*, **28**, 2061 (1968).
Glaudemans, C. P. J., and H. G. Fletcher, Jr., *J. Org. Chem.*, **29**, 3286 (1964).
Glaudemans, C. P. J., and H. G. Fletcher, Jr., *J. Org. Chem.*, **28**, 3004 (1963).
Hart, P. A., and J. P. Davis, *J. Amer. Chem. Soc.*, **91**, 512 (1969a).
Hart, P. A., and J. P. Davis, *Biochem. Biophys. Res. Commun.*, **34**, 733 (1969b).
Hubert-Habart, M., and S. S. Cohen, *Biochim. Biophys. Acta*, **59**, 468 (1962).
Kaplan, A. S., M. Brown, and T. Ben-Porat, *Mol. Pharmacol.*, **4**, 131 (1968).
Kaufman, H. E., and E. D. Maloney, *Arch. Ophthalmol.*, **69**, 626 (1963).
Kaufman, H. E., J. A. Capella, E. D. Maloney, J. E. Robbins, G. M. Cooper, and M. H. Uotila, *Arch. Ophthalmol.*, **72**, 535 (1964).
Keller, F., and A. R. Tyrrill, *J. Org. Chem.*, **31**, 1289 (1966).
Kihlman, B. A., and G. Odmark, *Hereditas*, **56**, 71 (1966).
Kim, J. H., and M. L. Eidinoff, *Cancer Res.*, **25**, 698 (1965).
Klee, A. and S. H. Mudd, *Biochemistry*, **6**, 988 (1967).
Kornberg, A., *Science*, **163**, 1410 (1969).
Koyama, G., K. Maeda, H. Umezawa, and Y. Iitaka, *Tetrahedron Letters*, 1966, 597.
Kurtz, S. M., R. A. Fisken, D. H. Kaump, and J. L. Schardein, *Antimicrobial Agents Chemotherapy*, 180 (1968).
Lee, W. W., A. Benitez, L. Goodman, and B. R. Baker, *J. Amer. Chem. Soc.*, **82**, 2648 (1960).
LePage, G. A., *Can. J. Biochem.*, **48**, 75 (1970).
LePage, G. A., and I. G. Junga, *Cancer Res.*, **23**, 739 (1963).
LePage, G. A., and I. G. Junga, *Cancer Res.*, **25**, 46 (1965).
Lepine, P., J. de Rudder, and P. de Garilhe, French Pat. M 3585; *Chem. Abstr.*, **64**, 11303a (1966).
Leung, H. B., A. M. Doering, and S. S. Cohen, *J. Bacteriol.*, **92**, 558 (1966).
Lindberg, B., H. Klenow, and K. Hansen, *J. Biol. Chem.*, **242**, 350 (1967).
Lucas-Lenard, J. M., and S. S. Cohen, *Biochim. Biophys. Acta*, **123**, 471 (1966).

Miles, D. W., M. J. Robins, R. K. Robins, M. W. Winkley, H. Eyring, *J. Amer. Chem. Soc.*, **91**, 824 (1969a); 831 (1969b).

Miller, F. A., G. J. Dixon, J. Ehrlich, B. J. Sloan, and I. W. McLean, Jr., *Antimicrobial Agents Chemotherapy*, 1968, 136.

Momparler, R. L , *Biochem. Biophys. Res. Commun.*, **34**, 465 (1969).

Montgomery, J. A., G. J. Dixon, E. A. Dulmage, H. J. Thomas, R. W. Brockman, and H. E. Skipper, *Nature*, **199**, 769 (1963).

Moore, E. C., and S. S. Cohen, *J. Biol. Chem.*, **242**, 2116 (1967).

Nagyvary, J., *Federation Proc.*, **26**, 292 (1967a).

Nagyvary, J., 21st Congr. of IUPAC, Prague, 1967b, Abstract N-48.

Nagyvary, J., *J. Am. Chem. Soc.*, **91**, 5409 (1969).

Nagyvary, J., and R. G. Provenzale, *Biochemistry*, **8**, 4769, (1969).

Nagyvary, J., and C. M., Tapiero, *Tetrahedron Letters*, No. 40, 3481, (1969).

Nagyvary, J., R. Provenzale, and J. M. Clark, Jr., *Biochem. Biophys. Res. Commun.*, **31**, 508 (1968).

Naito, T., M. Hirata, Y. Nakai, *Chem. Abstr.*, **69**, 107013h (1968a).

Naito, T., M. Hirata, Y. Nakai, *Chem. Abstr.*, **69**, 107014j (1968b).

Nichols, W. W., *Cancer Res.*, **24**, 1502 (1964).

Nishimura, T., and B. Shimizu, *Chem. Pharm. Bull. (Tokyo)*, **13**, 803 (1965).

Otter, B. A., E. A. Falco, and J. J. Fox, *J. Org. Chem.*, **33**, 3593 (1968).

Parke, Davis and Company, Belgium Pat. 671,557 (1967).

Pierre, K. J., A. P. Kimball, and G. A. LePage, *Can. J. Biochem.*, **45**, 1619 (1967).

Pizer, L. I., and S. S. Cohen, *J. Biol. Chem.*, **235**, 2387 (1960).

Pollman, W., and G. Schramm, *Biochim. Biophys. Acta*, **80**, 1 (1964).

Provenzale, R. G., and J. Nagyvary, *Biochemistry*, **9**, 1744 (1970).

Rao, R. N., and A. T. Natarajan, *Cancer Res.*, **25**, 1764 (1965).

Reichard, P., *The Biosynthesis of Deoxyribose*, Wiley, New York, 1967.

Reist, E. J., A. Benitez, L. Goodman, B. R. Baker, and W. W. Lee, *J. Org. Chem.*, **27**, 3274 (1962).

Reist, E. J., J. H. Osiecki, L. Goodman, and B. R. Baker, *J. Amer. Chem. Soc.*, **83**, 2208 (1961).

Reist, E. J., V. J. Bartuska, and L. Goodman, *J. Org. Chem.*, **29**, 3725 (1964).

Reist, E. J., D. F. Calkins, and L. Goodman, *J. Med. Chem.*, **10**, 130 (1967).

Reist, E. J., D. F. Calkins, L. V. Fisher, and L. Goodman, *J. Org. Chem.*, **33**, 1600 (1968).

Renis, H. E., C. A. Hollowell, and G. E. Underwood, *J. Med. Chem.*, **10**, 777 (1967).

Renis, H. E., and H. G. Johnson, *Bacteriol. Proc.*, *Microbiology*, 1962, 140.

Rockwell, M., and M. H. Maguire, *Mol. Pharmacol.*, **2**, 574 (1966).

Saslaw, L. D., N. A. Chaney, and V. S. Waravdekar, *Toxicol. Appl. Pharmacol.*, **12**, 455 (1968).

Schabel, F. M., Jr., *Chemotherapy*, **13**, 321 (1968).

Schabel, F. M., Jr., *Ann. Symp. Fundamental Cancer Res.*, **21**, 379 (1967).

Schabel, F. M., Jr., H. E. Skipper, M. W. Trader, and W. S. Wilcox, *Cancer Chemotherapy Rept.*, **48**, 17 (1965).

Schaeffer, H. J., and H. J. Thomas, *J. Amer. Chem. Soc.*, **80**, 3738 (1958).

Schardein, J. L., and R. W. Sidwell, *Antimicrobial Agents Chemotherapy*, 1968, 158.

Schnebli, H. P., D. L. Hill, and L. L. Bennett, Jr., *J. Biol. Chem.*, **242**, 1997 (1967).
Schramm, G., and I. Ulmer-Schürnbrand, *Biochim. Biophys. Acta*, **145**, 7 (1967).
Schramm, G., H. Grotsch, and W. Pollmann, *Angew. Chem. Intern. Ed. Engl.*, **1**, 1 (1962).
Schrecker, A. W., and M. J. Urshel, *Proc. Amer. Assoc. Cancer Res.*, **8**, 58 (1967).
Sidwell, R. W., G. J. Dixon, F. M. Schabel, Jr., and D. H. Kaump, *Antimicrobial Agents Chemotherapy*, 1968a, 148.
Sidwell, R. W., G. J. Dixon, S. M. Sellers, and F. M. Schabel, *Appl. Microbiol.*, **16**, 370 (1968b).
Sidwell, R. W., S. M. Sellers, and G. J. Dixon, *Antimicrobial Agents Chemotherapy*, 1966, 483.
Skipper, H. E., F. M. Schabel, Jr., and W. S. Wilcox, *Cancer Chemotherapy Rept*, **51**, 125 (1967).
Sloan, B. J., F. A. Miller, J. Ehrlich, I. W. McLean, Jr., and H. E. Machamer, *Antimicrobial Agents Chemotherapy*, 1968, 161.
Smith, C. G., *3rd Intern. Pharmacol. Mtg.*, **5**, 33 (1968).
Smith, C. G., H. H. Buskirk, and W. L. Lummis, *Proc. Amer. Assoc. Cancer Res.*, **6**, 60 (1965).
Smith, C. G., H. H. Buskirk, W. L. Lummis, *J. Med. Chem.*, **10**, 774 (1967).
Stevens, M. A., D. I. Magrath, H. W. Smith, and G. B. Brown, *J. Amer. Chem. Soc.*, **80**, 2755 (1958).
Toji, L., and S. S. Cohen, *Proc. Natl. Acad. Sci.* (*U. S.*) **63**, 871 (1969).
Tono, H., and S. S. Cohen, *J. Biol. Chem.*, **237**, 1271 (1962).
Ts'o, P. O. P., N. S. Kordo, M. P. Schweizer, and D. P. Hollis, *Biochemistry*, **8**, 997 (1969).
Upjohn Co., French Pat. 1,396,003 (1965), *Chem. Abstr.* **63**, 13392a (1965).
Underwood, G. E., *Proc. Soc. Exptl. Biol. Med.*, **111**, 660 (1962).
Walwick, E. R., W. K. Roberts, and C. A. Dekker, *Proc. Chem. Soc.*, 1959, 84.
Ward, D. C., and E. Reich, *Proc. Natl. Acad. Sci.* (*U. S.*), **61**, 1494 (1968).
Wechter, W. J., *J. Med. Chem.*, **10**, 762 (1967).
York, J. L , and G A. LePage, *Can. J. Biochem.*, **44**, 19 (1966a).
York, J. L., and G. A. LePage, *Can. J. Biochem.* **44**, 331 (1966b).
Young, R. S. K., and G. A. Fischer, *Biochem. Biophys. Res. Commun.*, **32**, 23 (1968).
Zelenzick, L. D., *Biochem., Pharmacol.*, **18**, 855 (1969).

CHAPTER 4

# 4-Aminohexose Pyrimidine Nucleosides

### 4.4. BAMICETIN 211



### 4.5. PLICACETIN 214



### 4.6. AMICETIN A AND C 216



---

There are five naturally occurring pyrimidine nucleoside antibiotics in which a 4-aminohexose is linked to nitrogen 1 of a pyrimidine or which contain a 4-aminohexose linked through a disaccharide. They are: amicetin (allomycin or sacromycin), bamicetin, blasticidin S, gougerotin, and plicacetin (amicetin B) (Fig. 4.1). Four of these nucleosides have (*1*) the cytosine chromophore, (*2*) a 4-amino sugar, (*3*) at least one amino acid, and (*4*) a $\beta$ glycosyl linkage. These nucleoside antibiotics are all elaborated by the *Streptomyces*. Fox et al. (1966) and Korzybski et al. (1967) have recently reviewed amicetin, bamicetin, blasticidin S, gougerotin, and plicacetin.

The total chemical synthesis of this group of nucleosides has not yet been reported. Fox et al. (1968), Watanabe et al. (1969), and Watanabe et al., (1970) have reported on the synthesis of a series of derivatives of the nucleoside portion of gougerotin, including a uronic acid containing nucleoside (C substance) readily derived from gougerotin. Thus a partial synthesis of gougerotin has been achieved. The syntheses of the 3′-aminohexose nucleosides, as part of a program related to the synthesis of analogs of gougerotin and blasticidin S, have also been reported (Watanabe and Fox, 1964, 1966; Friedman et al., 1967). The compounds synthesized were 1-(3′-amino-3′-deoxy-$\beta$-D-mannopyranosyl)uracil; the 1-(3′-amino-3′-deoxy-$\beta$-D-glucopyranosyl) pyrimidines; and 1-($\beta$-deoxy-3-glycylamido-$\beta$-D-glucopyranosyl)-uracil and the 3′-sarcosylamido, L-alanylamido and D-phenylalanylamido analogs.

The two additional compounds that have been isolated by DeBoer and Hinman (1959a, 1959b) which appear to be nucleoside antibiotics are amicetin A and amicetin C. Their structures have not been elucidated.

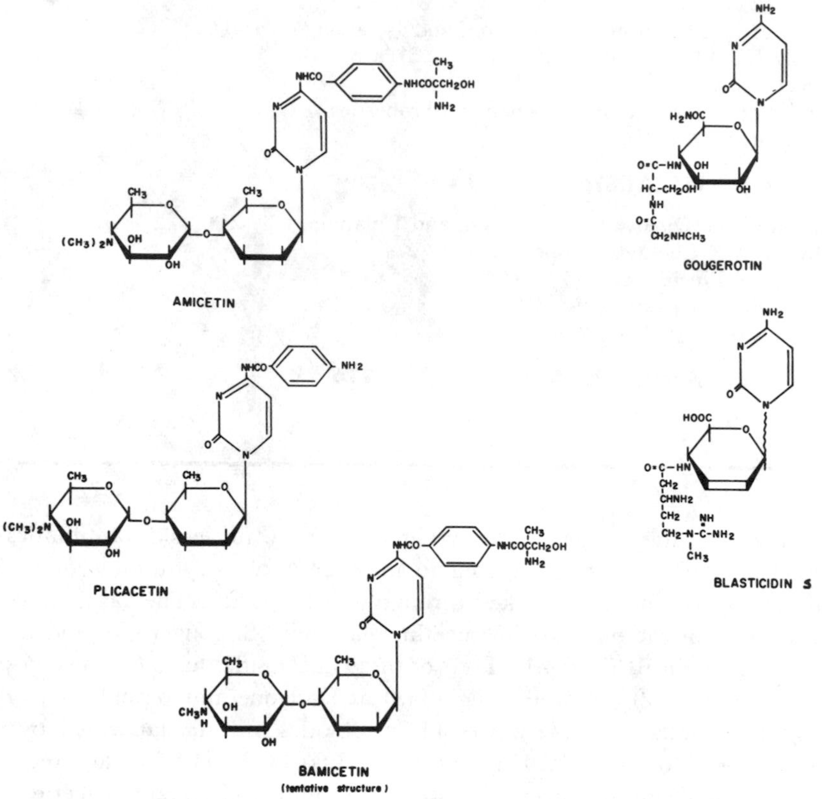

AMICETIN

GOUGEROTIN

PLICACETIN

BLASTICIDIN S

BAMICETIN
(tentative structure)

FIGURE 4.1 *Structures of pyrimidine nucleoside antibiotics.*

# 4.1. GOUGEROTIN

## INTRODUCTION

Gougerotin is a dipeptidyl pyrimidine nucleoside antibiotic isolated from *Streptomyces gougerotii*. The first structure for gougerotin reported by Iwasaki (1962) as a 1-cytosinyl-3′-amino-3′-deoxy-D-hexose derivative (Fig. 4.2*A*) was corrected by Fox et al. (1964) to a dipeptide cytosinyl nucleoside with a 4-amino-4-deoxyhexuronic acid amide of the galactopyranosyl configuration (Fig. 4.2*B*). This structure was most recently revised by Fox et al. (1968) (Fig. 4.2*C*). The correct structure assigned to gougerotin is 1-(cytosinyl)-4-sarcosyl-D-serylamino-1,4-dideoxy-β-D-glucopyranuronamide (Fig. 4.1 or Fig. 4.2*C*) (Fox et al., 1968). Gougerotin is a broad spectrum antibiotic and an inhibitor of protein synthesis.

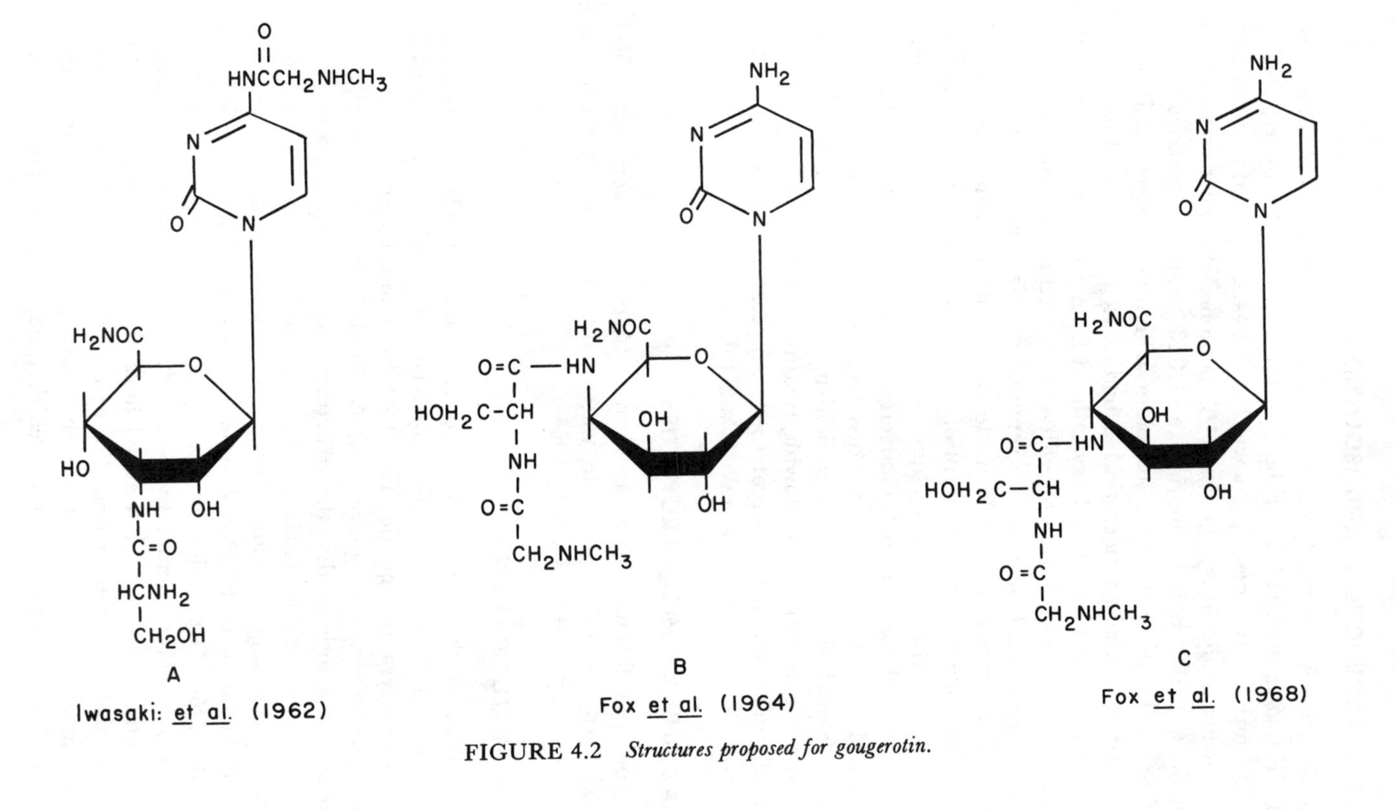


FIGURE 4.2 *Structures proposed for gougerotin.*

## DISCOVERY, PRODUCTION, AND ISOLATION

Gougerotin was first isolated from *S. gougerotii* from a soil sample in Kyoto, Japan by Kanzaki et al. (1962) at the Takeda Chemical Industries, Osaka, Japan. The initial culture description was No. 21544. Gougerotin is produced in the following medium: 3% glucose, 1% peptone, 1% meat extract, 1% soybean flour, 0.3% sodium chloride, and 0.3% calcium carbonate; pH 7.0, before sterilization. Since the antibiotic activity of gougerotin is weak and it is toxic to animals, gougerotin has had limited commercial use.

Gougerotin can be isolated and crystallized from the fermentation broth by the following method. The culture filtrate is added to a column of a weak cation-exchange resin. Gougerotin is retained on the resin. The column is washed with water and the crude nucleoside antibiotic eluted with acid. Following neutralization and adsorption on carbon, gougerotin is eluted with 80% acetone (pH 1.6). The eluant is neutralized with a weak anion-exchange resin and concentrated. Gougerotin is crystallized by the addition of acetone. The impure gougerotin is dissolved in water, added to an $Al_2O_3$ column, and eluted with water. The ninhydrin positive material (gougerotin) is collected and concentrated. Following treatment with carbon, gougerotin crystallized from methanol–water at room temperature and was recrystallized from methanol–water (Kanzaki et al., 1962).

## PHYSICAL AND CHEMICAL PROPERTIES

The molecular formula for gougerotin is $C_{16}H_{25}N_7O_8$; mp 200–215°C (decomp.); $[\alpha]_D^{21}$ +45° ($c$ = 1% in water); the ultraviolet spectrum resembles that of cytidine (Fox et al., 1964).

## STRUCTURAL ELUCIDATION

On the basis of the degradative studies, Iwasaki (1962) proposed the structure of gougerotin as 1-(*N*-sarcosyl-1-cytosinyl)-3-D-serylamino-1,3-dideoxy-β-D-allopyranuronamide (Fig. 4.2*A*). Fox and his co-workers reinvestigated the structure of gougerotin. A detailed chemical investigation demonstrated conclusively that the structure of gougerotin as proposed by Iwasaki was incorrect (Fox et al., 1964). The sarcosyl moiety was shown to be attached to the sugar portion of gougerotin and not to the amino group of cytosine. Isolation of DNP-sarcosylserine established a dipeptide bond (Fox et al., 1964). Reduction of the pyrimidine ring followed by acid hydrolysis permitted the isolation by Iwasaki (1962) of two crystalline aminohexose derivatives, namely, an *N*-acetyl hexosamine and a methyl glycosaminide tetraacetate. A comparison (Fox et al., 1964) of the physical properties of these crystalline derivatives with the known corresponding derivatives of the 2- and 3-aminohexose series showed that the amino group of the sugar moiety of gougerotin was not at the 2 or 3 position. From nmr and periodate oxidation studies on degradation products, they concluded that the sugar moiety of gougerotin was of the 4-amino-4-deoxy-β-D-*gluco*- or *galacto*-pyranosyl configuration. Since the physical properties of the methyl hexosaminide tetraacetate given by Iwasaki (1962) differed from those for the known anomers of methyl 4-amino-4-deoxy-D-glycopyranoside tetraacetate, Fox et al. (1964) assigned the *galacto*-configuration (Fig. 4.2*B*) to gougerotin. Periodate oxidation studies showed that the dipeptide sacrosyl-D-serine was linked to the 4-amino group of the sugar moiety by an amide linkage.

Subsequent degradative studies (Fox et al., 1968) showed that the *crystalline* methyl hexosaminide tetraacetate reported by Iwasaki was, in fact, a eutectic mixture of an anomeric pair of methyl 4-acetamido-2,3,6-tri-*O*-acetyl-4-deoxy-D-glucosides. On this basis Fox revised the structure of gougerotin to 1-(cytosinyl)-4-sarcosyl-D-serylamino-1,4-dideoxy-β-D-glucopyranuronamide (Fig. 4.2*C*). They concluded, further, that the nucleoside (C-substance) obtained by Iwasaki by acid hydrolysis of gougerotin in 6 *N* HCl under reflux is 1-(4-amino-4-deoxy-β-D-glucopyranosyluronic acid)cytosine.

Further confirmation of the *gluco* configuration of the sugar moiety of gougerotin and of the structure of C-substance were obtained from the following chemical syntheses:

*1.* The synthesis of 1-(4-acetamido-4-deoxy-β-D-glucopyranosyl)cytosine (**9**) and its polyacetylated derivative (**8**) and their identity with compounds obtained by reduction of the COOH function of C-substance derived from gougerotin (Watanabe et al., 1969, 1970).

*2.* The total synthesis of the 4-amino-uronic acid nucleoside 1-(4-amino-4-deoxy-β-D-glucopyranosyluronic acid)cytosine and its identity with C-substance (Fig. 4.3, Watanabe et al., 1970).

*3.* The synthesis (Fig. 4.4) of methyl (methyl 4-acetamido-4-deoxy-2,3-di-*O*-acetyl-α-D-glucopyranosid)uronate (**31**) (Kotick et al., 1969), a derivative of the sugar moiety of gougerotin.

The syntheses of the acetylated, C-6′-reduced C-substance (**9**) and C-substance are shown in Figure 4.3. The first study involved the chemical synthesis of the nucleosides, 1-(4-acetamido-2,3,6-tri-*O*-acetyl-4-deoxy-β-D-glucopyranosyl)-$N^4$-acetylcytosine (**8**) and 1-(4-acetamido-4-deoxy-β-D-glucopyranosyl)cytosine (**9**) from methyl 2,3,6-tri-*O*-benzoyl-4-*O*-mesyl-α-D-galactoside (**1**) (Fig. 4.3) (Watanabe et al., 1970). Gougerotin was converted to this nucleoside (**9**) and then peracetylated to **8**. The nucleosides obtained from gougerotin were identical with compounds **8** and **9** synthesized from D-galactose.

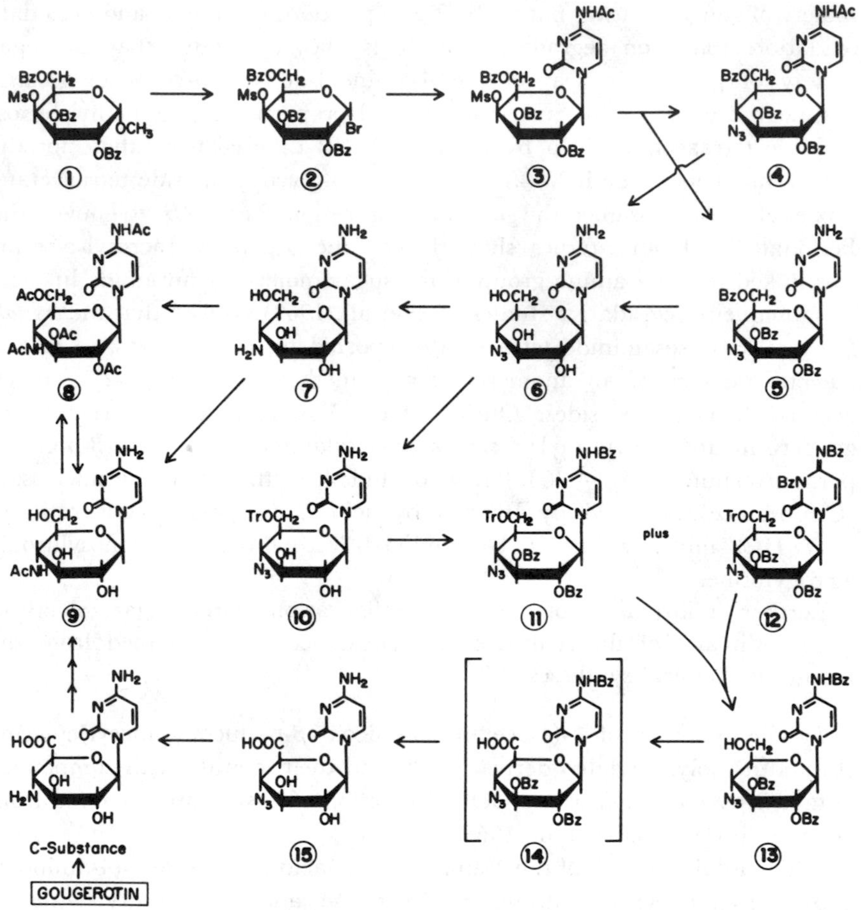

FIGURE 4.3 *The total syntheses of* 1-(4-*amino*-4-*deoxy*-β-D-*glucopyranosyluronic acid*) *cytosine, or C-substance, and the acetylated C-*6′-*reduced C-substance nucleosides,* 8 *and* 9. (*From Watanabe et al.*, 1970.)

The total chemical synthesis of C-substance (the nucleoside from gougerotin) is also shown in Figure 4.3 (Watanabe, et al. 1970). The 4-azido nucleoside (**6**) was converted to 1-(4-azido-2,3-di-*O*-benzoyl-4-deoxy-β-D-glucopyranosyl)-$N^4$-benzoylcytosine (**13**) by tritylation, benzoylation, and detritylation. Compound **13** was oxidized and debenzoylated to 1-(4-azido-4-deoxy-β-D-glucopyranosyluronic acid)cytosine (**15**). Reduction of **15** gave 1-(4-amino-4-deoxy-β-D-glucopyranosyluronic acid)cytosine, which is identical with C-substance obtained by acid hydrolysis of gougerotin.

The first total chemical synthesis of methyl(methyl-4-acetamido-4-deoxy-2,3-di-*O*-acetyl-α-D-glucopyranosid)uronate has now been achieved (Fig.

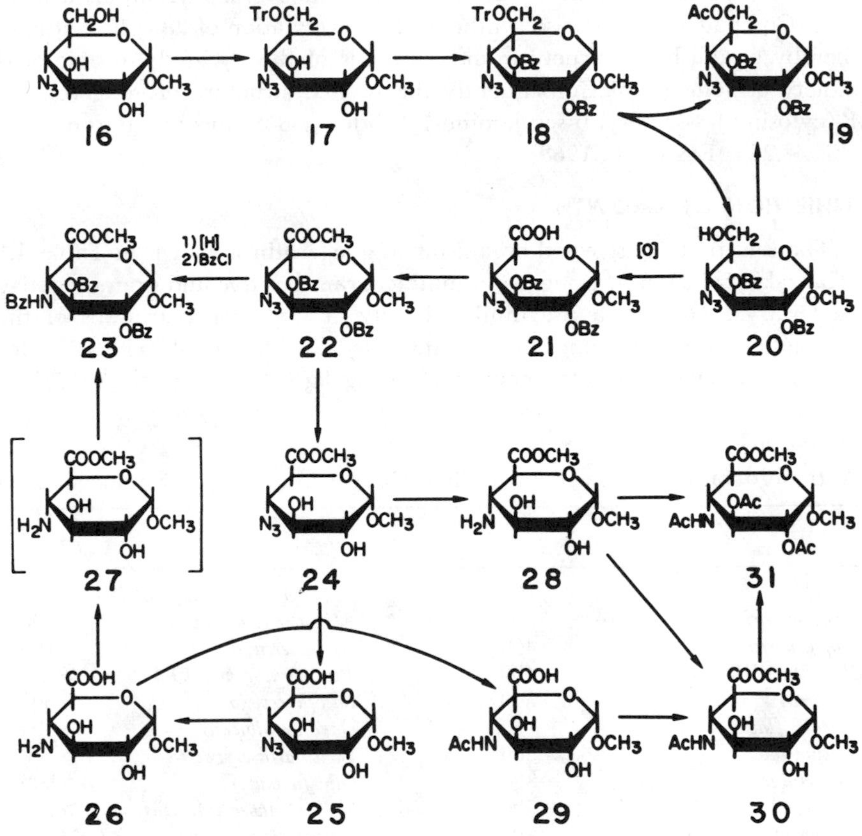

Figure 4.4 *Synthesis of methyl (methyl 4-acetamido-4-deoxy-2,3-di-O-acetyl-α-D-glucopyranosid)uronate* (31). (*From Kotick et al.*, 1969).

4.4, **31**) (Kotick et al., 1969). This synthesis was accomplished by tritylation of methyl 4-azido-4-deoxy-α-D-glucopyranoside (**16**) to the 6-tritylate (**17**). This was converted to the 2,3-di-*O*-benzoate (**18**) and detritylated to methyl 4-azido-4-deoxy-2,3-di-*O*-benzoyl-α-D-glucopyranoside (**20**), which on oxidation gave the glucuronic acid derivative (**21**). Methyl (methyl 4-benzamido-4-deoxy-2,3-di-*O*-benzoyl-α-D-glucopyranosid)uronate (**23**) was synthesized by esterification of **21** to **22**, followed by reduction and benzoylation of **22** to **23**. Debenzoylation of methyl (methyl-4-azido-4-deoxy-2,3-di-*O*-benzoyl-α-D-glucopyranosid)uronate (**22**) with sodium methylate to **24** followed by deesterification and hydrogenation gave crystalline methyl 4-amino-4-deoxy-α-D-glucopyranosiduronic acid (**26**). *N*-Acetylation of **26** yielded methyl 4-acetamido-4-deoxy-α-D-glucopyranosiduronic acid (**29**), which was then esterified and acetylated to the peracetylated derivative methyl-

(methyl 4-acetamido-4-deoxy-2,3-di-*O*-acetyl-α-D-glucopyranosid)uronate (**31**). Compound **31** was also synthesized by acetylation of **28**. These studies thereby establish the structure and synthesis of the pyrimidyl-uronic acid moiety of gougerotin. Consequently the revised structure of gougerotin is 1-(cytosinyl)-4-sarcosyl-D-serylamino-1,4-dideoxy-β-D-glucopyranuronamide (Fig. 4.2*C*) (Fox et al., 1968).

## INHIBITION OF GROWTH

The sensitivity of selected organisms to gougerotin is shown in Table 4.1 (Kanzaki et al., 1962). Gougerotin inhibits gram-positive and gram-negative bacteria. Mycobacteria are inhibited only at high concentrations of the antibiotic. Yeast and fungi are not inhibited at 500 μg/ml. The $LD_{50}$ for mice given gougerotin intravenously is 57 mg/kg (Kanazki et al., 1962).

**Table 4.1**

Antimicrobial Spectrum of Crystalline Gougerotin by Agar-Streak Method

| | μg/ml | | μg/ml |
|---|---|---|---|
| *E. coli* | 200 | *Mycobacterium smegmatis* | >800 |
| *Prot. vulgaris* | 800 | *Mycobacterium phlei* | 800[a] |
| *Staph. aureus* 209 P | 400 | *M. tuberculosis* H 37 Rv | 100 |
| *B. subtilis* PCl 219 | 400 | *Pen. chrysogenum* Q 176 | >500 |
| *B. cereus* | >800 | *Sacc. cerevisiae* | >500 |
| *Microc. flavus* | 40 | *Candida albicans* | >500 |
| *Sarcina lutea* | 800 | *Piricularia oryzae* | >500 |
| *Ps. aeruginosa* | 800 | *Gib. fujikuroi* | >500 |
| *B. brevis* | 200 | *Phytophythora infestans* | >500 |
| *Mycobacterium avium* | 800 | *Colleto, lagernarium* | >500 |
| *M. avium* streptomycin-fast | 800 | *Glomerella cingulata* | >500 |
| *Mycobacterium* 607 | 800 | *Alternaria kikuchiana* | >500 |

From Kanzaki et al., 1962.
[a] Serial broth dilution method.

## BIOCHEMICAL PROPERTIES

The data available up through 1968 indicate that gougerotin blocks protein synthesis by acting as a nonfunctional analog of aminoacyl-tRNA by inhibiting peptidyl transferase. However, more recent studies by Coutsogeorgopoulos, Pestka, Monro, and Yukioka and Monisawa indicate a more complex participation of this nucleoside antibiotic and other inhibitors of protein synthesis.

In the recent review on gougerotin, Clark (1967) stated that the similarity between the structures of gougerotin and puromycin led them to investigate the effect of gougerotin on protein synthesis in *E. coli*. The inhibition of protein synthesis in cell-free extracts of *E. coli* by gougerotin was first reported by Clark and Gunther (1963). This aminoacylnucleoside antibiotic was shown to inhibit protein synthesis at the stage of amino acid transfer from aminoacyl-tRNA to polypeptide. Gougerotin did not inhibit protein synthesis by blocking the formation of aminoacyl-tRNA, nor did it inhibit the ATP generating system. A more detailed study of the mode of action of gougerotin was reported by Clark and Chang (1965). A comparison of the inhibition of protein synthesis in cell-free systems of *E. coli* and rabbit reticulocytes was made with puromycin, gougerotin, chloramphenicol, cycloheximide, neomycin *β*, and tetracycline. All compounds inhibited this bacterial protein synthesis system except cycloheximide. In contrast, only cycloheximide inhibits, cell-free protein synthesis in rabbit reticulocytes.

Additional proof that gougerotin inhibits protein synthesis following the binding of aminoacyl-tRNA to ribosomes was shown by experiments involving both puromycin and gougerotin. Puromycin is known to catalyze the release of incomplete polypeptide chains from peptidyl-tRNA by acting as an analog of tRNA. The product of this reaction is peptidyl-puromycin. Casjens and Morris (1965) and Clark and Chang (1965) reasoned that if gougerotin acts in a manner similar to puromycin, then gougerotin should also catalyze the release of peptides from the ribosome complex. When puromycin and gougerotin were added to cell-free systems from *E. coli* or rabbit reticulocytes containing prelabeled polypeptides bound to the ribosomes, gougerotin inhibited the puromycin-dependent release of polypeptides from the ribosomes. Therefore, gougerotin is incapable of catalyzing the release of peptidyl materials. From these data it appeared that gougerotin inhibited ribosomal peptidyl transferase by acting as a nonfunctional analog of aminoacyl-tRNA. However, gougerotin cannot serve as an acceptor for transfer of the peptidyl group. The lack of ability to form peptidyl-gougerotin may probably be attributed to the differences in structures of gougerotin and puromycin.

Additional evidence that gougerotin acts as a specific inhibitor of the peptide bond-forming reaction was shown by the studies in which gougerotin does not inhibit the release of completed proteins from ribosomes (Casjens and Morris, 1965). Whereas protein synthesis in cell-free systems normally is accompanied by an orderly depolymerization of polysomes (Noll et al., 1963), polysome structures are maintained by gougerotin (Casjens and Morris, 1965). In contrast, puromycin accelerates the breakdown of polysomes (Williamson and Schweet, 1964). Thus, these initial observations supported the notion that gougerotin blocks peptide bond formation by inhibiting peptidyl transferase. Recent studies have ruled out this idea.

Finally, Casjens and Morris (1965) and Goldberg and Mitsugi (1967) studied the competition between gougerotin and puromycin. Casjens and Morris concluded that gougerotin is a strict competitive inhibitor of puromycin in the peptide-release reaction and suggested that these two antibiotics have a common site of action. Goldberg and Mitsugi also studied the effect of gougerotin on the puromycin-induced release of polylysine from sRNA bound to ribosomes. Their kinetic data suggest that the competition between gougerotin and puromycin belongs to the "mixed" type inhibitors. The results of Goldberg and Mitsugi differ from those of Casjens and Morris.

Sinohara and Sky-Peck (1965) studied the effect of gougerotin on protein synthesis in cell-free systems from mouse liver. They reported that gougerotin inhibits protein synthesis in cell-free systems of mouse liver by inhibiting the transfer of amino acids from aminoacyl-tRNA to protein. Sinohara and Sky-Peck compared the structure of gougerotin with the structural analogs of puromycin. Nathans and Neidle (1963) showed that the puromycin analogs with aromatic L-amino acids linked to the 3′-amino group of the 3′-deoxyribose were good inhibitors of protein synthesis. Gougerotin did not meet these structural requirements to inhibit protein synthesis in the same manner.

Coutsogeorgopoulos (1967) used gougerotin to determine if this aminoacylnucleoside antibiotic interfered with aminoacyl-tRNA at either an enzyme site or a ribosomal site. The system used was the poly-U-directed polyphenylalanine synthesis with cell-free extracts from *E. coli.* Gougerotin is a stronger inhibitor of polyphenylalanine synthesis during the initial stages of the reaction at 25°C than it is at 37°C. The inhibition of peptide synthesis by gougerotin was not reversed by increasing amounts of phenylalanyl-tRNA (see Table 4.6, Sect. 4.2).

Goldberg and Mitsugi (1967) reported on the inhibition of the puromycin reaction with the antibiotics sparsomycin, gougerotin and chloramphenicol. Sparsomycin ($1 \times 10^{-7}$ *M*) blocks polypeptide formation at a point beyond the attachment of aminoacyl-tRNA to ribosomes (Goldberg and Mitsugi, 1966, 1967). Sparsomycin is about 200 times better an inhibitor in blocking the puromycin reaction than is gougerotin or chloramphenicol with cell-free extracts of *E. coli* using poly A as the mRNA. Goldberg and Mitsugi also showed that gougerotin was most effective as an inhibitor when poly C was the mRNA. Gougerotin was less effective against poly A or poly U. More recent studies by Herner, Goldberg and Cohen (1969) showed that gougerotin acts on the 50-S subunit and competes with sparsomycin for the stimulation of binding and formation of the 70-S ribosomes. However, much higher concentrations of gougerotin are needed for stimulation of binding of *N*-acetyl-L-phenylalanyl-tRNA to ribosomes (Figure 4.5).

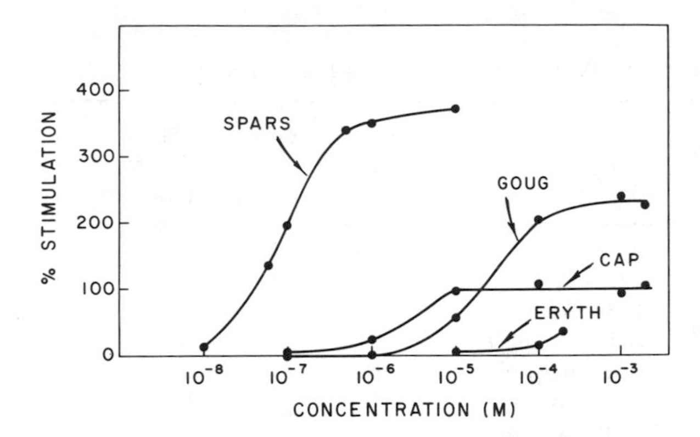


FIGURE 4.5 *Concentration curves for stimulation of binding of N-acetyl-*L*-phenylalanyl-tRNA to ribosomes. Sparsomycin, SPARS; gougerotin, GOUG; chloramphenicol, CAP; erythromycin, ERYTH.* (*From Herner, Goldberg and Cohen,* 1969.)

The role of gougerotin in elucidating peptide synthesis using the "fragment reaction," in which hexanucleotide-formylmethionine or CAACCA-formylmethionine serves as a formylmethionine donor has been reported (Marcker, 1965; Monro and Marcker, 1967). Gougerotin inhibits the puromycin dependent "fragment reaction" in a manner similar to its inhibition of the aminoacyl-tRNA-puromycin reaction.

A possible common site of action of gougerotin and sparsomycin are indicated by further work with the "fragment reaction." Whereas gougerotin inhibited the puromycin-reaction with the aminoacyl-tRNA system, it only partially inhibited the sparsomycin-induced complex with the fragment reaction (Monro et al., 1969) (Table 4.2). Under the same experimental conditions, but in the absence of sparsomycin, gougerotin stimulated the binding of CACCA-leucyl-acetate to ribosomes by 35%. The authors stated that the mode of action of gougerotin may be related to that of sparsomycin in that they compete for the same site on the ribosome complex. However, the inert complexes formed with gougerotin are less stable than the sparsomycin complexes. The data in Table 2 strongly suggest that the action of chloramphenicol, in peptide bond formation, may differ from the action of gougerotin or amicetin. Now that the structure of sparsomycin is known (see page 26 for structure) it will be interesting to compare the inhibition of protein synthesis by gougerotin and sparsomycin.

Initial studies with amicetin, gougerotin, blasticidin S, chloramphenicol, or sparsomycin indicated that they acted in protein synthesis by inhibiting the formation of the peptide bond per se (peptidyl transfer reaction). It was suggested that these antibiotics (see p. 26 for sparsomycin) were structurally similar to the aminoacyl end of amino acyl-tRNA. The most recent

**Table 4.2**

Effects of Various Antibiotics on Complex Formation

| Addition | Final conc., m$M$ | Percentage of control sparsomycin-stimulated binding |
|---|---|---|
| None | — | 100 |
| Chloramphenicol | 1 | 3 |
| Carbomycin | 0.1 | 0 |
| Spiramycin III | 0.1 | 0 |
| Streptogramin A | 0.1 | 2 |
| Lincomycin | 1 | 5 |
| Amicetin | 1 | 27 |
| Gougerotin | 1 | 42 |

From Monro et al., 1969.

studies of Pestka (1969b; 1970c) support the earlier suggestion of Coutsogeorgopoulos (1967) that chloramphenicol may also act in protein synthesis as an analog of the amino acyl adenylyl terminus of amino acyl tRNA or of the peptidyl adenylyl terminus of peptidyl tRNA.

Pestka (1970c) has reported that the binding of a $^{14}$C-L-phenylalanyl-oligonucleotide by washed ribosomes from *E. coli* at 20 m$M$ $Mg^{2+}$ is inhibited by chloramphenicol, gougerotin, amicetin, or sparsomycin. Thus, it has been proposed (Pestka) that the mechanism by which these antibiotics inhibit peptide bond formation may be the specific inhibition of the binding of the amino acyl adenylyl terminus of amino acyl-RNA or of the peptidyl-adenylyl terminus of peptidyl tRNA to the ribosome (presumably at the 50-S subunit, in the case of *E. coli*) prior to the peptidyl transfer reaction (peptide bond formation per se). This proposal could explain the inhibition of the peptidyl transfer (peptide bond formation per se) by these antibiotics in the reactions between puromycin and model peptidyl-tRNAs. However, such a mechanism of action would not easily explain the stimulation of the enzymatic binding of L-phenylalanyl-tRNA to the ribosomes by gougerotin, as observed by Yukioka and Morisawa (1969c) (Table 4.3), if this binding is to be considered as part of the mechanism by which gougerotin inhibits peptide bond formation. The differences in the inhibition of tRNA binding to ribosomes by gougerotin as reported by Pestka (1969a, 1969b, 1969c, 1970a, 1970b, 1970c, private communication) and Yukioka and Morisawa (1969a, 1969b, 1969c) may be attributed to different experimental procedures used. For example, Pestka used aminoacyl-oligonucleotides while Yukioka and Morisawa used intact aminoacyl-tRNA. Pestka (1970b) has shown that gougerotin does not inhibit the binding of intact aminoacyl-tRNA to ribosomes but does inhibit the binding of aminoacyl-oligonucleotides to *E. coli* and human (placenta) ribosomes (Fig. 4.6 and 4.7) (Pestka, private communication).

**Table 4.3**

Effect of Antibiotics on the $^{14}C$-Phenylalanyl-tRNA Binding to Ribosomes

| System | $^{14}C$-Phenylalanyl-tRNA bound: Experiment 1 in the presence of T and GTP, cpm | Experiment 2 in the absence of T and GTP, cpm |
|---|---|---|
| Complete | 3344 | 1021 |
| +Gougerotin ($4.5 \times 10^{-4}$ *M*) | 6016 | 1025 |
| +Blasticidin S ($4.7 \times 10^{-7}$ *M*) | 5455 | 1038 |
| +Puromycin ($2.1 \times 10^{-3}$ *M*) | 5080 | 1030 |
| +Chloramphenicol ($6.2 \times 10^{-4}$ *M*) | 3332 | 1004 |
| +Sparsomycin ($3.3 \times 10^{-7}$ *M*) | 3025 | 1007 |
| +Tetracycline ($2.3 \times 10^{-3}$ *M*) | 1637 | 585 |

From Yukioka and Morisawa, 1969c.

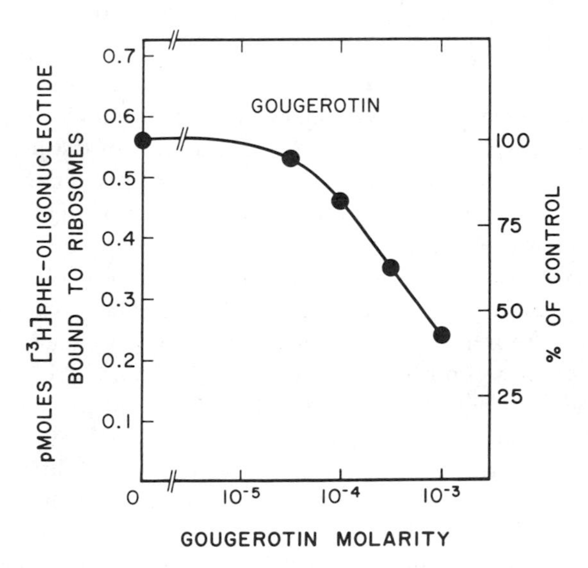


FIGURE 4.6 *Binding of phenylalanyl-oligonucleotide to ammonium chloride washed ribosomes isolated from E. coli. Each 0.05-ml reaction mixture contained the following components:* 20% *ethanol;* 0.05 *M Tris-chloride, pH* 7.2; 0.4 *M KCl;* 0.04 *M* $Mg^{++}$; 0.06 *M* $NH_4Cl$; 4.1 $A_{260}$ *units of ribosomes;* 1.5 *pmoles* [$^3H$]*-phenylalanyl-oligonucleotide; reaction time*—20 *min,* 24° *(From Pestka, private communication).*

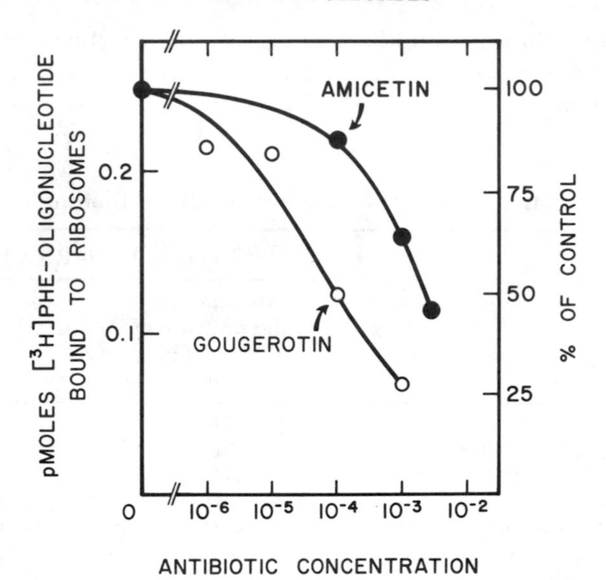


FIGURE 4.7 *Binding of phenylalanyl-oligonucleotide to ammonium chloride washed ribosomes isolated from human placenta. Each 0.050-ml reaction mixture contained the following components:* 0.050 *M Tris-chloride, pH* 7.2; 0.4 *M KCl;* 0.06 *M* $NH_4Cl$; 0.04 *M magnesium acetate;* 4.1 $A_{260}$ *units of ribosomes;* 3.0 *pmoles* [$^3H$] *phenylalanyl-oligonucleotide; reaction*-20 *min*, 24° (*From Pestka, private communication*).

The scheme which visualizes chloramphenicol, amicetin, gougerotin, blasticidin S, and sparsomycin as specific inhibitors of the peptidyl transfer reaction cannot easily explain the formation of phenylalanine-containing oligopeptides (Oligophe) from L-phenylalanyl-tRNA (Table 4.4) in the presence of chloramphenicol, blasticidin S, gougerotin, amicetin, or sparsomycin, at concentrations at which polyphenylalanine (Polyphe) formation is inhibited (Coutsogeorgopoulos, 1969, 1970). The formation of small peptide chains from L-phenylalanyl-tRNA in the presence of gougerotin ($4.5 \times 10^{-4}$ *M*) has also been observed by Yukioka and Morisawa (1969a). When the $Mg^{2+}$ concentration is lowered from 11.2 m*M* to 5.0 m*M*, L-phenylalanyl-tRNA polymerization is possible only if *N*-Ac-phenylalanyl-tRNA is also added as a peptide chain initiator. Under these conditions (Table 4.5, left-hand side) the formation of the phenylalanine-containing oligopeptide fraction was inhibited to various extents by chloramphenicol, blasticidin S, gougerotin, or sparsomycin. However, if at this $Mg^{2+}$ concentration the polymerization conditions are changed, namely, UTP is used in place of GTP, the observed inhibition of the phenylalanine-containing oligopeptide fraction (Oligophe) is reversed by 20–30%, except in the case of sparsomycin (Table 4.5, right-hand side). From these data of Coutsogeorgopoulos (1970), it appears that the mechanism of initiating the polypeptide chains or the state of the ribosome-poly-U-phenylalanyl-tRNA complex at the time of peptide chain initiation determines the degree of the observed inhibition of peptide bond formation by these antibiotics. At 11.2 m*M* $Mg^{2+}$ L-phenylalanyl-tRNA can initiate the peptide chains, whereas at 5.0 m*M* $Mg^{2+}$ chain initiation is possible only when *N*-Ac-phe-RNA is also used. On the other hand, pre-incubation with UTP creates conditions under which the initiation mechanism at 5.0 m*M* $Mg^{2+}$ is changed, as evidenced by the fact

**Table 4.4**

Inhibition of Phe-RNA Polymerization in the Presence of the Ribosomal Wash at 11.2 m*M* $Mg^{2+}$ [a]

| Inhibitor (*M*) | Polyphe (% input) | Ibn, % | Oligophe (% input) | Ibn, % |
|---|---|---|---|---|
| No inhibitor | 32 | — | 26 | — |
| Chloramphenicol $10^{-3}$ | 26 | 19 | 28 | Stim. |
| Blasticidin S $4 \times 10^{-6}$ | 3 | 91 | 31 | Stim. |
| Gougerotin $4 \times 10^{-4}$ | 6 | 81 | 39 | Stim. |
| Amicetin $4 \times 10^{-5}$ | 26 | 19 | 28 | Stim. |
| Sparsomycin $10^{-5}$ | 10 | 69 | 39 | Stim. |
| Fusidin $10^{-3}$ | 3 | 91 | 21 | 19 |
| Tetracycline $2.4 \times 10^{-5}$ | 22 | 31 | 12 | 54 |

From Coutsogeorgopoulos, 1970.

[a] The complete system (no inhibitor) contained in 0.25 ml: 25 μmoles of Tris-HCl buffer (pH 7.2); 25 μmoles of ammonium chloride, magnesium acetate to a final concentration 11.2 m*M*; 10 μg poly U; 0.2 μmole of GTP; 8.0 $A_{260}$ units of ribosomes (*E. coli* B) washed with 0.5 *M* $NH_4Cl$-buffer three times; 0.1 mg (protein) of the ribosomal wash as a source of transfer and initiation factors; 3.2 $A_{260}$ units of $^{14}C$-L-phe-RNA charged with 50 μμmoles of L-phenylalanine (25,000 cpm total). The reagents were added in the order cited. The inhibitors were added after the poly U and before the ribosomes. After incubation at 25°C for 4 min, 25 μl of 4 *N* KOH was added. Reincubation followed at 37°C for 20 min. The whole mixture was then applied on a 3 × 56.9 cm 3MM Whatman paper strip and developed in *n*-butanol/acetic acid/water (5:2:3v/v). The paper strip is first scanned in a Tracerlab 4 π paper strip radioactivity scanner. The radioactive areas, as well as the rest of the paper strip, are cut and counted in a Scintillation Spectrometer. The radioactivity remaining at the origin, which separates well from the rest of the radioactivity in the other areas of the paper, is taken as a measure of Polyphe formation. Free $^{14}C$-phenylalanine, produced mainly from the unutilized Phe-RNA, moves with an *Rf* of about 0.6 and separates well from a radioactive fraction which moves towards the front of the paper. The latter fraction is taken as a measure of Oligophe formation. "% input" represents the fraction of the total radioactivity measured in all the areas of the paper strip. Abbreviations: Phe-RNA, phenylalanyl-RNA; Polyphe, polyphenylalanine; Oligophe, oligophenylalanine; Ibn, inhibition.

**TABLE 4.5**

Inhibition of Phe-RNA Polymerization in the Presence of the Ribosomal Wash at 5.0 m*M* $Mg^{2+}$

| | In the presence of GTP[a] | | | | In the presence of UTP | | | |
|---|---|---|---|---|---|---|---|---|
| Inhibitor (M) | Polyphe (% input) | Ibn, % | Oligophe (% input) | Ibn, % | Polyphe (% input) | Ibn, % | Oligophe (% input) | Ibn, % |
| No inhibitor | 30 | — | 31 | — | 20 | — | 50 | — |
| Chloramphenicol $10^{-3}$ | 12 | 60 | 22 | 29 | 7 | 65 | 50 | 0 |
| Blasticidin S $4 \times 10^{-6}$ | 2 | 93 | 15 | 52 | 1 | 95 | 33 | 34 |
| Gougerotin $4 \times 10^{-4}$ | 1 | 97 | 19 | 39 | 1 | 95 | 45 | 10 |
| Amicetin $4 \times 10^{-5}$ | 9 | 70 | 30 | 3 | 8 | 60 | 59 | stim. |
| Sparsomycin $10^{-5}$ | 5 | 83 | 17 | 45 | 2 | 90 | 25 | 50 |
| Fusidin $10^{-3}$ | 2 | 93 | 9 | 71 | 3 | 85 | 33 | 34 |
| Tetracycline $2.4 \times 10^{-5}$ | 5 | 83 | 7 | 78 | 10 | 50 | 23 | 55 |
| UMPPCP $8 \times 10^{-4}$ | 1 | 97 | 4 | 87 | 15 | 25 | 28 | 44 |

From Coutsogeorgopoulos, 1970.

[a] "In the presence of GTP" the complete system was identical to that described in Table 4.4 with the following differences: Final $Mg^{2+}$ concentration, 5.0 m*M*; 1.6 $A_{260}$ units of *N*-Ac-$^{12}$C- phe-RNA, charged with 24 $\mu\mu$moles of $^{12}$C-L-phenylalanine, was added along with the $^{14}$C-phe-RNA. "In the presence of UTP," the incubation conditions were changed as follows: The complete system (no inhibitor) was first preincubated at 25°C for 4 min in a volume of 0.17 ml containing 25 $\mu$moles of Tris-HCl buffer, pH 7.2, 25 $\mu$moles of ammonium chloride, magnesium acetate to a final concentration of 5.0 m*M*, 10 $\mu$g poly U, 0.2 $\mu$moles of UTP, 8.0 $A_{260}$ units of washed ribosomes, and 0.1 mg (protein )of ribosomal wash. After this pre-incubation the mixture was cooled in ice and the following additions were made: distilled water to a final volume of 0.25 ml, magnesium acetate to a final concentration of 5.0 m*M*, inhibitors as indicated, and finally 1.6 $A_{260}$ units of *N*-Ac-$^{12}$C-phe-RNA and 3.2 $A_{260}$ units of $^{14}$C-L-phe-RNA. No GTP was added. The mixture was reincubated at 25°C for 4 min. The assay for Oligophe and Polpyhe formation was the same as given in Table 4.4. Abbreviations: *N*-Ac-phe-RNA, *N*-acetylphenylalanyl-RNA; UMPPCP, 5′-uridylylmethylenediphosphonate.

that the inhibition of the formation of the phenylalanine-containing oligopeptide fraction (Oligophe) by tetracycline was lowered when UTP was used in place of GTP (Table 4.5). Tetracycline is considered a specific inhibitor of the binding of aminoacyl-tRNA at the acceptor site (Cundliffe and McQuillen, 1967).

This new evidence indicates that modifications are necessary in the current scheme which visualizes these antibiotics as specific inhibitors of the peptidyl transfer reaction. Coutsogeorgopoulos (1970) suggests that chloramphenicol, blasticidin S, gougerotin, amicetin, or sparsomycin may be involved in and interfere with substeps (in the overall peptide bond formation step) which are distinct from the formation of the actual amide bond (peptidyl transfer reaction).

It will be extremely interesting to see if there are indeed substeps that precede or follow the formation of the peptide bond per se. If so, this would be another example of the utilization of the nucleoside antibiotics as biochemical tools in elucidating complex biological reactions.

Finally, Capecchi (1969) has used gougerotin and sparsomycin to study chain termination in protein synthesis. He has isolated, purified, and studied the release factors $R_1$ and $R_2$. Factor $R_1$ mediates the release of *N*-formylmethionine from the 50-S ribosome when UAG or UAA is the terminating code. Gougerotin or sparsomycin blocks the release of peptide by release factor $R_1$. This inhibition might be due to gougerotin acting as a steric inhibitor (Capecchi and Klein, 1970).

Gougerotin has been reported to inhibit pseudorabies, vaccinia, Newcastle disease, fowl plague, and Western equine encephalomyelitis viruses. Gougerotin is an effective inhibitor of uracil-rich virus messenger RNA, while sparsomycin is more effective on the activity of virus messenger RNA that is rich in guanine and cytosine (Thiry, 1968).

Gougerotin does not have any effect on respiration, phosphorylation, adenosine triphosphatase activity, or energized transport of alkali metal cations into isolated rat liver mitochondria (Lardy, private communication).

## Summary

Gougerotin, a broad spectrum antibiotic, is a dipeptidylpyrimidine nucleoside antibiotic in which the structure of the 4-amino carbohydrate moiety has been shown to have the *gluco* configuration. The nucleoside degradation products of gougerotin, C-substance, and the sugar moiety of gougerotin have now been synthesized.

Gougerotin is an acylaminoacylnucleoside antibiotic that blocks protein synthesis by acting as an inhibitor of peptide chain elongation. Gougerotin inhibits the puromycin reaction, but does not form a covalent bond with nascent peptides. The most recent evidence on oligopeptide formation in the presence of gougerotin shows that peptide bond(s) can form in the presence of gougerotin. At high concentrations of $Mg^{2+}$ (20 m*M*), gougerotin inhibits the binding of aminoacyloligonucleotides to the ribosomes. It may well be that peptide bond formation per se consists of several substeps that either precede or follow the formation of the peptide bond.

The sparsomycin-induced complex is only partially inhibited by gougerotin. Gougerotin was a most effective inhibitor when poly C was the messenger RNA. It was less effective against poly A or poly U and has no effect on respiration of phosphorylation in isolated rat liver mitochondria.

## References


Capecchi, M. R., and H. A. Klein, *Cold Spring Harbor Symp.*, **34**, 469 (1970).

Capecchi, M. R., 158th National Meeting, American Chemical Society, New York, 1969, Abstracts Biol. 037.

Casjens, S. R., and A. J. Morris, *Biochim. Biophys. Acta*, **108**, 677 (1965).

Clark, Jr., J. M., *Antibiotics*, *I*, Springer-Verlag, N. Y., p. 278 (1967).

Clark, Jr., J. M., and A. Y. Chang, *J. Biol. Chem.*, **240**, 4734 (1965).

Clark, Jr., J. M., and J. K. Gunther, *Biochim. Biophys. Acta*, **76**, 636 (1963).

Coutsogeorgopoulos, C., *Biochemistry*, **6**, 1704 (1967).

Coutsogeorgopoulos, C., *Federation Proc.*, **28**, 844 (1969).

Coutsogeorgopoulos, C., *Proc. 6th Intern. Congr. Chemotherapy*, *Tokyo*, 1970.

Cundliffe, E , and K. McQuillen, *J. Mol. Biol.*, No. **30**, 137 (1967).

DeBoer, C. and J. W. Hinman, U.S. Pat. 2,909,463 (1959a); *Chem. Abstracts*, **54**, 2669g (1960).

DeBoer, C. and J. W. Hinman, U.S. Pat. No. 2,909,464 (1959b); *Chem. Abstr.*, **54**, 2669i (1960).

Fox, J. J., Y. Kuwada, K. A. Watanabe, T. Ueda, and E. B. Whipple, Antimicrobial Agents Chemotherapy, 1964, 518.

Fox, J. J., K. A. Watanabe, and A. Bloch, *Progr. Nucleic Acid Res. Mol. Biol.*, **5**, 251 (1966).

Fox, J. J., Y. Kuwada, and K. A. Watanabe, *Tetrahedron Letters*, 1968, 6029.

Friedman, H. A., and K. A. Watanabe, and J. J. Fox, *J. Org. Chem.*, **32**, 3775 (1967).

Goldberg, I. H., and K. Mitsugi, *Biochem. Biophys. Res. Commun.*, **23**, 453 (1966).

Goldberg, I. H., and K. Mitsugi, *Biochemistry*, **6**, 372 (1967).

Herner, A. E., I. H. Goldberg, and L. B. Cohen, *Biochemistry*, **8**, 1335 (1969).

Iwasaki, H., *Zasshi Yakugaku*, **82**, 1358 (1962).

Kanzaki, T., E., Higashide, H. Yamamoto, M. Shibata, K. Nakazawa, H. Iwasaki, T. Takewaka, and A. Miyake, *J. Antibiotics* (*Tokyo*), **15A**, 93 (1962).

Korzybski, T., Z. Kowszyk-Gindifer, and W. Kurylowicz, Eds., *Antibiotics*, Vol. 1, Pergamon Press, New York, 1967.

Kotick, M. P., R. S. Klein, K. A. Watanabe, and J. J. Fox, *Carbohydrate Res.*, **11**, 369 (1969).
Marcker, K., *J. Mol. Biol.*, **14**, 63 (1965).
Monro, R. E., and K. A. Marcker, *J. Mol. Biol.*, **25**, 347 (1967).
Monro, R. E., M. L. Celma, and D. Vazqez, *Nature*, **222**, 356 (1969).
Nathans, D., and A. Neidle, *Nature*, **197**, 1076 (1963).
Noll, H., T. Staehelin, and F. O. Wettstein, *Nature*, **198**, 632 (1963).
Pestka, S., *Biochem. Biophys. Res. Commun.*, **36**, 589 (1969a).
Pestka, S., *Proc. Natl. Acad. Sci. (U.S.)*, **64**, 709 (1969b).
Pestka, S., *J. Biol. Chem.*, **244**, 1533 (1969c).
Pestka, S., *Arch. Biochem. Biophys.*, **136**, 80 (1970a).
Pestka, S., *Arch. Biochem. Biophys.*, **136**, 89 (1970b).
Pestka, S., *Cold Spring Harbor Symp.*, **34**, 395 (1970c).
Sinohara, H., and H. H. Sky-Peck, *Biochem. Biophys. Res. Commun.*, **18**, 98 (1965).
Thiry, L., *J. Gen. Virol.*, **2**, 143 (1968).
Watanabe, K. A., and J. J. Fox, *Chem. Pharm. Bull.*, **12**, 975 (1964).
Watanabe, K. A., and J. J. Fox, *J. Org. Chem.*, **31**, 211 (1966).
Watanabe, K. A., M. P. Kotick, and J. J. Fox, *Chem. Pharm. Bull.*, **17**, 416 (1969).
Watanabe, K. A., M. P. Kotick, and J. J. Fox, *J. Org. Chem.* **35**, 231, (1970).
Williamson, A. R., and R. Schweet, *Nature*, **202**, 435 (1964).
Yukioka, M. and S. Morisawa, *J. Biochem.*, **66**, 225 (1969a).
Yukioka, M. and S. Morisawa, *J. Biochem.*, **66**, 233 (1969b).
Yukioka, M. and S. Morisawa, *J. Biochem.*, **66**, 241 (1969c).

## 4.2. BLASTICIDIN S

### INTRODUCTION

The amino acylaminonucleoside antibiotic blasticidin S (Fig. 4.8, **1**; Fig. 4.10) specifically inhibits protein synthesis in intact cells and cell-free systems (Pestka, 1969, 1970a, 1970b, 1970c; Yamaguchi et al., 1965; Yamaguchi and Tanaka, 1966; Coutsogeorgopoulos, 1969, 1967a, 1967b). The final revised structure assigned to blasticidin S is 1-(1′-cytosinyl)-4-[L-3′-amino-5′-(1″-*N*-methylguanidino)-valerylamino]-1,2,3,4-tetradeoxy-2,3-dehydro-β-D-*erythro*-hex-2-ene uronic acid. Blasticidin S is very inhibitory against the pathogenic fungus *Piricularia oryzae*. While most fungi are not inhibited by blasticidin S at low concentrations, this nucleoside antibiotic is a protective and therapeutic agent against *P. oryzae* in concentrations as low as 5 ppm (Misato et al., 1959).

### DISCOVERY, PRODUCTION, AND ISOLATION

Takeuchi et al. (1958) described the isolation of blasticidin S from the culture filtrates of *Streptomyces griseochromogenes*. Blasticidin S has also been isolated from *S. moro-okaensis* SF337 (Tsuruoka and Niida, 1963). 8-Aza-

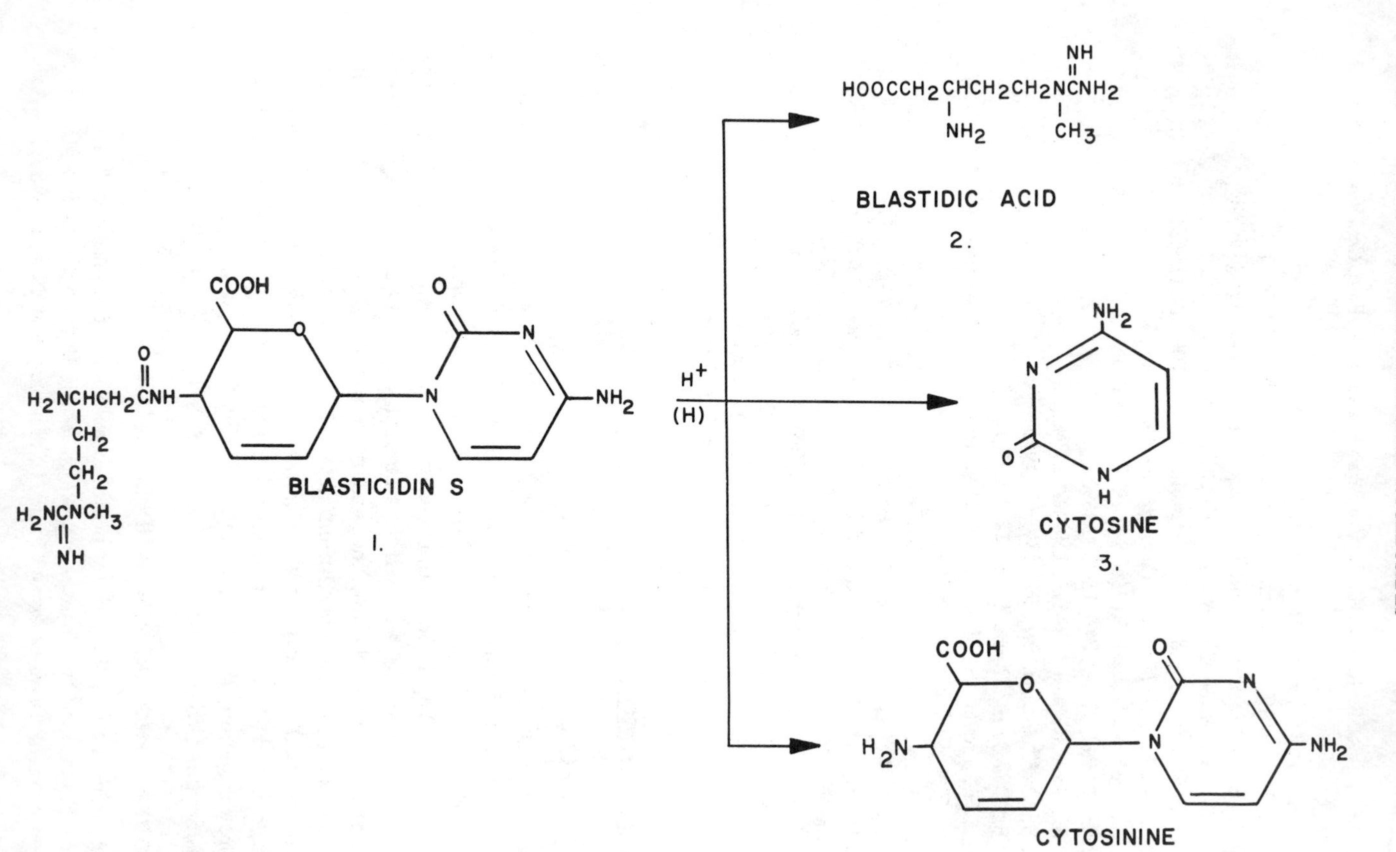


FIGURE 4.8 *Degradation of blasticidin (Ōtake et al., 1965b, 1966b; Yonehara et al., 1963).*

guanine was also isolated from this same organism. Three other members of the blasticidin family have been isolated. They are blasticidin A, blasticidin B, and blasticidin C (Fukunaga et al., 1955; Kono et al., 1968).

Blasticidin S is isolated from the culture filtrate by absorption on a cation-exchange resin (IRC-50-$Na^+$) and elution with 0.5 *N* HCl. The eluant is concentrated *in vacuo* (pH 6.0) and stored at 0°C. Crude blasticidin S monohydrochloride crystallizes. The impure blasticidin S is dissolved in water and passed through an anion-exchange resin (IR-4B-$OH^-$) to remove the excess HCl. Blasticidin S is crystallized by storage at 0°C and recrystallized from water as white needles (Yonehara et al., 1963).

## PHYSICAL AND CHEMICAL PROPERTIES

The molecular formula for blasticidin S is $C_{17}H_{26}N_8O_5$; mp 253–255°C (decomp.); $[\alpha]_D^{11}$ + 108.4° ($c$ = 1% in water); $\lambda_{max}$ 274 mμ ($\epsilon$ = 13,400) in 0.1 *N* HCl; $\lambda_{max}$ 266 mμ ($\epsilon$ = 8850) in 0.1 *N* NaOH. Blasticidin S is soluble in water and acetic acid, but insoluble in organic solvents (Ōtake et al., 1965a, 1965b, 1966a, 1966b). Yonehara et al. (1963) reported that Blasticidin S had a melting point of 235–236°C.

## STRUCTURAL ELUCIDATION

Since blasticidin S was used as an antibiotic in controlling rice blast disease, Yonahara and his co-workers proceeded to study the structure of this complex antibiotic. In 1963 Yonahara et al. reported that blasticidin S is a 1-substituted cytosine nucleoside. The antibiotic exhibits an ultraviolet absorption spectrum similar to that of cytidine. On refluxing blasticidin S (**1**) in acid and catalytic hydrogenation of cytosinine the compounds isolated were: blastidic acid (**2**), cytosine (**3**), and cytosinine (**4**) (Fig. 4.8) (Ōtake et al., 1965b, 1966b). Ōtake et al. (1965b) reported that the structure of blastidic acid was $\epsilon$-*N*-methyl-$\beta$-arginine. The absolute configuration of blastidic acid was shown to be of the L-configuration (Yonehara and Ōtake, 1966). This conclusion was based on the positive Cotton effect exhibited by blastidic acid. They also established the $\beta$ configuration for the anomeric center of cytosinine (**4**) (Fig. 4.9). This was accomplished by converting the methyl ester of cytosinine (**5**) to the aminotriol (**6**). Treatment of **6** with periodate gave rise to a dialdehyde (**7**) that was identical to the dialdehyde formed following periodate oxidation of cytidine (**8**). Fox and Watanabe (1966) reinvestigated the structure of the carbohydrate linkage of blasticidin S as proposed by Ōtake et al. (1965a) and reported that the aminohexose moiety of cytosinine is of the D-configuration. Yonehara and Ōtake (1966) also provided unequivocal evidence for C-4 (*S*) and C-5 (*S*) configurations for the 4-aminohexose and assigned the D-configuration to the sugar moiety.

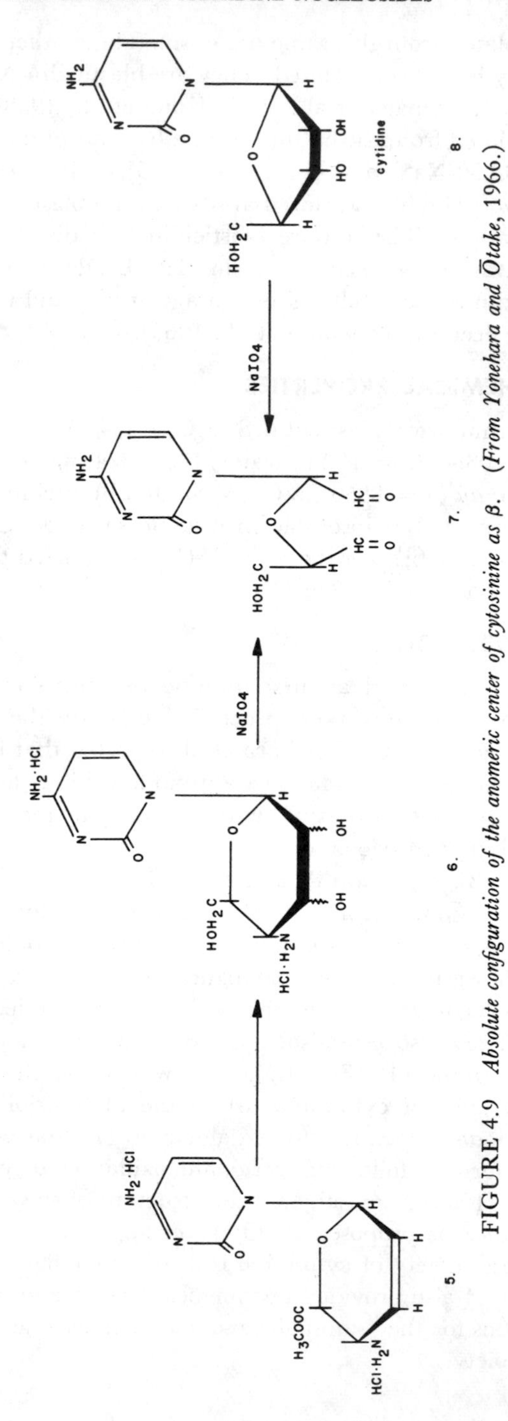


FIGURE 4.9 *Absolute configuration of the anomeric center of cytosinine as β.* (*From Yonehara and Ōtake*, 1966.)

Yonehara et al. (1963) hydrolyzed blasticidin S in 0.4 *N* NaOH at 27°C for 2 weeks and isolated cytosinine. They also isolated cytomycin, the deamination product, by using a milder alkaline treatment (0.1 *N* NaOH, room temperature, 2 weeks). Ōtake et al., (1965b, 1966b) hydrolyzed cytomycin in acid to cytosinine. The amide linkage between blastidic acid and cytosinine was shown to involve the amino group of the dehydropyran moiety of cytosinine and the carboxyl group of blastidic acid. From the nmr spectral evidence, Ōtake et al. (1966a) concluded that the protons at C-4 and C-5 of the sugar moiety were diaxial. One of the oxidation products of ozonolysis of the carbohydrate moiety of the substituted cytosine was shown to be β-hydroxyaspartic acid. Ōtake et al. (1965a) concluded that this amino acid was *erythro*-β-hydroxy-D-aspartic acid. Ōtake et al. (1966a, 1966b) and Yonehara and Ōtake (1966) subsequently revised their original structure for blasticidin S. The final structure assigned blasticidin S by Yonehara and Ōtake (1966) is 1-(1′-cytosinyl)-4-[L-3′-amino-5′-(1″-*N*-methylguanidino)-valerylamino]-1,2,3,4-tetradeoxy-β-D-*erythro*-hex-2-eneuronic acid (Fig. 4.10). Finally, the structural interpretation for blasticidin *S* hydrobromide has been provided by X-ray analysis (Onuma et al., 1966).

The total chemical synthesis of the unsaturated amino sugar of blasticidin S, methyl 4-amino-2,3,4-trideoxy-α-D-*erythro*-hex-2-enopyranosiduronic acid, has been reported by Goody, Watanabe and Fox (1970). This is the first synthesis of this new group of carbohydrates, the unsaturated amino sugars. Fox and his collaborators have also described the synthesis of the carbohydrate fragment of gougerotin (see Fig. 4.4).

The occurrence of 2,3-unsaturated sugars in nature prompted new studies to synthesize unsaturated carbohydrates and nucleosides with unsaturated sugars. For example, the chemical synthesis of a 2′,3′-unsaturated pyranosyl guanine nucleoside has been reported by Leutzinger, Robins, and Townsend (1968). The method for the synthesis of this nucleoside involved the condensation of the unsaturated sugar 3,4,6- tri-*O*-acetyl-D-glucal with 2-acetamido-6-chloropurine. The product isolated was 9-(2′,3′-didehydro-2′,3′-dideoxy-D-*erythro*-hexopyranosyl)guanine. The endocyclic double bond at the 2′ and 3′ positions of the hexose is similar to the 2′,3′-unsaturated

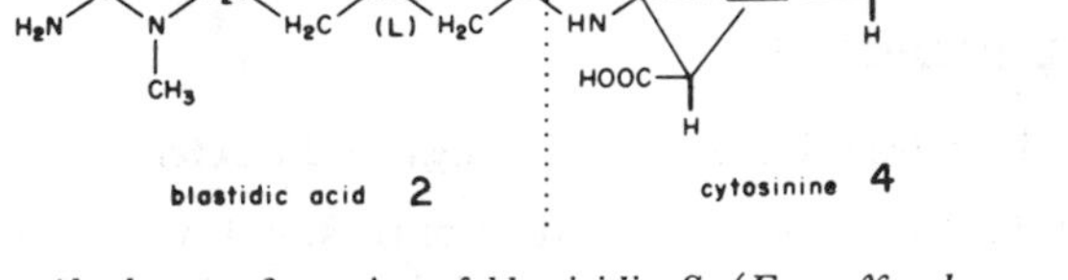

blastidic acid 2 | cytosinine 4

FIGURE 4.10 *Absolute configuration of blasticidin S. (From Yonehara and Ōtake,* 1966.)

hexose in blasticidin S. Since unsaturated sugars have been shown to occur in nature and play an important role in nucleoside antibiotics, Lemieux, Fraga, and Watanabe (1968) described a method for the preparation of methyl 4,6-*O*-benzylidene-D-hex-2-enopyranosides.

## BIOSYNTHESIS

Yonehara and Ōtake (1965) reported that cytosinine and blastidic acid were taken up by washed cells of *S. griseochromogenes*, resulting in an increased formation of blasticidin S. The per cent conversion from a 30-min incubation of washed cells was half the theoretical value. More recently, Seto et al. (1966b, 1968a, and 1968b) studied the biosynthesis of blastidic acid using carbon-14 labeled suspected precursors. Carbon-14-labeled compounds were added to cultures of *S. griseochromogenes* 48 hr after inoculation. Blasticidin S was isolated 24 hr later and degraded. The results obtained showed that cytosine $^{14}C$ was the precursor for the pyrimidine ring of blasticidin S. The carbon-14 from cytidine-(U)-$^{14}C$ was also incorporated, but all the radioactivity resided in cytosine. The ribose moiety of cytidine is incorporated into blasticidin S only after hydrolysis of cytidine to the base and the sugar.

The *N*-methyl group of blastidic acid was shown to arise from the methyl group of methionine. Arginine served as the carbon–nitrogen source for blastidic acid. The mechanism for $\beta$-amino acid formation from the $\alpha$-amino acid arginine has not been elucidated.

## INHIBITION OF GROWTH

Blasticidin S inhibits both gram-positive and gram-negative bacteria, and several members of the genus *Pseudomonas* and fungi. Tanaka et al. (1960) reported that blasticidin S is bactericidal in action. Misato et al. (1959) reported that blasticidin S is particularly effective in controlling rice infections by *P. oryzae*. The antibiotic exerts a greater effect on the mycelium rather than the spores. When phenylmercuric acetate was mixed with blasticidin S, the inhibitory properties of the antibiotic against rice blast disease were markedly increased. Blasticidin S exhibited antitumor activity against Walker adenocarcinoma 256, Ehrlich carcinoma, and sarcoma 180. The $LD_{50}$ dose for blasticidin intravenously is 2.8 mg/kg, while the peroral dose is 39.5 mg/kg. Blasticidin S is not toxic to fish and can be used near fish ponds and in rice paddies.

## BIOCHEMICAL PROPERTIES

### I. Inhibition of Protein Synthesis in Fungi and Bacteria

Huang et al. (1964a) studied the effect of blasticidin S in a cell-free system of *P. oryzae*. They reported that blasticidin S markedly inhibited the incorporation of carbon-14 labeled amino acids into protein. Yamaguchi et al. (1965) and Yamaguchi and Tanaka (1966) reported that blasticidin S inhibited growing cultures of *E. coli* and *B. megeaterium* at $2.3 \times 10^{-4}$ *M*. The addition of blasticidin S to cell-free protein-synthesizing systems of extracts of these two organisms strongly inhibited the incorporation of leucine and phenylalanine into polypeptide. Blasticidin S inhibition was markedly reversed with increased concentrations of sRNA. This report is interesting since Coutsogeorgopoulos (1967a) showed that increasing concentrations of L-phenylalanyl-tRNA did not reverse the inhibition of polyphenylalanine synthesis by blasticidin S. To determine if blasticidin S interfered with the polypeptide synthesis of the ribosome-messenger complex, the ribosomes were charged with poly U before the addition of blasticidin S (Fig. 4.11). The results show that blasticidin S does not interfere with the formation of the ribosome-messenger complex (Fig. 4.12), but did inhibit polypeptide synthesis. Apparently, some step following the formation of this complex is affected by blasticidin S (Yamaguchi and Tanaka, 1966).

Blasticidin S greatly interfered with the puromycin reaction, but was unlike the puromycin reaction in that blasticidin S prevented the release of nascent peptide chains from the ribosome to the supernatant. These data suggest that blasticidin S acts on some site of peptide bond forming reaction

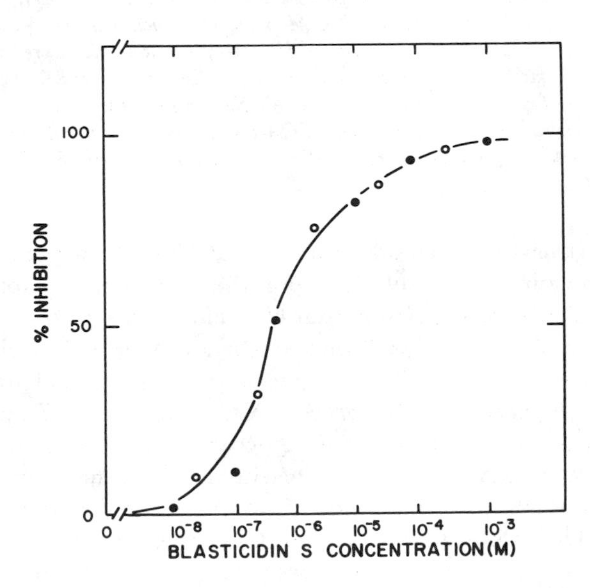


FIGURE 4.11 *Sensitivity of poly U-charged ribosomes to blasticidin S. inhibition.* (○) *Precharged ribosomes;* (●) *noncharged ribosomes.* (*From Yamaguchi and Tanaka, 1966.*)

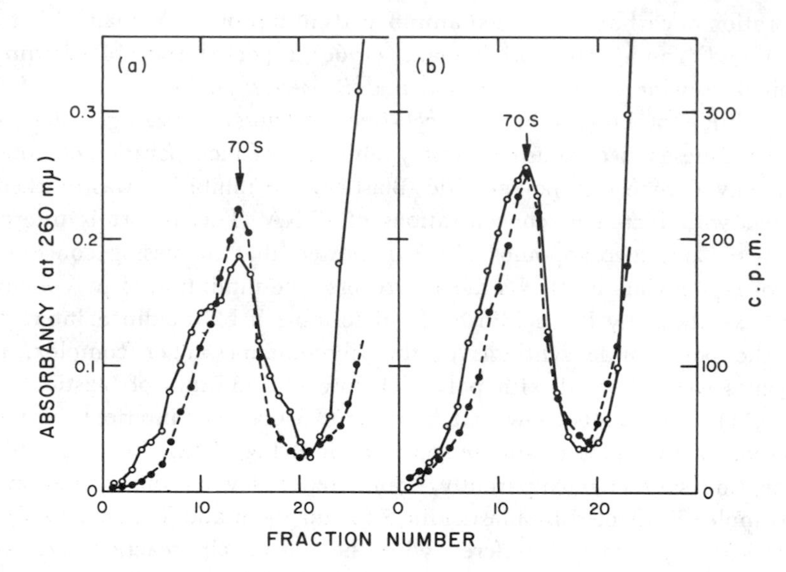


FIGURE 4.12 *Effect of blasticidin S on the binding of* $^{14}$*C-phenylalanyl-sRNA to ribosome-poly U complexes. Binding of* $^{14}$*C-phenylalanyl-sRNA with poly U to E. coli ribosomes was examined by the density gradient centrifugation, with and without the addition of blasticidin S. The reaction mixture contained, in a volume of* 0.4 *ml:* 2 *mg of washed ribosomes,* 20 *μg of poly U, 200 μg of* $^{14}$*C-phenylalanyl-sRNA* (22,400 *cpm*), *0.05 M Tris, pH 7.0, 0.1 M KCl,* 0.02 *M* $MgCl_2$, 0.006 *M mercaptoethanol, and, where indicated below,* $10^{-3}$ *M blasticidin S. All components except* $^{14}$*C-phenylalanyl-sRNA were incubated for 5 min at* 25°*C, after which* $^{14}$*C-phenylalanine-sRNA was added and the reaction was run for* 10 *min at* 0°*C. Centrifugation was carried out at* 40,000 *rpm for* 60 *min. Twelve-drops of fractions were collected and assayed for cold TCA-precipitable radioactivity* (○) *and absorbancy at* 260 *mμ* (●). (*a*) *Without blasticidin S;* (*b*) *with blasticidin S.* (*From Yamaguchi and Tanaka*, 1966.)

involving the transfer of peptide from peptidyl-tRNA to puromycin. Apparently blasticidin S can substitute for the incoming aminoacyl-tRNA. Additional evidence for the notion that blasticidin S is localized on the 50-S subunit was supplied by Weisblum and Davies (1968); they showed that blasticidin S also inhibited the binding of chloramphenicol to ribosomes isolated from *Bacillus stearothermophilus*. Yamaguchi and Tanaka (1966) suggested that the action of blasticidin S occurs in the step of peptide transfer from peptidyl-tRNA on the donor site with the incoming aminoacyl-tRNA (on the acceptor site), but does not affect the preceding steps in protein biosynthesis. The inhibition of leucine-$^{14}$C into polypeptide in the *in vitro* protein-synthesizing system with ribosomal particles of mouse Ehrlich carcinoma by blasticidin S was much greater than for the same *in vitro* system using normal rat liver ribosomal particles (Yamaguchi et al., 1965).

Coutsogeorgopoulos (1967a, 1967b) studied the aminoacyl nucleosides puromycin, chloramphenicol, gougerotin, amicetin, and blasticidin S as inhibitors of protein synthesis. The inhibition of blasticidin S was higher at the initial stages of the reaction. Also, the inhibition of protein synthesis by blasticidin S was greater at 25°C than at 37°C. The inhibition of polypeptide synthesis by blasticidin S could not be reversed when the concentration of phenylalanyl-tRNA was increased (Table 4.6). A graphical analysis of this inhibition with respect to L-phenylalanyl-RNA showed that the inhibition by blasticidin S is not competitive (Fig. 4.13). In a related study, Coutsogeorgopoulos (1967b) studied the effect of amicetin and blasticidin S on the inhibition of the puromycin reaction with ribosomes from *E. coli* and L-polylysyl-tRNA. At $1 \times 10^{-4}$ *M*, amicetin and blasticidin S inhibited the puromycin reaction 75 and 76%, respectively. The inhibition of the puromycin reaction by blasticidin S was similar to that of chloramphenicol. It appears that the linkage of the carboxyl group of amino acids or peptides to an amino group of a nucleoside, such as the exocyclic amino group of cytosine, produces compounds that are able to interact with the ribosome complex to inhibit the puromycin reaction (Coutsogeorgopoulos, 1967b). These data strongly suggest that the amino acylamino nucleoside antibiotics act as inhibitors of peptidyl transferase. However the recent data of Coutsogeorgopoulos (1969) in which he reports experiments that show the accumulation of oligophenylalanine from phenylalanyl-tRNA when blasticidin ($10^{-5}$ *M*) is added to $NH_4Cl$-washed *E. coli* ribosomes argues against this idea.

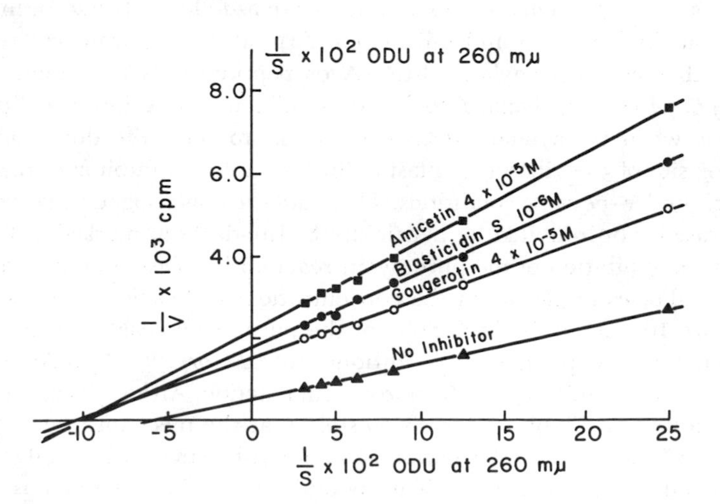


FIGURE 4.13 Nature of the inhibition of polyphenylalanine synthesis: (▲) *Control;* (●) *gougerotin;* (○) *blasticidin S;* (■) *amicetin.* (*From Coutsogeorgopoulos,* 1967*a.*)

**Table 4.6**

Effect of Increasing Concentrations of Phenylalanyl-RNA on the Inhibition of Polyphenylalanine Synthesis

| | ODU at 260 mμ of Phe-RNA (% inhibition) | | | | | | | | |
|---|---|---|---|---|---|---|---|---|---|
| Inhibitor, *M* | 4.0 | 8.0 | 12.0 | 16.0 | 20.0 | 24.0 | 32.0 | 36.0 | 40.0 |
| Puromycin ($2.4 \times 10^{-5}$) | 67 | 65 | 66 | 63 | 62 | 62 | 61 | 60 | 56 |
| Chloramphenicol ($10^{-3}$) | 56 | 53 | 53 | 53 | 50 | 50 | 47 | 47 | 46 |
| Gougerotin ($4 \times 10^{-5}$) | 52 | 52 | 56 | 57 | 58 | 60 | 61 | | |
| Blasticidin S ($10^{-6}$) | 58 | 61 | 65 | 66 | 64 | 67 | 67 | | |
| Amicetin ($4 \times 10^{-5}$) | 65 | 68 | 70 | 70 | 72 | 72 | 72 | | |
| Streptomycin ($2 \times 10^{-5}$) | 32 | 47 | 48 | 52 | 53 | 54 | 55 | | |
| Tetracycline ($1.2 \times 10^{-5}$) | 57 | 61 | 62 | 63 | 64 | 65 | 63 | | |

From Coutsogeorgopoulos, 1967a.

Tanaka (private communication) also argues against the notion that the mode of action of blasticidin S is related to the inhibition of peptidyl transferase. His experiments used a combination of blasticidin S and mikamycin A to support his arguments. His experiments are as follows: It had been established that blasticidin S and mikamycin A inhibit the puromycin reaction when either acetyl-phenylalanyl-tRNA or peptidyl-tRNA is bound to the ribosome. However, Tanaka found that mikamycin A inhibits dipeptide synthesis when phenylalanyl-tRNA is bound to both the donor site and acceptor site of the ribosome. Blasticidin S is not an inhibitor of dipeptide synthesis under the same conditions. These data strongly suggest that peptidyl transferase is not inhibited by blasticidin S. In addition, marked differences exist in the inhibition of the puromycin reaction by blasticidin S and mikamycin A. For example, when the concentrations of blasticidin S and puromycin are $10^{-4}$ *M* or $10^{-5}$ *M*, 50% of the puromycin reaction is inhibited. Under identical experimental conditions, $10^{-6}$ *M* or $10^{-7}$ *M* mikamycin A produces a 50% inhibition of the puromycin reaction. Although experimental data is not yet available, it is also possible that the mechanism of action of blasticidin S involves the formation of peptidyl-blasticidin S (similar to the peptidyl-puromycin reaction). This peptidyl-blasticidin S remains bound to the ribosomes instead of being released as in the case of the puromycin reaction. This notion, for the mode of action of blasticidin S, would be a substitute for the proposed ideas that blasticidin S acts either by inhibiting peptidyl transferase or as an inhibitor for the binding of the aminoacyl end of charged aminoacyl-tRNA or peptidyl-tRNA to the ribosomes.

The recent studies of Pestka (1969, 1970a, 1970b, 1970c) show that while blasticidin S does not inhibit the binding of the intact aminoacyl-tRNA to the ribosomes, it does inhibit the binding of phenylalanyl-oligonucleotide to ribosomes from human placenta and *E. coli* (Fig. 4.14, 4.15). Pestka reported similar findings with gougerotin (Fig. 4.6, 4.7) and amicetin (p. 209). These studies with charged tRNA strongly suggest that the aminoacyl nucleoside antibiotics did inhibit protein synthesis by interfering with the binding of the aminoacyl-end of aminoacyl-tRNA to the 50-S subunit of ribosomes.

Huang et al. (1964b) studied the effect of blasticidin S on sensitive and tolerant strains of *P. oryzae*. They reported that the incorporation of carbon-14 labeled amino acids into microsomal protein in cell-free extracts was markedly inhibited by blasticidin S from tolerant and sensitive cells. However, amino acid incorporation into protein of resistant cells growing in the presence of blasticidin S was not inhibited. Therefore, the resistance of *P. oryzae* seems to be attributed to differences in the permeability of the tolerant strain compared with that of the normal sensitive strain of *P. oryzae*. On the other hand, blasticidin S had little effect on the incorporation of amino acids into polypeptides with cell-free systems from *Pellicularia sasakii*. Apparently

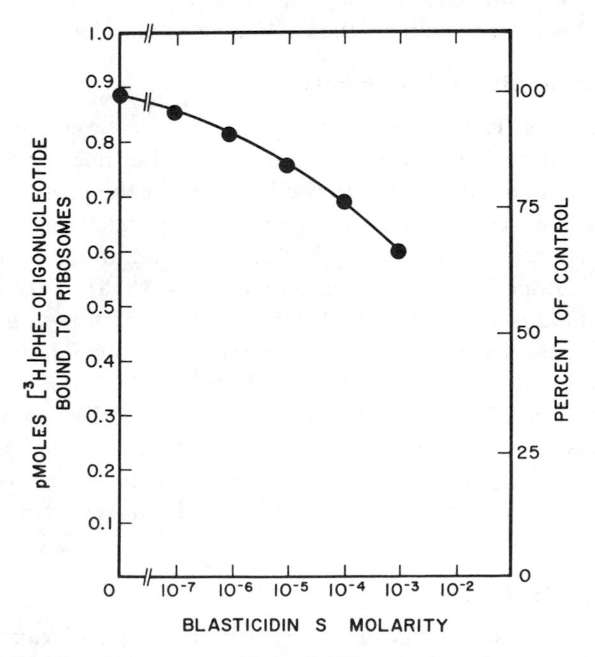


FIGURE 4.14 *Same legend as Fig.* 4.6 (*From Pestka, private communication*).

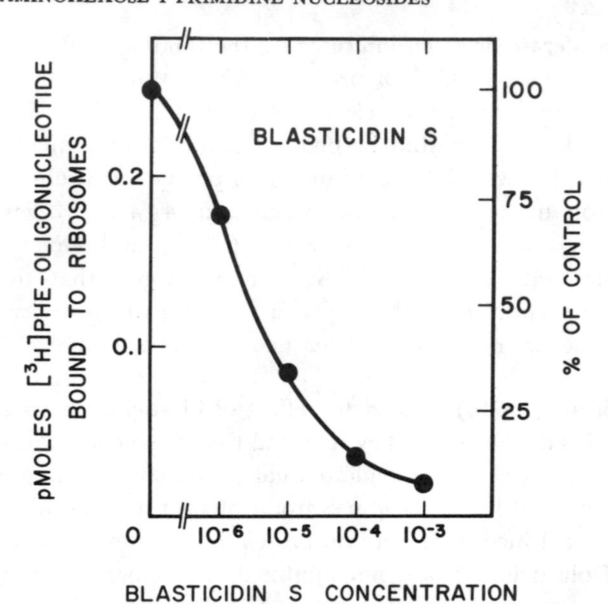


FIGURE 4.15 *Same legend as Fig.* 4.7 (*From Pestka, private communication*).

this insensitivity of *P. sasakii* to blasticidin S is not due to a difference in permeability as is the case of *P. oryzae*, but probably may be attributed to a different pathway in protein synthesis (Huang et al., 1964b).

## II. Effects of Blasticidin S on Plants

Blasticidin S is translocated to the leaves of the broad bean when cut stems are immersed in solutions containing the nucleoside antibiotic. Blasticidin S is not translocated to the leaves from the roots (Misato et al., 1958). When blasticidin S is added to the sweet potato root, phenylalanine ammonia-lyase activity is severely inhibited. In addition, polyphenol synthesis is also stopped (Minamikawa and Uritani, 1965). Takeo studied the effect of blasticidin S on polyphenoloxidase in tea leaf discs and showed that blasticidin S inhibited this enzyme. Apparently the increase in polyphenoloxidase during withering of the leaf is attributed to the *de novo* synthesis of this enzyme. Blasticidin S also suppressed *O*-diphenoloxidase in sweet potato slices and was more active than puromycin. Chlorotic spots are produced on the leaves of rice plants when sprayed with blasticidin S at concentrations exceeding 40 $\mu$g/ml. Phytotoxicity has also been reported to occur in numerous other plants.

## III. Effect on Plant Virus

Blasticidin S greatly reduced the rate of infection of rice plants with stripe virus (Hirai et al., 1968; Hirai and Shimomura, 1965). The leafhopper is the vector for transmitting rice stripe virus. When a leafhopper feeds on rice stems infiltrated with blasticidin S, the ability of progeny larvae to transmit the virus is greatly reduced. Carbon-14 labeled blasticidin S, introduced into the rice through the roots, and fed on by hoppers was distributed in sufficient concentrations to inhibit the virus in the bodies of the leafhoppers, especially in the salivary gland, gut, and fat body. Blasticidin S was easily degraded into compounds that were inert against the virus in rice. However, 2 days after spray treatment, blasticidin S was still found in effective concentrations in rice shoots to be inhibitory to the stripe virus. Blasticidin S was also found to inhibit tobacco mosaic virus (TMV) on bean and tomato plants (Hirai and Shimomura, 1965; Hirai et al., 1966).

### IV. Metabolism of Blasticidin S

*Aspergillus fumigatus* rapidly transforms blasticidin S into four compounds. One of the compounds, deaminoblasticidin S, was shown to be the result of deamination of the pyrimidine nucleus such that the cytosine ring is changed to the uracil ring (Seto et al., 1966a; Yonehara, 1968). Deaminoblasticidin S was 1/100 as active as blasticidin S against Walker adenocarcinoma 256, carcinoma 755 and Bashfold tumors. It was not toxic to mice when injected intravenously at 800 mg/kg/day for 7 days. In addition, the uracil derivative is not phytotoxic, but it did reduce disease spots in rice infected with *P. oryzae*. Three additional compounds were isolated as metabolic products following the addition of blasticidin S to flask cultures of *A. fumigatus*. These three compounds were biologically inactive.

## Summary

Blasticidin S is an amino acylamino nucleoside antibiotic that has the cytosine chromophore. The structure and configuration have been established. Blasticidin S inhibits protein synthesis by blocking peptide elongation. It has been suggested that this nucleoside antibiotic either inhibits the peptide transfer of peptidyl-tRNA from the donor site to the aminoacyl-tRNA on the acceptor site or it is identified with certain processes related to peptidyl transferase. Blasticidin S allows the formation of oligophenylalanine from phenylalanyl-RNA. With increasing concentrations of blasticidin S, polyphenylalanine formation is inhibited while oligophenylalanine accumulates. It does not inhibit the preceding steps in protein biosynthesis. While blasticidin S blocks the puromycin reaction, it does not cause a release of nascent peptide chains from the ribosome-mRNA complex to the supernatant. Blasticidin S markedly inhibits the binding of aminoacyl-oligonucleotides to ribosomes from human placenta and *E. coli*.

Cytosine, but not cytidine, was incorporated into blasticidin S. The *N*-methyl group of blastidic acid came from methionine.

## References


Coutsogeorgopoulos, C., *Biochemistry*, **6**, 1704 (1967a).
Coutsogeorgopoulos, C., *Biochem. Biophys., Res. Commun.*, **27**, 46 (1967b).
Coutsogeorgopoulos, C., *Federation Proc.*, **28**, 844 (1969).
Fox, J. J., and K. A. Watanabe, *Tetrahedron Letters*, 1966, 897.
Fukunaga, K., T. Misato, I. Ishii, and M. Asakawa, *Bull. Agr. Chem. Soc.* (*Japan*), **19**, 181 (1955).
Goody, R. S., K. A. Watanabe and J. J. Fox, *Tetrahedron Letters*, 293 (1970).
Hirai, T., and T. Shimomura, *Phytopathology*, **55**, 291 (1965).
Hirai, T., A. Hirashima, T. Otah, T. Takahashi, T. Shimomura, and Y. Hayashi, *Phytopathology*, **56**, 1236 (1966).
Hirai, T., T. Saito, H. Onda, K. Kitani, and A. Kiso, *Phytopathology*, **58**, 602 (1968).
Huang, K. T., T. Misato, and H. Asuyama, *J. Antibiotics* (*Tokyo*), **17A**, 65 (1964a).
Huang, K. T., T. Misato, and H. Asuyama, *J. Antibiotics* (*Tokyo*), **17A**, 71 (1964b).
Hyodo, H., and I. Uritani, *Agr. Biol. Chem.* (*Tokyo*), **30**, 1083 (1966).
Kono, Y., S. Takeuchi, and H. Yonehara, *J. Antibiotics* (*Tokyo*), **21A**, 433 (1968).
Lemieux, R. U., E. Fraga, and K. A. Watanabe, *Can. J. Chem.*, **46**, 61 (1968).
Leutzinger, E. E., R. K. Robins, and L. B. Townsend, *Tetrahedron Letters*, 1968, 4475.
Minamikawa, T., and I. Uritani, *Agr. Biol. Chem.* (*Tokyo*), **29**, 1021 (1965).
Misato, T., M. Asakawa, and K. Fukunaga, *Ann. Phytopath. Soc.* (*Japan*), **23**, 97 (1958).
Misato, T., I. Ishii, M. Asakawa, Y. Okimoto, K. Fukunaga, *Ann. Phytopath. Soc.* (*Japan*), **24**, 302 (1959).
Onuma, S., Y. Nawata, and Y. Saito, *Bull. Chem. Soc.* (*Japan*), **39**, 1091 (1966).
Ōtake, N., S. Takeuchi, T. Endō, and H. Yonehara, *Tetrahedron Letters*, 1965a, 1405.
Ōtake, N., S. Takeuchi, T. Endō, and H. Yonehara, *Tetrahedron Letters*, 1965b, 1411.
Ōtake, N., S. Takeuchi, T. Endō, and H. Yonehara, *Agr. Biol. Chem.* (*Tokyo*), **30**, 126 (1966a).
Ōtake, N., S. Takeuchi, T. Endō, and H. Yonehara, *Agr. Biol. Chem.* (*Tokyo*), **30**, 132 (1966b).
Pestka, S., *Proc. Natl. Acad. Sci.* (*U. S.*), **64**, 709 (1969).
Pestka, S., *Arch. Biochem. Biophys.*, **136**, 80 (1970a).
Pestka, S., *Arch. Biochem. Biophys.*, **136**, 89 (1970b).
Pestka, S., *Cold Spring Harbor Symp.*, **34**, 395 (1970c).
Seto, H., N. Ōtake, and H. Yonehara, *Agr. Biol. Chem.* (*Tokyo*), **30**, 877 (1966a).
Seto, H., I. Yamaguchi, N. Ōtake, and H. Yonehara, *Tetrahedron Letters*, 1966b, 3793.
Seto, H., I. Yamaguchi, N. Ōtake, and H. Yonehara, *Agr. Biol. Chem.*, (*Tokyo*), **32**, 1292 (1968a).
Seto, H., N. Otake, and H. Yonehara, *Agr. Biol. Chem.* (*Tokyo*), **32**, 1299 (1968b).
Takeo, T., *Agr. Biol. Chem.* (*Tokyo*), **30**, 1211 (1966).
Takeuchi, S., K. Hirayama, K. Ueda, H. Sakai, and H. Yonehara, *J. Antibiotics* (*Tokyo*), **11A**, 1 (1958).
Tanaka, N., T. Nishimura, N. Miyairi, Y. Sakagami, Y. Hsu, H. Yonehara, and H. Umezawa, *Gann. Suppl.*, **50**, 7 (1960).

Tsuruoka, T., and T. Niida, *Meiji Shika Kenkyu Nempo*, 1963, 23.
Weisblum, B., and J. Davies, *Bacteriol. Rev.*, **32**, 493 (1968).
Yamaguchi, H., and N. Tanaka, *J. Biochem. (Tokyo)*, **60**, 632 (1966)
Yamaguchi, H., C. Yamamoto, and N. Tanaka, *J. Biochem. (Tokyo)*, **57**, 667 (1965).
Yonehara, H., Japan Pat. No. 21,760 (1967); *Chem. Abstr.*, **68**, 28467n (1968).
Yonehara, H., and N. Ōtake, *Antimicrobial Agents Chemotherapy*, 1965, 855.
Yonehara, H., and N. Ōtake, *Tetrahedron Letters*, 1966, 3785.
Yonehara, H., S. Takeuchi, N. Ōtake, T. Endō, Y. Sakagami, and Y. Sumiki, *J. Antibiotics (Tokyo)*, **16A**, 195 (1963).

## 4.3. AMICETIN

### INTRODUCTION

Amicetin (allomycin or sacromycin) (Fig. 4.16, **1**, Fig. 4.17) was the first pyrimidine aminoacyl nucleoside antibiotic isolated. The total structure and stereochemistry has been established. The dipeptide $\alpha$-methylseryl-*p*-aminobenzoate is attached to the amino group of cytosine.

Amicetin inhibits gram-positive and gram-negative bacteria and is a potent inhibitor of protein synthesis *in vitro*.

### DISCOVERY, PRODUCTION, AND ISOLATION

Amicetin was first isolated by DeBoer et al. (1953) and Hinman et al. (1953) from a soil sample near Kalamazoo, Michigan. McCormick and Hoehn (1953) also isolated amicetin from *S. fasciculatis* NOV. *sp.* Tatsuoka et al. (1955) and Hinuma et al. (1955) isolated amicetin and assigned the names "allomycin" and "sacromycin." Haskell et al. (1958) subsequently isolated amicetin, bamicetin, and plicacetin from *S. plicatus*. Sensi et al. (1957) also isolated plicacetin and assigned the name amicetin B.

Amicetin is produced by submerged culture or shake flasks in soybean and yeast medium maintained at 26°C for 4–6 days. DeBoer et al. (1953) isolated and crystallized amicetin from the culture medium by extraction with 1-butanol (pH 7.5). The water–butanol was removed and the amicetin was dissolved in water, lyophilized, and purified by countercurrent distribution between methylene chloride and water. It was finally crystallized from water.

The procedure of McCormick and Hoehn (1953) for the isolation of amicetin involved extraction at pH 8.5 with 1-butanol and conversion to the picrate. The picrate was decomposed in ethanol saturated with dry HCl. Amicetin was crystallized by adding acetone to a water solution containing the antibiotic.

### PHYSICAL AND CHEMICAL PROPERTIES

The molecular formula for amicetin is $C_{29}H_{42}N_6O_9$; mp 244–245°C; $[\alpha]_D^{24} = +116.5°$ ($c = 0.5\%$ in 0.1 *N* HCl); $\lambda_{max}$ 305 m$\mu$ (pH. 7.0);

$\lambda_{max}$ 316 m$\mu$ (0.1 *N* HCl); $\lambda_{max}$ 322 m$\mu$ (0.1 *N* NaOH). It is soluble in water and dilute acid, but insoluble in organic solvents.

## STRUCTURAL ELUCIDATION

The structural assignment of amicetin (Fig. 4.16, **1**) was established in large part by Flynn, Haskell, Hanessian, and Stevens. Amicetin (**1**) is unstable in alkali and yields cytosamine (**2**) and a dipeptide. Acid hydrolysis

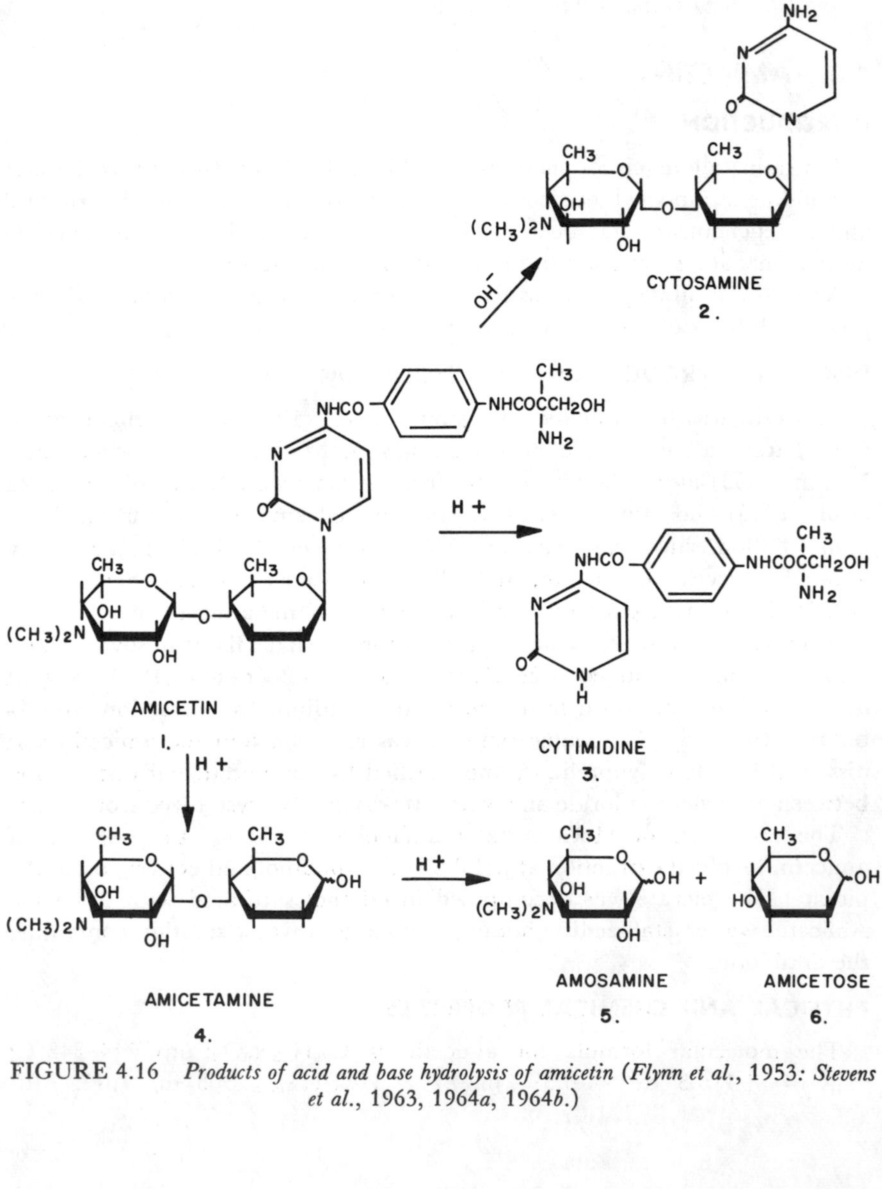

FIGURE 4.16 *Products of acid and base hydrolysis of amicetin (Flynn et al., 1953: Stevens et al., 1963, 1964a, 1964b.)*

produced the water-insoluble base cytimidine (**3**), which is a cytosine peptide of α-methylseryl-*p*-aminobenzoic acid (Flynn et al., 1953). Stevens et al. (1956) reported that amicetin could be hydrolyzed to amicetamine (**4**) and finally to amosamine (4,6-dideoxy-4-dimethylamino-D-glucose) (**5**) and amicetose (2,3,6-trideoxy-D-*erythro*-aldohexose) (**6**). The absolute proof of structure of these two sugars was established by their chemical syntheses (Stevens et al., 1963, 1964a, 1964b). Sensi et al. (1957) reported that cytosamine is a 1-substituted cytosine nucleoside. The anomeric configuration of the disaccharide has been shown to be α and the *N*-glycosylic linkage β (Hanessian and Haskell, 1964). The total structure and stereochemistry of amicetin (**1**) is shown in Figure 4.17. Fox et al. (1966) and Korzybski et al. (1967) have recently completed a review of amicetin.

## SYNTHESIS OF PYRIMIDINES RELATED TO AMICETIN

Noell and Chang (1966) synthesized the 1-(tetrahydro-2-pyranyl)pyrimidines, since a number of 9-substituted tetrahydropurine derivatives and the aminoacyl nucleosides were known to be potent biological inhibitors. These compounds were prepared by the Hibert-Johnson method from 2-chlorotetrahydropyran and the corresponding 2,4-dimethoxypyrimidines.

## INHIBITION OF GROWTH

Amicetin inhibits gram-positive and gram-negative organisms, *Mycobacterium tuberculosis*, and *Mycoplasma pneumoniae* (DeBoer et al., 1953; Tatsuoka et al., 1955; Araki et al., 1955; Arai et al., 1966; Brosbe et al., 1964). Smith et al. (1959) reported that amicetin inhibited the KB strain of human epidermoid carcinoma cells. The survival time of mice with leukemia 82 was also increased with amicetin (Burchenal et al., 1954). Amicetin was inactive against acute leukemia in children who had developed resistance to methotrexate, 6-mercaptopurine, and steroids (Tan and Burchenal, 1956). Renis (1965) reported that amicetin inhibited the methylation of DNA

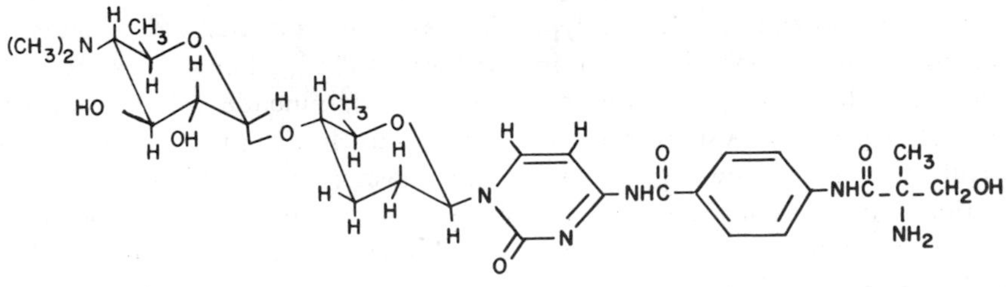

$(CH_3)_2N$ ... HO ... OH ... $CH_3$ ... O ... N ... NHC ... $NHC-C-CH_2OH$ ... $CH_3$ ... $NH_2$

FIGURE 4.17 *The total structure and stereochemistry of amicetin* (*Hanessian and Haskell,* 1964.)

viruses (herpes, vaccinia, and pseudorabies). The RNA viruses were not as sensitive. Cytimidine and cytosamine were not inhibitory.

Haskell (1958) reported on the importance of the structure of amicetin for antibacterial activity. Cytimidine and cytosamine were not inhibitory to bacteria, while the $N^6$-acetylated cytosamine derivatives were similar to amicetin as antibacterial compounds.

## BIOSYNTHESIS

The biosynthesis of the amosamine moiety (**5**) of amicetin has been studied both *in vitro* and *in vivo* (Edmundowicz, private communication). Sonic extracts of *S. plicatus* contained the enzyme cytidine diphosphoglucose pyrophosphorylase, which catalyzes the formation of cytidine diphosphate-D-glucose. CDP-glucose may be an important intermediate in the biosynthesis of CDP-amosamine. The incorporation of the methyl groups of labeled methionine into D-amosamine was taken as evidence that cytidine diphosphate D-amosamine is formed. It was postulated that CDP-D-amosamine reacted with an acceptor to form amicetin plus CDP.

## BIOCHEMICAL PROPERTIES

The first indication that amicetin was involved in the inhibition of protein synthesis was the report of Brock (1963). When amicetin (10 $\mu$g/ml) was added to *Streptococcus pyogenes*, there was an inhibition of growth and synthesis of *M* protein.

The structural resemblance of amicetin to the aminoacyladenylyl terminus of aminoacyl-tRNA led Bloch and Coutsogeorgopoulos (1966) to speculate that amicetin might inhibit protein synthesis in a manner similar to that of puromycin. Amicetin is a nucleoside antibiotic with an aminoacyl moiety attached to the amino group of cytosine. The authors reported that amicetin blocked protein synthesis, but RNA and DNA synthesis continued in *E. coli*. In cell-free extracts of *E. coli*, amicetin blocked the poly-U directed incorporation of L-phenylalanine into polyphenylalanine. Amicetin did not block the formation of phenylalanyl-tRNA but did interfere with the transfer of tRNA-bound amino acid to polypeptide. At a concentration of 1 $\mu$mole/ml, amicetin inhibited polyphenylalanine synthesis by 56%. The same structural relationship holds for puromycin or homocitrullylaminoadenosine. Amicetin inhibits protein synthesis in a manner similar to that of puromycin, homocitrullylaminoadenosine, gougerotin and blasticidin S. They all block peptide synthesis following aminoacyl-tRNA formation. This need not mean that amicetin has the same mechanism of action as other nucleosides that inhibit protein synthesis. When growing cultures of *E. coli* were supplemented with purines or pyrimidines, the inhibition of growth was not reversed. In a subsequent study, Coutsogeorgopoulos (1967b) studied protein synthesis with puromycin and amicetin to determine if amicetin blocked the puromycin reaction, as had been reported earlier for chloramphenicol and blasticidin S. When amicetin or blasticidin S was added to the *in vitro* protein-synthesizing systems, the puromycin reaction was inhibited 75 and 76%, respectively. Amicetin, in the absence of puromycin, did not release radioactive polylysine attached to tRNA. At a constant concentration of puromycin, increasing concentrations of amicetin blocked the puromycin reaction. When the molar concentration of amicetin was constant while the molar concentration of puromycin was increased, the inhibition of the puromycin reaction by amicetin decreased. Amicetin and blasticidin S are, respectively, 5 and 10 times as potent as chloramphenicol as inhibitors of the puromycin reaction. Similarly, Weisblum and Davies (1968) showed that ribosomes from *B. stearothermophilus* were 2.5 times as sensitive to amicetin as ribosomes from *E. coli*. As a result of this increased sensitivity of amicetin to *B. Stearothermophilus* ribosomes, Chang, Siddhikol, and Weisblum (1969) have now shown that the 50-S subunit is where several antibiotics exert their mode of action. Figure 4.18 shows the dose dependence of several antibiotics for

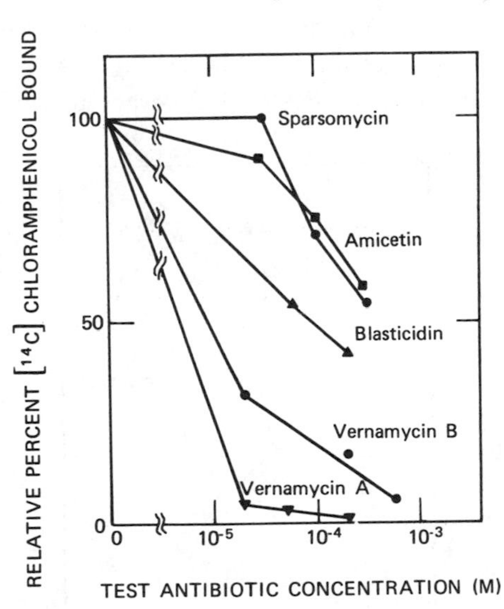


FIGURE 4.18 *Inhibition of* [$^{14}$C] *chloramphenicol binding to B. stearothermophilus ribosomes by several test antibiotics—dose response of inhibition. B. stearothermophilus ribosomes* (110 *pmoles*, 5 $A_{260}$ *m*μ *units*) *were incubated in a medium containing* 0.1 *M Tris HCl* (*pH* 7.7), *0.12 M* $MgCl_2$, *0.08 M KCl, 0.08 M* $NH_4Cl$, 0.007 *M* β*-mercaptoethanol*, 0.2 *mM* [$^{14}$C] *chloramphenicol* (*specific activity, 18* μ*C/*μ*mole*)*; unlabeled test antibiotic, concentration as noted on abscissa; total volume, 100* μ*l. After incubation for* 15 *min at* 37°*C, the mixture was worked up as in the tRNA binding assay described by Ravel* (1967). (*From Chang, Sidd, Weisblum*, 1969.)

inhibiting [$^{14}$C]chloramphenicol binding. Vernamycin A and B (streptogramin types A and B antibiotics) are the most effective antibiotics for inhibiting the binding of chloramphenicol, while blasticidin S, amicetin, and sparsomycin are least inhibitory. Pactomycin, bottromycin, fusidic acid, and rifamycin do not inhibit the binding of chloramphenicol to *B. stearothermophtlus* 50-S subunits (Fig. 4.19). Coutsogeorgopoulos (1967b) concluded that:

> A generalized requirement for the inhibition of the puromycin reaction, and by inference of the analogous reaction of aminoacyl-RNA, appears to be the presence of amino acids or peptides linked through their carboxyl to an amino function of a suitable structure containing either cytosine, as has been noted (Fox et al. 1966), or a *p*-nitrophenyl group.

In a related study using the amino acylaminonucleoside inhibitors of protein synthesis, Coutsogeorgopoulos (1967a) examined the question of whether the aminoacyl nucleoside inhibitors interfered with aminoacyl-tRNA at either an enzyme or ribosomal site. The inhibition exerted by amicetin was not reversed by increased amounts of phenylalanyl-tRNA (see Fig. 4.13 and Table 4.6, Sect. 4.2). Similar results were obtained with the pyrimidine nucleoside antibiotics gougerotin and blasticidin S. Although it appears that amicetin, gougerotin, and blasticidin S interact at the same site, conclu-

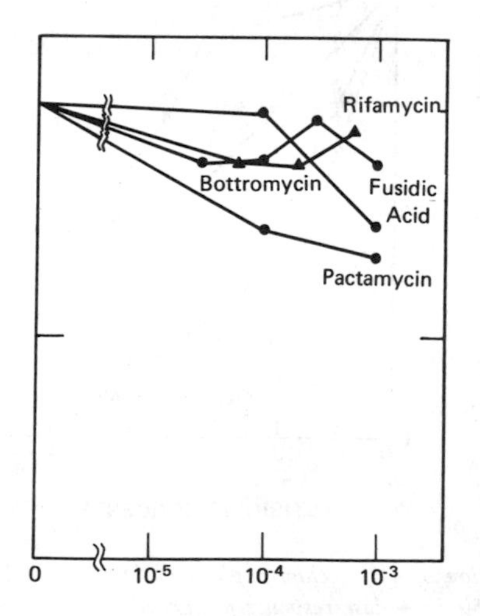


FIGURE 4.19 *Inhibition of [$^{14}$C]chloramphenicol binding to B. stearothermophilus ribosomes by several test antibiotics—dose response of inhibition. Experimental conditions were the same as those used in Fig. 4.14, except that other antibiotics were tested, as indicated. (From Chang, Siddhikol, and Weisblum,* 1969.)

sions on the mechanism of action of these antibiotics must await further investigation. When the level of phenylalanyl-tRNA was kept constant, the inhibition of peptide synthesis by amicetin was much stronger during the initial stages of the reaction. Whereas puromycin, gougerotin, and blasticidin S gave higher inhibitions at lower incubation temperatures, the inhibition exerted by amicetin was almost independent of temperature. Coutsogeorgopoulos (1969), using $NH_4Cl$-washed *E. coli* ribosomes, showed that while amicetin ($4 \times 10^{-5}$ *M*) inhibited the puromycin reaction over 50%, oligophenylalanine (but not polyphenylalanine) formation occurred. As the concentration of amicetin was increased, there was a decrease in polyphenylalanine formation and an accumulation of oligophenylalanine. Tetracycline inhibits oligophenylalanine and polyphenylalanine.

The effect of amicetin on the ribosome-catalyzed peptidyl transfer was studied by Monro and Vazquez (1967). For this study they used the formylmethionyl-hexanucleotide fragment. This fragment is prepared by the treatment of *N*-formylmethionyl-tRNA with ribonuclease $T_1$ (Monro and Marcker, 1967; Monro, 1967). Since the evidence indicates that the fragment reaction is catalyzed by the same mechanism as peptidyl transfer in protein synthesis, the aminoacyloligonucleotide fragment eliminates other reactions which are normally interlinked with peptidyl transfer through other moieties of the tRNA molecule. This type assay provides a means for determining the effect of specific inhibitors of protein synthesis. When amicetin was added to incubation mixtures, the fragment reaction was strongly inhibited (Monro and Vazquez, 1967). It is of interest that gougerotin, sparsomycin, and amicetin inhibit the fragment reaction, but do not effect chloramphenicol binding. These data strongly indicate that peptidyl transfer may be inhibited by the action upon at least two sites. Presumably, both of these sites are located on the 50-S subunit. Amicetin partially inhibits the formation of the sparsomycin-induced complex (Monro et al., 1969). In the absence of sparsomycin, amicetin stimulated the binding of the oligonucleotide fragment (CACCA-leucyl-acetate) to ribosomes. The authors concluded that the mode of action of amicetin is related to that of sparsomycin in that they compete for the same site. The inert complex formed in the presence of amicetin may be less stable than the sparsomycin complex. Pestka (1969, 1970a, 1970b, 1970c) has also reported extensive studies in which he has compared the effect of many inhibitors of protein synthesis on the binding of intact aminoacyl-tRNA with that of amino-acyl-oligonucleotide. Whereas the binding of phenylalanyl-oligonucleotide to ribosomes from *E. coli* and human placenta (Fig. 4.6; 4.7) is markedly inhibited, the binding of intact phenylalanyl-tRNA to *E. coli* ribosomes is not inhibited (Pestka, 1970b). These data had been interpreted by Pestka to support the notion that the aminoacyl nucleoside antibiotics inhibit the binding of the aminoacyladenylyl terminus of tRNA to the ribosomes. Amicetin does not inhibit respiration or phosphorylation of isolated liver mitochondria with glutamate as the substrate (Lardy et al., 1958).

## Summary

Amicetin is a cytosine nucleoside antibiotic with an aminoacyl group attached to the amino group of cytosine. The total structure and stereochemistry of amicetin have been established. It is an inhibitor of bacteria, human carcinoma cells, and viruses. Amicetin selectively inhibits the ribosome-catalyzed peptidyl transfer with extracts of *E. coli*. It also inhibits the puromycin reaction. It does not inhibit respiration or phosphorylation of mitochondria. It appears that amicetin is sufficiently similar in structure to the aminoacyl adenylyl terminus of tRNA to be bound to the ribosome-messenger RNA-peptidyl complex. Oligophenylalanine accumulates with increasing concentrations of amicetin, while the formation of polyphenylalanine is inhibited.

## References

Arai, S., K. Yoshida, A. Izawa, K. Kumagai, and N. Ishida, *J. Antibiotics (Tokyo)*, **19A**, 118 (1966).

Araki, T., K. Tsuchiya, T. Tsuchiya, and F. Hiraiwa, *Ann. Rept. Takeda Res. Lab.*, **14**, 112 (1955).

Bloch, A., and C. Coutsogeorgopoulos, *Biochemistry*, **5**, 3345 (1966).

Brock, T. D., *J. Bacteriol.*, **85**, 527 (1963).

Brosbe, E. A., P. T. Sugihara, C. R. Smith, and L. Hyde, *Antimicrobial Agents Chemotherapy*, 1964, 733.

Burchenal, J. H., M. Yuceoglu, M. Y. Dagg, and C. C. Stock, *Proc. Soc. Exptl. Biol. Med.*, **86**, 891 (1954).

Chang, N., C. Siddhikol, B. Weisblum, *Biochim. Biophys. Acta*, **186**, 396 (1969).

Coutsogeorgopoulos, C., *Biochemistry*, **6**, 1704 (1967a).

Coutsogeorgopoulos, C., *Biochem. Biophys. Res. Commun.*, **27**, 46 (1967b).

Coutsogeorgopoulos, C., *Federation Proc.*, **28**, 844 (1969).

DeBoer, C., E. L. Caron, and J. W. Hinman, *J. Amer. Chem. Soc.*, **75**, 499 (1953).

Flynn, E. H., J. W. Hinman, E. L. Caron, and D. O. Woolf, Jr., *J. Amer. Chem. Soc.*, **75**, 5867 (1953).

Fox, J. J., K. A. Watanabe, and A. Bloch, *Progr. Nucl. Acid Res. Mol. Biol.*, **5**, 252 (1966).

Hanessian, S., and T. H. Haskell, *Tetrahedron Letters*, 1964, 2451.

Haskell, T. H., *J. Amer. Chem. Soc.*, **80**, 747 (1958).

Haskell, T. H., A. Ryder, R. P. Frohardt, S. A. Fusari, Z. L. Jakubowski, and Q. R. Bartz, *J. Amer. Chem. Soc.*, **80**, 743 (1958).
Hinman, J. W., E. L. Caron, and C. DeBoer, *J. Amer. Chem. Soc.*, **75**, 5864 (1953).
Hinuma, Y., M. Kuroya, T. Yajima, K. Ishihara, S. Hamada, K. Watanabe, and K. Kikuchi, *J. Antibiotics (Tokyo)*, **8A**, 148 (1955).
Korzybski, T., Z. Kowszyk-Gindifer, and W. Kurylowicz, *Antibiotics: Origin, Nature and Properties*, Vol. 1, Pergamon Press, Warsaw, 1967.
Lardy, H. A., D. Johnson, and W. C. McMurray, *Arch. Biochem. Biophys.*, **78**, 587 (1958).
McCormick, M. H., and M. M. Hoehn, *Antibiot. Chemotherapy*, **3**, 718 (1953).
Monro, R. E., and K. A. Marcker, *J. Mol. Biol.*, **25**, 347 (1967).
Monro, R. E., *J. Mol. Biol.*, **26**, 147 (1967).
Monro, R. E., and D. Vazquez, *J. Mol. Biol.*, **28**, 161 (1967).
Monro, R. E., M. L. Celma, and D. Vazquez, *Nature*, 222, 356 (1969).
Noell, C. W., and C. C. Cheng, *J. Heterocyclic Chem.*, **3**, 5 (1966).
Pestka, S., *Proc. Natl. Acad. Sci. (U.S.)*, **64**, 709 (1969).
Pestka, S., *Arch. Biochem. Biophys.*, **136**, 80 (1970a).
Pestka, S., *Arch. Biochem. Biophys.*, **136**, 89 (1970b).
Pestka, S., *Cold Spring Harbor Symp.*, **34**, 395 (1970c).
Ravel, J., *Proc. Natl. Acad. Sci. (U.S.)*, **57**, 1811 (1967).
Renis, H. E., *Antimicrobial Agents, Chemotherapy*, 1965, 567.
Sensi, P., A. M. Greco, G. G. Gallo, and G. Rolland, *Antibiot. Chemotherapy*, **7**, 645 (1957).
Smith, C. G., W. L. Lummis, and J. E. Grady, *Cancer Res.*, **19**, 847 (1959).
Stevens, C. L., P. Blumbergs, and F. A. Daniher, *J. Amer. Chem. Soc.*, **85**, 1552 (1963).
Stevens, C. L., R. J. Gasser, T. K. Mukherjee, and T. H. Haskell, *J. Amer. Chem. Soc.*, **78**, 6212 (1956).
Stevens, C. L., N. A. Nielsen, P. Blumbergs, and K. G. Taylor, *J. Amer. Chem. Soc.*, **86**, 5695 (1964b).
Stevens, C. L., P. Blumbergs, and D. L. Wood, *J. Amer. Chem. Soc.*, **86**, 3592 (1964a).
Tan, C. T. C., and J. H. Burchenal, *Antibiot. Med. Clin. Therapy*, **3**, 126 (1956).
Tatsuoka, K., K. Nakazawa, M. Inoue, and S. Fujii, *Yakugaku Zasshi*, **75**, 1206 (1955).
Weisblum, B., and J. Davies, *Bacteriol. Rev.*, **32**, 493 (1968).

## 4.4. BAMICETIN

### INTRODUCTION, DISCOVERY, PRODUCTION, AND ISOLATION

The pyrimidine nucleoside antibiotic bamicetin is one of the three pyrimidine nucleoside antibiotics isolated from the fermentation cultures of *S. plicatus* (Haskell, et al., 1958 )(Fig. 4.20, **1**). This nucleoside antibiotic is closely related in structure to amicetin and plicacetin.

Bamicetin and amicetin were separated by countercurrent distribution using 1-butanol-0.1 *M* phosphate buffer (pH 6.9). The partition coefficients of amicetin and bamicetin are sufficiently different to permit a good separation by this technique. Amicetin and bamicetin are also isolated from the culture filtrates by extraction with 1-butanol followed by reextraction into aqueous acid and finally precipitation of both antibiotics as the mixed hydrochloride salts. The mixed hydrochloride salt is stirred in water–chloroform (pH 8.2). Amicetin is distributed in the chloroform fraction and the bamicetin is found in the aqueous fraction. When the pH is adjusted to 5.5, followed by concentration, a change of pH to 8.5, and extraction with 1-butanol, bamicetin partitions in the organic layer. Bamicetin hydrochloride is formed by addition of acetone and concentrated HCl to the 1-butanol. Bamicetin recrystallizes from methanol–water and gives a white microcrystalline precipitate.

## PHYSICAL AND CHEMICAL PROPERTIES

The molecular formula of bamicetin is $C_{28}H_{40}N_6O_9$; mp 240–241°C (decomp.); $[\alpha]_D^{26}$ +121° ($c$ = 0.5 in 0.1 $N$ HCl); $\lambda_{max}$ 314 m$\mu$ (in 0.1 $N$ HCl); $\lambda_{max}$ 302 m$\mu$ (phosphate buffer, pH 7.0); $\lambda_{max}$ 322 m$\mu$ (in 0.1 $N$ NaOH).

## STRUCTURAL ELUCIDATION

The tentative structure proposed for bamicetin is attributed to Haskell (1958). Acid hydrolysis of bamicetin (**1**) yields bamicetamine (**2**) and cytimidine (**3**) (Fig. 4.20). The aminodisaccharide bamicetamine (**2**) is a C-13 hydrolysis product. Whereas periodate oxidation of amicetin afforded dimethylamine, a similar oxidation of bamicetin gave monomethylamine. This was the only observable difference between the two molecules aside from biological potency and solubility. If this methyl group is the only difference in structure between amicetin and bamicetin, it may be that bamicetin is the precursor of amicetin.

## INHIBITION OF GROWTH AND BIOCHEMICAL PROPERTIES

Bamicetin is active against experimental mouse tuberculosis. The test organism to assay bamicetin during isolation was *E. coli* (Haskell, 1958). Although bamicetin is an aminoacyl aminopyrimidine nucleoside antibiotic and is structurally related to amicetin, it has not been studied in detail like amicetin. If this nucleoside can be classified with the other acylaminonucleoside antibiotics puromycin, gougerotin, blasticidin S, and amicetin, then bamicetin would probably interfere with protein synthesis by acting as an analog of aminoacyl-tRNA in the peptidyl transfer step of the transfer reaction in protein synthesis. Brosbe et al. (1963) reported that bamicetin was less potent than amicetin in inhibiting the growth of *M. fortuitum*.

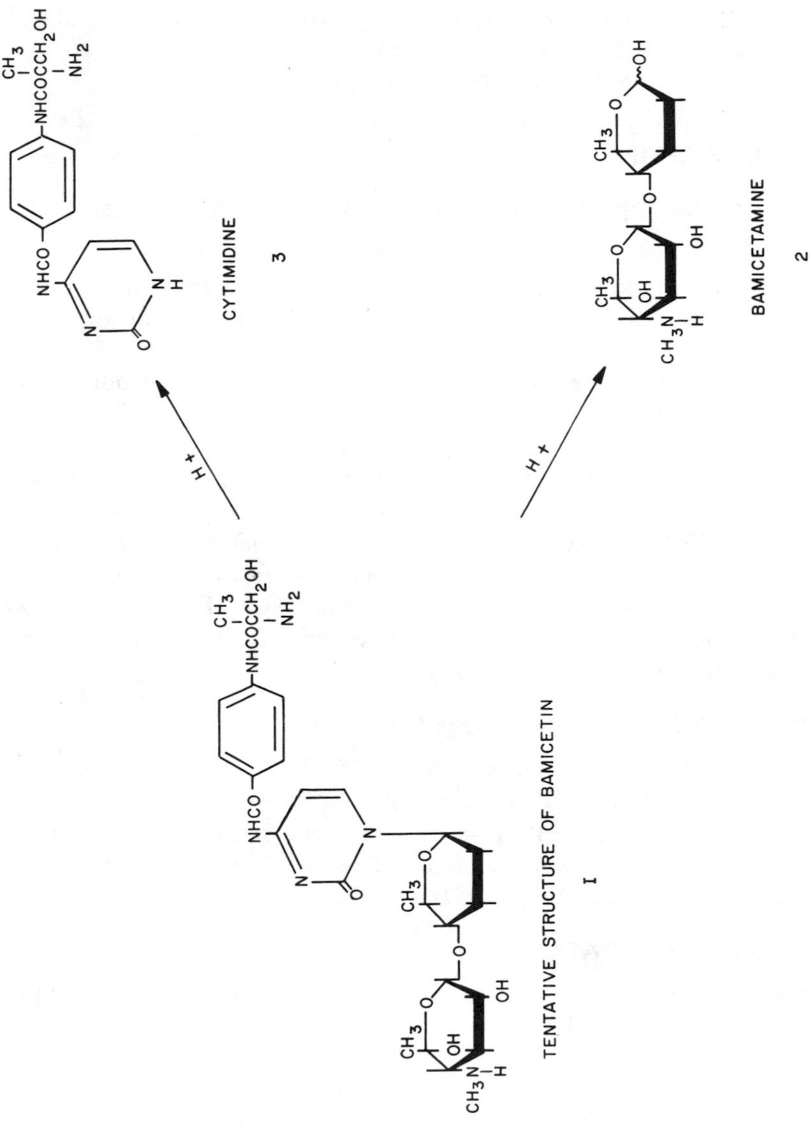

TENTATIVE STRUCTURE OF BAMICETIN

I

$H^+$

CYTIMIDINE

3

$H^+$

BAMICETAMINE

2

FIGURE 4.20 *Hydrolysis products of bamicetin (Haskell, 1958).*

## References

Brosbe, E. A., P. T. Sugihara, C. R. Smith, and L. Hyde, *Amer. Rev. Resp. Diseases*, **88**, 112 (1963).
Haskell, T. H., A. Ryder, R. P. Frohardt, S. A. Fusari, Z. L. Jakubowski, and Q. R. Bartz, *J. Amer. Chem. Soc.*, **80**, 743 (1958).
Haskell, T. H., *J. Amer. Chem. Soc.*, **80**, 747 (1958).

# 4.5. PLICACETIN

## INTRODUCTION, DISCOVERY, PRODUCTION, AND ISOLATION

Plicacetin (Fig. 4.21) is isolated from the culture filtrates of *S. plicatus* along with amicetin and bamicetin (Haskell et al., 1958). Plicacetin has also been isolated independently from a sample of soil in Chile from an unclassified strain of the *Actinomyces*, *Streptomyces sp.* 285 by Sensi et al. (1957). Sensi assigned the name "amicetin B." As a result of the chemical studies on the structure of plicacetin, Haskell et al. (1958) and Haskell (1958) reported that the structure of plicacetin is identical with that of amicetin B.

Plicacetin is isolated from the culture filtrates (pH 7.1) by extraction with 1-butanol. The butanol is evaporated and the plicacetin is extracted into 0.01 *N* HCl. The acid extract is passed through an anion column (amberlite IR-4B) and concentrated to dryness. Plicacetin was extracted with absolute methanol. Ether was added to precipitate plicacetin. This procedure was repeated. Plicacetin was crystallized from water (pH. 9.0) and finally from aqueous methanol as colorless needles.

## PHYSICAL AND CHEMICAL PROPERTIES

The molecular formula for plicacetin is $C_{25}H_{35}N_5O_7$; molecular weight (by Rast method) 520 to 522; mp 160–163°C (when crystallized from ethyl acetate); $[\alpha]_D^{26}$ + 181° ($c$ = 2.7% in methanol); $\lambda_{max}$ 257 mμ and 311.5 mμ in 0.1 *N* HCl; $\lambda_{max}$ 249 mμ and 321 mμ in phosphate buffer, pH 7.0; and $\lambda_{max}$ 329 mμ in 0.1 *N* NaOH.

## STRUCTURAL ELUCIDATION

Haskell et al. (1958) and Haskell (1958) reported the assigned structure to plicacetin as shown in Figure 4.22. Alkaline hydrolysis of plicacetin (**1**) produced cytosamine (**2**) and *p*-aminobenzoic acid (**3**). Acid hydrolysis of plicacetin yields cytosine (**4**) and *p*-aminobenzoic acid (**3**). α-Methylserine was not a product of acid hydrolysis as it was in the case of mild acid hydrolysis of amicetin or bamicetin. Since the only structural difference between amicetin and plicacetin is the α-methylserine group, Haskell et al. (1958) suggested that plicacetin is probably a precursor of amicetin. Plicacetin

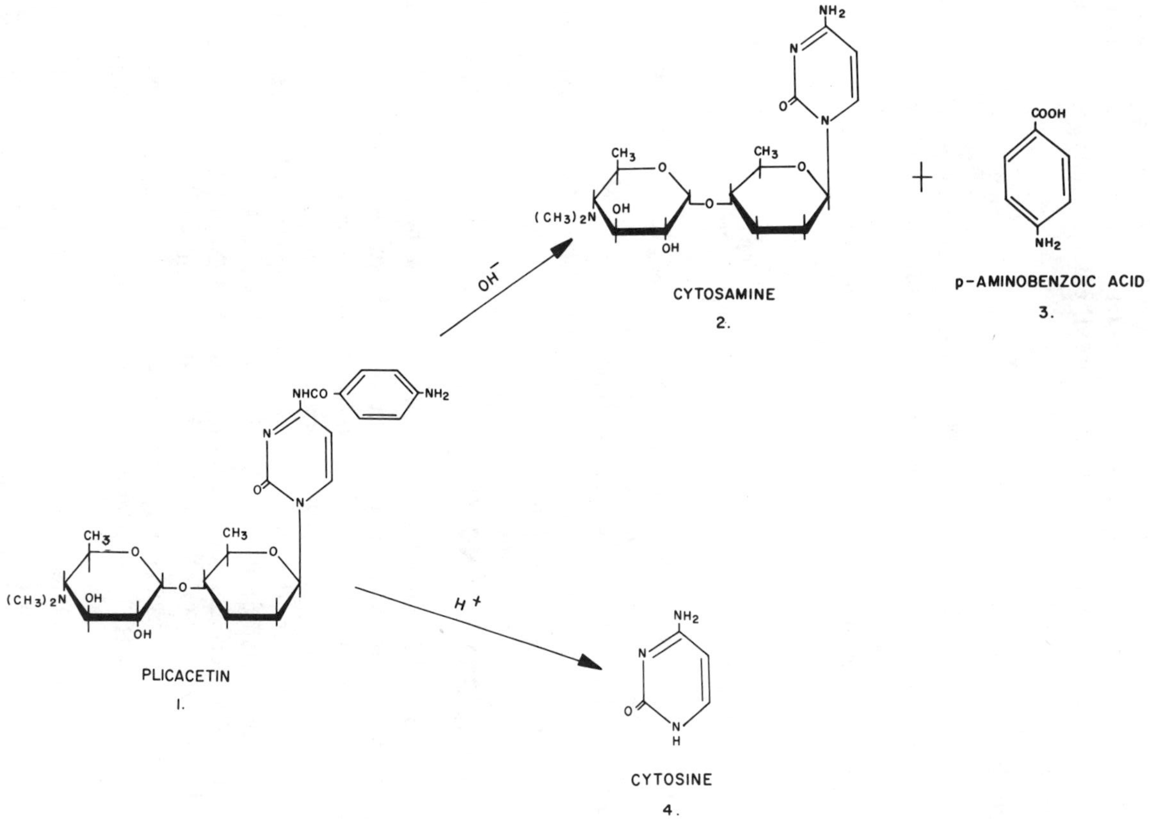

PLICACETIN
1.

CYTOSAMINE
2.

p-AMINOBENZOIC ACID
3.

CYTOSINE
4.

FIGURE 4.21 *Structural determination for plicacetin. (From Haskell et al., 1958; Haskell, 1958).*

is probably not a precursor for the biosynthesis of bamicetin, since the amino group on the sugar moiety of plicacetin has two methyl groups, while bamicetin has only one. If such a biogenetic relationship did exist, then the conversion of plicacetin to bamicetin would require the removal of one of the *N*-methyl groups on the aminohexose moiety. Plicacetin can be synthesized from cytosamine and *p*-nitrobenzoyl chloride followed by the reduction of the nitro group.

## INHIBITION OF GROWTH

Plicacetin is less effective against *Mycobacterium tuberculosis* H 37R than is amicetin or bamicetin. Since plicacetin is a cytosine nucleoside antibiotic, it probably inhibits protein synthesis in a manner similar to that reported for puromycin, gougerotin, blasticidin S, amicetin, and bamicetin. However, detailed experimental evidence is not available.

### References

Haskell, T. H., A Ryder, R. P. Frohardt, S. A. Fusari, Z. L. Jakubowski, and Q. R. Bartz, *J. Amer. Chem. Soc.*, **80**, 743 (1958).
Haskell, T. H., *J. Amer. Chem. Soc.*, **80**, 747 (1958).
Sensi, P., A. M. Greco, G. G. Gallo, and G. Rolland, *Antibiot. Chemotherapy*, **7**, 645 (1957).

# 4.6. AMICETIN A AND C

DeBoer and Hinman (1959a) reported on the isolation of amicetin A and amicetin C from the culture filtrates of *Streptomyces vinaceus-drappus*. Although the structure of both of these antibiotics is not known, their chemical and physical properties strongly suggest that they are nucleoside antibiotics. However, this has not been rigorously established (Hinman, private correspondence). Both antibiotics are effective against mycobacteria. The molecular formula for amicetin A is $C_{31}H_{42}N_6O_8$; mp 209–212°C; $\lambda_{max}$ 251 mμ ($E^{1\%}_{1\,cm}$ 290); 320 mμ ($E^{1\%}_{1\,cm}$ 540); 230 mμ ($E^{1\%}_{1\,cm}$ 236); 272 mμ ($E^{1\%}_{1\,cm}$ 176); $\lambda_{max}^{0.1\,N\,HCl}$ 256 mμ ($E^{1\%}_{1\,cm}$ 341); 312 mμ ($E^{1\%}_{1\,cm}$ 288); $[\alpha]_D^{22}$ = +138.8° (in 0.1 *N* HCl). Amicetin A is slightly soluble in water and methyl alcohol.

Amicetin C was isolated from the same organism (DeBoer and Hinman, 1959b). The molecular formula is $C_{56}H_{94}N_{10}O_{19}S$; mp 160–163°C; $\lambda_{max}^{H_2O}$ 257 mμ ($E^{1\%}_{1\,cm}$ 289); 304 mμ ($E^{1\%}_{1\,cm}$ 250); 229 mμ ($E^{1\%}_{1\,cm}$ 132.5); 283 mμ ($E^{1\%}_{1\,cm}$ 194.9); $\lambda_{max}^{0.1\,N\,HCl}$ 255 mμ ($E^{1\%}_{1\,cm}$ 183); 313 mμ ($E^{1\%}_{1\,cm}$ 332); 230 mμ ($E^{1\%}_{1\,cm}$ 141.3); 278 mμ ($E^{1\%}_{1\,cm}$ 138.5); $[\alpha]_D^{22}$ = +123.1° (in 0.1 *N* HCl).

## References

DeBoer, C., and J. W. Hinman, U. S. Pat. No. 2,909,463 (1959a). *Chem. Abstr.*, **54**, 2669g (1960).

DeBoer, C., and J. W. Hinman, U. S. Pat. No. 2,909,464 (1959b), *Chem. Abstr.*, **54**, 2669i (1960).

CHAPTER 5

# Peptidyl-Pyrimidine Nucleosides



## 5.1. POLYOXINS

### INTRODUCTION

The polyoxins represent a new group of peptidyl-pyrimidine nucleoside antibiotics that are antifungal in their action. They are produced by *Streptomyces cacaoi* var. *asoensis*. Twelve polyoxins (A–L) have been isolated and their structures have been elucidated. The twelve polyoxins include α-L-amino acids, 5-aminofuranuronoside, and the 2,4-dioxy chromophore. All the polyoxins except C and I are unique in that they are extremely toxic towards phytopathogenic fungi, but have no biological activities against other organisms. The inhibition of uptake of glucosamine suggests that the site of action may be related to cell-wall chitin biosynthesis (Eguchi et al., 1968). Dipeptides reverse the inhibition of polyoxin B and D. The general formula for the polyoxins is shown in Figure 5.1. All polyoxins are made up of combinations of $R_1$, $R_2$, and $R_3$.

### DISCOVERY, PRODUCTION, AND ISOLATION

The isolation, physical properties, chemical degradations, structural elucidations, and biochemical properties of this unique group of nucleoside antibiotics may be in large part attributed to the elegant work of Suzuki, Isono, and their co-workers. Three strains of *Streptomyces* which belong to *S. cacaoi* were isolated from soil samples collected at Aso district of Kumamoto prefecture by Isono et al. (1965).

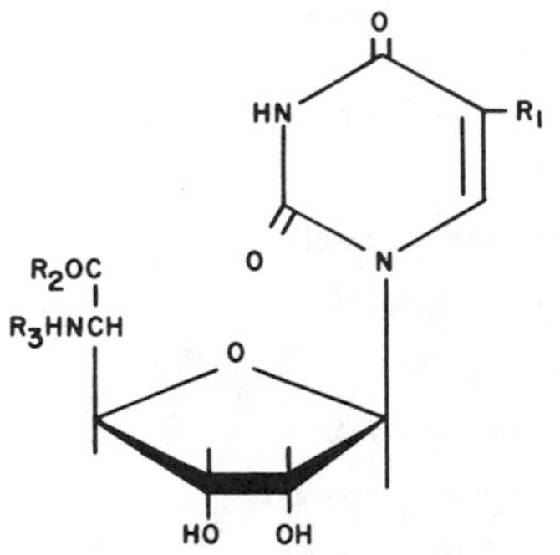

$R_1$ = $-CH_2OH$, $-COOH$, $-CH_3$, $-H$

$R_2$ = 3-ETHYLIDENE-L-AZETIDINE-2-CARBOXYLIC ACID

$R_3$ = 5-O-CARBAMOYL-2-AMINO-2-DEOXY-L-XYLONIC ACID OR THE 3-DEOXY DERIVATIVE

FIGURE 5.1 *The general formula for the polyoxins.* (*From Isono et al.*, 1968*b*.)

The medium used for production of the polyoxin complex is as follows: 2% glucose, 1% meat extract, and 1% peptone (Isono et al., 1967). The seed culture was transferred to medium A, which was then used to inoculate medium B and finally medium C (Table 5.1). Fermentation in medium C was carried out for 72 hr at 28°C with agitation and aeration. The isolation of polyoxins A–I is shown in Figure 5.2. More recently, polyoxins, J, K, and L have been isolated and characterized by Isono et al. (1968) and Isono and Suzuki (1968b; 1968d). Polyoxins A, B, G, H, and I were crystallized from aqueous ethanol. Polyoxins A and H were obtained as needle crystals,

**Table 5.1**

Composition of Culture Media

| | A | B | C |
|---|---|---|---|
| Glycerol | — | — | 1.0% |
| Glucose | 1.5% | 0.5% | 1.5 |
| Soluble Starch | 3 | 5 | — |
| Soybean meal | 2 | 0.5 | 1.5 |
| Ammonium sulfate | — | — | 0.5 |
| Dried yeast | 3 | 1 | 0.5 |
| Wheat embryo | — | 2 | — |
| Sodium chloride | — | — | 0.5 |
| Calcium carbonate | 0.4 | 0.1 | 0.5 |

From Isono et al., 1967.

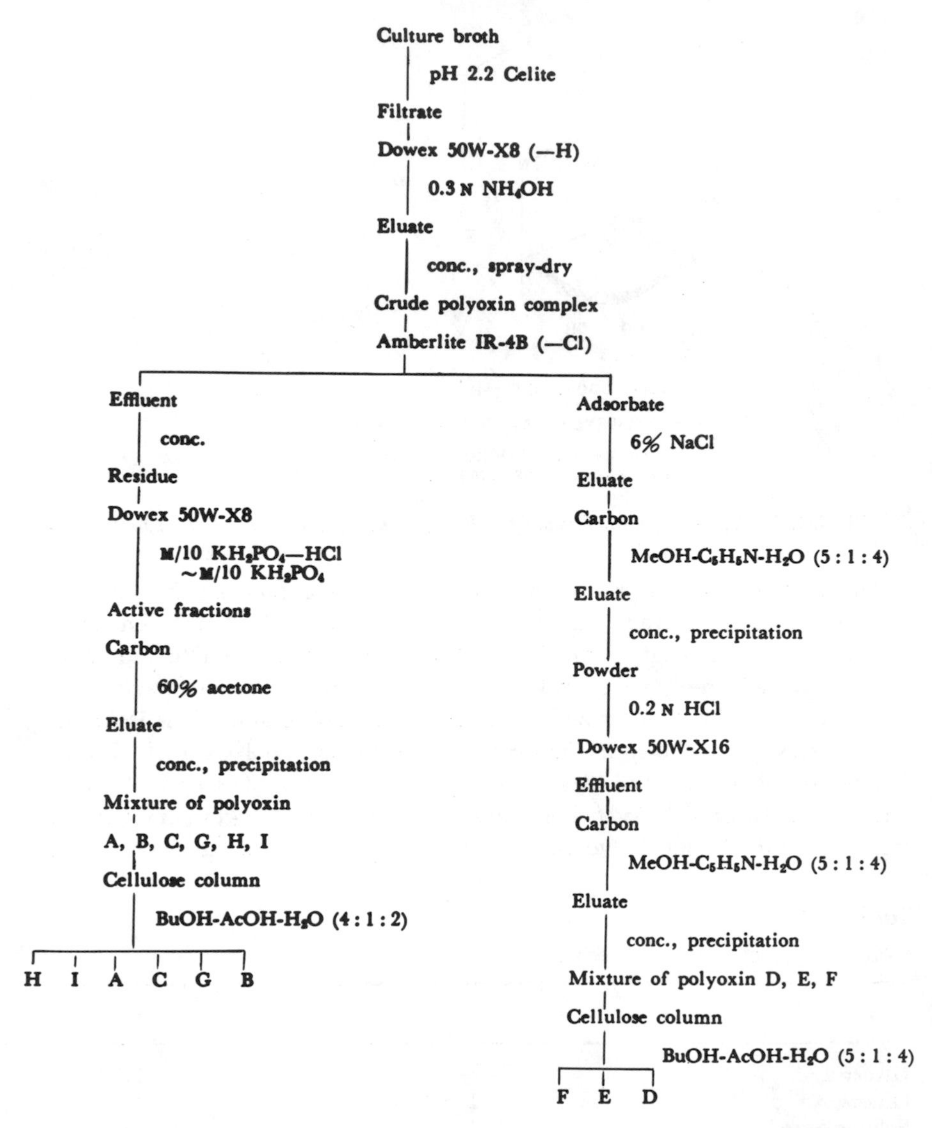


FIGURE 5.2 *Isolation procedure for polyoxins.* (*From Isono et al.*, 1967.)

whereas polyoxins B, G, and I were obtained as amorphous powders. Polyoxin C was also crystallized from water to give needle crystals. Polyoxins D, E, and F were crystallized from aqueous ethanol. Polyoxin D was obtained as a crystalline powder, while polyoxins E and F were amorphous. For a complete review of the polyoxins, see Isono et al. (1969).

**Table 5.2**
Physicochemical Properties of Polyoxins

| Polyoxin | Formula | p*Ka'* | | | | $[\alpha]_D^{20°}$ (*c*-l, water), deg. | 0.05 *N* HCl $\lambda_{max}$, log $\epsilon$, m$\mu$ | 0.05 *N* NaOH $\lambda_{max}$, log $\epsilon$, m$\mu$ |
|---|---|---|---|---|---|---|---|---|
| A | $C_{23}H_{32}N_6O_{14}$ | 3.0 | — | 7.3 | 9.6 | −30 | 262 (3.94) | 264 (3.80) |
| B | $C_{17}H_{25}N_5O_{13}$ | 3.0 | — | 6.9 | 9.4 | +34 | 262(3.94) | 264 (3.82) |
| C | $C_{11}H_{15}N_3O_8$ | 2.4 | — | 8.1 | 9.5 | +11 | 262(3.97) | 264 (3.87) |
| D | $C_{17}H_{23}N_5O_{14}$ | 2.6 | 3.7 | 7.3 | 9.4 | +30 | 276 (4.05) | 271 (3.85) |
| E | $C_{17}H_{23}N_5O_{13}$ | 2.8 | 3.9 | 7.4 | 9.3 | +19 | 276 (4.00) | 271 (3.81) |
| F | $C_{23}H_{30}N_6O_{15}$ | 2.7 | 3.9 | 7.2 | 9.3 | −18 | 276 (4.06) | 271 (3.87) |
| G | $C_{17}H_{25}N_5O_{12}$ | 3.2 | — | 7.3 | 9.3 | +37 | 262 (3.92) | 264 (3.82) |
| H | $C_{23}H_{32}N_6O_{13}$ | 3.3 | — | 7.2 | 9.4 | −38 | 265 (3.88) | 266 (3.79) |
| I | $C_{17}H_{22}N_4O_9$ | 2.7 | — | 6.2 | 9.3 | −35 | 262.5 (3.94) | 264 (3.78) |
| J | $C_{17}H_{25}N_5O_{12}$ | 3.0 | — | 7.1 | 9.9 | +31.7 | 264 (3.91) | 267 (3.81) |
| K | $C_{22}H_{30}N_6O_{13}$ | 3.0 | — | 7.2 | 9.3 | −16.5 | 259 (3.95) | 262 (3.86) |
| L | $C_{16}H_{23}N_5O_{12}$ | 3.0 | — | 7.1 | 9.4 | +34.4 | 259 (3.96) | 262 (3.85) |

From Isono et al., 1967; Inoso and Suzuki, 1968d; Isono et al., 1969.

## PHYSICAL AND CHEMICAL PROPERTIES

The physicochemical properties of the twelve polyoxins are summarized in Tables 5.2 and 5.3.

The $R_f$ values for the polyoxin complex were also reported for paper chromatography, thin-layer chromatography, and paper electrophoresis (Isono et al., 1967).

**Table 5.3**

Molecular Constituents of Polyoxins

| Polyoxin | Chromophore | Polyoximic acid (XV) | Polyoxamic acid (XVI) | Desoxy-polyoxamic acid |
|---|---|---|---|---|
| A | 5-Hydroxymethyluracil | + | + | − |
| B | 5-Hydroxymethyluracil | − | + | − |
| C | 5-Hydroxymethyluracil | − | − | − |
| D | Uracil-5-carboxylic acid | − | + | − |
| E | Uracil-5-carboxylic acid | − | − | + |
| F | Uracil-5-carboxylic acid | + | + | − |
| G | 5-Hydroxymethyluracil | − | − | + |
| H | Thymine | + | + | − |
| I | 5-Hydroxymethyluracil | + | − | − |
| J | Thymine | − | + | |
| K | Uracil | + | + | |
| L | Uracil | − | + | |

From Isono et al., 1967, 1968, 1969.

## STRUCTURAL ELUCIDATION

The structure of polyoxin C was first determined by Isono and Suzuki (1968a). Polyoxin C is the smallest molecule of the polyoxins. Although it lacks biological activity, it was a key compound in determining the structure of the polyoxins since hydrolysis of polyoxins A, B, G, and I gives rise to polyoxin C (Isono and Suzuki, 1966a; Isono and Suzuki, 1968c). In addition, polyoxin C acid is also a product of hydrolysis of polyoxins D, E, and F (Isono et al., 1967; Suzuki et al., 1966; Isono and Suzuki, 1966b).

Isono and Suzuki (1968a) used a combination of physical and chemical techniques to assign the structure 1-(5′-amino-5′-deoxy-β-D-allofuranuronosyl)-5-hydroxymethyluracil for polyoxin C (Fig. 5.3, I). Polyoxin C is an amphoteric compound having $pK_a$ values of 2.4 ($-COOH$), 8.1 ($-NH_2$), and 9.5 (uracil). It is ninhydrin positive and periodate-benzidine positive. Preliminary evidence establishing the furanose structure and the position

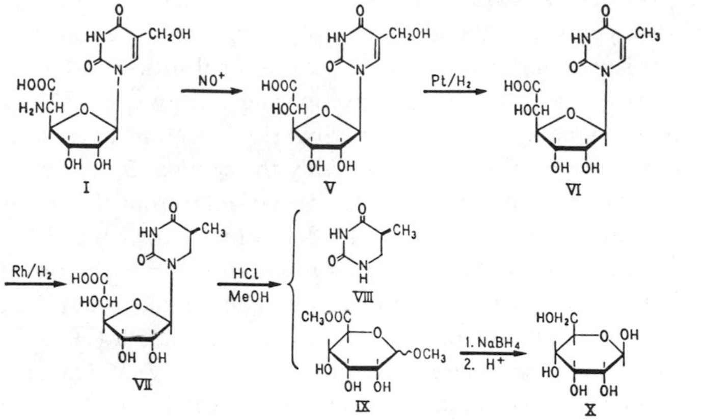

FIGURE 5.3 *Determination of the structure of the amino-uronic acid moiety of polyoxin C.* (*From Isono and Suzuki, 1968*a.)

of the amino group in polyoxin C was obtained by hydrogenolysis of *N*-acetylpolyoxin C (III) to form *N*-acetyldeoxypolyoxin C (IV) (Fig. 5.4). The nmr spectrum of compound IV with its assignments is shown in Figure 5.4. The C-5′ proton (quartet) coupled with the AcNH-proton ($J$ = 8.0 cps) becomes a resolved doublet ($J$ = 5.5 cps) on a spin decoupling experiment. These data indicate that the AcNH position is C-5′. The addition of deuterium oxide gave a similar doublet. The positive color reaction for $\alpha$-amino acids supports the notion that the amino group is on the C-5′ position of polyoxin C. The furanose structure for polyoxin C was supported by three lines of evidence. First, polyoxin C appears to be an $\alpha$-amino acid.

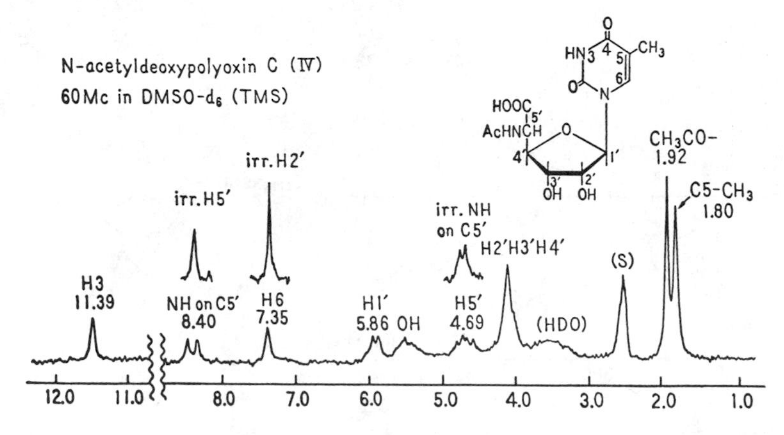


FIGURE 5.4 *The spectrum of N-acetyldeoxypolyoxin C.* (*From Isono and Suzuki*, 1968*a*.)

Second, *N*-acetylpolyoxin C rapidly consumed 1 mole of periodate. Third, *N*-acetyldeoxypolyoxin C had a value of 5.0 cps for the anomeric proton.

Evidence suggesting that polyoxin C is a *N*-1 substituted 5-hydroxymethyluracil nucleoside was obtained by comparing the uv spectra of polyoxin C with 5-hydroxymethyluridine, 5-methyluridine, and thymidine. The uv spectrum of polyoxin C (I) was essentially the same as 5-methyluridine, thymidine and 1-methyluracil, but differed markedly from the uv spectrum of 3-methyluracil (Isono and Suzuki, 1968a). Finally, prolonged hydrolysis of polyoxin C with 3 *N* HCl resulted in the isolation of 5-hydroxymethyluracil.

The proof of structure of the aminouronic acid moiety and the configuration of the nucleoside was reported by Isono and Suzuki (1968a) (Fig. 5.3). Polyoxin C (I) was deaminated with nitrous acid to give V. Compound V was then hydrogenated over platinum and then by rhodium on alumina to give compound VII. Treatment of compound VII with HCl and methanol resulted in the isolation of dihydrothymine (VIII) and the methylglycoside of the ester (IX). Reduction of compound IX with sodium borohydride followed by hydrolysis with Dowex-50-$H^+$ resulted in the isolation of β-D-allose (X). The allose isolated was identical to synthetic allose as determined by thin-layer chromatography and X-ray powder diffraction.

The configuration of the 5′ carbon was confirmed by the positive Cotton effect at 359 mμ of the derivative *N*-dithiocarbethoxy-2′,3′-*O*-isopropylidenedeoxypolyoxin C (XI) (Fig. 5.5) (Isono and Suzuki, 1968a). The β configuration of polyoxin C would be expected since a normal uracil ring and furanose ring comprise this pyrimidine nucleoside. The positive Cotton effects and CD maxima of polyoxin C and desaminodesoxypolyoxin C indicate that the configuration of this nucleoside is β.

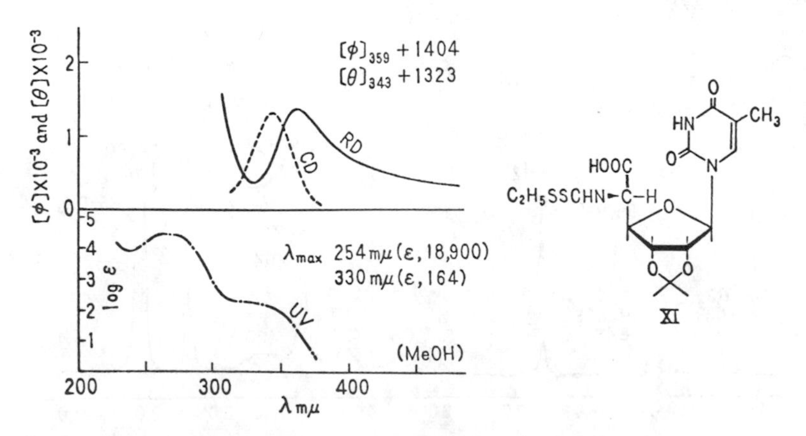


FIGURE 5.5 *Configuration of the 5′ carbon of polyoxin C.* (*From Isono and Suzuki*, 1968*a*.)

The presence of 5-amino-hexoses and aminohexuronic acids in nature are very rare. Inouye et al. (1966) reported on the occurrence of the 5-amino sugar nojirimycin, which they proposed as D-glucopiperidinose (Inouye et al., 1968). The identification of 5-amino-hexuronic acid (5-amino-5-deoxy-D-allofuranuronic) in polyoxin C is the first example of a naturally occurring sugar of this type. The chemical synthesis and isolation of *gluco-*, *galacto-*, and *manno-*2-amino-2-deoxyuronic acids has been reported (Williamson and Zamenhof, 1963; Haskell and Hanessian, 1963; Heyns et al., 1959; Perkins, 1963). Watanabe, Kotick, and Fox (1969) and Fox et al. (1968) have recently shown that the carbohydrate moiety of gougerotin is the 4-amino-4-deoxy-hexuronic acid, 4-amino-4-deoxy-D-glucuronic acid (see Chap. 4, Sect. 4.1). Fox and his co-workers have now described the first chemical synthesis of the 4-aminohexuronic acid 4-amino 4-deoxy-D-glucuronic acid moiety of gougerotin (Kotick et al., 1969) and the unsaturated amino sugar of blasticidin S, methyl 4-amino-2,3,4-trideoxy-$\alpha$-D-*erythro*-hex-2-eno-pyranosiduronic acid (Goody et al., 1970).

The biosynthesis of polyoxin C would be extremely interesting. Asahi et al. (1968) reported on the configuration of polyoxin C from X-ray analysis. Once the structure of polyoxin C was elucidated, Isono and Suzuki (1968c) presented evidence assigning the structures XII and XIII to polyoxins A and B, respectively (Fig. 5.6). Isono and Suzuki refer to these unique nucleoside antibiotics as "peptidic nucleosides" because they are pyrimidine nucleosides that contain tri- or dipeptides with unique $\alpha$-L-amino acids. The hydrolysis of polyoxin A is shown in Figure 5.7.

Mild alkaline hydrolysis (0.5 *N* NaOH, 65°C, 4 hr) of polyoxin A gave rise to 5-hydroxymethyluracil (XIV), polyoxin C (I), polyoximic acid (XV), and polyoxamic acid (XVI). One equivalent each of ammonia and carbon dioxide was also formed. When polyoxin A was hydrolyzed (3 *N* HCl

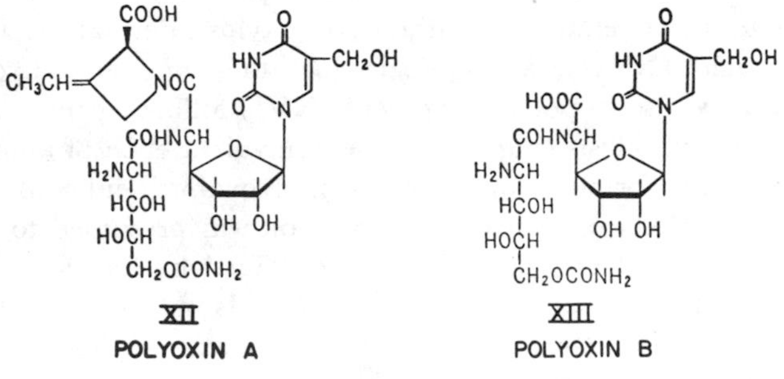

XII
POLYOXIN A

XIII
POLYOXIN B

FIGURE 5.6 *Structures of polyoxin A (XII) and polyoxin B. (XIII). (From Isono and Suzuki,* 1968*c.)*

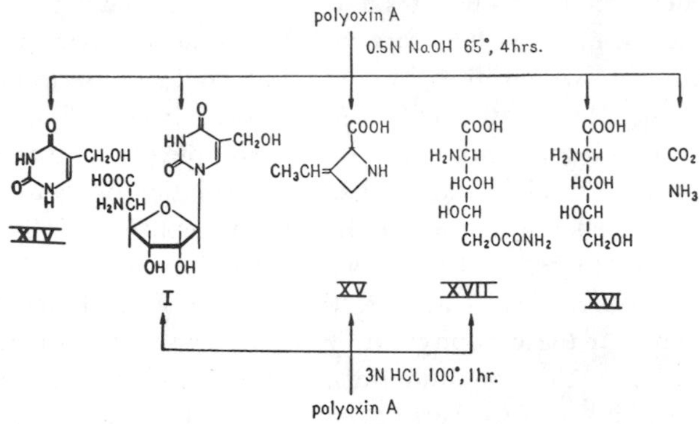

polyoxin A

0.5N NaOH 65°, 4hrs.

XIV I XV XVII XVI $CO_2$ $NH_3$

3N HCL 100°, 1hr.

polyoxin A

FIGURE 5.7 *Hydrolysis products of polyoxin A.* (*From Isono and Suzuki*, 1968*c*.)

at 100°C for 1 hr), three products were isolated: polyoxin C (I), 3-ethylidine-L-azetidine-2-carboxylic acid (XV), and carbamylpolyoxamic acid (XVII). The structure of polyoxin C had been shown to be 1-(5′-amino-5′deoxy-β-D-allofuranuronosyl)-5-hydroxymethyluracil (Fig. 5.3, I) (Isono and Suzuki, 1968a). The structure of polyoximic acid was shown to be 3-ethylidene-L-azetidine-2-carboxylic acid (XV) as established earlier by Isono and Suzuki (1966b). The structure of polyoxamic acid (XVI) was determined by the conversion to the *N*-acetyl-γ-lactone (Isono and Suzuki, 1968c). Whereas polyoxamic acid was oxidized with periodate to give one equivalent each of carbon dioxide, ammonia, and formaldehyde and three equivalents of formic acid, the *N*-acetyl-γ-lactone was not oxidized with periodate. These data suggested that polyoxamic acid (XVI) was a straight-chain trihydroxyamino acid. The *N*-acetyl-γ-lactone of XVI was converted to the five-membered unsaturated lactone by exhaustive acetylation and subsequent β elimination. The reduction of the *N*-acetyl-γ-lactone of XVI gave rise to 2-acetamido-2-deoxy-β-L-xylose. The physical and chemical properties of this sugar were identical with the chemically synthesized D-xylose derivative. Carbamylpolyoxamic acid (XVII), a major product following acid hydrolysis of polyoxin A, gave rise to polyoxamic acid (XVI) following mild alkaline or prolonged acid hydrolysis (Fig. 5.7). One equivalent each of ammonia and carbon dioxide was formed. Since carbamylpolyoxamic acid consumed three equivalents of periodate, the carbamyl position was presumed to be at C-5. Proof of the carbamyl group on C-5 was obtained by formation of a five-membered unsaturated lactone on acetylation. Isono and Suzuki (1968c) assigned the structure 5-*O*-carbamyl-2-amino-2-deoxy-L-xylonic acid (XVII) for carbamylpolyoxamic acid.

To determine the peptide sequence of these three unusual α-L-amino acids (I, XV, and XVII), polyoxin A was deaminated with nitrous acid. Following alkaline hydrolysis, compounds I and XV were isolated. However, polyoxamic acid (XVI) was not isolated. These data indicate that polyoxamic acid (XVI) is the *N*-terminal amino acid. When polyoxin A was treated with alkali under mild conditions, polyoxamic acid (XVI) was removed and compound XVIII was isolated (Fig. 5.8). Treatment of compound XVIII with nitrous acid resulted in the isolation of 3-ethylidene-L-azetidine-2-carboxylic acid (XV). Polyoxin C (I) was not isolated. These data indicated that polyoxin C is *N*-terminal. The structure proposed for polyoxin A (XII) is consistent with the physical and chemical properties of this peptide nucleoside antibiotic. The $pK_a$ values of 3.0, 7.3, and 9.6 correspond to the carboxyl, amino group, and the uracil chromophore, respectively. The presence of the furanose structure is strongly indicated by the $J$ value of an anomeric proton at 5.5 cps. The rapid reaction of polyoxin A with three equivalents of periodate is in agreement with the structure as proposed.

Hydrolysis of the nucleosidic bond of polyoxin A in alkali was explained by the reactivity of a peptide $\alpha$-hydrogen to hydroxide ion (Fig. 5.9). The *N*-glycoside bond of polyoxin B or polyoxin C is not labile to alkaline hydrolysis, since the carboxylate anion prevents the dissociation of the $\alpha$-hydrogen. Therefore, alkaline hydrolyses of polyoxin B and C do not yield any significant amounts of 5-hydroxymethyluracil. The same type mechanism was postulated for the alkaline hydrolysis of adenosylmethionine (Baddiley et al., 1962) and the cyanide-catalyzed decomposition of vitamin $B_{12}$ (Johnson and Shaw, 1961).

The structure of polyoxin B (XIII) was determined by acid and alkaline hydrolysis and comparison of the products of hydrolysis with those of the hydrolysis of polyoxin A. Similar products were obtained, except that polyoximic acid (XV) was not isolated and very little 5-hydroxymethyluracil was detected.

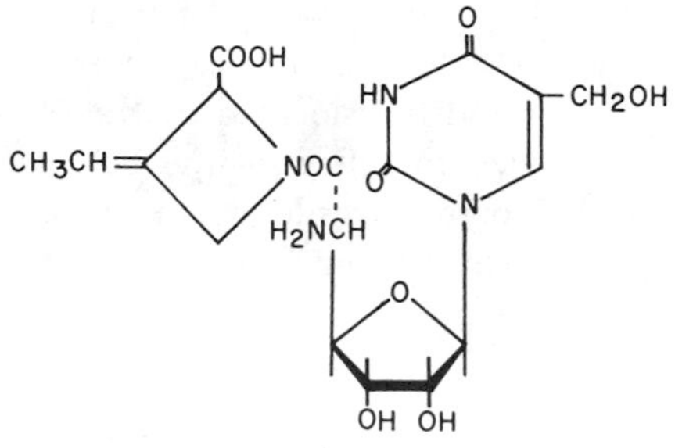

FIGURE 5.8 *Mild alkaline hydrolysis of Polyoxin A produces compound XVIII.* (*From Isono and Suzuki*, 1968*c*.)

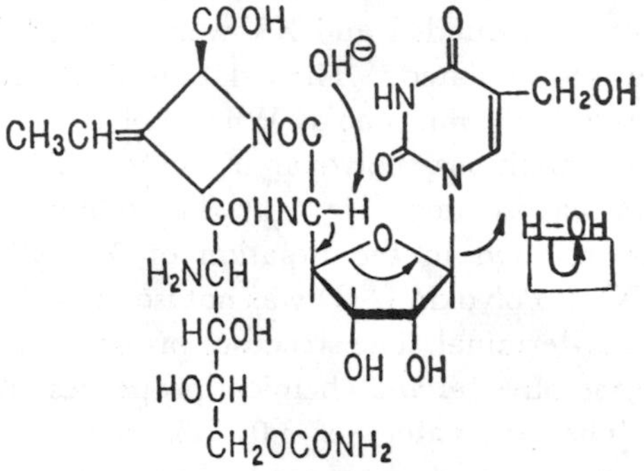

FIGURE 5.9 *Hydrolysis of nucleosidic bond of polyoxin A in alkali.* (*From Isono and Suzuki*, 1968c.)

5-Aminofuranuronoside is common to all the polyoxins. In addition, they all possess the uracil ring and are unique in that they are the first examples of peptide nucleosides (Isono and Suzuki, 1968b; Isono et al., 1969).

## INHIBITION OF GROWTH

The biological activities of the polyoxins are very characteristic because of their specific action against phytopathogenic fungi and lack of activity against the gram-positive and gram-negative bacteria and *Mycobacteria*. The polyoxins have low or no toxicity toward mice, fish, and plants. They are not inhibitory to *Aspergillus*, *Pencillium*, *Mucor*, *Trichophyton*, or *Candida*. The antifungal activities of polyoxins A–I are summarized in Table 5.4.

Sasaki et al. (1968) reported on the fungicidal activity of the polyoxins against sheathblight of rice *Pellicularia sasakii* (Shirai) S. Ito. The effects of the polyoxins in plant tissue were also studied. Polyoxin D was the most effective in controlling this disease. It appears that the polyoxins penetrated the leaves. In addition, when the stem of kidney bean plants or damaged root of rice plants was immersed in solutions of the polyoxins, a large amount of the antibiotic was absorbed and translocated by the plant.

Eguchi et al. (1968) studied the effect of the polyoxins against *Alternaria mali* and *A. kikuchianna*. Polyoxin B was found to be the most effective of the polyoxins in inhibiting the development of mycelium, spore germination, and sporulation.

## BIOCHEMICAL PROPERTIES

Isono and Suzuki (1968b) suggested that the site of action of the polyoxins may be related to cell wall chitin biosynthesis. This was based on the swelling out phenomenon of the fungal cell wall after treatment with polyoxins (Eguchi et al., 1968). Ohta et al. (1969) have reported that the uptake of glucosamine-$^{14}C$ in the plant pathogen, *Cochliobolus miyabeanus*, was inhibited by polyoxin D. More recently, Endo and Misato (1969) have reported that polyoxin D at 100 $\mu g/ml$ inhibited the incorporation of glucosamine-$^{14}C$ into the cell wall fraction of *Neurospora crassa*. There was no effect on the incorporation of $^{32}P$ into nucleic acids and of labeled glutamic acid into protein. In addition, polyoxin D caused an accumulation of UDP-*N*-acetyl-glucosamine (UDP-GlcNAc). These results indicate that polyoxin D was inhibiting the chitin synthetase reaction since glucosamine must be converted to UDP-GlcNAc before incorporation into chitin. The inhibition of chitin synthetase of polyoxin D is shown in Table 5.5. The kinetics of the inhibition is of the competitive type. The $K_m$ for UDP-GlcNAc is $1.43 \times 10^{-3}$ $M$; the $K_i$ (for polyoxin D) is $1.40 \times 10^{-6}$ $M$. Polyoxin D does not inhibit UDP-GlcNAc pyrophosphorylase. It was proposed that polyoxins D and L may indeed be structural analogs of UDP-*N*-acetylglucosamine. Figure 5.10 compares the structures of polyoxin L and UDP-*N*-acetylglucosamine. The polyoxins are not inhibitors of protein synthesis.

A recent study on the effect of polyoxin A on plant viruses was reported by Yun and Hirai (1968). They reported that polyoxin A was an inhibitor of TMV and was more effective than blasticidin S.

Mitani and Inoue (1968) reported their studies on the reversal of the antifungal activity of polyoxin B and D. The influence of nitrogen compounds on reversing the inhibition against *Rhizoctonia solani* and *Pellicularia sasaki* is shown in Tables 5.6 and 5.7. When nitrogenous-containing substances such

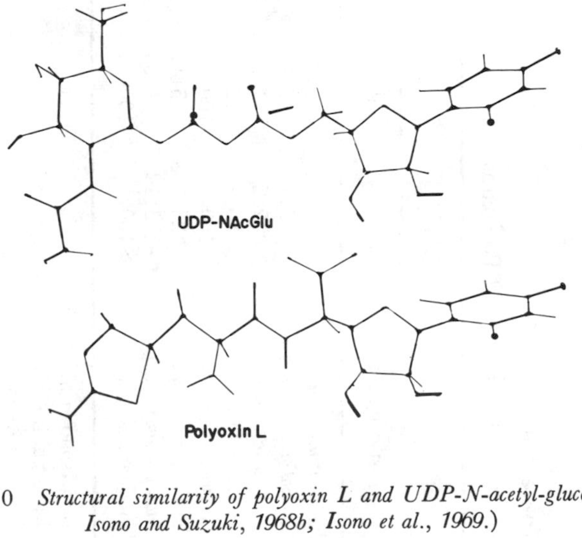

UDP-NAcGlu

Polyoxin L

FIGURE 5.10 *Structural similarity of polyoxin L and UDP-N-acetyl-glucosamine.* (*From Isono and Suzuki, 1968b; Isono et al., 1969.*)

**Table 5.4**
Antifungal Activities of Polyoxins

| Test-organism | MIC[a], mcg/ml | | | | | | | | |
|---|---|---|---|---|---|---|---|---|---|
| | A | B | C | D | E | F | G | H | I |
| *Piricularia oryzae* | 3.12 | 6.25 | >100 | 3.12 | 12.5 | 25 | 6.25 | 3.12 | >100 |
| *Cochliobolus miyabeanus* | 3.12 | 3.12 | >100 | 6.25 | 12.5 | 6.25 | 3.12 | 25 | >100 |
| *Pellicularia sasakii* | 12.5 | 1.56 | >100 | >1.56 | 1.56 | 50 | 1.56 | 50 | >100 |
| *Alternaria kikuchiana* | 50 | 12.5 | >100 | 50 | 50 | >100 | 6.25 | 12.5 | >100 |
| *Glomerella cingulata* | >100 | >100 | >100 | >100 | >100 | >100 | >100 | >100 | >100 |
| *Physalospora laricina* | 25 | 3.12 | >100 | 100 | 50 | >100 | 6.25 | 12.5 | >100 |
| *Cladosporium fulvum* | 3.12 | 1.56 | >100 | 100 | 25 | 25 | 3.12 | 6.25 | >100 |
| *Fusarium oxysporum* | >100 | >100 | >100 | >100 | >100 | >100 | >100 | >100 | >100 |

From Isono et al., 1967.
[a] Minimal inhibitory concentration.

**Table 5.5**

Effect of Polyoxin D on the Chitin Synthetase of *N. crassa*[a]

| Variation of complete system | UDP-GlcNAc incorporated (mμmoles/5 min) | Per cent inhibition |
|---|---|---|
| None | 16.0 | |
| + 4.7 μM polyoxin D | 4.3 | 73 |
| + 23.5 μM polyoxin D | 1.5 | 91 |

From Endo and Misato, 1969.

[a] Ten grams of 48-hr mycelia of *N. crassa* were ground with sand and 30 ml of 0.05 *M* Tris-HCl, pH 7.5 in a mortar at 0°C. The mixture was centrifuged at 20,000 × *g* for 30 min. The supernatant solution was then centrifuged at 100,000 × *g* for 2 hr. The particles obtained were dispersed in 8 ml of 0.05 *M* Tris-HCl, pH 7.5. The complete reaction mixture contained per 0.2 ml final volume: 37.5 m*M* Tris-HCl, pH 7.5, 1.8 m*M* UDP-GlcNAc (46,800 cpm), 5 m*M* $MgCl_2$, 5m*M* GlcNAc, 0.05 ml of the particle suspension. After 5 min of incubation at 24°C the reaction was stopped by heating for 1 min at 100°C. The precipitate was collected by centrifugation, washed twice with cold water, and counted.

as peptone, yeast extract, and casein were added, the antifungal activity was completely reversed (Table 5.6). Glycylglycine, glycylglycylglycine, and glycylglycylglycylglycine partially reversed polyoxin inhibition. Glycine was not antagonistic to polyoxin. Dipeptides were the most effective nitrogenous sources in reversing the inhibition of polyoxin (Table 5.7).

**Table 5.6**

The Influence of Nitrogen Compounds on the Polyoxin Inhibitory Action

| Nitrogen compound | Per cent polyoxin inhibition | | | | |
|---|---|---|---|---|---|
| | 1.0% | 0.5% | 0.1% | 0.05% | 0.01% |
| Peptone | 0 | 10 | 60 | 95 | — |
| Yeast extract[a] | — | 36 | 64 | 95 | 100 |
| Casein[a] (Hammarstein) | — | 31 | 70 | 95 | 100 |
| | 20 m*M* | 10 m*M* | 2 m*M* | 1 m*M* | |
| Glycine[b] | 100 | 100 | 100 | 100 | |
| Glycylglycine[b] | 32 | 57 | 88 | 100 | |
| Glycylglycylglycine[b] | 72 | 95 | 100 | 100 | |
| Glycylglycylglycylglycine[b] | 53 | 66 | 72 | 95 | |

From Mitani and Inoue, 1968.

[a] *Rhizoctonia solani* was used as the test organism.

[b] *Pellicularia sasakii* was used as the test organism.

**Table 5.7**

Antagonistic Action of Nitrogen Compounds on the Antifungal Activity of Polyoxin

| Nitrogen compounds, m*M* | Per cent polyoxin inhibition |
|---|---|
| None | 100 |
| Glycylglycine 10 | 30 |
| Glycyl L-adenine 10 | 0 |
| Glycyl DL-valine 10 | 0 |
| Glycylglycylglycine 10 | 95 |
| Glycylglycylglycylglycine 10 | 55 |
| DL-Alanyl DL-alanine 10 | 100 |
| DL-Alanyl glycine 10 | 0 |
| DL-Alanyl DL-methionine 10 | 64 |
| DL-Alanyl DL-valine 10 | 100 |
| DL-Leucyl DL-leucine 10 | 95 |
| γ-Aminobutyrate 100 | 100 |
| L-Asparagine 100 | 100 |
| L-Arginine 10 | 72 |
| L-Cysteine 100 | 47 |
| L-Glutamic acid 100 | 100 |
| D-Glutamic acid 100 | 100 |
| Glycine 200 | 100 |
| L-Glutamine 100 | 95 |
| DL-α-Alanine 100 | 100 |
| DL-β-Alanine 100 | 100 |
| L-Isoleucine 100 | 95 |
| DL-Methionine 100 | 86 |
| DL-Valine 50 | 95 |
| L-Phenylalanine 100 | 100 |
| L-Lysine[a] | 79 |
| L-Ornithine[a] 10 | 92 |
| Ascorbic acid 10 | 100 |
| Adenine 10 | 100 |
| Uracil 10 | 100 |
| Xanthine 10 | 95 |
| Thymine 10 | 100 |
| Cytosine 10 | 100 |

From Mitani and Inoue, 1968.
[a] Polyoxin D: 1 μg/ml.

DL-Alanyl-DL-alanine and DL-alanyl-DL-valine did not show any antagonism to polyoxin inhibition. Amino acids could reverse some of the polyoxin inhibition, but the concentration needed was much greater than with the dipeptides. There was no direct relationship between the length of the peptide chain and the reversal of inhibition.

## Summary

A new family of antifungal peptide nucleosides has been isolated. Polyoxins C and I are not inhibitors of fungi. They all contain the pyrimidine chromophore, α-L-amino acids, and the 5-aminofuranuronoside moiety. The peptide linkage and structures have been determined.

Peptone, yeast extract, or milk casein reverse the antifungal activity of the polyoxins. Of the di-, tri-, and tetrapeptides studied to reverse polyoxins B and D inhibition, the dipeptides were most effective. The polyoxins block cell wall chitin biosynthesis by inhibiting chitin synthetase. Polyoxins D and L exert an effect in fungi by acting as competitive inhibitors of UDP-*N*-acetylglucosamine.

## References

Asahi, K., T. Sakurai, K. Isono, and S. Suzuki, *Agr. Biol. Chem.* (*Tokyo*), **32**, 1046 (1968).

Baddiley, J., W. Frank, N. A. Hughes, and J. Wieczorkowski, *J. Chem. Soc.*,1962, 1999.

Eguchi, J., S. Sasaki, N. Ohta, T. Akashiba, T. Tsuchiyama, and S. Suzuki, *Ann. Phytopathol. Soc.*, **34**, 280 (1968).

Endo, A. and T. Misato, *Biochem. Biophys. Res. Commun.*, **37**, 718 (1969).

Fox, J. J., Y. Kuwada, and K. A. Watanabe, *Tetrahedron Letters*, 1968, 6029.

Goody, R. S., N. A. Watanabe and J. J. Fox, *Tetrahedron Letters*, **293** (1970).

Haskell, T. H., and S. Hanessian, *Nature*, **199**, 1075 (1963).

Heyns, K., G. Kiessling, W. Lindenberg, H. Paulsen, and M. E. Webster, *Chem. Ber.*, **92**, 2435 (1959).

Inouye, S., T. Tsuruoka, and T. Niida, *J. Antibiotics* (*Tokyo*), **19***A*, 288 (1966).

Inouye, S., T. Tsuruoka, T. Ito, and T. Niida, *Tetrahedron*, **24**, 2125 (1968).

Isono, K., J. Nagatsu, Y. Kawashima, and S. Suzuki, *Agr. Biol. Chem.*, **29**, 848 (1965).

Isono, K., and S. Suzuki, *Agr. Biol. Chem.*, **30**, 813 (1966a).

Isono, K., and S. Suzuki, *Agr. Biol. Chem.*, **30**, 815 (1966b).

Isono, K., J. Nagatsu, K. Kobinata, K. Sasaki, and S. Suzuki, *Agr. Biol. Chem.*, **31**, 190 (1967).

Isono, K., K. Kobinata, and S. Suzuki, *Agr. Biol. Chem.*, **32**, 792 (1968).

Isono, K., and S. Suzuki, *Tetrahedron Letters*, 1968a, 203.

Isono, K., and S. Suzuki, 156th Meeting, American Chemical Society, Atlantic City, N.J., September, 1968b, Abstracts Medi., 35.

Isono, K., and S. Suzuki, *Tetrahedron Letters*, 1968c, 1133.

Isono, K., and S. Suzuki, *Agr. Biol. Chem.*, **32**, 1193 (1968d).

Isono, K., K. Asahi, and S. Suzuki, *J. Am. Chem. Soc.*, **91**, 7490 (1969).

Johnson, A. W., and N. Shaw, *Proc. Chem. Soc.*, 1961, 447.

Kotick, M. P., R. S. Klein, K. A. Watanabe, and J. J. Fox, *Carbohydrate Res.*, **11**, 369 (1969).

Mitani, M., and Y. Inoue, *J. Antibiotics* (*Tokyo*), **21A**, 492 (1968).

Ohta, N., T. Akashiba, S. Sasaki, S. Kuroda and T. Misato, Abstracts, Annual Meeting, Agricultural Chemical Society of Japan, Tokyo, April 1, 1969, p. 15.

Perkins, H. R., *Biochem. J.*, **86**, 475 (1963).

Sasaki, S., N. Ohta, J. Eguchi, Y. Furukawa, T. Akashiba, T. Tsuchiyama, and S. Suzuki, *Ann. Phytopathol. Soc.*, **34**, 272 (1968).

Suzuki, S., K. Isono, J. Nagatsu, Y. Kawashima, K. Yamagata, K. Sasaki, and K. Hashimoto, *Agr. Biol. Chem.*, **30**, 817 (1966).

Watanabe, K. A., M. P. Kotick, and J. J. Fox, *Chem. Pharm. Bull.*, **17**, 416 (1969).

Williamson, A. R., and S. Zamenhof, *J. Biol. Chem.*, **238**, 2255 (1963).

Yun, T. G., and T. Hirai, *Nippon Shokubutsu Byori Gakkaiho*, **34**, 109 (1968).

CHAPTER 6

# Adenine and Purine Nucleosides

---

The five substituted purines to be reviewed in this chapter are aristeromycin, nucleocidin, septacidin, nebularine, and crotonoside (isoguanosine). They are all inhibitors of cellular reactions except crotonoside. Crotonoside is included in this chapter since it is a naturally occurring nucleoside. Septacidin, although not a true $N^9$ nucleoside, is also included since it is an adenine-glycosyl antibiotic in which the sugar moiety is substituted on the amino group of adenine. These five substituted purines have modifications in the base or carbohydrate moieties. Aristeromycin is another example of a nucleoside in which the chemical synthesis was reported prior to the isolation from natural sources. The biochemical effects of these compounds on protein synthesis, nucleic acid synthesis, and cellular reactions will be discussed.

## 6.1. ARISTEROMYCIN

### INTRODUCTION

Aristeromycin (Fig. 6.1, II) is a carbocyclic analog of adenosine (Fig. 6.1, I). The synthesis of the racemic compound was described by Shealy and Clayton (1966, 1969). Kusaka et al. (1968) subsequently isolated the same nucleoside as a naturally occurring, optically active (−) compound from the culture filtrates of *Streptomyces citricolor* and assigned the name "aristero-

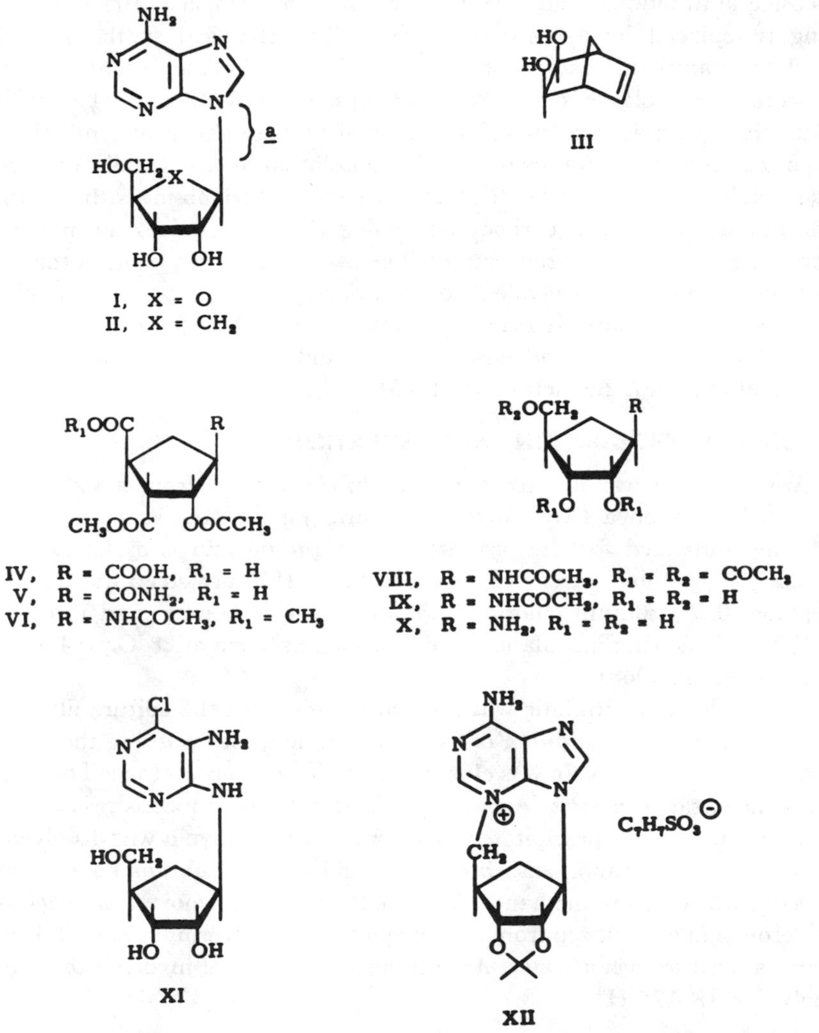

I, X = O
II, X = $CH_2$

III

IV, R = COOH, $R_1$ = H
V, R = $CONH_2$, $R_1$ = H
VI, R = $NHCOCH_3$, $R_1$ = $CH_3$

VIII, R = $NHCOCH_3$, $R_1$ = $R_2$ = $COCH_3$
IX, R = $NHCOCH_3$, $R_1$ = $R_2$ = H
X, R = $NH_2$, $R_1$ = $R_2$ = H

XI

XII

FIGURE 6.1 *Chemical synthesis of* 9-[β-(2′α,3′α-*dihydroxy*-4′β-(*hydroxymethyl*)*cyclopentyl*]*adenine* (*II*). (*From Shealy and Clayton*, 1966, 1969.)

mycin."[1] The structure of the carbocyclic analog of adenosine as synthesized by Shealy and Clayton (1966, 1969) and the structure assigned by Kishi et al. (1967) to aristeromycin isolated from nature (Kusaka et al., 1968) is 9-[β-(2′α,3′α-dihydroxy-4′β-(hydroxymethyl)cyclopentyl)]adenine. This nu-

[1] The name assigned to aristeromycin by Kusaka et al. (1968) was derived from its optical activity; *aristero* means levus in Latin.

cleoside antibiotic is unique in that the oxygen atom in the ribofuranosyl ring is replaced by a methylene group. The chemical synthesis of this nucleoside antibiotic was reported by Shealy and Clayton (1966) just prior to its reported isolation from *Streptomyces citricolor* by Kishi et al. (1967). This is another example of a biological antagonist whose chemical synthesis was reported almost simultaneously with the isolation as a naturally occurring product. The replacement of the furanose oxygen of adenosine with a methylene group produced a carbocyclic analog that antagonizes the naturally occurring nucleosides or nucleotides. The carbon–nitrogen bond of this type nucleoside should be equivalent to alkyl derivatives and be less susceptible to enzymatic cleavage. It is highly cytotoxic to H. Ep. #2 cells in culture and is a substrate for adenosine kinase and adenosine aminohydrolase (Allan et al., 1967; Bennett et al., 1968).

## DISCOVERY, PRODUCTION, AND ISOLATION

Aristeromycin was isolated from *S. citricolor nov. sp.* from a soil sample collected in Nagoya City, Aichi Prefecture, Japan (Kusaka et al., 1968). The medium used for the production of aristeromycin is as follows: 2% glucose, 3% soluble starch, 1% soybean flour, 1% corn steep liquor, 0.5% peptone, 0.3% sodium chloride, 0.5% calcium carbonate, and tap water (pH 7.0). Following inoculation, the flasks were shaken at 28°C for 4–6 days (Kusaka et al., 1968).

The nucleoside antibiotic was isolated by adjusting the culture filtrate to pH 9. Following the addition of carbon, filtration, and washing the carbon with water, aristeromycin was eluted with 80% aqueous acetone. The eluate was concentrated *in vacuo*. Acetone was added to the aqueous residue, and crude aristeromycin precipitated. This crude aristeromycin was dissolved in 55% aqueous methanol, adsorbed on an acid-washed alumina column, and eluted with aqueous methanol. The methanol was removed *in vacuo* and aristeromycin crystallized from the aqueous fraction overnight at 0°C. It was recrystallized from hot water. An alternative procedure involved the use of Amberlite IR-120-$H^+$.

## PHYSICAL AND CHEMICAL PROPERTIES

The empirical formula is $C_{11}H_{15}N_5O_3$. The molecular weights as determined by mass spectroscopy and X-ray analyses are 265 and 269.7, respectively (Kusaka et al., 1968). The melting point of aristeromycin is 213–215°C (decomp.) (Kusaka et al., 1968). The melting point of the chemically synthesized racemic aristeromycin is 238–242°C (Shealy and Clayton, 1966). The optical rotation is $[\alpha]_D^{25}$ $-52.5°$ ($c$ = 1% in dimethylformamide). It is soluble in acetic acid, dimethylsulfoxide, dimethylformamide, ethylene glycol, water, aqueous methanol, and aqueous acetone. It is only slightly soluble or insoluble in methanol, ethanol, ethyl acetate, chloroform, diethyl ether, and benzene. The ultraviolet spectral properties are as follows: $\lambda_{max}$ 258 m$\mu$ ($\epsilon = 14.5 \times 10^3$) and 212 m$\mu$ ($20.6 \times 10^3$) in 0.1 *N* HCl (Shealy and Clayton, 1969). The $R_f$ of aristeromycin in three solvents for paper chromatography and one system for thin-layer chromatography have been described by Kusaka et al. (1968). Unlike the purine nucleosides, aristeromycin is stable to acid.

## STRUCTURAL ELUCIDATION

The nmr spectra of aristeromycin showed the presence of two aromatic hydrogens at 8.17 and 8.10 ppm as singlets (each 1H) and an amino group at 7.2 ppm (2H) which disappeared following the addition of deuterium oxide (Kishi et al., 1967). These data and the ultraviolet spectra suggested that aristeromycin may be an adenosine derivative. When aristeromycin was refluxed with dilute sulfuric acid, there was no hydrolysis, which suggested the absence of a *N*-glycoside linkage. On the basis of the acid stability of aristeromycin and the spin-decoupling studies in nmr spectroscopy, Kishi et al. (1967) proposed the structure as shown in Figure 6.1 (II) for this nucleoside antibiotic. Further confirmation of the suggested structure, as well as the determination of the absolute configuration, was obtained by use of X-ray analysis. The absolute structure for aristeromycin isolated from *S. citricolor* by Kusaka et al. (1968) was established by Kishi et al. (1967) as (1′*R*, 2′*S*, 3′*R*, 4′*R*) (−)-9-[β-2′α, 3′α-dihydroxy-4′β-(hydroxymethyl)-cyclopentyl]adenine (Fig. 6.1, II). The physical and chemical properties of aristeromycin isolated from nature are identical to those of the synthetic compound.

## CHEMICAL SYNTHESIS OF ARISTEROMYCIN

Shealy and Clayton (1966, 1969) reported on the chemical synthesis of this carbocyclic analog of adenosine as a continuation of their studies on the synthesis of analogs related to adenosine such that their biochemical and therapeutic properties could be studied. It was anticipated that the *N*-glycoside-like bond of this carbocyclic analog was less susceptible to enzymatic cleavage than would be the corresponding furanosyl heterocycles. The synthesis of the cyclopentane analog of thymidine had been reported by Murdock and Angier (1962). The synthesis of the carbocyclic analog of adenosine as reported by Shealy and Clayton (1966, 1969) is shown in Figure 6.1. The cyclopentane 2α,3α-diacetoxy-1β,4β-cyclopentanedicarboxylic acid (IV) was synthesized from *exo,cis*-5-norbornene-2,3-diol (III). Treatment of the cyclic anhydride of compound IV with ammonia gave 2α,3α-diacetoxy-4β-carbamoyl-1β-cyclopentanecarboxylic acid (V). Compound V was converted to the cyclopentylamine (X) by two routes starting with compound V. One route involved the conversion of the carboxamide group of compound V to the amino group by use of the Hoffman hypobromite reaction. This reaction is known to occur with retention of configuration. The product of this reaction, after esterification and acetylation, was methyl 4β-acetamido-2α,3α-diacetoxy-1β-cyclopentanecarboxylate (VI). Compound VI was reduced with lithium borohydride and acetylated to give the tetraacetyl derivative, compound VIII. Deacetylation of compound VIII with base gave crystalline *N*-(2α,3α-dihydroxy-4β-hydroxymethyl-1β-cyclopentyl)acetamide (IX). Complete deacetylation of compound VIII gave rise to the requisite cyclopentylamine (X). Treatment of compound X with 5-amino-4,6-dichloropyrimidine gave the trihydroxycyclopentylaminopyrimidine (XI). Compound XI was converted to the 6-chloropurine derivative with triethyl orthoformate. Treatment of the 6-chloropurine derivative with ammonia gave the carbocyclic adenosine analog (±)-9-[β-(2α,3α-dihydroxy-4β-(hydroxymethyl)cyclopentyl)]adenine (II).[2] The synthetic compound was shown to have the 2′- and 3′-hydroxyl groups in the *cis* form, and the 5′-hydroxymethyl group *cis* to the purine ring was shown by the conversion of the 5′-*p*-toluenesulfonate to the cyclonucleoside derivative (XII).

Shealy and Clayton (1969) and Shealy and O'Dell (1969) have also reported on the chemical synthesis of the racemic forms of the carbocyclic (cyclopentane) analogs of inosine, 6-mercaptopurine ribonucleoside, 6-(methylthio)purine ribonucleoside, 2′-deoxyadenosine, and 3′deoxyadenosine (cordycepin).

## INHIBITION OF GROWTH

Kusaka et al. (1968) reported that aristeromycin was inhibitory against *Xanthomonas oryzae* and *Piricularia oryzae*. It was not inhibitory against yeast, pathogenic fungi, and bacteria except for the acid-fast bacteria. A very interesting contribution is the recent report that aristeromycin regulates the growth of plants (Hagimoto et al., private communication). Some of the organisms studied for inhibition by aristeromycin by Kusaka et al. (1968) are shown in Table 6.1. Additional organisms were also studied and reported. Aristeromycin was effective against the blast disease of rice plants in green houses. The inhibiting effect against spore formation of *P. oryzae* on the leaves of rice plants was about the same as that required for inhibition by blasticidin S.

[2] The DL nomenclature was originally used by Shealy and Clayton (1966) to convey the fact that their chemical synthesis is racemic. In their subsequent full-length publication (Shealy and Clayton, 1969), the (±) nomenclature was substituted to show both enantiomers of a racemic form. The structure (Fig. 6.1, II) represents only one enantiomer. The naturally occurring, optically active compound aristeromycin, as described by Kishi et al. (1967), is the (−) isomer.

**Table 6.1**

Antimicrobial Spectra of Aristeromycin

| Test organisms | Minimum inhibitory concentration, mcg/ml | | | Condition for pre-incubation | Assay condition |
|---|---|---|---|---|---|
| | pH 6 | pH 7 | pH 8 | | |
| *Bacillus subtilis* | >100 | >100 | >100 | Bouillon | Bouillon |
| *Bacillus brevis* | >100 | >100 | >100 | Agar | Agar |
| *Bacillus cereus* | >100 | >100 | >100 | 37°C | 37°C |
| *Staphylococcus aureus* | >100 | >100 | >100 | 1−2 days | 18 hrs |
| *Micrococcus flavus* | >100 | >100 | >100 | | |
| *Escherichia coli* | >100 | >100 | >100 | | |
| *Proteus vulgaris* | >100 | >100 | >100 | | |
| *Pseudomonas aeruginosa* | >100 | >100 | >100 | | |
| *Serratia marcescens* | — | >100 | — | | |
| *Aerobacter aerogenes* | — | >100 | — | | |
| *Mycobacterium avium* | 100 | 100 | 100 | Glycerol | Glycerol |
| *M. avium*, streptomycin-resistant | 50 | 50 | 50 | Bouillon agar | Bouillon agar |
| *Mycobacterium smegmatis* | 100 | 100 | 50 | 37°C | 37°C |
| *Mycobacterium phlei* | 50 | 50 | 100 | 2−3 days | 42 hrs |
| *Mycobacterium* ATCC 607 | 50 | 100 | 50 | | |

From Kusaka et al., 1968.

Aristeromycin was not toxic to mice when injected intravenously at a concentration of 50 mg/kg. In addition, killifish in an aqueous solution of 10 μg/ml of aristeromycin did not die after 6 days (Kusaka et al., 1968). Daily treatment of mouse lymphoid leukemia L1210 or rat intramuscular Walker carcinosarcoma 256 showed that aristeromycin was toxic at doses of 50 mg/kg/day. At 25 mg/kg/day it was nontoxic (Shealy and Clayton, 1969).

## BIOCHEMICAL PROPERTIES

The most complete metabolic studies with carbocyclic analogs of purine nucleosides are the reports of Allan et al. (1967) and Bennett et al. (1968). They reported that the carbocyclic analog of adenosine is highly cytotoxic to H. Ep. #2 cells in culture. These workers hypothesized that the C—N linkage between the cyclopentane and purine rings should not be subject to the action of nucleoside phosphorylases or hydrolases. However, these analogs might be expected to resemble parent nucleosides and therefore be substrates for those enzymes, such as the deaminases and kinases, acting at positions in the molecule other than the glycosidic bond. The carbocyclic analogs of adenosine, inosine, 6-thiopurine ribonucleoside, and 6-methylthiopurine ribonucleoside were studied. Of these four carbocyclic analogs, only the adenosine analog was cytotoxic to H. Ep. #2 cells in culture. The cytotoxic concentration of the carbocyclic analog of adenosine was about the same as that of the ribonucleosides of 6-mercaptopurine and 6-methylthiopurine. Two sublines of H. Ep. #2 cells lacking adenosine kinase were not inhibited by the carbocyclic analog of adenosine up to concentrations 30 times as great as that used on the parent cells. Of interest was the observation that concentrations of the carbocyclic analog of adenosine of 20 $\mu M$ or greater did inhibit both resistant lines of cells. These same cell lines were reported earlier to have a very high resistance to many other adenosine analogs (Bennett et al., 1966). The inhibition of purine synthesis *de novo* with the carbocyclic analog of adenosine was studied.

The conversion of the tritium-labeled adenine nucleoside analog to nucleotides is shown in Table 6.2. As can be seen from the table, most of the carbocyclic analog of adenosine was converted to the 5′-mono-, di-, and triphosphates. Additional proof that the monophosphate of the carbocyclic analog is formed was obtained by incubation with ATP-$\gamma$-$^{32}$P, $Mg^{2+}$ and adenosine kinase. A $^{32}$P-labeled compound was observed in the position expected for AMP. The adenosine analog was deaminated to a small extent by the crude system from H. Ep. #2 cells and was also a substrate for calf intestinal adenosine deaminase. Very little or no radioactivity was found in the polynucleotides and there was no radioactive compound migrating with an $R_f$ similar to NAD.

However, Ikehara (private communication) compared the ability of aristeromycin 5′-triphosphate to replace ATP with the partially purified DNA-dependent RNA polymerase isolated from *E. coli* (Table 6.3). Aristeromycin 5′-triphosphate was able to replace ATP with several synthetic deoxypolynucleotide templates, as well as with calf thymus DNA. There was a selectivity of incorporation of aristeromycin 5′-triphosphate. When poly-dAT and poly-dTC·AG were the templates, there was no incorporation of the analog. Aristeromycin 5′-triphosphate replaced ATP most efficiently with four-letter alternating templates. The selectivity observed on the incorporation of aristeromycin 5′-triphosphate is similar to the earlier reports on the replacement of tubercidin 5′-triphosphate or formycin 5′-triphosphate for ATP (Nishimura et al., 1966; Ikehara et al., 1968; also see pp. 342 and 371 for discussions of these two references).

The $K_m$ for the deamination of the carbocyclic analog with partially purified adenosine kinase from H. Ep. #2 cells was $7.8 \times 10^{-5}$ $M$. The $K_m$ for the same compound with calf intestinal adenosine deaminase was $3.3 \times 10^{-3}$ $M$. The $K_m$ for adenosine from calf intestinal deaminase was $3.6 \times 10^{-5}$ $M$. The four carbocyclic analogs did not interfere with the phosphoryla-

**Table 6.2**

Metabolism of (±)-aristeromycin-$^3$H in H. Ep. #2/S Cells[a,b]

| Metabolite | cpm/$10^4$ cells | Percentage of total cpm |
|---|---|---|
| Di- and triphosphates of (±)-aristeromycin | 1412 | 75 |
| (±)-Aristeromycin monophosphate | 320 | 17 |
| (±)-Aristeromycin | 65 | 3.6 |
| Carbocyclic analog of inosine | 49 | 2.5 |
| Unidentified materials | 39 | 2.0 |

From Bennett et al. 1968.

[a] (±)-aristeromycin means that the racemic form of the synthetic nucleosides was used for these studies.

[b] Suspension cultures of H. Ep. #2/S cells were grown for 24 hr in medium containing (±) aristeromycin-$^3$H at a concentration of 3.6 $\mu M$. The di- and triphosphates are grouped together because they were incompletely separated and because some triphosphate was degraded to diphosphate during chromatography.

**Table 6.3**

Synthesis of RNA When Directed by Synthetic and Native DNA's with Either ATP or Aristeromycin 5′-Triphosphate

| Template | Product | ATP incorporated, μmoles/ml | ArTP incorporated, (% of ATP) |
|---|---|---|---|
| Poly-dAT[a] | Poly-AU | 4.35 | — |
| Poly-dTC·AG | Poly-AG | 7.5 | — |
| Poly-dTG·AC | Poly-AC | 2.3 | 15 |
| Poly-dATAG·TATC | Poly-CUAU | 4.5 | 31 |
| Poly-dATAG·TATC | Poly-GAUA | 4.3 | 51 |
| Poly-dAATG·TTAC | Poly-CAUU | 8.1 | 48 |
| Poly-dAATG·TTAC | Poly-GUAA | 4.3 | 71 |
| Calf thymus DNA | RNA | 11.9 | 27 |

From Ikehara, private communication.

[a] dAT stands for deoxypolynucleotide having T and A in alternating sequence.

tion or deamination of adenosine or the cleavage of inosine to hypoxanthine. The phosphorylation of 6-methylthiopurine ribonucleoside was inhibited by this carbocyclic analog.

Bennett et al. (1968) concluded that the cytotoxicity observed for the carbocyclic analog of adenosine is due, at least in part, to the formation of a phosphate derivative since the cell lines lacking adenosine kinase were much more resistant to the nucleoside analog. Since the resistant cells were still inhibited by increased concentrations of the adenosine analog, it appears that the nonphosphorylated form can also be toxic to H. Ep. #2 cells. Precedence for the toxicity of nucleosides to mammalian cells has been set by the reports that a number of 9-alkylpurines are toxic to cells, but are neither phosphorylated nor cleaved (Kelley et al., 1962; Kimball et al., 1964). Although a number of enzyme systems and biological reactions were examined by Bennett et al. (1968), they concluded that the metabolic block responsible for the cytotoxicity of the carbocyclic nucleoside analog of adenosine has not yet been found. It may be that this nucleoside analog acts by way of a feedback-type inhibition, as has been observed with other analogs of purines and nucleosides, or it could be that the nucleoside analog at the triphosphate level interferes with those reactions requiring ATP. It does not appear that the incorporation into polynucleotides would explain the mode of action. Finally, it appears that the oxygen atom of the ribofuranosyl ring is important, but not essential, for substrate activity for adenosine kinase and adenosine deaminase. The corresponding pyrimidine carbocyclic nucleoside analog of thymidine was without activity (Murdock and Angier, 1962). Similarly, a comparison of the inhibitory properties of the 9-cycloaliphatic derivatives (which cannot be phosphorylated) with the carbocyclic derivative of adenosine shows they need not be related even though there is a close structural similarity of the 9-position substituents on adenine. These types of analogs may well exhibit their inhibitory action by interacting at the catalytic or regulatory sites of any number of enzymes. For example, Zedeck et al. (1967) showed that 6-chloro-8-aza-9-cyclopentylpurine inhibited the steroid-induced synthesis of $\Delta^5$-3-ketosteroid isomerase in *Pseudomonas testeroni*, and, more recently, Zedeck et al. (1969) and Johnson et al. (1969) showed that this nucleoside analog also inhibits DNA, RNA, and protein synthesis and blocks the induction of $\beta$-galactosidase in *E. coli* B. DNA synthesis is more sensitive to inhibition than is the synthesis of either RNA or protein. The block in DNA synthesis appears to involve the formation of thymine nucleotides, while inhibition of RNA synthesis is due to the inhibition of RNA polymerase. The drug affects protein synthesis by inhibiting the polymerization of amino acids into polypeptide linkage.

The report by Hagimoto et al. (private communication) that aristeromycin regulates the growth of plants represents a most interesting contribution to plant biochemistry.

## Summary

The racemic mixture of the carbocyclic analog of adenosine has been synthesized and the (−)enantiomer, aristeromycin, has been isolated from the culture filtrate of *S. citricolor*. Aristeromycin is cytotoxic to H. Ep. #2 cells in culture and inhibits *X. oryzae* and *P. oryzae*. It is a substrate for adenosine kinase and adenosine aminohydrolyase. The 5′-mono, di-, and triphosphates have been isolated from the acid-soluble pool of H. Ep. #2 cells. Aristeromycin is not incorporated into NAD and does not interfere with the phosphorylation or deamination of adenosine. While the cleavage of inosine to hypoxanthine is not inhibited by aristeromycin, the phosphorylation of 6-methylthiopurine ribonucleoside is inhibited. Finally, *de novo* purine biosynthesis as determined by the reduction of formylglycinamide derivatives that accumulated in azaserine-treated cells is inhibited. Although aristeromycin is not incorporated into nucleic acids by H. Ep. #2 cells in culture, aristeromycin 5-triphosphate is incorporated into acid-insoluble material with partially purified RNA polymerase from *E. coli*.

Sublines of these cells that lacked adenosine kinase were resistant to this nucleoside. It appears that part of the inhibition of aristeromycin is due to the conversion to the 5′-phosphates. However, as has been observed with other nucleoside analogs, it appears that aristeromycin may act as an inhibitor by interfering with many processes in the cell.

Aristeromycin has been shown to regulate the growth of plants.

## References


Allan, P. W., D. L. Hill and L. L. Bennett, Jr., *Federation Proc.*, **26**, 730 (1967).

Bennett, L. L., Jr., P. W. Allan, and D. L. Hill, *Mol. Pharmacol.*, **4**, 208 (1968).

Bennett, L. L., Jr., H. P. Schnebli, M. H. Vail, P. W. Allan, and J. A. Montgomery, *Mol. Pharmacol.*, **2**, 432 (1966).

Ikehara, M., K. Murao, F. Harada, and S. Nishimura, *Biochim. Biophys. Acta*, **155**, 82 (1968).

Johnson, J. M., R. W. Ruddon, M. S. Zedeck, and A. C. Sartorelli, *Mol. Pharmacol.*, **5**, 271 (1969).

Kelley, G. G., G. P. Wheeler, and J. A. Montgomery, *Cancer Res.*, **22**, 329 (1962).

Kimball, A. P., G. A. LePage, and B. Bowman, *Can. J. Biochem.*, **42**, 1753 (1964).

Kishi, T., M. Muroi, T. Kusaka, M. Nishikawa, K. Kamiya, and K. Mizuno, *Chem. Commun.*, 1967, 852.

Kusaka, T., H. Yamamoto, M. Shibata, M. Muroi, T. Kishi, and K. Mizuno, *J. Antibiotics (Tokyo)*, **21A**, 255 (1968).

Murdock, K. C., and R. B. Angier, *J. Amer. Chem. Soc.*, **84**, 3758 (1962).

Nishimura, S., F. Harda, and M. Ikehara, *Biochim. Biophys. Acta*, **129**, 301 (1966).

Shealy, Y. F., and J. D. Clayton, *J. Amer. Chem. Soc.*, **88**, 3885 (1966).

Shealy, Y. F., and J. D. Clayton, *J. Amer. Chem. Soc.*, **91**, 3075 (1969).

Shealy, Y. F., and C. A. O'Dell, *Tetrahedron Letters*, 1969, 2231.

Zedeck, M. S., A. C. Sartorelli, P. K. Chang, K. Raška, Jr., R. K. Robins, and A. D. Welch, *Mol. Pharmacol.*, **3**, 386 (1967).

Zedeck, M. S., A. C. Sartorelli, J. M. Johnson, and R. W. Ruddon, *Mol. Pharmacol.*, **5**, 263 (1969).

## 6.2. NUCLEOCIDIN

### INTRODUCTION

Nucleocidin is an adenine glycoside antibiotic with a sulfamyloxy group and fluoride attached to the carbohydrate moiety. This is the first fluoro sugar to be isolated from natural sources. The structure originally proposed (Waller et al., 1957) (Fig. 6.2*A*) was revised to 9-(4-*O*-sulfamoylpentofuranosyl)-adenine. Patrick and Meyer (1968) and, most recently, Morton et al. (1969) have further revised the structure to 9-(4-fluoro-5-*O*-sulfamoylpentofuranosyl)adenine (Fig. 6.2*B*). Nucleocidin is a potent anti-trypanosomal antibiotic produced by *Streptomyces clavus nov. sp.* isolated from soil in India. Nucleocidin inhibits protein synthesis by preventing peptide elongation. It may form a complex with the ribosomes which then become inactive in peptide bond formation. Florini (1967) has written a recent review on nucleocidin. Since nucleocidin is an organic fluorine compound, the reader should also consult the recent review by Goldman (1969) on the properties, metabolic effects, structure–function relations, and enzymology of the carbon–fluorine bond in compounds of biological interest.

### DISCOVERY, PRODUCTION, AND ISOLATION

Nucleocidin was isolated from a soil sample from Dinepur, India and produced by *Streptomyces clavus*, Lederle strain T 3018 (ATCC No. 13,382) (Backus et al., 1957; Thomas et al., 1956/57; Hewitt et al., 1956/57). The cultural, morphological, and physiological characteristics of the nucleocidin-producing *Streptomyces* were studied in detail by Backus et al. (1957).

The fermentation medium for the production of nucleocidin is as follows:

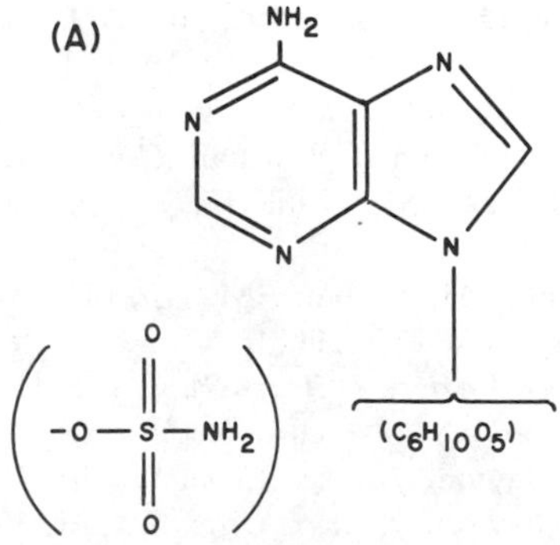

(A)

$NH_2$

N N N N

$-O-S(=O)(=O)-NH_2$

$(C_6H_{10}O_5)$

FIGURE 6.2 (*A*) *Partial structure for nucleocidin.* (*From Waller et al., 1957.*)

1.25% corn steep liquor, 1.0% mannitol, 0.2% sodium chloride, 0.2% dibasic ammonium phosphate, 0.05% dipotassium phosphate, 0.15% monopotassium phosphate, 0.025% magnesium sulfate, and trace elements. The pH of the medium is 6.9. The nucleoside is produced by submerged culture at 26–31°C and appears in the culture filtrates 96 hr after inoculation (Thomas et al., 1956/57; Thomas et al., 1959a, 1959b).

Nucleocidin is isolated from the culture medium at pH 7.0 by adsorption onto carbon (Darco G-60) and is eluted with acetone–water (95:5), lyophilized, dissolved in acetone–water (1:1), and passed through another charcoal column. The eluted material is rechromatographed, lyophilized, and dissolved in 0.06 *N* HCl and the pH is adjusted to pH 4.0. Nucleocidin crystallizes readily after seeding with pure crystalline nucleocidin.

## PHYSICAL AND CHEMICAL PROPERTIES

The molecular formula for nucleocidin is $C_{10}H_{13}N_6SO_6F$ (Morton et al., 1969). The melting point of the picrate of nucleocidin is 143–144°C; $\lambda_{max}$ 256 mμ ($\epsilon$ = 15,500) (neutral, aqueous solution); $[\alpha]_D^{24.5} = -33.3°$ ($c$ = 1.05 in ethanol–0.1 *N* HCl 1:1) (Thomas et al., 1959a). Nucleocidin is amphoteric and more readily soluble in acid than in alkali. It is soluble in methanol and acetone and insoluble in 1-butanol, ethyl acetate, benzene, and ether (Thomas et al., 1959a, 1959b). Nucleocidin is stable for 24 hr in aqueous solution at pH 3, 7, or 9. It is hydrolyzed by Dowex-50-$H^+$ to form free sulfamic acid (Waller et al., 1957). The mass spectrum of the tetra(trimethylsilyl) derivative of nucleocidin provides a molecular ion at *m/e* 652 and a prominent M-15 at *m/e* 637 (Morton et al., 1969). Computer analysis confirmed the presence of fluorine in nucleocidin. The 60-MHz nmr spectrum has been reported by Patrick and Meyer (1968). More recently, the $^1H$ nmr

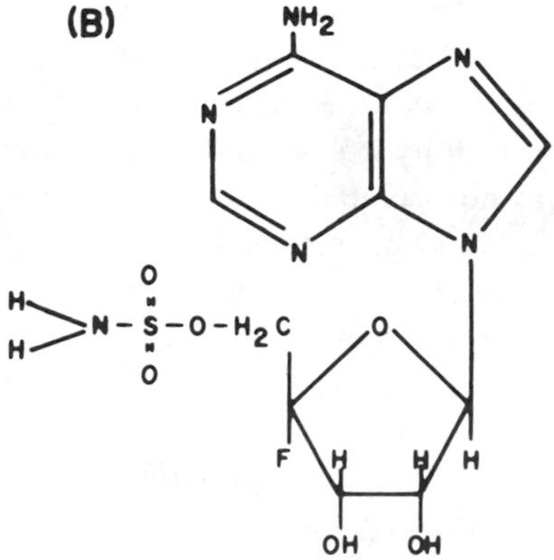

FIGURE 6.2 (*B*) *Revised structure for nucleocidin*, 9-(4-*fluoro*-5-*O*-*sulfamoylpentofuranosyl*)*adenine*. (*From* *Morton et al*., 1969.)

spectra at 60 and 100 MHz and the $^{19}F$ nmr spectrum at 56.4 MHz of nucleocidin have been reported by Morton et al. (1969).

## STRUCTURAL ELUCIDATION

The chemical structure of nucleocidin was first proposed by Waller et al. (1957). They stated that nucleocidin is a glycoside of adenine in which sulfamic acid is bound to the carboyhdrate moiety in an ester linkage. Nucleocidin exhibited an adenosine like ultraviolet absorption spectrum and was presumed to be a 9-adenyl compound. Acid hydrolysis of the antibiotic yielded a reducing sugar that gave a positive reaction with phenylhydrazine and a positive color test with aniline phthalate. The carbohydrate moiety was assumed to be a hexose ($C_6H_{12}O_6$). Adenine was also isolated following acid hydrolysis. When nucleocidin was hydrolyzed in 2 *N* HCl at 100°C for 30 min in a solution containing barium chloride, barium sulfate was isolated. The authors concluded that the sulfur atom in nucleocidin is hexavalent and is not linked directly to carbon since sulfones and alkane sulfonic acids are stable under the above conditions. When nucleocidin was hydrolyzed with Dowex-50-$H^+$, sulfamic acid ($HSO_3NH_2$) was identified by paper chromatography and by its specific reaction with nitrous acid–benzidine reagent. Nucleocidin has a p$K_a$ of 9.3. This p$K_a$ value excludes the possibility of a free sulfonic acid group, but is in agreement for a sulfonamide. To prove that nucleocidin is not an amine sulfonate, but is an ester of sulfamic acid, nucleocidin was treated with barium nitrite in dilute acid at room temperature. Barium sulfate was isolated. When barium chloride was added under the same experimental conditions, the antibiotic was unaffected. Waller et al. (1957) assigned the partial structure for nucleocidin as shown in Figure 6.2*A*.

More recently, Patrick and Meyer (1968) revised this structure to 9(4-*O*-sulfamoyloxypentofuranosyl)adenine ($C_{10}H_{14}N_6SO_7$). The adenyl and sulfamate moieties in nucleocidin were identified by the isolation of adenine and sulfamic acid following hydrolysis and by uv, ir, and nmr spectrometry. The 4-ketopentose was not isolated. However, the characteristic reductone transformations which would be expected of such a structure were observed and the products predicted by the oxidative degradation were obtained. The 2′-, 3′-diol system in the antibiotic was shown to be *cis*. The nmr studies indicate that the C-1′ proton is *trans* to the C-2′ proton. Four possible configurations were proposed: α-L- and β-D-*lyxo*- and α-L- and β-D-*ribo*-. There was no C-4′ proton observed in the nmr spectrum. The evidence presented by Patrick and Meyer (1968) strongly suggests that the sugar of nucleocidin is a 1,4-dicarbonyl pentose, and not an aldohexose as originally proposed (Fig. 6.2*A*).

Morton et al. (1969) have now more clearly defined the structure of nucleocidin by examination of the $^{1}H$ nmr spectrum at 100 MHz and the $^{19}F$ spectrum at 56.4 MHz. The 100 MHz spectrum clearly established extra spin–spin couplings of 18.0 and 9 cps which were due to fluorine interaction with the C-3′ H and the C-5′-methylene hydrogens, respectively. The 18 cps coupling constant between C-3′H and C-4′F is in agreement with the reported values for H—F vicinal coupling (Hall and Manville, 1965, 1967, 1968). Morton et al. stated that the C-4′F and the C-5′-methylene protons form an AA′X system with an apparent coupling of 9 cps for the H—F portion of the system (pyridine-$d_5$). Only two hydroxyl groups were observed in nucleocidin (C-2′OH, C-3′OH). Computer analysis of the mass spectrum of the tetra(trimethylsilyl) (TMS) derivative of nucleocidin provided positive evidence for the presence of fluorine.

The fragmentation pattern of this derivative of nucleocidin clearly showed a sulfamyloxy group on C-5′ and a fluorine in the sugar moiety attached at the C-4′ position. The fragmentation of the TMS is shown in Figure 6.3. A molecular ion at *m/e* 652 was reported, as well as a prominent M-15 (*m/e* 637).

On the basis of these new data, Morton et al. (1969) further revised the structure for nucleocidin to 9-(4′-fluoro-5′-*O*-sulfamoylpentofuranosyl)adenine (Fig. 6.2*B* or a stereoisomer). The stereochemistry of nucleocidin as possessing adenine and the 4′-sulfamoyloxymethyl as *cis* was established by forming the 5′,3-cyclonucleoside (Shuman, Robins, and Robins, 1969). Thus, in all probability nucleocidin has the *β*-configuration.

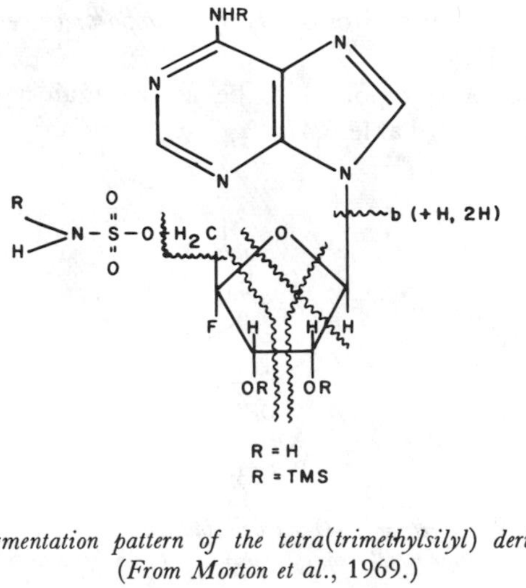

FIGURE 6.3 *Fragmentation pattern of the tetra(trimethylsilyl) derivative of nucleocidin. (From Morton et al.,* 1969.)

## SYNTHESIS OF ADENINE 5′-O-SULFAMOYL NUCLEOSIDES RELATED TO NUCLEOCIDIN

Shuman, Robins, and Robins (1969, 1970) have described the first synthesis of sulfomate esters of nucleosides that are direct structural models of nucleocidin. The compounds synthesized were 5′-*O*-sulfamoyladenosine (Fig. 6.4*A*) and 5′-*O*-sulfamoyl-2′-deoxyadenosine (Fig. 6.4*B*). The synthesis involved the treatment of 2′,3′-*O*-ethoxymethylideneadenosine or 3′-*O*-acetyl-2′-deoxyadenosine with sulfamoyl chloride. The circular dichroism curves for nucleocidin and 5′-*O*-sulfamoyladenosine in water were identical and were similar to that of adenosine. These data suggest that the adenine ring is above the sugar plane.

There was a 50% inhibition of *S. faecalis* with 5′-*O*-sulfamoyladenosine ($4 \times 10^{-6}$ *M*). At $5 \times 10^{-7}$ *M*, nucleocidin gave the same inhibition. *Trypanosoma rhodesiense* was also inhibited by 5′-*O*-sulfamoyladenosine.

## INHIBITION OF GROWTH

Hewitt et al. (1956/57) reported on the therapeutic properties of nucleocidin in controlling infections of mice with *Trypanosoma equiperdum*. Nucleocidin is about 20,000 times as strong as trypansamide and 4000 times as potent as puromycin. Stephen and Gray (1960) reported that a single intramuscular injection (0.25 mg/kg) cleared peripheral blood of *Trypanosomes* within 20 hr. However, the infection was not corrected and relapses occurred 18–33 days after treatment. Tobie (1957) reported that nucleocidin cured infections of *T. congolense*, *T. equinum*, and *T. gambiense* in rats and mice. It was the first antibiotic to be highly effective against *T. congolense*. Nucleocidin is very toxic to mammals (Hewitt et al., 1956/57). Arai et al. (1966) reported that nucleocidin was an inhibitor of *Mycoplasma pneumoniae* Mac, a pathogenic isolate of human origin.

Nucleocidin has been reported to be active against gram-positive and gram-negative bacteria (Table 6.4).

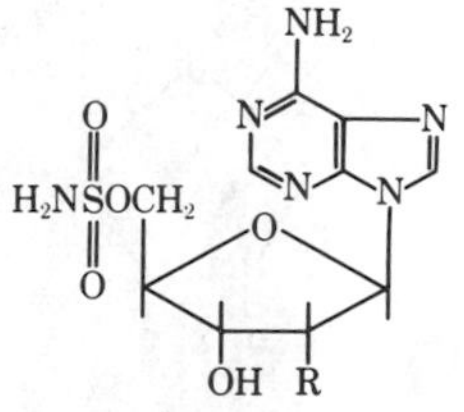

FIGURE 6.4 (*A*) *Structure for* 5′-*O-sulfamoyladenosine*. (*From Shuman et al., 1969*.) (*B*) *Structure for* 5′-*O-sulfamoyl-2′-deoxyadenosine*. (*From Shuman et al.,* 1969.)

**Table 6.4**

Inhibition by Nucleocidin

| Organism | μg/ml for complete inhibition of growth |
|---|---|
| Gram-positive | |
| *Bacillus cereus* | 4–8 |
| *Bacillus subtilis* | 2–4 |
| *Corynebacterium xerose* | 8–16 |
| *Corynebacterium flaccumfaciens* | 1–2 |
| *Micrococcus pyogenes* var. *albus* | 2–8 |
| *Micrococcus pyogenes* var. *aureus* | 2–16 |
| *Sarcina lutea* | 4–8 |
| *Streptococcus faecalis* | 1–2 |
| *Streptococcus mitia* | 2–8 |
| *Streptococcus pyogenes* var. *haemolyticus* | 0.05–0.1 |
| Gram-negative | |
| *Agrobacterium tumefaciens* | 0.25 |
| *Alcaligenes faecalis* | 128 |
| *Erwinia amylovora* | 0.5–1 |
| *Erwinia carotovora* | 4–16 |
| *Escherichia coli* | 8–16 |
| *Klebsiella pneumoniae* | 64 |
| *Proteus vulgaris* | 4–8 |
| *Pseudomonas aeruginosa* | 64–128 |
| *Pseudomonas glycinea* | 16 |
| *Pseudomonas solanacearum* | 4–16 |
| *Pseudomonas tabaci* | 0.5–1 |
| *Pseudomonas torelliana* | 4–16 |
| *Salmonella gallinearum* | 8–16 |
| *Salmonella pullorum* | 4–8 |
| *Salmonella typhosa* | 16 |
| *Serratia marcescens* | 4 |

From Thomas et al., 1956/57.

$LD_{50}$ for nucleocidin in mice is reported to be 0.2 mg/kg when administered intraperitoneally and 2.0 mg/kg when given orally (Hewitt et al., 1956/57). In their studies on the inhibition of protein synthesis by nucleocidin, Florini et al. (1966) reported that intraperitoneal injections of nucleocidin at 0.8 mg/kg killed all rats, whereas 0.4 mg/kg caused no deaths in 12 hr. When the dose was increased, animals died 2–4 hr after injection.

## BIOSYNTHESIS OF THE CARBON–FLUORINE BOND

Although no studies have been reported on the synthesis of the carbon–fluorine bond in nucleocidin, Peters et al. (1965) and Peters (1967) have reported experiments on the synthesis of fluoroacetate with cell-free extracts of *Acacia georginae*. The plant homogenates were prepared from seedlings of *A. georginae* in a cooled mortar. The incubation mixture consisted of fluoride, manganese, pyruvate, ATP, and phosphate. Fluoroacetate was identified by gas chromatography. The relevant peak was not present when no additions were made. It is not clear if ATP, $Mn^{2+}$, or pyruvate is required for the biosynthesis of fluoroacetate.

## BIOCHEMICAL PROPERTIES

The most significant study related to the effect of nucleocidin has been reported by Florini et al. (1966). Their studies involved the effect of nucleocidin on the incorporation of tritium-labeled leucine into rat liver protein *in vivo* and *in vitro*. Although nucleocidin *in vivo* inhibited the incorporation of leucine-$^3$H into rat liver protein, it had no effect on the incorporation of this labeled amino acid into brain protein. The effect of nucleocidin on the TCA-soluble pool and incorporation of leucine-$^3$H into rat liver protein is shown in Table 6.5. The data in Table 6.5 (experiment 1; 0.25 mg nucleocidin per kilogram body weight) show that the radioactivity in the protein

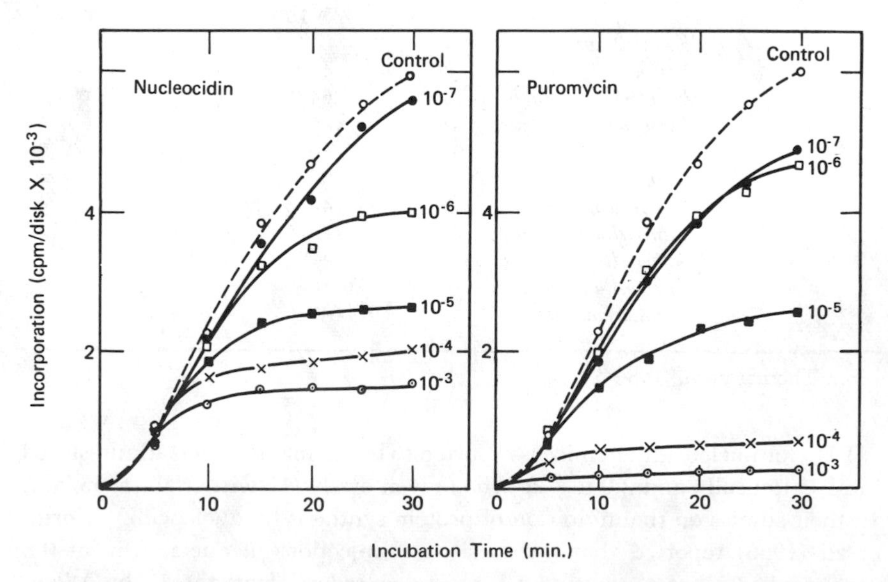


FIGURE 6.5 *Effect of nucleocidin and puromycin on the time course of $^3$H-leucine incorporation into protein by cell-free preparations from rat liver.* (*From Florini et al.*, 1966.)

was only 29.7% that of the control. Nucleocidin did not inhibit the uptake of leucine-$^3$H into the TCA-soluble fraction of liver. Therefore, the inhibition of protein synthesis cannot be attributed to a lower accumulation of this labeled amino acid. When puromycin was injected in a similar manner into rats at doses of 10 or 50 mg/kg, there was no effect on the incorporation of radioactive amino acids into liver proteins. Thus, nucleocidin is a much more potent inhibitor of protein synthesis *in vivo* than puromycin. However, *in vitro* studies on protein synthesis with nucleocidin and puromycin failed to show a marked difference in the inhibition of protein synthesis by these two nucleoside antibiotics (Fig. 6.5). The differences observed in the *in vitro* and *in vivo* studies relative to the inhibition by nucleocidin and puromycin might be attributed to the more rapid metabolism and/or excretion of puromycin. Very little nucleocidin was excreted or metabolized within 2 hr after injection. A comparison of the inhibition of amino acid incorporation into liver proteins *in vivo* by nucleocidin with that of cycloheximide showed that 50 mg/kg of cycloheximide was necessary to cause a 90% incorporation of labeled amino acids into protein of rabbit liver (Young et al., 1963).

**Table 6.5**

Effect of Nucleocidin on $^3$H-leucine Incorporation into Liver Protein *In Vivo*[a]

| | Radioactivity in liver fractions | | |
|---|---|---|---|
| | | Protein | |
| Nucleocidin injected, mg/kg | TCA-soluble, dpm/mg liver | Dpm/mg protein ($\times 10^{-5}$) | % of control |
| Experiment 1 | | | |
| None | 1420 ± 56 | 30.8 ± 0.6 | |
| 0.05 | 1370 ± 32 | 28.0 ± 2.4 | 90.9 |
| 0.25 | 1720 ± 105 | 0.2 ± 0.4 | 29.7 |
| Experiment 2 | | | |
| None | 2380 ± 41 | 82.0 ± 1.9 | |
| 0.20 | 3100 ± 41 | 12.4 ± 0.6 | 15.2 |
| 0.40 | 3785 ± 47 | 13.3 ± 47 | 16.3 |

From Florini et al., 1966.

[a] Nucleocidin was injected 2 hr before leucine-$^3$H. Experiment 1, protein was isolated 1 hr after the administration of leucine-$^3$H; experiment 2, protein was isolated 2 hours after administration of leucine-$^3$H.

The structural similarity of nucleocidin and puromycin led Florini et al. (1966) to examine the possibility that nucleocidin, like puromycin, might inhibit protein synthesis by virtue of its structural similarity to the 3′hydroxy-terminus of tRNA. Their studies on the two antibiotics, however, revealed different modes of action. A number of *in vitro* experiments were conducted on nucleocidin in parallel with those on puromycin to provide a direct comparison of the effects of these two antibiotics. The appreciable lag before nucleocidin exerted its inhibition on protein synthesis could be eliminated by a prior incubation of the antibiotic with the factors necessary for amino acid incorporation from tRNA. This prior incubation also enhanced the inhibition at low concentrations of the antibiotic. The formation of a ribosome–nucleocidin complex was shown by incubation of ribosomes with the nucleocidin. Ribosomes preincubated with nucleocidin were only about 40% as active as ribosomes preincubated without the antibiotic. The effect was reversible and the rate of reversal could be increased by the addition of the pH 5 fraction. Apparently the ribosome–nucleocidin complex is sufficiently strong to survive centrifugation through a 1.0 *M* sucrose layer. However, this complex dissociated when the ribosomes were again incubated with the pH 5 fraction, under the conditions required for amino acid incorporation into polypeptide.

To determine if nucleocidin affected the binding of tRNA to ribosomes, experiments were reported in which 0.1 m*M* nucleocidin was present in the protein-synthesizing systems with tRNA-$^{32}$P. Nucleocidin did not inhibit the binding of tRNA-$^{32}$P to ribosomes.

To determine if nucleocidin specifically affected the binding of ribosomes to synthetic messenger RNA, experiments were reported using copolymers of uridylic and guanylic acids. Nucleocidin did not interfere with the binding of ribosomes to synthetic messenger RNA. Since the sucrose gradient profiles of the reisolated ribosomes following incubation with nucleocidin were not changed, it was concluded that this nucleoside antibiotic did not inhibit the progression of ribosomes along the messenger RNA. Experiments were also reported to show that the adenine base of nucleocidin and the dimethyladenine base of puromycin do not interact with the same site on the ribosome. These data were obtained by prior incubation of ribosomes with nucleocidin. Nucleocidin had no effect on the puromycin-induced release of peptides from ribosomes. If nucleocidin and puromycin are bound at the same site on the ribosome, it may well be that puromycin is able to displace the bound nucleocidin very rapidly. Finally, Florini et al. (1966) showed that the specific activity of tRNA, in which a 50–70% inhibition of amino acid incorporation into protein was observed, was not altered in the presence of nucleocidin. Thus, it appears that nucleocidin inhibits amino acid incorporation into liver protein *in vitro* and *in vivo*. This inhibition occurs at a step following the formation of aminoacyl-tRNA. This inhibition of transfer of amino acids from aminoacyl-tRNA to protein requires the presence of the pH 5 fraction. Florini et al. (1966) provided excellent data to show that nucleocidin forms a complex with ribosomes which is then inactive in peptide bond formation. However, the exact manner in which protein synthesis is inhibited in the nucleocidin-ribosome complex is not known.

Although nucleocidin appears to be a structural analog of adenosine, the antibiotic did not inhibit RNA synthesis in cell-free preparations using rat liver nuclei.

## Summary

Nucleocidin is a unique adenine nucleoside antibiotic in that it contains C, H, N, S, O, and F. The fluoride is at position C-4′ and a sulfamyloxy group is at position C-5′. It is a very potent antitrypanosomal antibiotic. Nucleocidin is a much more potent inhibitor of protein synthesis *in vivo* than puromycin. However, *in vitro* studies showed that the inhibition of protein synthesis by nucleocidin and puromycin is the same. This difference in toxicity from *in vivo* studies may be attributed to a more rapid metabolism or a more rapid excretion by puromycin. Nucleocidin is excreted very slowly.

The *in vitro* inhibition of protein synthesis by nucleocidin appears to involve binding to the ribosomes and subsequent peptide formation. It does not block the binding of tRNA to ribosomes nor does it inhibit RNA synthesis. Nucleocidin does not inhibit the puromycin reaction. Apparently nucleocidin binds to the ribosome-mRNA-peptidyl complex at a site other than that occupied by puromycin or else puromycin displaces the bound nucleocidin very rapidly. The exact mechanism of binding of nucleocidin to the ribosome complex is not known.

## References

Arai, S., K. Yoshida, A. Izawa, K. Kumagai, and N. Ishida, *J. Antibiotics (Tokyo)*, **19A**, 118 (1966).

Backus, E. J., H. D. Tresner, and T. H. Campbell, *Antibiot. Chemotherapy*, **7**, 532 (1957).

Florini, J. R., in *Antibiotics* I, D. Gottlich and P. D. Shaw, Eds., Springer-Verlag, New York, 1967, p. 427.

Florini, J. R., H. H. Bird, and P. H. Bell, *J. Biol. Chem.*, **241**, 1091 (1966).

Goldman, P., *Science*, **164**, 1123, (1969).

Hall, L. D., and J. F. Manville, *Chem. Ind. (London)*, 1965, 991.

Hall, L. D., and J. F. Manville, *Can. J. Chem.*, **45**, 1299 (1967).

Hall, L. D., and J. F. Manville, *Chem. Commun.*, 1968, 37.
Hewitt, R. I., A. R. Gumble, L. H. Taylor, and W. S. Wallace, *Antibiot. Ann.*, 722, 1956/57.
Morton, G. O., J. E. Lancaster, G. E. Van Lear, W. Fulmor, and W. E. Meyer, *J. Amer. Chem. Soc.*, **91**, 1535 (1969).
Patrick, J. B., and W. E. Meyer, 156th Meeting, American Chemical Society, Atlantic City, N. J., September, 1968, Abstracts Medi. 24.
Peters, R. A., M. Shorthouse, and P. F. V. Ward, *Life Sci.*, **4**, 749 (1965).
Peters, R. A., *Rev. Roumanian Biochem.*, **4**, 79 (1967).
Shuman, D. A., R. K. Robins, and M. J. Robins, *J. Amer. Chem. Soc.*, **91**, 3391 (1969).
Shuman, D. A., and M. J. Robins, *J. Amer. Chem. Soc.*, **91**, 3434 (1970).
Stephen, L. E., and A. R. Gray, *J. Parasitol.*, **46**, 509 (1960).
Thomas, S. O., J. A. Lowery, and V. L. Singleton, British Pat., No. 815,381 (1959a).
Thomas, S. O., V. L. Singleton, J. A. Lowery, R. W. Sharpe, L. M. Pruess, J. N. Porter, J. H. Mowat, and N. Bohonos, *Antibiot. Ann.*, 1956/57, 716.
Thomas, S. O., J. A. Lowery, and V. L. Singleton, U.S. Pat. No. 2,914,525 (1959b).
Tobie, E. J., *J. Parasitol.*, **43**, 291 (1957).
Waller, C. W., J. B. Patrick, W. Fulmor, and W. E. Meyer, *J. Amer. Chem. Soc.*, **79**, 1011 (1957).
Young, C. W., P. F. Robinson, and B. Sacktor, *Biochem. Pharmacol.*, **12**, 855 (1963).

## 6. 3. SEPTACIDIN

### INTRODUCTION

Although septacidin is not a true $N^9$-adenine nucleoside antibiotic, it is included in this review since it is a $N^6$-glycosyl-adenine derivative. The structure of the major component is shown in Figure 6.6. Septacidin is a cytotoxic, antifungal, and antitumor antibiotic. Hydrolysis of septacidin yielded adenine, glycine, and a 4-amino-aldoheptose and fatty acid fraction containing $C_{12}$ to $C_{18}$ fatty acids (mostly branched). The predominant fatty acid is iso-palmitic acid. Septacidin is active against tumor cells and mammalian cells in culture. It is not antibacterial, but is inhibitory for several filamentous fungi.

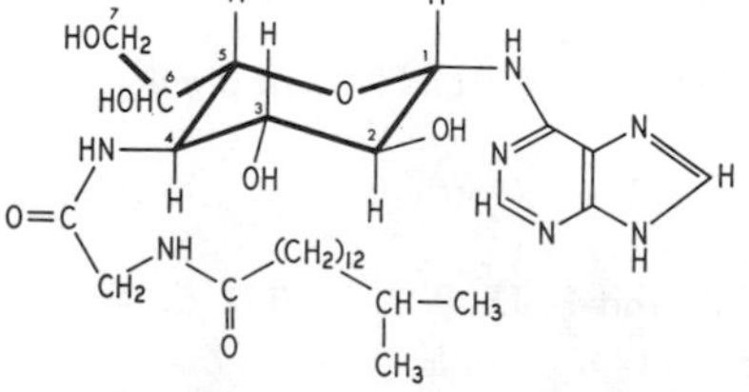

FIGURE 6.6 *Septacidin*

Another example of $N^6$-substitution of the adenine moiety is the work of Hall et al. (1966), in which they showed the natural occurrence of $N^6$-($\Delta^2$-isopentenyl) adenosine in the tRNA of yeast.

## DISCOVERY, PRODUCTION, AND ISOLATION

Septacidin was first isolated from *Streptomyces fimbriatus* by Dutcher et al. (1963). The culture of *S. fimbriatus* (ATCC 15051) was maintained on the following seed medium: 15% soybean meal, 15% dehydrated mashed potatoes, 5% glucose, 0.005% cobalt chloride, 1% calcium carbonate, and 25% agar. The cultures were maintained on an incubator shaker for 72–96 hr at 25°C. A 10% inoculum was made into 125-ml portions of the same medium (minus agar) and incubated for 7 days. The fermentation medium was as follows: 3% soybean meal, 5% glucose, 0.7% calcium carbonate, and 0.25% cornsteep liquor. Four days were required for maximum production of septacidin (Dutcher et al., 1963, 1964).

Septacidin is isolated by extracting the mycelium with 1-butanol. The butanol is removed by evaporation *in vacuo* and the water removed to give crude septacidin. Septacidin is extracted with dimethylformamide, filtered, and treated with an equal volume of methyl alcohol, and crystallized.

## PHYSICAL AND CHEMICAL PROPERTIES

The molecular formula for septacidin where the fatty acid is iso-palmitic acid is $C_{30}H_{51}N_7O_7$; mp 215–220°C (decomp.); $[\alpha]_D^{23} = +6.6°$ ($c = 1.0\%$ in dimethylformamide); $\lambda_{max} = 264$ m$\mu$ ($E_{1\,cm}^{1\%}$ 253) in methanol; $\lambda_{max}$ 272 m$\mu$ ($E_{1\,cm}^{1\%}$ 275) in 0.1 *N* NaOH (Dutcher et al., 1963).

## STRUCTURAL ELUCIDATION

Acid hydrolysis of septacidin resulted in the isolation of adenine, 4-amino-4-deoxy-L-glycero-L-glycoheptose, glycine, and a fatty acid fraction (Dutcher et al., 1964) (Fig. 6.7). Three families of fatty acid components were found. The major component (about 40%) was identified as iso-palmitic acid. Since Fox et al. (1966) have recently reviewed the structural elucidation of septacidin, only a brief treatment has been given here. The utlraviolet spec-spectrum of septacidin is consistent with N-6 substitution rather than N-9 or N-7. The elucidation of the structure and the stereochemistry of the aminoaldoheptose moiety are discussed by von Saltza et al. (1964) and Agahigian et al. (1965). They showed that periodate oxidation of a derivative of the monoaminoheptose with production of 1 mole of formaldehyde resulted in the isolation of a monoaminohexose ($C_6H_{13}NO_5$). The position of the amino group and the configuration of this sugar were established by proton decoupling studies and comparison of the physical and chemical properties of the $\alpha$-methyl glycoside tetraacetate derivative. This derivative

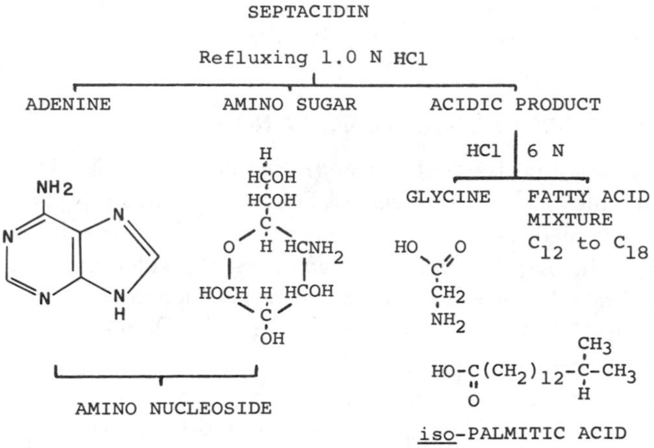

SEPTACIDIN

Refluxing 1.0 N HCl

ADENINE — AMINO SUGAR — ACIDIC PRODUCT

HCl 6 N

GLYCINE — FATTY ACID MIXTURE $C_{12}$ to $C_{18}$

AMINO NUCLEOSIDE

iso-PALMITIC ACID

FIGURE 6.7 *Products from hydrolysis of septacidin* (*From Dutcher et al.*, 1963).

was identical with authentic methyl-4-acetamido-4-deoxy-2,3,6-tri-*O*-acetyl-α-D-glucopyranoside except for the sign of rotation. The nmr spectrum of the pentaacetate derivative of this aminohexose in chloroform-*d* was similar to the nmr spectrum of β-D-glucopyranose pentaacetate reported by Lemieux et al. (1957, 1958). By comparing the chemical shifts and coupling constant in the spectra of various aminohexose derivatives, von Saltza et al. (1964) and Agahigian et al. (1965) were able to determine whether adjacent methine protons on carbons 1 through 5 were coupled axial–axial or axial–equatorial. In addition, the nmr spectrum indicated that the methyl glycoside tetraacetate derivative existed in the pyranose form.

The evidence that the glycosidic linkage binding the aminoaldoheptose to the 6-amino group of adenine has a β configuration was obtained in two ways (von Saltza, 1964; private communication). First, a comparison was made of the $[M]_D^{22°}$ +10,400 (0.1 *N* NaOH) of the *N*-acetyl aminonucleoside, containing the aminohexose, with the $[M]_D^{22°}$ values of the *N*-acetyl methyl-α- and β-glycosides of 4-amino-4-deoxy-L-glucose, which are −35,750 (water) and +8,000 (methanol), respectively. This clearly indicates a β-glycosidic linkage in the aminonucleoside. Second, the nmr spectrum of the aminonucleoside dihydrochloride in deuterium oxide has a doublet at $4.47\tau$ ($J$ = 8.5 cps) assigned to the C-1 proton of the pyranose ring. This indicates that the C-1 proton is coupled in an axial–axial manner with the C-2 proton.

To establish the L-configuration of the asymmetry at the C-6 position in the aminoheptose moiety of septacidin, the 6′,7′-*O*-isopropylidene derivative of the aminonucleoside was prepared (Fig. 6.8, **1**). The dialdehydo-nucleoside derivative resulting from periodate oxidation (**2**) was oxidized with sodium

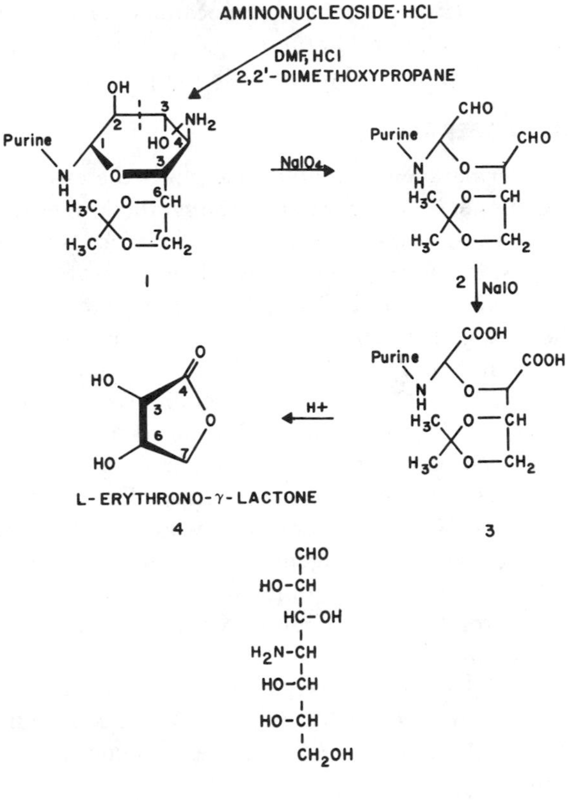

FIGURE 6.8 *Degradative scheme to isolate* L-*erythrono-γ-lactone* (4) *to establish the asymmetry at carbon 6 of the aminoheptose.* (*From von Saltza, et al.*, 1964, *private communication.*)

hypoiodite to a dicarboxylic acid derivative (**3**), which on acid hydrolysis yielded L-erythrono-γ-lactone (**4**) (mp 102.5–103°C; $[\alpha]_D^{22°}$ +69° ($H_2O$). This lactone was identical except for the sign of rotation with the lactone obtained by hypoiodite oxidation of D-erythrose. Since this lactone is derived from carbons 4, 5, 6, and 7 of the aminoheptose (**5**), the asymmetric center at C-6 has the L-configuration and the aminoheptose moiety of septacidin is 4-amino-4-deoxy-L-glycero-L-glucoheptose.

## INHIBITION OF GROWTH

Dutcher et al. (1963) reported that septacidin exhibited cytotoxic activity against Earle's L cells (NCTC 929) in tissue culture. It had antifungal activity against *Trichophyton mentagrophytes* and *Fusarium bulbigenum*. Septacidin exhibited antitumor activity against adenocarcinoma (CA 755) in mice. It was inactive against Walker 256 tumor and leukemia 1210.

The $LD_{50}$ in mice was approximately 0.88 mg per kg body weight.

## BIOCHEMICAL PROPERTIES

Very limited information is available concerning the mechanism of action of septacidin. Aszalos et al. (1965) studied the synthesis of septacidin analogs in an effort to determine the structural requirement for toxicity.

Analogs of septacidin were prepared as follows: Treatment of septacidin with 0.3 *N* methanolic HCl at 25°C for 72 hrs resulted in methanolysis of the molecule at the linkage between the amino sugar and the glycine moieties. This technique permitted the isolation of the aminonucleoside portion of septacidin. The synthetic analogs of septacidin were synthesized by the attachment of various groups by amide linkage to the amino group of the amino nucleoside. The analogs were tested for cytotoxic activity against Earle's L cells (NCTC 929) in suspension culture. The lauroylglycyl analog of septacidin had a therapeutic index of 6 which made it acceptable for further testing as an anticancer agent. This analog was considerably less cytotoxic than septacidin. The data obtained showed that structural modification of septacidin resulted in compounds with a therapeutic index greater than that of septacidin.

In an attempt to demostrate a synergistic activity with vernamycin A, Meyers et al. (1965) reported that vernamycin A and septacidin were synergistic against Earle's L cells growing in suspension-culture in a calf serum-containing medium.

The aminonucleoside portion [$N^6$-(4-amino-4-deoxyaldoheptosyl)adenine] alone does not possess inhibitory or cytotoxic activity compared to septacidin (Dutcher et al., 1963). Septacidin is not an inhibitor of protein synthesis (von Saltza, private communication).

## Summary

Septacidin is a $N^6$-glycosyl-adenine antibiotic. The 4-amino-aldoheptose is covalently bound to the $N^6$-amino group of adenine by a $\beta$-glycosidic linkage. The configuration of the hydroxyl group at C-6 of this aminoheptose has been shown to be L. The structure assigned this sugar is 4-amino-4-deoxy-L-glycero-L-glucoheptose. Glycine and a mixture of fatty acids make up the remainder of this antibiotic. Septacidin inhibits filamentous fungi, Earle's L cells (NCTC 929) in tissue, and adenocarcinoma (CA 755) in mice. $N^6$-(4-Amino-4-deoxyaldoheptosyl)adenine, the aminonucleoside portion, is not inhibitory. Septacidin does not inhibit protein synthesis.

## References

Agahigian, H., G. D. Vickers, M. H. von Saltza, J. Reid, A. I. Cohen, and H. Gauthier, *J. Org. Chem.*, **30**, 1085 (1965).

Aszalos, A., P. Lemanski, B. Berk, and J. D. Dutcher, *Antimicrobial Agents Chemotherapy*, 1965, 845.

Dutcher, J. D., M. H. von Saltza, and F. E. Pansy, *Antimicrobial Agents Chemotherapy*, 1963, 83.

Dutcher, J. D., F. E. Pansy, and M. H. von Saltza, U.S. Pat. No. 3,155,647 (1964); *Chem. Abstr.*, **62**, 9235f (1965).

Fox, J. J., K. A. Watanabe, and A. Bloch, in Progress in Nucleic Acid Research and Molecular Biology, Vol. V, J. N. Davidson and W. E. Cohn, Eds., Academic Press, New York, 1966, p. 251.

Hall, R. H., M. J. Robins, L. Stasiuk, and R. Thedford, *J. Amer. Chem. Soc.*, **88**, 2614 (1966).

Lemieux, R. U., R. K. Kullnig, H. J. Bernstein, and W. G. Schneider, *J. Amer. Chem. Soc.*, **79**, 1005 (1957).

Lemieux, R. U., R. K. Kullnig, H. J. Bernstein, and W. G. Schneider, *J. Amer. Chem. Soc.*, **80**, 6098 (1958).

Meyers, E., D. A. Smith, and D. Perlman, *Antimicrobial Agents Chemotherapy*, 1965; 256; *Chem. Abstr.*, **65**, 11135 (1966).

von Saltza, M. H., J. D. Dutcher, and J. Reid, Abstracts, 148th National Meeting American Chemical Society, Chicago, Illinois, September, 1964, p. 15A.

## 6.4. NEBULARINE

### INTRODUCTION

Nebularine (9-β-D-ribofuranosylpurine) (Fig. 6.9) is a naturally occurring nucleoside antibiotic elaborated by the mushroom and *Streptomyces yokosukanensis*. The structure and chemical synthesis of this nucleoside antibiotic have been described. This compound is toxic to *Mycobacterium*, animal cells in culture, and tumors. It is very toxic to mice. Fox et al. (1966) have recently reviewed this nucleoside antibiotic.

### DISCOVERY, PRODUCTION AND ISOLATION

Nebularine was first isolated by Ehrenberg et al. (1946a) from the mushroom *Agaricus* (*clitocybe*) *nebularis* Batsch. More recently, Isono and Suzuki (1960) reported on the isolation of nebularine from a Streptomyces species. Nakamura (1961) named the organism "*S. yokosukanensis n sp.*"

Two procedures are described here for the isolation of nebularine.

A. The isolation of nebularine from the mushroom was accomplished by adding 3 volumes of acetone to the extract of the fungi (Löfgren et al., 1954). This mixture was allowed to stand at 0°C for 3 hr. The insoluble material

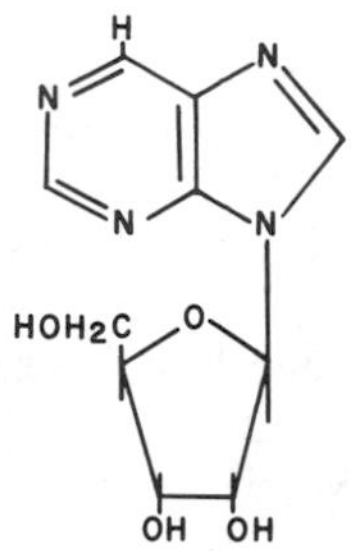

FIGURE 6.9 *Structure of nebularine (9-β-D-ribofuranosylpurine).*

was removed by filtration and the acetone distilled off under vacuum. The aqueous solution was washed with chloroform and ether. Charcoal was added to the aqueous solution and allowed to stand overnight at 0°C. Nebularine was eluted from the charcoal with water–ethanol–2-octanol. The eluant, containing nebularine, was concentrated *in vacuo*, and aluminum oxide was added with stirring. The aluminum oxide was separated and washed with water. The solution was evaporated and extracted with chloroform–ethanol, and the organic fraction was concentrated *in vacuo*. Butanol was added to the residue, followed by water. The aqueous solution containing the nucleoside was evaporated. The residue was dissolved in propyl alcohol and added slowly to ether. Nebularine crystallized on standing overnight at 0°C. It was further purified by countercurrent distribution between 1-butanol and water.

B. The isolation of nebularine from *S. yokosukanensis* was performed as described by Isono and Suzuki (1960). The fermentation medium was as follows: 6% molasses, 2% soybean meal, 1% dipotassium phosphate, and 0.6% calcium carbonate. The fermentation flasks were aerated at 27°C. Nebularine was isolated by adsorption and elution from charcoal, followed by absorption and elution from C-1 resin with 7% aqueous acetone. The acetone was removed and the aqueous portion extracted with 1-butanol (pH 8.0) and reextracted into acid–water (pH 2.0). The aqueous layer was concentrated and lyophilized. Nebularine was crystallized from ethanol.

## PHYSICAL AND CHEMICAL PROPERTIES

The molecular formula for nebularine is $C_{10}H_{12}N_4O_4$; mp 181° to 182°C; $[\alpha]_D^{25}$ −48.6 ($c$ = 1% in water); $\lambda_{max}$ in water, 0.1 $N$ HCl, and 0.1 $N$ NaOH is 262.5 mμ (Brown and Weliky, 1953). The ORD curve was negative plain in the range 300–700 mμ. The synthesis and chemical and enzymatic properties of the $N$-oxide of nebularine were reported by Stevens et al. (1962). Nebularine is readily hydrolyzed in acid to form purine and D-ribose (Löfgren et al. 1954). The purine ring is unstable in alkali (Gordon et al., 1957a).

## STRUCTURAL ELUCIDATION

Löfgren et al. (1954) hydrolyzed nebularine in HCl and isolated purine and D-ribose. When nebularine was treated with periodate, the spectral properties of the dialdehyde indicated that the sugar was linked to the purine nucleus at nitrogen-9 (Löfgren and Lüning, 1953).

## CHEMICAL SYNTHESIS

Brown and Weliky (1953) provided evidence for the proof of the structure of nebularine by reporting on the synthesis of this nucleoside antibiotic. The physical and chemical properties of nebularine and the chemically synthesized purine ribonucleoside were identical. The synthesis involved the condensation of the monochloromercuri derivative of 6-chloropurine and tri-*O*-acetyl-D-ribofuranosylchloride. The acetyl groups were removed with methanolic ammonia to give 6-chloro-9-β-D-ribofuranosylpurine, which was reduced catalytically to nebularine. The absolute proof of the β configuration of nebularine was established by the amination of 6-chloro-9-β-D-ribofuranosylpurine to adenosine. The chemical synthesis of nebularine has also been reported by several other methods. Fox et al. (1958) reported on the conversion of 6-mercapto-9-β-D-ribofuranosyl to nebularine by removal of the sulfur with Raney nickel. Hashizume and Iwamura (1966) and Iwamura and Hashizume (1968) described the synthesis of nebularine by a modified fusion method. The synthesis of nebularine-8-$^{14}$C has been described by Gordon and Brown (1956). While studying the properties of several purine derivatives, Gordon et al. (1957a) observed that nebularine was extremely labile to dilute alkali at room temperature. Two products were isolated following the opening of the imidazole ring. They also reported that nebularine gave rise to formic acid (carbon 8) when treated with *p*-toluenesulfonic acid or sulfuric acid. Adenosine did not give rise to formic acid. Apparently the amino group on the purine ring stabilizes the imidazole ring. Treatment of nebularine 5′-monophosphate with dilute alkali also resulted in a loss of carbon 8 as formic acid. The compound isolated was tentatively identified as the 5-amino derivative 4-(5′-phosphoribofuranosylamino)-pyrimidine.

The synthesis of the isopropylidine derivative, 2′,-3′-*O*-isopropylidine-9-β-D-ribofuranosylpurine was reported by Hampton and Magrath (1957). This compound was used as an intermediate in the synthesis of the 5′-phosphate of nebularine by Magrath and Brown (1957). The isopropylidine derivative of nebularine was treated with dibenzylphosphorochloridate in pyridine to give 2′,3′-*O*-isopropylidine-9-β-D-ribofuranosyl purine-5′-dibenzylphosphate. Hydrogenation removed the benzyl groups, and 9-β-D-ribofuranosylpurine 5′-phosphate was isolated.

The chemical synthesis of 7-deazanebularine has been reported by Gerster et al. (1967). The synthesis of this nucleoside was accomplished by the conversion of acetylated 7-deazanebularine to 4-chloro-7-(β-D-ribofuranosyl)pyrrolo[2,3-*d*]pyrimidine. The chloride was removed by catalytic hydrogenation. Fox et al. (1966) suggested that the biochemical behavior of nebularine should be compared with that of 7-deazanebularine.

## INHIBITION OF GROWTH

Nebularine inhibits the growth of human, bovine and avian strains of *M. tuberculosis* in Dorest's medium (Ehrenberg et al., 1946a; Löfgren et al., 1949). Ehrenberg et al. subsequently (1946b) reported on the effect of nebularine on other organisms. *Brucella abortus* was inhibited by nebularine, but *Staphylococcus aureus* and *E. coli* were not. Nebularine is a very toxic riboside of purine (Löfgren and Lüning, 1953; Löfgren et al., 1954; Biesele et al., 1955; Isono and Suzuki, 1960; Nakamura, 1961). Nebularine is very toxic to mice and rats, while the algycone (purine) is relatively nontoxic (Brown and Weliky, 1953). Twenty milligrams of nebularine per kilogram of body weight for a 3-day period was lethal to mice. The corresponding base was not toxic at 500 mg/kg/day. The $LD_{50}$ of nebularine when injected subcutaneously into mice is 100 mg/kg.

Biesele et al. (1955) studied the toxicity of purine and nebularine to cells of embryonic mouse skin and mouse sarcoma 180. Nebularine was toxic to sarcoma 180 cells, embryonic fibroblasts, and epithelial tissue of embryonic skin. Nebularine appeared to cause mitotic aberrations as well as chromosome breaks on root tips of *Allium cepa* (Löfgren et al., 1954).

## BIOCHEMICAL PROPERTIES

A limited number of studies have been reported on the biological effects of nebularine in bacteria and mammalian tissues.

### I. Effect on Glucose Dehydrogenase

Brink (1953) reported on the inhibition constant of nebularine as a competitive inhibitor of NAD with beef liver glucose dehydrogenase. The inhibition constant for nebularine was $2 \times 10^{-2}$ *M*. This compares with a $K_m$ of $4.3 \times 10^{-6}$ *M* for NAD.

### II. Effect on Animal Cells in Culture and Tumor Cells

Biesele et al. (1955) reported that nebularine was a very toxic ribonucleoside against sarcoma-180 cells in culture. Nebularine was more toxic to these cells than to embryonic fibroblasts, and more toxic to fibroblasts than to epithelial cells of embryonic skin. Adenosine, AMP, and NAD partially blocked the toxicity of nebularine. This nucleoside antibiotic caused mitotic aberrations and chromosome breaks in root tips of *Allium cepa* (Löfgren et al., 1954).

Sugiura and Creech (1956) studied the inhibition of ascites tumor growth by nebularine in mice. This nucleoside antibiotic was injected intraperitoneally and the degree of inhibition of ascites tumor growth was measured. The authors reported that nebularine exerted a moderate inhibition of sarcoma-180 ascites tumor cells, but did not inhibit Ehrlich ascites carcinoma or Krebs 2 ascites carcinoma.

## III. Metabolism of Nebularine in Animals

Gordon and Brown (1956) studied the metabolism of nebularine-8-$^{14}$C. The nucleoside antibiotic was injected intraperitoneally into male rats. The only urinary compound detected with sufficient radioactivity was allantoin. Very little radioactive carbon dioxide was detected as respiratory carbon dioxide. Hydrolysis of the nucleic acid resulted in the isolation of $^{14}$C-labeled adenine and guanine. These data suggest that amination at position 6 of the purine ring of nebularine had occurred. Nebularine was not incorporated into the nucleic acids. Analysis of the soluble pool indicated the presence of three nucleotide-containing purines. These nucleotides appeared to be the 5′-mono-, di-, and triphosphates of nebularine. The metabolism of purine-8-$^{14}$C was shown to be markedly different from that of its nucleoside (Gordon et al., 1957c).

## IV. Phosphorylation

Lindberg et al. (1967) and Schnebli et al. (1967) reported that nebularine was a substrate for rabbit liver adenosine kinase. The $K_m$ for nebularine and adenosine were $7.8 \times 10^{-5}$ *M* and $1.6 \times 10^{-6}$ *M*, respectively. Nebularine 5′-monophosphate was converted to its corresponding triphosphate in the presence of myokinase. Shigeura and Sampson (1967) studied the structural requirements for the 3′-hydroxyl groups of nebularine for phosphorylation. 3′-Deoxynebularine was not phosphorylated.

Gordon and Brown (1956) and Gordon et al. (1957b) reported on the isolation of the 5′-monophosphate and 5′-diphosphate of nebularine from the perchloric acid extracts of rat liver following the administration of nebularine-8-$^{14}$C.

## V. Effect on Viruses

The effect of nebularine as an inhibitor of influenza B virus multiplication was studied by Tamm et al. (1956). Nebularine was extremely inhibitory to influenza B virus, but it showed a low degree of selectivity.

## VI. Effect on Mengovirus Polymerase

Kapuler et al. (1969) tested nebularine as a replacement for ATP in the Mengovirus polymerase reaction. Nebularine triphosphate could not replace ATP. Nebularine can form only one hydrogen bond with template UMP residues. Similarly, the corresponding 7-deazapurine riboside 5′-triphosphate was not a substrate.

## VII. Resistance of Animal Cells to Adenosine Analogs

Bennett et al. (1966) reported that sublines of H. Ep. #2 cells that had lost adenosine kinase activity and could not phosphorylate cytotoxic nucleosides were resistant to these analogs. They showed that nebularine was not toxic to the sublines of cells that were missing adenosine kinase. These data strongly indicate that adenosine kinase is an essential enzyme for the activation of cytotoxic nucleosides and that those cells devoid of this enzyme activity may represent a mechanism of resistance to these nucleosides. Caldwell et al. (1967) reported a tumor subline of Ehrlich ascites carcinoma cells which was deficient in purine in ribonucleoside kinase. This subline of cells was not inhibited by nucleoside analogs. In addition, these cells were no longer susceptible to feedback inhibition by nebularine. It was proposed that the absence of the purine ribonucleoside kinase may be related to the resistance of these cells to nucleoside analogs.

## Summary

Nebularine is inhibitory towards bacteria, mammalian cells in culture, tumor cells, and influenza B virus. It is very toxic to mice. The corresponding free base has little inhibitory properties. The experimental evidence obtained indicates that nebularine must be phosphorylated to the 5′-nucleotide before it becomes cytotoxic. Sublines of tumor cells in culture devoid of adenosine kinase were not inhibited by nebularine. Nebularine 5′-triphosphate could not replace ATP in the Mengovirus polymerase reaction.

## References

Bennett, L. L., Jr., H. P. Schnebli, M. H. Vail, P. W. Allan, and J. A. Montgomery, *Mol. Pharmacol.*, **2**, 432 (1966).

Biesele, J. J., M. C. Slautterback, and M. Margolis, *Cancer*, **8**, 87 (1955).

Brink, N. G., *Acta Chem. Scand.*, **7**, 1081 (1953).

Brown, G. B., and V. S. Weliky, *J. Biol. Chem.*, **204**, 1019 (1953).

Caldwell, I. C., J. F. Henderson, and A. R. P. Paterson, *Can. J. Biochem.*, **45**, 735 (1967).
Ehrenberg, L., H. Hedström, N. Löfgren, and B. Takman, *Svensk Kem. Tidsskr.*, **58**, 269 (1946a).
Ehrenberg, L., H. Hedström, N. Löfgren, and B. Takman, *Svensk Farm. Tidsskr.*, **50**, 645 (1946b).
Fox, J. J., I. Wempen, A. Hampton, and I. L. Doerr, *J. Amer. Chem. Soc.*, **80**, 1669 (1958).
Fox, J. J., K. A. Watanabe, and A. Bloch, in Progress in Nucleic Acid Research and Molecular Biology, Vol. 5, J. N. Davidson and W. E. Cohn, Eds., Academic Press, New York, 1966, p. 251.
Gerster, J. F., B. Carpenter, R. K. Robins, and L. B. Townsend, *J. Med. Chem.*, **10**, 326 (1967).
Gordon, M. P., and G. B. Brown, *J. Biol. Chem.*, **220**, 927 (1956).
Gordon, M. P., V. S. Weliky, and G. B. Brown, *J. Amer. Chem. Soc.*, **79**, 3245 (1957a).
Gordon, M. P., D. I. Magrath, and G. B. Brown, *J. Amer. Chem. Soc.*, **79**, 3256 (1957b).
Gordon, M. P., O. M. Intrieri, and G. B. Brown, *J. Biol. Chem.*, **229**, 641 (1957c).
Hampton, A., and D. I. Magrath, *J. Amer. Chem. Soc.*, **79**, 3250 (1957).
Hashizume, T., and H. Iwamura, *Tetrahedron Letters* (1966), 643.
Isono, K., and S. Suzuki, *J. Antibiotics* (*Tokyo*), **13A**, 270 (1960).
Iwamura, H., and T. Hashizume, *J. Org. Chem.*, **33**, 1796 (1968).
Kapuler, A., D. C. Ward, N. Mendelsohn, H. Klett, and G. Acs, *Virol.*, **37**, 701 (1969).
Lindberg, B., H. Klenow, and K. Hansen, *J. Biol. Chem.*, **242**, 350 (1967).
Löfgren, N., B. Takman, and H. Hedström, *Svensk Farm. Tidsskr.*, **53**, 321 (1949).
Löfgren, N., and B. Lüning, *Acta Chem. Scand.*, **7**, 225 (1953).
Löfgren, N., B. Lüning, and H. Hedström, *Acta Chem. Scand.*, **8**, 670 (1954).
Magrath, D. I., and G. B. Brown, *J. Amer. Chem. Soc.*, **79**, 3252 (1957).
Nakamura, G., *J. Antibiotics* (*Tokyo*), **14A**, 94 (1961).
Schnebli, H. P., D. L. Hill, and L. L. Bennett, Jr., *J. Biol. Chem.*, **242**, 1997 (1967).
Shigeura, H. T., and S. D. Sampson, *Nature*, **215**, 419 (1967).
Stevens, M. A., A. Giner-Sorolla, H. W. Smith, and G. B. Brown, *J. Org. Chem.*, **27**, 567 (1962).
Sugiura, K., and H. J. Creech, *Ann. N.Y. Acad. Sci.*, **63**, 962 (1956).
Tamm, I., K. Folkers, and C. H. Shunk, *J. Bacteriol.*, **72**, 59 (1956).

## 6.5. CROTONOSIDE

### INTRODUCTION

Crotonoside (isoguanosine; 2-hydroxy-9-β-D-ribofuranosyladenine) (Fig. 6.10) was first isolated from the seeds of *Croton tiglium L* (Cherbuliez and Bernhard, 1932a). This purine nucleoside cannot serve as a sole source of purine in the growth of *Lactobacillus casei,* nor is it incorporated into mammalian nucleic acids. Although crotonoside is not a nucleoside antibiotic, it is included since it is a naturally occurring structural analog of guanosine.

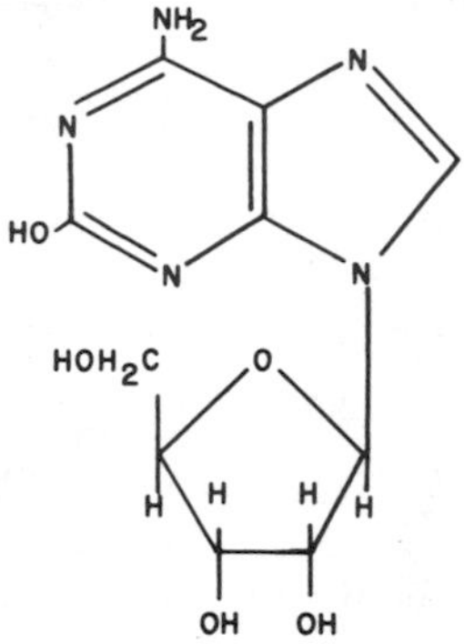

FIGURE 6.10 *Structure of crotonoside (isoguanosine).*

## DISCOVERY AND ISOLATION

Cherbuliez and Bernhard (1932a) discovered crotonoside as a naturally occurring nucleoside in *Croton tiglium* seeds. Crotonoside was isolated by extraction of ground seeds with methanol followed by filtration. The filtrate was treated with lead acetate and hydrogen sulfide, concentrated, and recrystallized from water.

## PHYSICAL AND CHEMICAL PROPERTIES

The molecular formula of crotonoside is $C_{11}H_{13}O_5N_5 \cdot 2H_2O$; $[\alpha]_D$–60.38° (in 0.1 *N* NaOH); mp 237–241°C (decomp.) (Cherbuliez and Bernhard, 1932a). Cherbuliez and Bernhard (1932b) and Falconer et al. (1939) reported that the ultraviolet absorption spectrum of crotonoside resembled that of 9-methylisoguanine and concluded that the attachment of the ribose to the base was at position 9. Falconer et al. (1939) reported that treatment of crotonoside with nitrous acid gave xanthosine. Spies (1939) showed that isoguanine is resistant to deamination with nitrous acid. D-Ribose was identified as the sugar moiety of crotonoside (Spies and Drake, 1935).

## CHEMICAL SYNTHESIS

Davoll (1951) described the chemical synthesis of crotonoside by treating 2,6-diamino-9-β-D-ribofuranosylpurine with nitrous acid to yield 9-β-D-ribofuranosylisoguanine. This product was identical with crotonoside that was isolated from Croton beans according to Falconer's procedure (1939). More recently, Montgomery and Hewson (1968) reported on the synthesis of crotonoside by starting with 9-(2′,3′,5′-tri-*O*-acetyl-β-D-ribofuranosyl)-2,6-dichloropurine. Treatment of this compound with sodium azide followed by catalytic hydrogenation and treatment with 48% fluoroboric acid and sodium nitrite gave the triacetate of crotonoside. Deacetylation was accomplished by treatment with ethanolic ammonia to give crotonoside.

Davoll (1951) reported that the deamination of 2,6-diaminopurine riboside resulted in the formation of a "gelatinous mass." He did not further characterize the aggregated state of this nucleoside. It was also known that dilute aqueous solutions of crotonoside become viscous on cooling, and at higher concentrations they form thick opalescent gels. Ravindranathan and Miles (1965) studied this phenomenon and presented evidence that crotonoside exists as an asymmetric, regular, ordered structure, presumably helical. By the use of viscosity, ultraviolet spectra, infrared spectra, and rotatory dispersion curves, the authors presented data that suggest that the structure of the gel is highly asymmetric and probably helical.

### SYNTHESIS OF AGLYCONE DERIVATIVES

The synthesis of several derivatives of the aglycone of crotonoside has been reported. The synthesis of isoguanine-1-*N*-oxide was reported by Parham et al. (1967). Cresswell and Brown (1963) reported on the synthesis of the 2-mercaptoadenine 1-*N*-oxide.

### BIOCHEMICAL PROPERTIES

Brown and his co-workers studied the utilization of crotonoside in nucleic acid synthesis in the rat and *L. casei* (Ballis et al., 1952; Lowy et al., 1952; Elion, 1953). Crotonoside-2-$^{14}$C was synthesized from radioactive 2,6-diamino-9-$\beta$-D-ribofuranosylpurine. The nucleoside was administered intraperitoneally into adult rats. A very small amount of crotonoside was incorporated into nucleic acids (Lowy et al., 1952). With *L. casei*, Ballis et al. (1952) reported that crotonoside-2-$^{14}$C was not incorporated into the nucleic acids. Crotonoside was unable to serve as a sole source of purine for the growth of a diaminopurine-resistant strain of *L. casei* (Elion et al., 1953).

While studying the deamination and vasodepressor effects of adenine analogs, Clarke et al. (1952) reported that crotonoside was 15–19 times as active as adenosine. Mihich et al. (1954) studied the effect of adenosine analogs on isolated intestine of rabbits. They reported that the 2-substituted adenosine analogs cause a transient reduction of tone and spontaneous motility in isolated intestine of rabbit.

Schnebli et al. (1967) showed that crotonoside did not inhibit phosphorylation of adenosine, but it did have a slight inhibition on the phosphorylation of 6-methylthiopurine riboside.

## Summary

Crotonoside (isoguanosine) is a naturally occurring nucleoside analog of guanosine. It cannot serve as a sole source of purine for the growth of a diaminopurine-resistant strain of *L. casei*. Crotonoside did not inhibit phosphorylation of adenosine, but did exert a slight inhibition of the phosphorylation of 6-methylthiopurine riboside.

## References

Balis, M. E., D. H. Levin, G. B. Brown, G. B. Elion, H. Vanderwerff, and G. H. Hitchings, *J. Biol. Chem.*, **199**, 277 (1952).

Cherbuliez, E., and K. Bernhard, *Helv. Chim. Acta*, **15**, 464 (1932a).

Cherbuliez, E., and K. Bernhard, *Helv. Chim. Acta*, **15**, 878 (1932b).

Clarke, D. A., J. Davoll, F. S. Philips, and G. B. Brown, *J. Pharmacol. Exptl. Therapy.*, **106**, 291 (1952).

Cresswell, R. M., and G. B. Brown, *J. Org. Chem.*, **28**, 2560 (1963).

Davoll, J., *J. Amer. Chem. Soc.*, **73**, 3174 (1951).

Elion, G. B., H. Vanderwerff, G. H. Hitchings, M. E. Balis, D. H. Levin, and G. B. Brown, *J. Biol. Chem.*, **200**, 7 (1953).

Falconer, R., J. M. Gulland, and L. F. Story, *J. Chem. Soc.*, 1939, 1784.

Lowy, B. A., J. Davoll, and G. B. Brown, *J. Biol. Chem.* **197**, 591 (1952).

Mihich, E., D. A. Clarke, and F. S. Philips, *J. Pharmacol. Exptl. Therapy.*, **111**, 335 (1954).

Montgomery, J. A., and K. Hewson, *J. Org. Chem.*, **33**, 432 (1968).

Parham, J. C., J. Fissekis, and G. B. Brown, *J. Org. Chem.*, **32**, 1151 (1967).

Ravindranathan, R. V., and H. T. Miles, *Biochim. Biophys. Acta*, **94**, 603 (1965).

Schnebli, H. P., D. L. Hill, and L. L. Bennett, Jr., *J. Biol. Chem.*, **242**, 1997 (1967).

Spies, J. R., *J. Amer. Chem. Soc.*, **61**, 350 (1939).

Spies, J. R., and N. L. Drake, *J. Amer. Chem. Soc.*, **57**, 774 (1935).

CHAPTER 7

# Azapyrimidine Nucleosides



## 7.1. 5-AZACYTIDINE

### INTRODUCTION

The nucleoside antibiotic 5-azacytidine (4-amino-1-β-D-ribofuranosyl-1,3,5-triazin-2-one) is a *s*-triazine nucleoside analog of cytidine (Fig. 7.1). It is a powerful bacteriostatic, antitumor, and mutagenic agent. Studies on the inhibitory action of this nucleoside related to bacteria, tumor cells, calf thymus nuclei, mutations, inducible enzymes, protein synthesis, RNA synthesis, DNA synthesis, ribosomal synthesis, bacteriophage, and the biochemical mechanisms of resistant cells are described in this section. This nucleoside antibiotic is primarily directed against lymphoid leukemia (Šorm

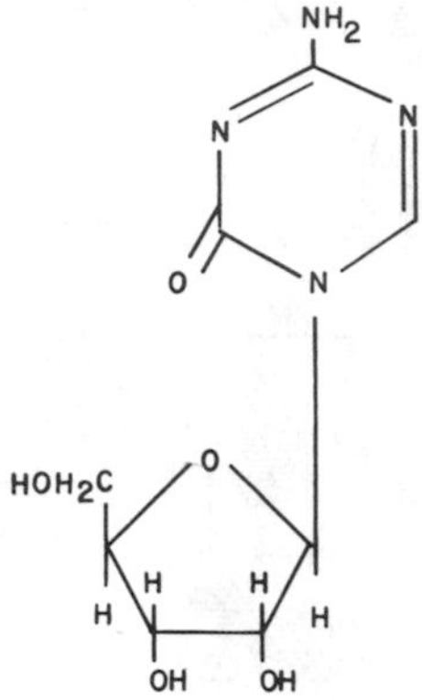

FIGURE 7.1 *Structure of 5-azacytidine.* (*From Pịhova et al.*, 1965*a*.)

and Veselý, 1964; Evans and Haňka, 1968). Šorm and co-workers and Haňka et al. have studied the metabolism and mechanism of action of 5-azacytidine in a number of biochemical reactions with bacterial and mammalian cells. 5-Azacytidine is readily phosphorylated and is a strong inhibitor of protein synthesis. 5-Azacytidine is incorporated into RNA and DNA, but nucleic acid synthesis is not inhibited (Doskočil et al., 1967). 5-Azacytidine does not inhibit DNA synthesis in uninfected cells, but does inhibit the synthesis of bacteriophage DNA following infection of *E. coli* with bacteriophage $T_4$ (Doskočil and Šorm, 1967). Cytidine reverses the inhibition of 5-azacytidine. This nucleoside antibiotic also brings about chromosomal and gene mutations (Fučík et al., 1965b, 1965c). Four mechanisms have been proposed to define the interference of 5-azacytidine with nucleic acid synthesis: (*1*) competition for the kinases (Raška et al., 1966); (*2*) inhibition of oroditylic acid decarboxylase (Veselý et al., 1967); (*3*) incorporation into ribonucleic acids (Jurovčík et al., 1965), and (*4*) the incorporation into DNA (Zadražil et al., 1965). Since the symmetrical triazine molecule is not stable and decomposes rapidly on slight changes in pH towards the alkaline region, it was postulated that a similar type of degradation of 5-azacytidine in the nucleic acids would also take place (Čihak et al., 1964; Pičhová et al., 1965b). However, Pačes et al. (1968a) subsequently isolated and identified intact 5-azacytidine following the incorporation into polynucleotide. These data tend to eliminate the idea that 5-azacytidine forms faulty nucleic acids by degradation following incorporation into the nucleic acids.

The first chemical synthesis of 5-azacytidine was reported by Pískala and Šorm (1964). Hanka et al. (1966) reported on the isolation of 5-azacytidine from the culture filtrates of *Streptoverticillium ladakanus*. 5-Azacytidine and its derivatives, with substituents on position 1, exists in the tautomeric form shown in Fig. 7.1 (Pičhová et al., 1965a).

## DISCOVERY, PRODUCTION, AND ISOLATION

Stock cultures of *S. ladakanus* are maintained under sterile soil. The seed medium is as follows: 2.5% glucose and 2.5% pharmamedia. Following inoculation of this seed culture with spores, the seed flasks were maintained on an incubator–shaker for 48 hr at 28°C. The production medium for 5-azacytidine was as follows: 4% black strap molasses, 3% pharmamedia, 1% dried whole yeast, and 2% dextrin. The flasks were shaken at 28°C. Maximum formation of the nucleoside antibiotic occurs between 70 and 90 hr following inoculation. The yield of 5-azacytidine was 920 $\mu$g/ml (Hanka et al., 1966). Production of 5-azacytidine is assayed by the disc-plate method against *E. coli*, ATCC 26.

The isolation of 5-azacytidine from the culture filtrates was first described by Bergy and Herr (1966). The culture filtrate was filtered at pH 7.8, followed by the addition of Darco G-60. The 5-azacytidine was eluted with 50% acetone and the eluants concentrated to aqueous solution. Five volumes of acetone were added and the inactive materials that precipitated were removed by filtration. The filtrate was concentrated to aqueous solution and freeze dried. Further purification was achieved by partition chromatography using diatomaceous earth with the solvent system 1-butanol–ethyl acetate–buffer (pH 6.0) (75:25:35, v/v/v). The fractions containing the nucleoside antibiotic were combined and two volumes of petroleum hexane were added. The solvent phases were separated and the aqueous phase was freeze dried. The residue was further chromatogramed on silica gel buffered with $Na_2HPO_4$ and $KH_2PO_4$. Chloroform–methanol (7:3) was used as a solvent. 5-Azacytidine crystallized from the concentrate.

While studying the biosynthesis of 5-azacytidine, Suhadolnik (manuscript in preparation) isolated a second nucleoside from the culture filtrates from *S. ladakanus*. This nucleoside has been unequivocally shown to be pseudouridine. Approximately 188–220 mg of crystalline pseudouridine have been isolated per liter of culture filtrate. Since this unusual C—C nucleoside has been found only in tRNA, the presence of pseudoridine in such large amounts in the culture filtrates of *S. ladakanus* poses a most fascinating question as to its biosynthesis and function as a metabolic product. With exception of the isolation of pseudouridine from the medium of *Agrobacterium tumefaciens* (Suzuki and Hochster, 1964), this is the first example of the isolation of the β-D-isomer of pseudouridine in such large amounts from the culture filtrate of a *Streptomyces*. While uracil-2-$^{14}$C is an effective precursor for the base moiety of psuedouridine, uridine-U-$^{14}$C is only incorporated into pseudouridine following hydrolysis of the *N*-riboside bond; and then only the uracil portion is utilized in the biosynthesis of pseudouridine (Suhadolnik and Uematsu, manuscript in preparation). Pseudouridine could not be isolated following incubation of uracil and ribose-5-phosphate with cell-free extracts of *S. ladakanus*. This *in vitro* assay is essentially the same as that reported by Suzuki and Hochster (1966) or Heinrikson and Goldwasser (1964).

## PHYSICAL AND CHEMICAL PROPERTIES

The molecular formula for 5-azacytidine is $C_8H_{12}N_4O_5$. The molecular weight is 252; mp 228–230°C; $[\alpha]_D^{25} = +39°$ ($c = 1\%$ in water). The ultraviolet absorption is as follows: $\lambda_{max}$ 241 m$\mu$ ($\epsilon = 8767$) in water; $\lambda_{max}$ 249 m$\mu$ ($\epsilon = 3077$) in 0.01 *N* HCl; $\lambda_{max}$ 223 m$\mu$ ($\epsilon = 24{,}200$) in 0.01 *N* KOH (Bergy and Herr, 1966). The infrared and nuclear magnetic resonance spectra strongly supported a nucleoside structure. The mass spectrum of 5-azacytidine did not show a molecular ion peak. However, a major peak at 111 mass units corresponded to the aglycone fragment of the formula, $C_3H_3N_4O$ (Bergy and Herr, 1966). Acid hydrolysis resulted in the isolation of the aglycone and ribose. The physical and chemical properties of the 5-azacytidine isolated from the culture filtrates of *S. ladakanus* and the chemically synthesized 5-azacytidine were the same (Pískala and Šorm, 1964). Bergy and Herr (1966) concluded that the compound isolated from the culture filtrates was 5-azacytidine or 1-$\beta$-D-ribofuranosyl-5-azacytosine (Fig. 7.1).

## CHEMICAL SYNTHESIS

The first total chemical syntheses of 5-azacytidine, 1-glycosyl derivatives of 5-azauracil, and 5-azacytosine were reported by Pískala and Šorm (1964). Their synthesis for 5-azacytidine is shown in Figure 7.2. When 1-chloro-2,3,5-tri-*O*-acetyl-D-ribofuranose (**1**) was allowed to react with silver cyanate, 2,3,5-tri-*O*-acetyl-$\beta$-D-ribofuranosyl isocyanate (**2**) was isolated. The $\beta$-configuration at the glycosidic center was ascribed on the basis of Baker's rule (Baker, 1957). Treatment of the isocyanate with 2-methylisourea gave the crystalline 1-(2,3,5-tri-*O*-acetyl-$\beta$-D-ribofuranosyl)-4-methylisobiuret (**3**). When this product was treated with ethylorthoformate, 1-(2,3,5-tri-*O*-acetyl-$\beta$-D-ribofuranosyl)-4-methoxy-2-oxo-1,2-dihydro-1,3,5-triazine (**4**) was isolated. Treatment of this ribofuranosyl derivative (**4**) with methanolic ammonia afforded crystalline 1-$\beta$-D-ribofuranosyl-5-azacytosine (5-azacytidine, **5**). Treatment of compound **4** with sodium methoxide and Dowex-50-$H^+$ gave 5′,6-anhydro-6-hydroxy-5,6-dihydro-5-azauridine (**6**).

The most recent synthesis of 5-azacytidine is that described by Winkley and Robins (1970). Their synthesis is the first example of the direct glycosylation of the 1,3,5-triazine ring. This was accomplished by treatment of the trimethylsilyl derivative of 4-amino-1,3,5-triazin-2-one (5-azacytosine) with 2,3,5-tri-*O*-acetyl-D-ribofuranosyl bromide in acetonitrile. The synthesis of 2′-deoxy-5-azacytidine was also described. The $\alpha$- and $\beta$-anomers of 2′-deoxy-5-azacytidine were clearly established for the first time by nmr. Pliml

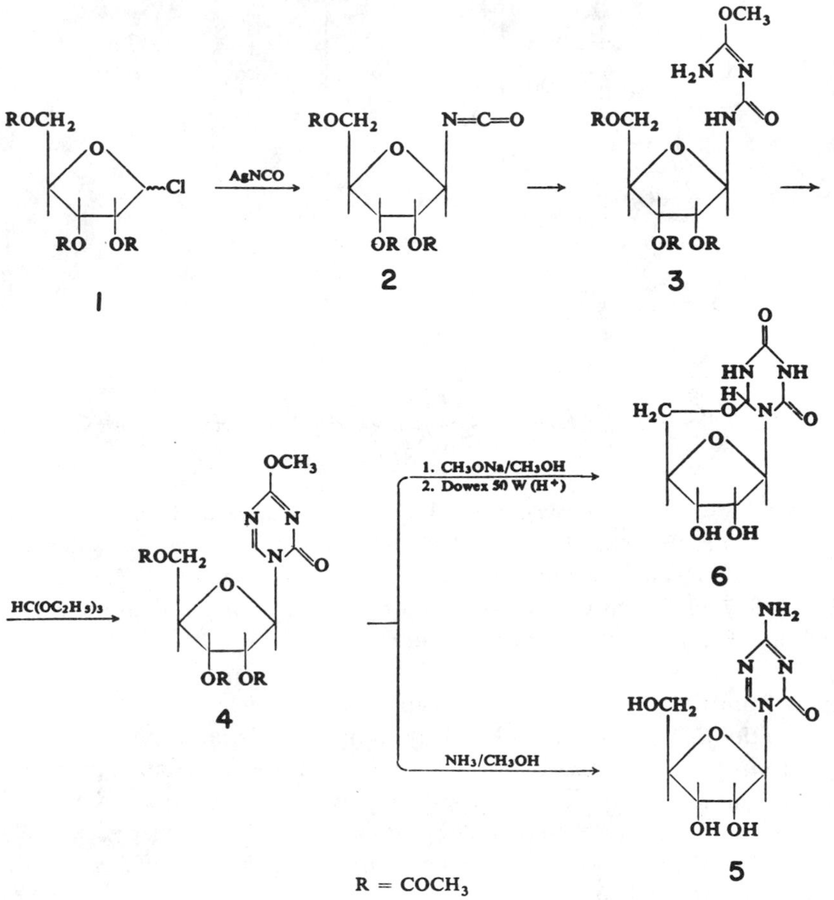

FIGURE 7.2 *Synthesis of* 5-*azacytidine.* (*From Pískala and Šorm*, 1964.)

and Šorm (1964) described the synthesis of 5-aza-2′-deoxycytidine, but no proof of anomeric structure was given. This nucleoside is an inhibitor of *E. coli* (Raška et al., 1965a). The chemical synthesis of 1-β-D-ribopyranosyl and glucopyranosyl-5- azacytosine has also been reported (Pískala and Šorm, 1964).

Píthová et al. (1965b) studied the hydrolysis of 5-azacytidine. The hydrolysis of 5-azacytidine in acid, neutral, and basic media is shown in Figure 7.3. Under acidic conditions, the glycosidic bond of 5-azacytidine (**5**) is cleaved to give the aglycone (**7**), the deaminated product (**8**) and D-ribose. In alkaline media, there is an attack at position 6 and the subsequent isola-

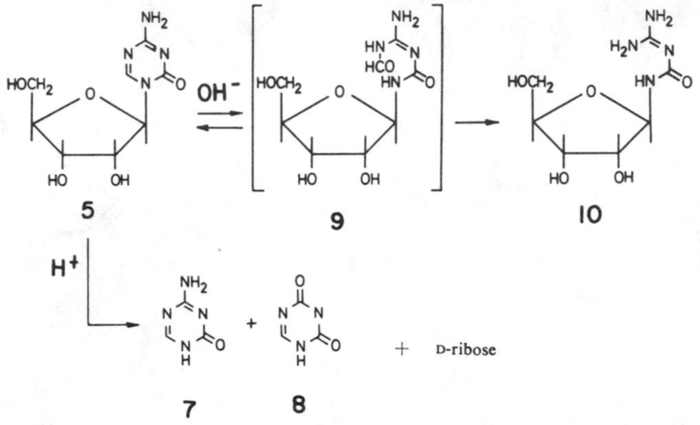

FIGURE 7.3 *Chemical behavior of* 5-*azacytidine in alkali and acid.* (*From Piťhova et al.,* 1965*b.*)

tion of 1-β-D-ribofuranosyl-3-guanylurea (**10**). Compound **10** retained the furanoid structure as determined by its electrophoretical behavior in a borate buffer. The β configuration at the glycosidic center did not change. The *N*-formyl derivative (**9**) is assumed to be an intermediate in the alkaline hydrolysis of 5-azacytidine to compound **10**.

The chemical properties of the 5-aza analogs of pyrimidine nucleosides differ from the pyrimidine and 6-azapyrimidine nucleosides. For example, 5-azacytidine is readily hydrolyzed by mineral acids, and the triazine ring is extremely labile in alkali (Pískala and Šorm, 1964). A detailed description of the synthesis of the 1-glycosyl-5-azacytosine can be obtained from the patent by Šorm and Pískala (1967). Pískala (1967) and Bergy and Herr (1966) described a procedure for the synthesis of the aglycone of 5-azacytidine and its methyl derivatives.

## INHIBITION OF GROWTH

Šorm et al. (1964), Šorm and Veselý (1964), Hanka et al. (1966) and Evans and Hanka (1968) reported that 5-azacytidine is a good therapeutic agent in the treatment of experimental leukemia in mice, T-4 lymphoma, and leukemia L1210. Šorm and Veselý (1964) reported that 5-azacytidine was more effective as an inhibitor of Ehrlich ascitic tumor than of Ehrlich solid tumor. The minimal inhibitory concentrations of 5-azacytidine against several bacteria is shown in Table 7.1. The inhibition of *E. coli* by 5-azacytidine was also reported by Čihák and Šorm (1965) and Šorm et al. (1964). Complete inhibition of growth of *E. coli* occurs at $1.2 \times 10^{-6}$ *M* (Čihák and Šorm, 1965). At a ratio of 1:100 (μg/ml) (5-azacytidine: pyrimidine precursor), uridine, deoxyuridine, cytidine, deoxycytidine and thymidine reversed the inhibition of growth of *E. coli* (Čihák and Šorm, 1965). When the molar ratio of 5-azacytidine to pyrimidine precursor was 1:50, uridine and cytidine reversed the inhibition of 5-azacytidine in *E. coli*; thymidine did not reverse the inhibition (Table 7.2). 5-Azacytidine also inhibits the production of viable phage particles by bacteria infected with bacteriophage $T_4$ (Doskočil and Šorm, 1967). Since Svatá et al. (1966) reported that 5-azacytidine administered intraperitoneally interrupted pregnancy, Seifer-

**Table 7.1**

Minimal Inhibitory Concentrations (MIC) of 5-Azacytidine against Several Bacteria

| Microorganism | MIC (μg/ml) at 24 hr in | |
|---|---|---|
| | Synthetic broth | Nutrient Broth |
| *Escherichia coli* ATCC 26 | 0.010 | 50 |
| *Salmonella gallinarum* USDA 8410 | 0.08 | 200 |
| *S. schottmuelleri* ATCC 9149 | 0.01 | 100 |
| *Pseudonomas mildenbergii* ATCC 795 | >200 | >200 |
| *Proteus vulgaris* ATCC 8427 | 0.005 | 200 |
| *Staphylococcus aureus* FDA 209P | | >200 |

From Hanka et al., 1966.

**Table 7.2**

Reversal of Inhibition by 5-Azacytidine (2.0 μM) of *Escherichia coli* by Several Purines and Pyrimidines

| Reversing compound[a] | Growth[b] |
|---|---|
| Orotidylic acid | 0 |
| Uridine | 92 |
| Cytidine | 95 |
| Thymidine | 5 |
| Adenosine | 4 |
| Guanosine | 0 |
| Inosine | 0 |
| Xanthosine | 0 |
| Control 1 (no reversing compound) | 0 |
| Control 2 (no 5-azacytidine) | 100 |

Hanka et al., 1966.
[a] All compounds were at a concentration of 100 μM.
[b] Expressed in per cent of control 2.

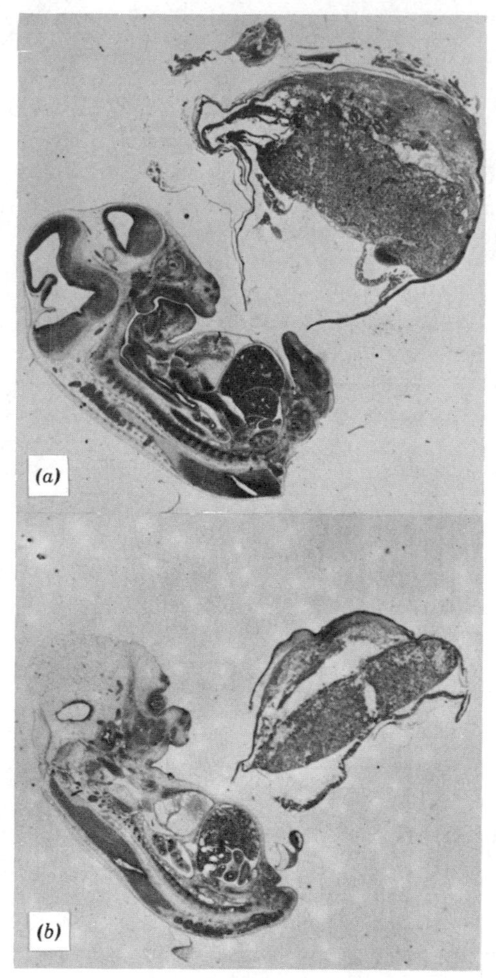


FIGURE 7.4 *Effect of 5-azacytidine on liver, placenta, and brain of 14-day-old mouse embryos: (a) control; (b) after administration of 3 doses of 5-azacytidine. (From Seifertová et al., 1968.)*

tová et al. (1968) studied the effect of 5-azacytidine on the developing mouse embryo (Fig. 7.4). Gross developmental abnormalities were observed in the cranial parts and in the livers of treated foetuses. The nervous tissue was mostly destroyed and only a few pycnotic cells remained. Likewise, hepatocytes greatly diminished in number, and dividing cells were absent. Similar changes were present in the placentae, which were considerably smaller than in the control. The $LD_{50}$ of 5-azacytidine administered intraperitoneally in mice is 115 mg/kg (Šorm et al., 1964).

## BIOCHEMICAL PROPERTIES

5-Azacytidine has been used as a biochemical tool for a large spectrum of studies in biological tissues. The earliest studies were concerned with fundamental information about the inhibitory effect of 5-azacytidine towards *E. coli* (Čihák and Šorm, 1965). They showed that in a bacterial medium, as well as with cell-free extracts of *E. coli*, 5-azacytidine was readily metabolized to ribosyl *N*-formylbiuret and ribosylbiuret. These two compounds are formed as decomposition products of the unstable nucleoside 5-azauridine. The inhibition of cytosine deaminase by 5-azacytidine was assumed to be attributed to 5-azauracil. This latter compound is formed from 5-azacytidine during incubation. 5-Azacytidine had a pronounced inhibition on the phosphorolytic cleavage of uridine. The synthesis of uridine from uracil and ribose 1-phosphate was not affected. Short term incubations with 5-azacytidine did not inhibit uridine phosphorylase. The authors suggested that the actual inhibitor for this enzyme was not 5-azacytidine, but some metabolic product such as 5-azauridine. This view was confirmed by preincubating 5-azacytidine with the cell-free extract of *E. coli* before estimating uridine phosphorylase inhibition. When the cell-free extract was preincubated with 5-azacytidine, there was a marked increase in the per cent inhibition of uridine phosphorylase. Uridine and cytidine reverse the inhibition of *E. coli* by 5-azacytidine. In addition, the formation of filamentous cell forms in the presence of 5-azacytidine was also reversed by the addition of uridine or cytidine, but not of uracil or cytosine (Čihák and Šorm, 1965).

To determine if the inhibitory effect of 5-azacytidine on the hematopoietic and lymphatic systems was due to the formation of the nucleotide, Jurovčík et al. (1965) reported that Ehrlich ascites tumor cells were able to phosphorylate 5-azacytidine to the 5′-mono-, di-, and triphosphates. Cell-free extracts of mouse tissues phosphorylated 5-azacytidine to the 5′-monophosphate. They also reported that 5-azacytidine-4-$^{14}$C was incorporated into RNA. The incorporation of 5-azacytidine into RNA was especially significant since it had not been possible to demonstrate the incorporation of pyrimidine aza-analogs into RNA of animal tissues. The radioactivity in DNA was not studied further. However, Pithová et al. (1965c) reported that 5-azacytidine was incorporated into DNA in the board bean (*Vicia faba*). The 2′-deoxy-5-azacytidine isolated from the DNA was identical with chemically synthesized 2′-deoxyazacytidine.

### I. Cytologic Effects of 5-Azacytidine in Normal Mice

Veselý and Šorm (1965) studied the effects of 5-azacytidine at the cellular level using tritium-labeled cytidine and thymidine with AK inbred mice. The autoradiographic experiments were performed 24 hr after three doses of 5-azacytidine on three consecutive days. They showed that RNA synthesis was diminished in the myeloid cells, while DNA synthesis was affected only in the mature elements. The largest drop in RNA and DNA synthesis in the mice treated with 5-azacytidine occurred in the thymus. The spleen and lymph nodes were involved to a smaller degree.

## II. Metabolism of 5-Azacytidine and 5-Aza-2′-deoxycytidine in Mice

Raška et al. (1965a, 1966) studied the distribution of the cancerostatic agents 5-azacytidine and 5-aza-2′-deoxycytidine following intraperitoneal administration to AKR mice. The two nucleoside analogs were distributed in the animal tissues, except for the central nervous system. Neither nucleoside penetrated the central nervous system. One-half of both substances was excreted within 6–8 hr after injection. The urine contained metabolic products as well as the original nucleosides. The data showed that mice could deaminate 5-azacytidine and 5-aza-2′-deoxycytidine and that 5-azacytidine substantially increased the amount of orotic acid and orotidine in the urine of AKR mice. 5-Azacytosine or its deoxyribonucleoside did not give the same results. The phosphorylation of 5-azacytidine and 5-aza-2′-deoxycytidine in normal and leukemia tissues of AKR mice has been studied (Šorm et al., 1966b). Cell-free extracts of all normal tissue phosphorylate 5-azacytidine, while only extracts of spleen and thymus glands phosphorylate 5-aza-2′-deoxycytidine. Leukemia tissues phosphorylate 5-aza-2′-deoxycytidine better than 5-azacytidine. There is no difference in the phosphorylation of 5-azacytidine in normal and leukemia tissues. Although these data indicate that one of the sites of action of 5-azacytidine would interfere with the decarboxylation of orotidylic acid, the authors concluded that this observation did not explain the marked biological effects of 5-azacytidine.

## III. Effect on Protein Synthesis

The inhibition of protein synthesis and of inducible enzymes by 5-azacytidine has been studied in several biological systems. The data reviewed in this section show that 5-azacytidine completely inhibits the synthesis of total protein and inducible enzymes concurrently, while the overall rate of synthesis of RNA remains nearly normal.

Raška et al. (1965b) reported on the effect of 5-azacytidine on the synthesis of protein and RNA in isolated nuclei of calf thymus. Figure 7.5 shows the changes in the incorporation of various precursors into protein and RNA in the presence of varying concentrations of 5-azacytidine. Protein synthesis, as indicated by the uptake of leucine-1-$^{14}C$ (curve 1) into thymus nuclei, is not inhibited (Fig. 7.5). 5-Azacytidine ($1 \times 10^{-3}$ *M*) blocked the incorporation of cytidine into RNA, while the incorporation of adenosine was reduced only to one-half. These observations on the inhibition of RNA synthesis were

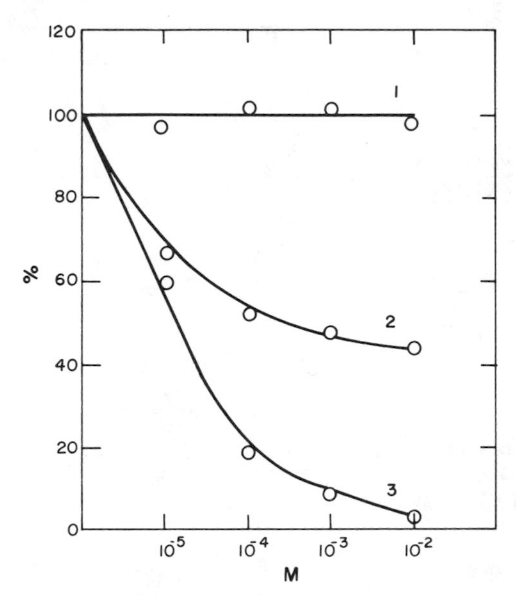


FIGURE 7.5 *Inhibitory effect of 5-azacytidine on the incorporation of precursors into ribonucleic acids and proteins of isolated nuclei of calf thymus.* (*Curve 1*) *incorporation of leucine*-1-$^{14}$*C;* (*curve 2*) *incorporation of adenosine*-8-$^{14}$*C;* (*curve 3*) *incorporation of cytidine*-$^{14}$*C.* (*From Raška et al.*, 1965*b*.)

in agreement with the earlier findings, in which the addition of 5-azacytidine resulted in a significant decrease in RNA synthesis in mouse thymus (Veselý and Šorm, 1965).

Since the mechanism of inhibition of 5-azacytidine appeared to be related to the formation of abnormal RNA and DNA following the incorporation of this nucleoside antibiotic, Čihák et al. (1967a; 1967b) studied the changes that occur in the formation of the inducible enzyme tryptophan pyrrolase. 5-Azacytidine completely inhibited the hydrocortisone-induced tryptophan pyrrolase in rat liver when administered either simultaneously or prior to the addition of the inducer (Fig. 7.6). 5-Azacytidine did not inhibit the activity of tryptophan pyrrolase when added to liver homogenates *in vitro*. The inhibition of the hormonal induction of tryptophan pyrrolase by 5-azacytidine was reversed by the addition of cytidine. The authors concluded that the inhibition of tryptophan pyrrolase induction by 5-azacytidine was due to the incorporation of this nucleoside antibiotic into newly synthesized messenger RNA. However, in a more detailed study, Čihák et al. (1969) showed that the inhibitory effect of 5-azacytidine on the substrate-induced tryptophan pyrrolase depends on the time between the administration of the pyrimidine analog and tryptophan. When 5-azacytidine and tryptophan are adminis-

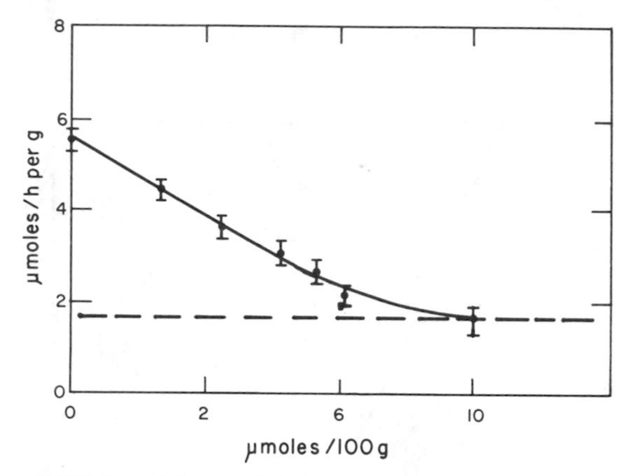


FIGURE 7.6. *Inhibition of hormonal induction of tryptophan pyrrolase by 5-azacytidine. Abscissa: µmoles* 5-*azacytidine applied per* 100 *g body weight; ordinate: activity of tryptophan pyrrolase. Dashed line is enzyme activity in absence of hormone.* (*From Čihák et al.*, 1967*a*.)

tered simultaneously, there is a 60% inhibition of tryptophan pyrrolase. When 5-azacytidine is administered 18 or 24 hr prior to the administration of tryptophan, there is a marked increase in enzyme activity when compared to the control. When tryptophan pyrrolase is induced by the hormone, there is no influence by the 24-hr pretreatment with 5-azacytidine. 5-Azacytidine is incorporated into the nucleic acids of the liver. This incorporation causes changes in the distribution pattern of the liver polyribosomes.

Levitan and Webb (1969) investigated the mechanism of inhibition of hepatic protein biosynthesis by 5-azacytidine. They studied the effect of this nucleoside analog on the induction of tyrosine transaminase, tryptophan pyrrolase, and hepatic polyribosome breakdown. By using adrenalectomized rats, they showed that the incorporation of radioactive amino acids into hepatic protein was inhibited by 5-azacytidine (Table 7.3). Although 5-azacytidine inhibited the degradation of the induced tyrosine transaminase and the induction of tryptophan pyrrolase, it did not inhibit the induction of tyrosine transaminase (Fig. 7.7). This differential effect on protein biosynthesis indicates that the drug exhibits a high degree of selectivity. According to current concepts this selectivity could only occur through the production of defective messengers. Those of short half-life or those newly produced would be affected, while preformed stable messenger RNA would not. The latter accounts for control at the translational level of protein synthesis. The data show that the induction of tryptophan pyrrolase was inhibited by the base analog. Levitan and Webb (1969) suggested that the messenger RNA affected may code for an enzyme which degrades or inactivates tyrosine transaminase (Kenney, 1967). The possibility that 5-azacytidine stabilizes

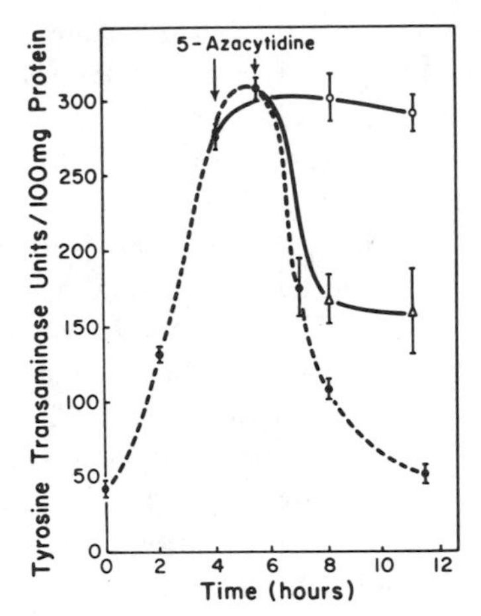


FIGURE 7.7. *Prevention by 5-azacytidine of tyrosine transaminase inactivation following its hormonal induction. Rats received* (●) 40 *mg of hydrocortisone sodium succinate per kilogram or hydrocortisone followed by* 20 *mg of* 5-*azacytidine per–kilogram at* (○) *4 or* (△) 5.5 *hr.* (*From Levitan and Webb*, 1969.)

tyrosine transaminase by a manner other than protein synthesis has not been eliminated. Čihák et al. (1970) also reported that 5-azacytidine inhibits the dietary induction and the corticosteroid induction of serine dehydratase.

In a subsequent report, Doskočil et al. (1967) elucidated the effect of 5-azacytidine on the induced synthesis of β-galactosidase in *E. coli*. The syn-

**Table 7.3**

Incorporation of Amino Acids into Soluble Hepatic Proteins[a]

| Amino acid | Specific radioactivity of soluble proteins, counts/min/mg[b] | | Inhibition by 5-azacytidine, % |
|---|---|---|---|
| | Control | 5-Azacytidine | |
| [$^{14}$C] Leucine | 202 | 76 | 63 |
| [$^{3}$H] Lysine | 531 | 156 | 71 |
| [$^{3}$H] Tyrosine | 157 | 55 | 65 |

From Levitan and Webb, 1969.

[a] Rats received saline or 20 mg of 5-azacytidine per kilogram of body weight, 4 hrs before removal of the liver. They were given [$^{14}$C] leucine (5 μc), [$^{3}$H]lysine (50 μc), or [$^{3}$H]tyrosine (30 μc) 15 min before liver removal. The specific radioactivity of the soluble hepatic proteins was measured by liquid scintillation counting.

[b] The data represent the average of 2 rats.

thesis of this enzyme was compared with the concurrent synthesis of RNA and DNA. 5-Azacytidine (10 μg/ml) completely and permanently inhibited the formation of β-galactosidase (Fig. 7.8). Under similar conditions, when the synthesis of β-galactosidase was completely inhibited, the total synthesis of protein was equally reduced (Fig. 7.9). The inducible enzyme, thymidine phosphorylase, was also inhibited. While 5-azacytidine completely inhibited the incorporation of leucine into total cell protein and the synthesis of two inducible enzymes, the rate of RNA synthesis was only slightly inhibited and the synthesis of DNA was not inhibited at all. The inhibition of β-galactosidase synthesis was reversed by either cytidine or uridine, and the activity of preformed thymidine phosphorylase was not inhibited by 5-azacytidine. The authors concluded that 5-azacytidine inhibited protein synthesis by selectively affecting the function of some species or portion of RNA that is essential for protein synthesis. While emphasizing the absence of any inhibitory effect on the synthesis of DNA in *E. coli*, Doskočil and Šorm (1967) stated that 5-azacytidine is a very strong inhibitor of the synthesis of DNA of bacteriophage $T_4$. However, this inhibitory action was shown to be direct and did not result from a preceding inhibition of enzymes specific for phage DNA synthesis. Doskočil and Šorm also stated that the inhibition of protein synthesis occurring in the uninfected host is of minor importance in the $T_4$-infected bacteria. They concluded that the inhibition of synthesis of phage DNA is probably related to the synthesis of 5-hydroxymethyldeoxycytidylic acid and is limited to those phage containing this nucleotide in their DNA. The notion that 5-azacytidine could interfere with the synthesis of the terminal sequence of tRNA was discounted since the turnover of the terminal sequence in *E. coli* is too slow (Rosset and Monier, 1965).

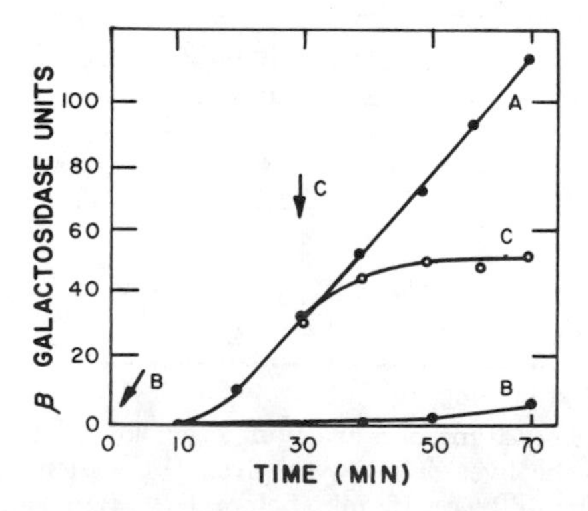


FIGURE 7.8 *Inhibition of synthesis of β-galactosidase by* 5-*azacytidine: (A) control; (B)* 5-*azacytidine (10 μg–ml) added simultaneously with the inducing lactose; (C)* 5-*azacytidine (10μg–ml) added 30 min later. (From Doskočil et al.*, 1967.)

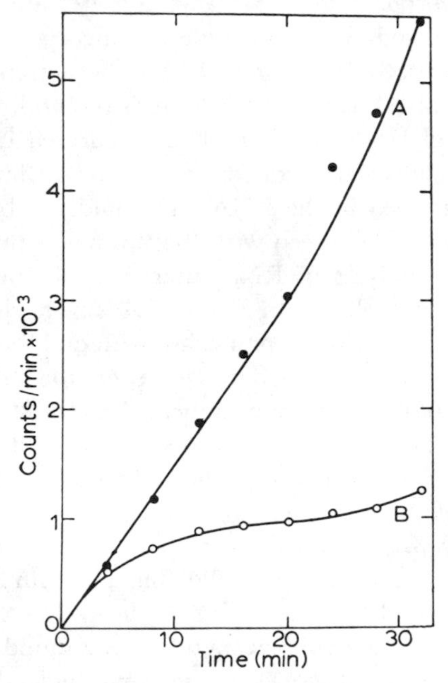


FIGURE 7.9. *Incorporation of leucine-$^{14}$C by a culture of E. coli: (A) control; (B) 5-azacytidine* (10-$\mu$*g/ml*). (*From Doskočil et al.*, 1967.)

Although Doskočil et al. (1967) stated that 5-azacytidine probably does not interfere with the inhibition of $\beta$-galactosidase synthesis due to the incorporation of this nucleoside antibiotic into the terminal sequence of *E. coli* tRNA, it should be noted that Kalousek et al. (1966) studied the effect of 5-azacytidine on the acceptor ability of tRNA, its physical properties ($T_m$), hyperchromic effect, and sedimentation analysis following the incorporation of 5-azacytidine.

Doskocil and Sorm (1970) have now compared the effects of 5-azacytidine and 5-azauridine on protein synthesis in *E. coli*. They observed that in *E. coli* deficient in cytidine deaminase, 5-azacytidine is a weak inhibitor of protein synthesis. However, the nucleoside antibiotic is extensively incorporated into RNA. With wild type strains, there was strong inhibition of protein synthesis. The authors attributed this inhihition to the formation of 5-azauridine that is formed by the deamination of 5-azacytidine. They also stated that the blocking of replication of phage T4 shown previously to be due to primary inhibition of replication of phage DNA is a function of 5-azacytidine itself.

Pačes et al. (1968a) subsequently studied the incorporation of 5-azacytidine into nucleic acids of *E. coli* and the inhibition of synthesis of $\beta$-galactosidase.

The kinetics of incorporation of 5-azacytidine-$^{14}$C into different species of nucleic acids was studied. The nucleoside was incorporated with a relatively high efficiency into both RNA and DNA. The action of 5-azacytidine is different from that of 5-fluorouracil in that 5-azacytidine does not interfere with the synthesis of DNA and is itself incorporated into DNA with high efficiency. When cytidine was added, the carbon-14 labeled RNA from 5-azacytidine-$^{14}$C remained in the RNA and could not be chased efficiently. The sedimentation profile agreed with the normal population of cell RNA. The sedimentation analyses of RNA after a 1-min pulse with carbon-14 labeled 5-azacytidine is shown in Figure 7.10*a*. The highest radioactivity was in the region around 15-S. Upon chase with cytidine, the sedimentation profile changed and the radioactivity followed the absorbance profile of total RNA (Fig. 7.10*b*). These results indicate an almost uniform distribution of 5-azacytidine in both 23-S and 16-S RNA, as well as in tRNA. Long-term labeling with 5-azacytidine-$^{14}$C showed the same sedimentation characteristics (Fig. 7.10*c*).

Twenty to thirty per cent of the cytidine in RNA and deoxycytidine in DNA was replaced by 5-azacytidine following a 25 min labeling experiment. Some of the 5-azacytidine was recovered from RNA following enzymic degradation. A large amount of 5-azacytidine was found in the tRNA. Since the turnover rate of the end sequence CpCpA is high, the authors expected that most of the 5-azacytidine would replace cytidine in this sequence. However, when tRNA, isolated from the inhibited cells, was tested *in vitro* for amino acid acceptor activity, only weak inhibitory effects were observed. The RNA pulse labeled in the presence of 5-azacytidine was capable of forming hydrids with homologous DNA. The authors concluded that the effects of 5-azacytidine upon components of the protein-synthesizing system other than mRNA were relatively insignificant. It appears then that mRNA containing 5-azacytidine does not function properly for protein synthesis. The notion that ring-opening of the 5-azacytidine in RNA was responsible for the biological effects of this nucleotide, once it was incorporated into ribopolynucleotide, was not supported in the experiments reported. Therefore, it appears most likely that the inhibition of protein synthesis may be attributed to the presence of intact 5-azacytidine in mRNA.

In subsequent experiments, Čihák et al. (1968) studied the effect of 5-azacytidine on thymidine kinase synthesis in regenerating rat liver. This study was performed since regenerating liver appears to be mediated by short-lived RNA molecules; therefore, any changes in the RNA or DNA would result in marked effects on protein synthesis. 5-Azacytidine prevented the formation of thymidine kinase in partially hepatectomized animals (i.e., in the regenerating liver remnant after removal of two-thirds of the liver to induce regeneration). 5-Aza-2′-deoxycytidine was not inhibitory. These

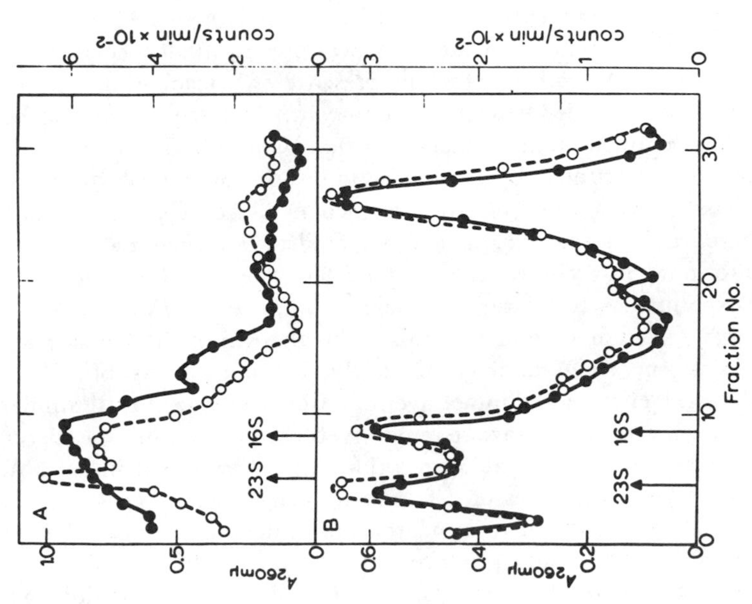


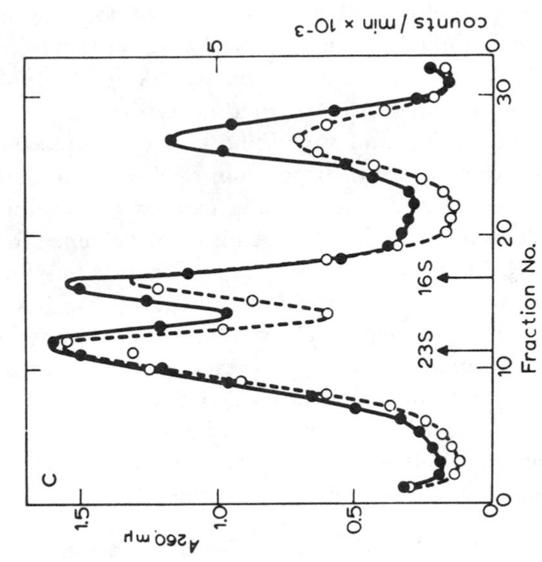


FIGURE 7.10. *Sedimentation profiles of RNA labeled with 5-azacytidine-*$^{14}$*C.* (*A*) *Cells were labeled for 1 min;* (*B*) *cells were labeled for 1 min and cytidine added subsequently for a 10-min period;* (*C*) *cells were labeled for 30 min.* (*From Pačes et al.,* 1968*a.*)

results provided evidence for the different inhibitory mechanisms of these two nucleoside analogs. When the patterns of polyribosomes in regenerating rat liver were studied, it was observed that 5-azacytidine completely abolished the heavy polyribosome region (Čihák et al., 1968). 6-Azauridine, a specific inhibitor of *de novo* pyrimidine synthesis, had no effect on polyribosome distribution and thymidine kinase. It appears that the observed changes are due to the incorporation of 5-azacytidine into newly synthesized messenger RNA's, as was the case in the inhibition of tryptophan pyrrolase (Čihák et al., 1967a). Additional evidence for this proposal was supplied by Čihák and Veselý (1969). They found that the induction of thymidine kinase is temporarily inhibited by 5-azacytidine administered 12–14 hr after partial hepatectomy. Cytidine could not reverse the inhibition of thymidine kinase caused by 5-azacytidine in partially hepatectomized rats. Sucrose-gradient analysis revealed that there was a loss of absorbance in the heavy polyribosome region of the postmitochondrial liver supernatant fractions. This inhibitory process may be ascribed to the incorporation of 5-azacytidine into newly synthesized RNA.

## IV. Effect on RNA Synthesis and Ribosome Formation and Breakdown

A more detailed study on the effect of 5-azacytidine on the synthesis of ribosomes in *E. coli* was reported by Pačes et al. (1968b). When *E. coli* cells were exposed to 5-azacytidine for 10–20 min, the RNA formed during this period had a normal sedimentation profile. The profile consisted mainly of 23-S, 16-S, and 4-S material. 5-Azacytidine was present in all species. However, most of the RNA synthesized in the presence of 5-azacytidine was not found in the ribosomes. Instead, it was found in a heterogeneous fraction having a lower sedimentation coefficient. When the inhibitory action of 5-azacytidine was interrupted by adding cytidine, the 50-S peak appeared and two peaks at 25-S and 34-S were formed in place of the 30-S peak. Most of the particles formed in the cells exposed to 5-azacytidine were incapable of forming 70-S ribosomes even with high concentrations of $Mg^{2+}$. In control experiments, association was nearly complete. Čihák et al. (1968) showed that 5-azacytidine caused a massive breakdown of the heavy polyribosomes in regenerating rat liver. Similarly, Levitan and Webb (1969) observed the same effect in the intact liver of adrenalectomized rats. Similar findings were reported with 8-azaguanine (Webb, 1967; Kwan and Webb, 1967). These results suggest that 5-azacytidine is incorporated into RNA, which leads to the biosynthesis of defective messenger RNA. It is entirely possible that 5-azacytidylic acid is incorporated into the terminal-CpCpA sequence of tRNA. This could produce a RNA that would be biologically inert and might explain the greater inhibitory effect of 5-azacytidine as compared to that of 8-azaguanine. Levitan and Webb (1969) showed that

5-azacytidine caused an increase in the monomer–dimer concentration and that maximal polyribosome breakdown induced by 20 mg of 5-azacytidine/kg occurred after 4 hr. By pulse labeling of nascent proteins, they showed that polyribosome breakdown was not due to the action of nucleases. Since Veselý et al. (1968b) reported that 5-azacytidine inhibits *de novo* pyrimidine synthesis by inhibiting orotic acid decarboxylase, Levitan and Webb examined the effect of 5-azacytidine on RNA synthesis by using adenine-$^{14}C$ as a precursor. The kinetics of the incorporation of adenine-$^{14}C$ into the cytoplasmic ribosomes of control and 5-azacytidine-treated rats showed that RNA synthesis is inhibited after a lag of about 1.5 hr (Fig. 7.11). This lag was attributed to the time needed to deplete preexisting pyrimidines.

## V. Effect on DNA Synthesis

### 1. *In Rhesus Monkey Kidney Cells in Tissue Culture*

Šorm et al. (1966b) compared the effects of 5-azacytidine and 5-aza-2′-deoxycytidine on the incorporation of leucine into protein and cytidine into RNA and DNA in Rhesus monkey kidney cells in tissue culture. Both antimetabolites inhibited the growth of the primary Rhesus monkey kidney cells in tussue culture, but the effects of both analogs were different. 5-Azacytidine inhibited the formation of RNA and DNA, while the inhibitory effect of 5-aza-2′-deoxycytidine on ribonucleoside and deoxyribonucleoside incorporation in mammalian cells in culture was not as great as the effect of 5-azacytidine.

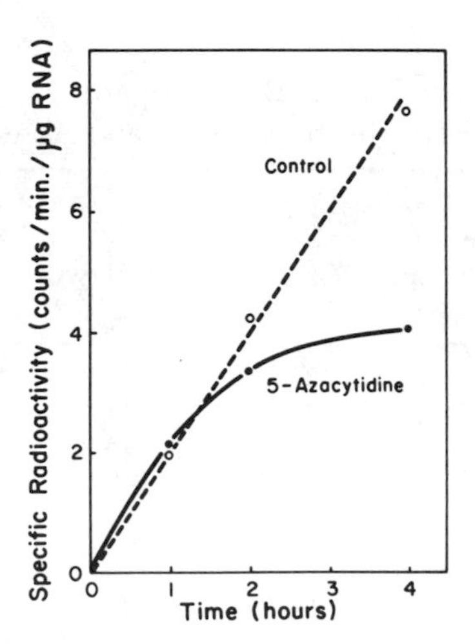


FIGURE 7.11. *Labeling of cytoplasmic ribosomes. Rats received 20µCi of adenine-$^{14}C$ with* (●) *5-azacytidine or* (○) *saline* (*From Levitan and Webb*, 1969.)

## 2. *In Normal and Phage-Infected Cells of* E. coli

Zadražil et al. (1965) studied the structure of DNA from *E. coli* grown in the presence of 5-azacytidine. The interest in studying this DNA resulted from earlier studies which showed that 5-azacytidine causes chromosomal (Fučík et al., 1965c) and gene mutations (Fučík et al., 1965b). It had been suggested that the DNA molecule containing 5-azacytidine would be less stable and subject to alteration of the secondary structure since this nucleoside is known to be extremely unstable (Piťhová et al., 1965b). The physicochemical properties of normal and 5-azacytidine-labeled DNA are shown in Table 7.4 (Zadražil et al., 1965). The sedimentation values and electron micrographs reveal great heterogeneity of the DNA preparation. This heterogeneity in molecular dimensions (guanine–cytosine content) was manifest by the gradual decrease of $T_m$ values of the DNAs isolated from the aging culture. The findings reported tend to support the notion that the incorporation and decomposition of 5-azacytosine in DNA can result in a labilization of the secondary structure, especially at the site of larger accumulation of guanine–cytosine pairs. The authors concluded that the observed changes in the structure of DNA containing 5-azacytosine were probably caused by the instability of the incorporated 5-azacytidine and might explain the biological effects of this compound in terms of its ability to bring about chromosomal and gene mutations that always result from some change or disturbance in DNA. However, Pačes et al. (1968a) showed that 5-azacytidine is stable in the polynucleotide and can be isolated following incorporation and subsequent hydrolysis.

**Table 7.4**

Physiochemical Characteristics of DNA Preparations Isolated from Different Phases of Growth of *E. coli* in the Presence and Absence of 5-Azacytidine

| Concentration of 5-azacytidine in the medium, μg/ml | Time of cultivation, hr | $s_{20,w}$ | Mol. wt. $\times 10^{-6}$ | $T_m$, °C | $\Delta A_{260m\mu}$ % |
|---|---|---|---|---|---|
| 0 | 1 | 25.7 | 12.8 | 95.7 | 21 |
| 0 | 2 | 24.3 | 11.1 | 94.2 | 22 |
| 0 | 3 | 27.9 | 15.5 | 93.6 | 27 |
| 0 | 4 | 23.2 | 10.0 | 93.0 | 28 |
| 0 | 18 | 23.2 | 10.0 | 90.5 | 30 |
| 100 | 1 | 3.7 | 0.3 | 93.5 | 12 |
| 100 | 2 | 18.6 | 5.3 | 93.5 | 17 |
| 100 | 3 | 26.2 | 13.5 | 90.2 | 23 |
| 100 | 4 | 19.2 | 5.5 | 89.3 | 25 |

From Zadražil et al., 1965.

Doskočil and Šorm (1967) studied the influence of 5-azacytidine on macromolecular syntheses induced by infecting *E. coli* cells with bacteriophage $T_4$. 5-Azacytidine at very low concentrations (0.5 $\mu$g/ml) is a strong inhibitor of replication of phage $T_4$. This inhibition was reversed by the simultaneous addition of cytidine. Deoxycytidine was not as effective in reversing this inhibition. When the incorporation of $^{32}P$ into the nucleic acids was studied, it was observed that the synthesis of RNA was not affected by 5-azacytidine, but the synthesis of DNA was strongly inhibited. The inhibition of DNA synthesis could not be reversed by deoxycytidine, but was reversed by cytidine. 5-Azacytidine did not inhibit DNA synthesis in uninfected cells. The addition of 5-azacytidine immediately following infection did not inhibit protein synthesis until the late period of phage development and the synthesis of late phage-specific proteins was therefore inhibited. Since DNA synthesis in the uninfected host cell is not inhibited by 5-azacytidine (Doskočil et al., 1967), there must be some reactions specific for phage DNA involved in this inhibition. The inhibition of 5-hydroxymethyldeoxycytidylic acid was studied. However, the deoxycytidylate hydroxymethylase was not inhibited. On the contrary, there was an overproduction of this enzyme in the 5-azacytidine-treated cultures. Studies on the effect of 5-azacytidine on genetic recombination of two amber mutants of $T_4$ showed that the frequency of wild-type recombinants of these two mutants is changed by 5-azacytidine. The cells of *E. coli* completely lose their ability to phosphorylate and incorporate 5-azacytidine upon infection with $T_4$ phage. The inhibition of replication of phage is more effective when 5-azacytidine is given before infection. Doskočil et al. (1967) speculated that very low amounts of 5-azacytidine (as the phosphorylated derivative) might be sufficient to inhibit the synthesis of phage DNA. This is in contrast to uninfected host cells which require higher amounts of 5-azacytidine to be incorporated into mRNA (Pačes et al., 1968b). An alternative explanation proposed that the inhibition of phage-DNA synthesis does not require the phosphorylation of 5-azacytidine. Phage ghosts selectively affect phosphorylation of 5-azacytidine in a manner like that of the whole phage (Doskočil and Šorm, 1969). The manner by which particles of phage protein can interact at the cell surface and destroy the ability to phosphorylate 5-azacytidine and not affect the phosphorylation of cytidine might be explained by a different mechanism for uptake and phosphorylation of 5-azacytidine compared to cytidine.

In another study, Doskočil and Pačes (1968) reported that the inducible enzyme thymidine phosphorylase, was strongly inhibited by 5-azacytidine in infected and uninfected *E. coli*.

### 3. On Root Meristem

Fučík et al. (1965b) reported that 5-azacytidine is highly mutagenic. It produced reversions of proline and tyrosine auxotrophic mutants to prototrophy. These reversions occurred to a much greater extent with 5-azacytidine than with 5-iodo-2′-deoxyuridine and suggested that 5-azacytidine was incorporated into DNA. Fučík, Šormová, and Šorm (1965a) studied the effect of 5-azacytidine on the primary roots of *Vicia faba* seedlings. They reported that 5-azacytidine led to a decrease in mitotic indices, but complete arrest of mitotic activity was not observed. The nucleoside antibiotic did not have a marked influence on the elongation growth of the root cell. When 5-azacytidine was removed, further growth was arrested for several days even though no signs of damage appeared. The addition of uridine or cytidine reversed the inhibitory action of 5-azacytidine. Uncoupling agents, such as 2,4-dinitrophenol, when added simultaneously with 5-azacytidine, blocked the inhibitory effect of this nucleoside. The authors concluded that energy was required for the phosphorylation of 5-azacytidine in order for it to be an inhibitor.

Mitotic depression and the occurrence of aberrant metaphases were observed only hours following the addition of 5-azacytidine. Apparently mitosis is not arrested due to the sudden deficiency of certain precursors for DNA synthesis.

## VI. Mutagenic Effects of 5-Azacytidine in Bacteria and Plants

5-Azacytidine brings about both chromosomal (Fučík et al., 1965c) and gene mutations (Fučík et al., 1965b) in auxotrophic strains of *E. coli* WP 14 Pro$^-$. Fučík et al. found that the number of gene mutations of this strain was substantially increased by 5-azacytidine (Table 7.5). The bacteria were maintained in a mineral medium enriched with casein hydrolyzate. It was

**Table 7.5**

The Effect of 5-Azacytidine on the Frequency of Revertants to Protrophy in *Escherichia coli* WP 14 Pro$^-$

| Experiment | Mutagen Concentration in the Medium, μg/ml | Number of Viable Cells per ml | Number of Mutants per $10^8$ Viable Cells |
|---|---|---|---|
| 1 | 0 | $5.2 \times 10^9$ | 15 |
| | 0.4 | $4.7 \times 10^9$ | 135 |
| | 4.0 | $3.8 \times 10^9$ | 2580 |
| 2 | 0 | $1.0 \times 10^{10}$ | 12 |
| | 0.4 | $7.8 \times 10^9$ | 38 |
| | 4.0 | $7.9 \times 10^9$ | 445 |

From Fučík et al., 1965b.

assumed that this mutagenic activity of 5-azacytidine was associated with the incorporation into the nucleic acids. Although the incorporation had not been demonstrated in bacteria, analogous experiments on the incorporation into DNA of plant material were reported. Halle (1968) reported that 5-azacytidine was mutagenic for abroviruses. Halle also suggested that the replicative form or the replicative intermediate species of viral RNA, or both, which are synthesized first, may be the primary sites of 5-azacytidine action.

## VII. Radioprotective Effect of 5-Azacytidine in AKR Mice

Since 5-azacytidine has a broad spectrum of activity in different biological systems, the effect on the survival of mice given a lethal dose of X-rays has been studied (Veselý et al., 1969b). When 5-azacytidine was administered to AKR mice prior to irradiation with a supra lethal dose of X-rays, there was a marked reduction in their mortality. The number of blood leukocytes and bone marrow nucleated cells of the 5-azacytidine treated mice is considerably higher when compared with those of animals that had only been irradiated. The authors concluded that the radioprotective effect of 5-azacytidine may influence the proliferation of the stem-cells since these cells are responsible for the repopulation of the bone marrow.

## VIII. Effect of 5-Aza-2′-Deoxycytidine against Mouse Leukemic and Mouse Hemopoietic Tissue

Šorm and Veselý (1968) showed that 5-aza-2′-deoxycytidine is an inhibitor of leukemic cells in AKR mice. The 2′-deoxy derivative is as effective as 5-azacytidine; however, the 2′-deoxy derivative appears to be more toxic on repeated administration. A partial reversal of inhibition of leukemic cells by 5-aza-2′-deoxycytidine occurred with the simultaneous addition of 2′-deoxycytidine. Thymidine did not reverse the inhibition of leukemic cells by 5-aza-2′-deoxycytidine. The simultaneous addition of both nucleoside analogs exerted an additive effect in prolonging the survival time of leukemic mice.

Leukemic cells resistant to 5-aza-2′-deoxycytidine remain sensitive toward 5-azacytidine.

## IX. Resistance of Cells to 5-Azacytidine

The incorporation of 5-azacytidine into the RNAs of leukemic mice sensitive and resistant to 5-azacytidine was first studied by Čihák et al. (1965). The uptake of 5-azacytidine-4-$^{14}C$ by leukemic cells of mice sensitive and resistant to this nucleoside showed that the proportion of labeled 5-azacytidine-resistant cells was 46% lower. When the soluble RNA and ribosomal RNA were examined, there were 43 and 61% decreases, respectively, in the amount of radioactivity in these nucleic acid fractions in the 5-azacytidine-resistant leukemic cells. Subsequently, Veselý et al. (1966, 1967) reported that uridine kinase and uridine phosphorylase activities were markedly decreased in the 5-azacytidine-resistant leukemic cells. It is generally accepted that resistance of cells to analogs of purines and pyrimidines is often associated with a decreased capacity of these resistant cells to form nucleotides. This leads to a block in the lethal synthesis of fraudulent nucleotides. These biochemical changes in the resistant cells appear to be related to the decreased ability to phosphorylate 5-azacytidine (Table 7.6) (Veselý et al., 1968b).

**Table 7.6**

*In Vivo* Phosphorylation of 5-Azacytidine in the Liver of Leukemic Mice Sensitive (AKR/s) and Resistant (AKR/r) to 5-Azacytidine[a]

| Expt. no. | Leukemic mice | Phosphorylation % | 5-Azacytidine 5′-phosphate, mμmole/g liver | Depression, % |
|---|---|---|---|---|
| 1 | AKR/s | 37.3 | 18.56 | |
| | AKR/r | 18.1 | 5.19 | 72.3 |
| 2 | AKR/s | 42.0 | 22.30 | |
| | AKR/r | 19.8 | 6.78 | 59.3 |

From Veselý et al., 1968b.

[a] 5-Azacytidine-4-$^{14}C$ (1 μc/μmole/mouse) was administered to two leukemic mice of both groups on the sixth day after inoculation and 1 hr before killing the animals.

Šorm et al. (1966a) reported that AKR leukemic cells resistant to 5-azacytidine differed markedly from the wild-type strain by the decreased incorporation *in vivo* of radioactive cytidine, thymidine, and 5-azacytidine into nucleic acids. The incorporation of thymidine into the acid-soluble pool was increased. Thymidine kinase activity was somewhat decreased in the resistant mutant cells and the degradation of thymidine and thymidine 5′-diphosphate and 5′-triphosphate was greater than in the 5-azacytidine-sensitive leukemic cells. There was no change in the activity of orotidylic acid decarboxylase in cell-free extracts of leukemic cells that were resistant to 5-azacytidine (Veselý et al., 1968b). Thymidine 5′-triphosphatase activity in particle-free supernatants of mouse leukemic cells (resistant to 5-azacytidine) was also increased (Veselý et al., 1968a). Veselý et al. (1968c) studied the characteristics of mouse leukemic cells resistant to 5-azacytidine and 5-aza-2′-deoxycytidine. The incorporation of labeled cytidine and 5-azacytidine into nucleic acids of AKR mouse leukemic cells during the development of resistance to 5-azacytidine was significantly lower than that of their wild-type counterparts. 5-Aza-2′-deoxycytidine-resistant mutant cells showed a decline in the uptake of 2′-deoxycytidine. Leukemic cells resistant to 5-aza-2′-deoxycytidine retained their sensitivity toward 5-azacytidine, but 5-azacytidine-resistant cells were cross-resistant to 5-aza-2′-deoxycytidine. Another biochemical change observed in 5-aza-2′-deoxycytidine-resistant leukemic cells was a 50% increase in the activity of deoxycytidine kinase (Veselý et al., 1969a). In addition, these resistant cells incorporate cytidine and adenosine *in vivo* into RNA and DNA much better than the sensitive cells. The incorporation of 2′-deoxycytidine into nucleic acids of mouse leukemic cells is extensively inhibited by 5-aza-2′-deoxycytidine (Veselý et al., 1969a).

The evidence presented indicates that the depressed anabolic conversion of 5-azacytidine to 5′-phosphates, i.e., to its biologically active form involves not only the inhibition by 5-azacytidine of orotidylic acid decarboxylase, but also the incorporation of the analog into nucleic acids.

## Summary

5-Azacytidine is a very potent antileukemic nucleoside antibiotic. It is also bacteriostatic and brings about chromosomal and gene mutations. Preliminary clinical tests indicate that 5-azacytidine may be useful in children with acute leukemia. It exerts a good therapeutic effect on transplanted leukemia of AKR mice. It is readily phosphorylated and incorporated into RNA and DNA and inhibits protein synthesis in bacterial, viral, and mammalian systems. It also inhibits orotidylic acid decarboxylase. 5-Azacytidine may prevent inactivation of tyrosine transaminase by inhibiting the synthesis of a degradative enzyme. By using 5-azacytidine as a biochemical tool it has been possible to demonstrate that the control of tyrosine transaminase takes place at the translational level of protein biosynthesis. The hormonal induction of tryptophan pyrrolase is inhibited by 5-azacytidine. 5-Azacytidine also inhibits the synthesis of phage-coat protein. The evidence presented to date indicates that mRNA containing 5-azacytidine does not function properly during protein synthesis. The notion that the ring-opening of the incorporated 5-azacytidine in the RNA is responsible for these biological effects has not been proven beyond all doubt. The physicochemical properties of DNA isolated following the incorporation of 5-azacytidine are markedly different from those of native DNA. 5-Azacytidine affects the synthesis of ribosomes in *E. coli* and causes a breakdown of hepatic polyribosomes to inactive monomers and dimers. Finally, the biochemical mechanism of resistance to 5-azacytidine appears to involve primarily uridine kinase, which is responsible for the first phosphorylation step of 5-azacytidine. This deficiency results in the impairment of the inhibitory action of 5-

azacytidine against orotidylic acid decarboxylase and in its decreased incorporation into RNA.

5-Aza-2′-deoxycytidine has potent antibacterial activity and remarkable antileukemic properties in AKR mice. Mouse leukemic cells that are resistant to 5-azacytidine show a decreased incorporation into the nucleic acids.

## References


Baker, B. R., *Ciba Found. Symp., Chem. Biol. Purines*, 1957, 120.

Bergy, M. E., and R. R. Herr, *Antimicrobial Agents Chemotherapy*, 1966, 625.

Čihák, A., and F. Šorm, *Collection Czech. Chem. Commun.*, **30**, 2091 (1965).

Čihák, A., J. Škoda, and F. Šorm, *Collection Czech. Chem. Commun.*, **29**, 300 (1964).

Čihák, A., J. Veselý, and F. Šorm, *Biochim. Biophys. Acta*, **108**, 516 (1965).

Čihák, A., J. Veselý, and F. Šorm, *Biochim. Biophys. Acta*, **134**, 486 (1967a).

Čihák, A., J. Veselý, and F. Šorm, *Collection Czech. Chem. Commun.*, **32**, 3427 (1967b).

Čihák, A., J. Veselý, and F. Šorm, *Biochim. Biophys. Acta*, **166**, 277 (1968).

Čihák, A., and J. Veselý, *Collection Czech. Chem. Commun.*, **34**, 910 (1969).

Čihák, A., J. Veselý, and F. Šorm, *Collection Czech. Chem. Commun.*, **34**, 1060 (1969).

Čihák, A., D. Wilkinson and H. C. Pitot, *Arch. Biochem.* Biophys., *in press* (1970).

Doskočil, J., and F. Šorm, *European J. Biochem.*, **8**, 75 (1969).

Doskočil, J., and V. Pačes, *Biochem. Biophys. Res. Commun.*, **30**, 153 (1968).

Doskočil, J., and F. Šorm, *Biochim. Biophys. Acta*, **145**, 780 (1967).

Doskočil, J., V. Pačes, and F. Šorm, *Biochim. Biophys. Acta*, **145**, 771 (1967).

Doskočil, J., and F. Sorm, *Biochem. Biophys. Res. Commun.*, **38**, 569 (1970).

Evans, J. S., and L. J. Haňka, *Experientia*, **24**, 922 (1968).

Fučík, V., Z. Šormová, and F. Šorm, *Biol. Plant. Akad. Sci. Bohemoslov*, **7**, 1 (1965a).

Fučík, V., S. Zadražil, Z. Šormová, and F. Šorm, *Collection Czech. Chem. Commun.*, **30**, 2883 (1965b).

Fučík, V., Z. Šormová, and F. Šorm, *Biol. Plant. Akad. Sci. Bohemoslov.* **7**, 58 (1965c).

Halle, S., *J. Virol.*, **2**, 1228 (1968).

Haňka, L. J., J. S. Evans, D. J. Mason, and A. Dietz, *Antimicrobial Agents Chemotherapy*, 1966, 619.

Heinrikson, R. L., and E. Goldwasser, *J. Biol. Chem.*, **239**, 1177 (1964).

Jurovčík, M., K. Raška, Jr., Z. Šormová, and F. Šorm, *Collection Czech. Chem. Commun.*, **30**, 3370 (1965).

Kalousek, F., K. Raška, Jr., M. Jurovčík, and F. Šorm, *Collection Czech. Chem. Commun.*, **31**, 1421 (1966).

Kenney, F. T., *Science*, **156**, 525 (1967).

Kwan, S. W., and T. E. Webb, *J. Biol. Chem.*, **242**, 5542 (1967).

Levitan, I. B., and T. E. Webb, *Biochim. Biophys. Acta*, **182**, 491 (1969).

Pačes, V., J. Doskočil, and F. Šorm, *Biochim. Biophys. Acta*, **161**, 352 (1968a).

Pačes, V., J. Doskočil, and F. Šorm, *FEBS Letters*, **1**, 55 (1968b).

Pískala, A., *Collection Czech. Chem. Commun.*, **32**, 3966 (1967).

Pískala, A., and F. Šorm, *Collection Czech. Chem. Commun.*, **29**, 2060 (1964).

Piťhová, P., A. Pískala, J. Piťha, and F. Šorm, *Collection Czech. Chem. Commun.*, **30**, 1626 (1965a).

Piťhová, P., A. Pískala, J. Piťha, and F. Šorm, *Collection Czech. Chem. Commun.*, **30**, 2801 (1965b).

Piťhová, P., V. Fučík, S. Zadražil, Z. Šormová, and F. Šorm, *Collection Czech. Chem. Commun.*, **30**, 2879 (1965c).

Pliml, J., and F. Šorm, *Collection Czech. Chem. Commun.*, **29**, 2576 (1964).

Raška, K., Jr., M. Jurovčík, Z. Šormová, and F. Šorm, *Collection Czech. Chem. Commun.*, **30**, 3001 (1965a).

Raška, K., Jr., M. Jurovčík, Z. Šormová, and F. Šorm, *Collection Czech. Chem. Commun.*, **30**, 3215 (1965b).

Raška, K., Jr., M. Jurovčík, V. Fučík, R. Tykva, Z. Šormová, and F. Šorm, *Collection Czech. Chem. Commun.*, **31**, 2809 (1966).

Rosset, R., and R. Monier, *Biochim. Biophys. Acta*, **108**, 385 (1965).

Seifertová, M., J. Veselý, and F. Šorm, *Experientia*, **24**, 487 (1968).

Šorm, F., and A. Pískala, U.S. Pat. No. 418,273 (1967).

Šorm, F., and J. Veselý, *Neoplasma*, **11**, 123 (1964).

Šorm, F., and J. Veselý, *Neoplasma*, **15**, 339 (1968).

Šorm, F., A. Pískala, A. Čihák, and J. Veselý, *Experientia*, **20**, 202 (1964).

Šorm, F., J. Veselý, and A. Čihák, *Acta Biochim. Pol.*, **13**, 385 (1966a).

Šorm, F., Z. Šormová, K. Raška, M. Jurovčík, *Rev. Roumaine Biochem.*, **3**, 139 (1966b).

Suzuki, T., and R. M. Hochster, *Can. J. Microbiol.*, **10**, 867 (1964).

Suzuki, T., and R. M. Hochster, *Can. J. Biochem.*, **44**, 259 (1966).

Svatá, M., K. Raška, Jr., and F. Šorm, *Experientia*, **22**, 53 (1966).

Veselý, J., and F. Šorm, *Neoplasma*, **12**, 1, (1965).

Veselý, J., J. Seifert, A. Čihák, and F. Šorm, *Intern. J. Cancer*, **1**, 31 (1966).

Veselý, J., A. Čihák, and F. Šorm, *Intern. J. Cancer*, **2**, 639 (1967).

Veselý, J., A. Čihák, and F. Šorm, *Collection Czech. Chem. Commun.*, **33**, 341 (1968a).

Veselý, J., A. Čihák, and F. Šorm, *Biochem. Pharmacol.*, **17**, 519 (1968b).

Veselý, J., A. Čihák, and F. Šorm, *Cancer Res.*, **28**, 1995 (1968c).

Veselý, J., A. Čihák, and F. Šorm, *Collection Czech. Chem. Commun.*, **34**, 901 (1969a).

Veselý, J., R. Gostof, A. Čihák, and F. Šorm, *Z. Naturforsch.*, **24b**, 318 (1969b).

Webb, T. E., *Biochim. Biophys. Acta*, **138**, 307 (1967).

Winkley, M. W., and R. Robins, *J. Org. Chem.*, **35**, 491 (1970).

Zadražil, S., V. Fučík, P. Bartl, Z. Šormová, and F. Šorm, *Biochim. Biophys. Acta*, **108**, 701 (1965).

CHAPTER 8

# Pyrrolopyrimidine Nucleosides

---

## ABBREVIATIONS

To, Toyocamycin; Tu, tubercidin; Sa, sangivamycin; F, formycin; TuTP, tubercidin 5′-triphosphate; ToTP, toyocamycin 5′-triphosphate; SaTP, sangivamycin 5′-triphosphate; TuMP, tubercidin 5′-monophosphate; ToMP, toyocamycin 5′-monophosphate; SaMP, sangivamycin 5′-monophosphate; poly (U,A), random copolymer of adenosine and uridine; ApApA, adenylyl-(3′,5′)-adenylyl-(3′,5′)-adenosine; poly d(T-C)·d(A-G) refers to the DNA-like polymer that contains alternating deoxythymidylate and deoxycytidylate units in one strand, and alternating deoxyadenylate and deoxyguanylate units in the complementary strand. Similarly, poly d(T-G)·d(A-C) refers to the polymer that contains alternating deoxythymidylate and deoxyguanylate units in one strand and deoxyadenylate and deoxycytidylate units in the complementary strand. Poly rTu.rU, poly rTor.U and poly rA.rU, 1:1 mixtures of the homopolymer of Tu, To or A with the homopolymer of uridine; Poly (A-G), poly (Tu-G), poly (Tu-U) and poly (Sa-U) are ribopolynucleotides containing repeating sequences of two ribonucleotides as indicated by the letters U, C, A, G, Tu or Sa; 3′-dATP, 3′-deoxyadenosine 5′-triphosphate; PRPP, 5-phosphoribosyl 1-pyrophosphate.

The structurally related pyrrolopyrimidine nucleoside antibiotics toyocamycin, tubercidin, and sangivamycin have been isolated from 13 *Streptomyces* cultures in independent laboratories. The structures (Fig. 8.1) of these three nucleosides have been elucidated and the total chemical syntheses have been reported. These three pyrrolopyrimidine ribosides represent a new group of nucleoside antibiotics that are highly cytotoxic to mammalian cell lines in culture and inhibitory to the growth of bacteria and fungi, as well as to RNA and DNA viruses. Modification of the 4-amino group and the cyano group at position 5 of toyocamycin or position 5 of tubercidin markedly changed the antineoplastic and antiviral activities of these pyrrolopyrimidine nucleoside antibiotics. The close structural relationship of tubercidin, toyocamycin, and sangivamycin to adenosine has made these nucleosides extremely valuable biochemical tools for studying the many cellular and enzyme reactions in which adenosine, AMP, ADP, or ATP has been replaced. A biosynthetic relationship between purines and the pyrrolopyrimidine nucleoside antibiotics has been reported.

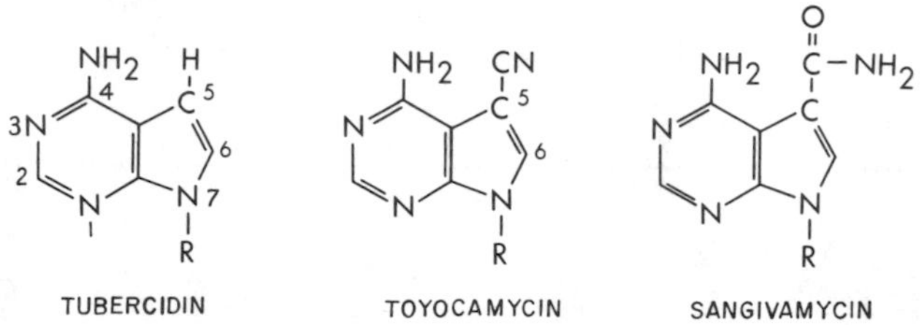

TUBERCIDIN TOYOCAMYCIN SANGIVAMYCIN

FIGURE 8.1. *Structures of the pyrrolopyrimidine nucleoside antibiotics.*

## 8.1. TOYOCAMYCIN

### DISCOVERY, PRODUCTION AND ISOLATION

Toyocamycin has been isolated in a number of independent laboratories. Nishimura et al. (1956) isolated this nucleoside from a culture of *S. toyocaensis* and Ohkuma (1960, 1961) isolated it from *Streptomyces* strain 1922 from the soil near Fuji City, Shizuoka Prefecture. Yamamoto et al. (1957) isolated toyocamycin from *Streptomyces* strain No. 1037 and assigned the name "antibiotic 1037." Antibiotic 1037 was shown to be the same as toyocamycin by Aszalos et al. (1966). Kikuchi (1955) and Katagiri et al. (1957) reported the isolation of an antibiotic from a culture of *Streptomyces sp.* no. E-212. Matsuoka and Umezawa (1960) and Matsuoka (1960) reported on the isolation of the antibiotic, unamycin B from the culture filtrates of *S. fungicidus*. They suggested that unamycin B was related to toyocamycin and antibiotic E-212. It may be that monilin, a compound isolated from *Streptomyces* 6633 (Fujii et al., 1955) and closely related to the pyrrolopyrimidine nucleosides, has the same structure as toyocamycin. The antibiotic vengicide was isolated from *S. vendargensis* (Stheeman and Struyk, 1953; Struyk and Stheeman, 1957, 1965). Rao et al. (1963) reported on the isolation of toyocamycin from a strain of *S. rimosus*. Arcamone et al. (1968) isolated toyocamycin from the mycelium of *Streptomyces* 86; mannosidohydroxystreptomycin and hydroxystreptomycin were also isolated.

The absolute proof that antibiotic 1037, unamycin B, vengicide, antibiotic E-212, and toyocamycin had the same structures was established by the total synthesis of the pyrrolo[2,3-*d*]pyrimidine nucleoside antibiotics toyocamycin, sangivamycin, and tubercidin by Robins and his co-workers (Tolman et al., 1968; 1969).

The references given above described the media that were used for the production of toyocamycin. The medium described here for the production of toyocamycin is that of Rao et al. (1963): 1 g glucose, 1.5 g soy bean meal, 0.2 g sodium chloride, 0.25 distillers' solubles, 0.25 g dibasic potassium phosphate, and 0.2 g calcium carbonate in 100 ml water. The flasks were shaken at a temperature of 26–36°C and maximum toyocamycin production occurred at 60–72 hr. Although most of the toyocamycin was found in the fermentation medium, some was also isolated from the mycelium. Toyocamycin has also been isolated by converting it to sangivamycin (Uematsu & Suhadolnik, 1970). The yield of crystalline sangivamycin by this method is 47mg/liter of medium.

Toyocamycin is isolated from the culture filtrates by adjusting the pH to 4–6, filtering, and extracting with 1-butanol (pH 8.0), following which the butanol is removed. The aqueous extract is passed through a weakly basic anion-exchange resin and the column washed with 1-butanol. The toyocamycin passes through the column while most of the impurities are absorbed. The fractions containing toyocamycin are combined and concentrated and the antibiotic crystallizes on storage. It is further purified by recrystallization from methanol–chloroform (1:1) (Rao et al., 1963).

## PHYSICAL AND CHEMICAL PROPERTIES

When toyocamycin is crystallized from ethanol, the molecular formula is $C_{12}H_{13}N_5O_4$. The molecular weight is 291. The molecular formula following crystallization from water is $C_{12}H_{13}N_5O_4 \cdot H_2O$ (Ohkuma, 1961). The melting points as reported by seven independent laboratories for toyocamycin ranged from 230 to 250°C. The highest melting point for toyocamycin was 247–250°C (Rao et al., 1963); $[\alpha]_D^{16} = -45.7°$ ($c = 1.05\%$ in 0.1 $N$ HCl). The ultraviolet spectral properties are as follows: $\lambda_{max}$ (ethanol) 231 m$\mu$ ($\epsilon = 9{,}300$) and 278 m$\mu$ ($\epsilon = 15{,}100$); $\lambda_{max}$ (in 0.1 N HCl) 232 m$\mu$ ($\epsilon = 16{,}000$) and 272 m$\mu$ ($\epsilon = 12{,}200$); $\lambda_{max}$ (pH 11) 227 m$\mu$ ($\epsilon = 10{,}200$) and 277 m$\mu$ ($\epsilon = 14{,}300$) (Tolman et al., 1969). Toyocamycin is slightly soluble in water, methanol, ethanol, acetone, dioxane, and 1-butanol. It is insoluble in petroleum ether, ethyl acetate, and chloroform.

## STRUCTURAL ELUCIDATION

The chemical structure of toyocamycin was in large part elucidated by Ohkuma (1960, 1961). Final proof for the structure of toyocamycin was the total chemical synthesis of the aglycones structurally related to 4-aminopyrrolo[2,3-*d*]pyrimidine (Taylor and Hendess, 1964, 1965). When Toyocamycin was refluxed with water and Dowex-50-H$^+$, followed by paper chromatography of the aqueous fraction, D-ribose was identified as the sugar moiety of this nucleoside antibiotic. While purine nucleosides are acid labile, the pyrrolo[2,3-*d*]pyrimidine nucleoside antibiotics are extremely stable to acid treatment (Ohkuma, 1961; Pike et al., 1964). Following the hydrolysis of toyocamycin with HCl, treatment with nitrous acid, and heating in a sealed tube with hydriodic acid and red phosphorus for 5 hr, 4-hydroxypyrrolo-

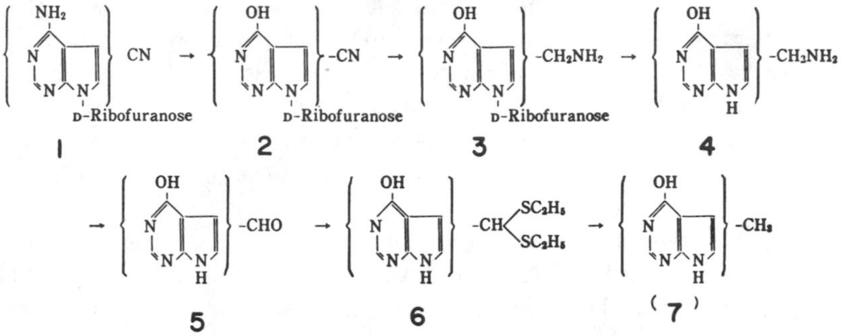

FIGURE 8.2. *Determination of the position of the cyano group of toyocamycin.* (*From Okhuma, 1961.*)

[2,3-*d*]pyrimidine was isolated. This compound was identical with authentic, chemically synthesized 4-hydroxypyrrolo[2,3-*d*]pyrimidine (Davoll, J., 1960; Wellcome Foundation Ltd., 1960). Treatment of toyocamycin by alkaline fusion at 250°C also yielded 4-hydroxypyrrolo[2,3-*d*]pyrimidine. Ohkuma (1961) determined the position of the cyano group in toyocamycin (Figure 8.2). The antibiotic (**1**) was deaminated with nitrous acid to yield 7-(D-ribosyl)-4-hydroxy-5-cyanopyrrolo[2,3-*d*]pyrimidine (**2**). Catalytic hydrogenation of the deaminated toyocamycin resulted in the isolation of desamino-hydroxy-descyano-aminomethyl toyocamycin (**3**). When compound **3** was treated with hydriodic acid and red phosphorus in a sealed tube, 150°C, 4 hrs, 4-hydroxy-5-aminomethylpyrrolo[2,3-*d*]pyrimidine (**4**) was isolated. Compound **4** was converted to 4-hydroxy-5-(diethylthio-)methyl-pyrrolo[2,3-*d*]pyrimidine (**6**). This compound was converted to 4-hydroxy-5-methylpyrrolo[2,3-*d*]pyrimidine (**7**) by refluxing with Raney nickel. Comparison of compound **7** with 2-methyl-4-hydroxypyrrolo[2,3-*d*]pyrimidine and 6-methyl-4-hydroxypyrrolo[2,3-*d*]pyrimidine showed that compound **7** was different from either of these two compounds. Ohkuma concluded that 5-methyl-4-hydroxypyrrolo[2,3-*d*]pyrimidine was the only remaining *C*-methyl derivative of 4-hydroxy-pyrrolo[2,3-*d*]pyrimidine. The total chemical synthesis of 5-methyl-4-hydroxypyrrolo[2,3-*d*]pyrimidine has been described by Ohkuma (1961). The physical and chemical properties of the chemically synthesized 4-hydroxy-5-methylpyrrolo[2,3-*d*]pyrimidine and compound **7** were identical. Consequently, Ohkuma (1961) unequivocally established the position of the cyano group in toyocamycin and proposed the structure of toyocamycin as 4-amino-5-cyano-7-(D-ribofuranosyl)-pyrrolo[2,3-*d*]pyrimidine. The configuration of the glycosidic linkage was not established. Additional proof for the structure of the aglycone of toyocamycin was obtained by Taylor and Hendess (1964, 1965). The synthesis of the pyrrolo-

[2,3-*d*]pyrimidine ring system utilized pyrrole intermediates for the synthesis of the aglycones of the pyrrolopyrimidine nucleoside antibiotics tubercidin and toyocamycin (Fig. 8.3). The synthesis of 2,5-diamino-3,4-dicyanothiophene (**8**) was accomplished by treatment of tetracyanoethylene with hydrogen sulfide. Rearrangement of compound **8** with alkali resulted in the formation of 5-amino-3,4-dicyano-2-mercaptopyrrole (**9**). Treatment of **9** with methyl orthoformate and alcoholic ammonia gave 4-amino-5-cyano-6-methylmercaptopyrrolo[2,3-*d*]pyrimidine (**10**). Treatment of compound **10** with Raney nickel in aqueous ammonium hydroxide gave 4-amino-5-cyanopyrrolo[2,3-*d*]pyrimidine (**11**). Compound **11** (the aglycone of toyocamycin) was converted to compound **12** by hydrolysis with 6 *N* HCl. The decarboxylation of compound **12** gave 4-aminopyrrolo [2,3-*d*]pyrimidine (**13**). Compound **13** is the aglycone of tubercidin.

Uematsu and Suhadolnik (1968) described a degradation procedure for the isolation of the aglycone and carbon atoms 1′ and 2′ of the pyrrolopyrimidine nucleosides. The degradation of these two nucleoside antibiotics is shown in Figure 8.4. The procedure as outlined in Figure 8.4 has been especially useful in studying the biosynthesis of the pyrrolopyrimidine nucleoside antibiotics since the aglycone and carbons 1′ and 2′ of the ribose moiety can be isolated. The aglycone of toyocamycin (**11**) was obtained in a 71% yield.

Most recently, the elegant work of Robins and co-workers has unequivocally established the structures of toyocamycin and sangivamycin by describing the total synthesis of these two pyrrolopyrimidine nucleoside antibiotics (Tolman et al., 1968). The synthesis is outlined in Figure 8.5. 4-Amino-6-bromo-5-cyanopyrrolo[2,3-*d*]pyrimidine (**14**) was synthesized by ring closure of 2-amino-5-bromo-3,4-dicyanopyrrole. Acetylation of **14** afforded compound **15**. When compound **15** and 1,2,3,5-tetra-*O*-acetyl-β-D-ribofuranose were heated at 175°C with catalytic amounts of bis(*p*-nitrophenyl) phosphate, the acetylated nucleoside (compound **16**) was formed. Deacetylation of compound **16** afforded 4-amino-6-bromo-5-cyano-7-(β-D-ribofuranosyl)pyrrolo[2,3-*d*]pyrimidine (compound **17**) in 92% yield. Proof that the D-ribofuranosyl moiety of compound **17** was at nitrogen 7 was supplied by Tolman et al. (1967). They reported that 4-amino-5-cyanopyrrolo[2,3-*d*]pyrimidines with substituents in the 6 position produced a 15–30 mμ bathochromic shift, while alkylation in the pyrrole ring produced only a small hypsochromic shift. The ultraviolet absorption maxima for compound **14** and the nucleoside (compound **17**) are identical. The 6-bromo group of compound **17** was removed to yield toyocamycin (compound **1**). The chemical and physical properties and mixed melting points with toyocamycin were identical. Proof of the β configuration of the nucleosides for toyocamycin, sangivamycin, and tubercidin was established by the appearance in the pmr of a sharp singlet at 6.65 δ (1H) for the anomeric proton of 2′,3′-*O*-isopropyli-

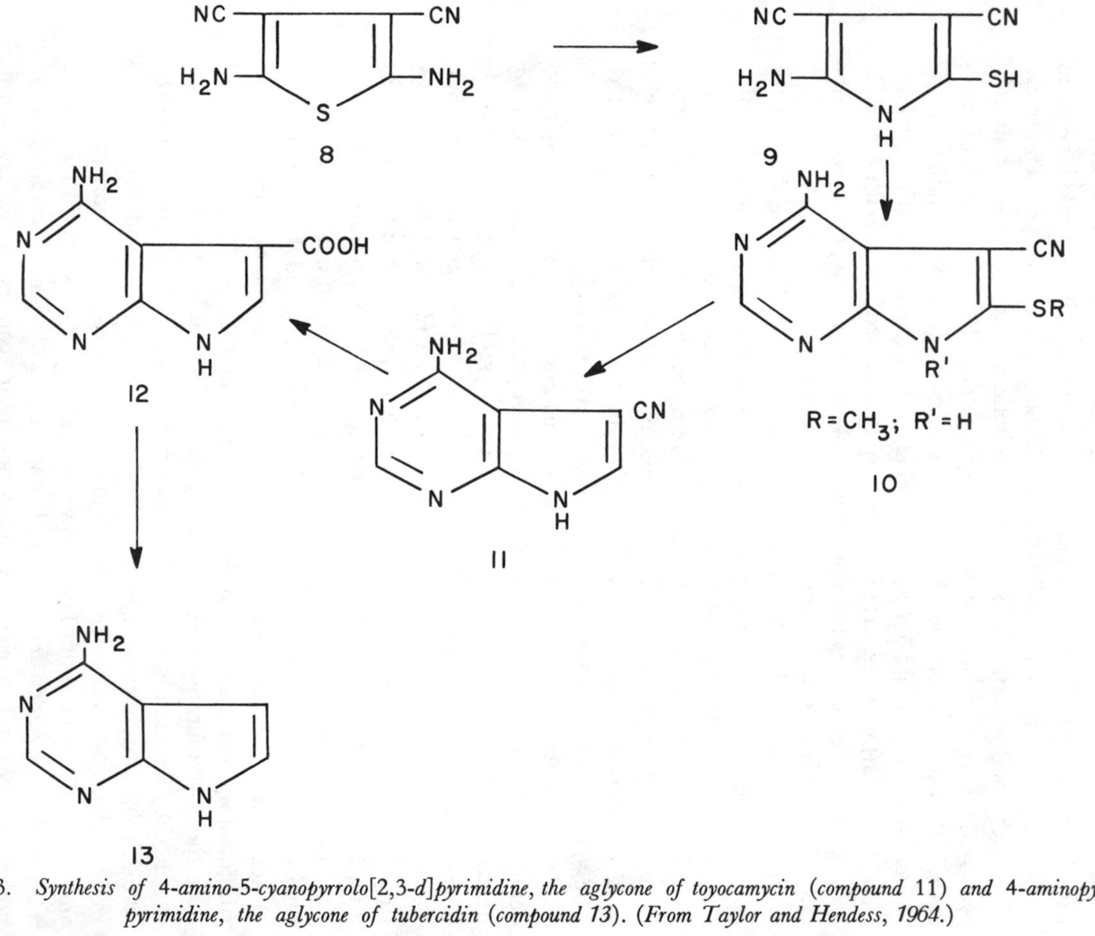


FIGURE 8.3. *Synthesis of 4-amino-5-cyanopyrrolo*[2,3-*d*]*pyrimidine, the aglycone of toyocamycin (compound 11) and 4-aminopyrrolo-*[2,3-*d*] *pyrimidine, the aglycone of tubercidin (compound 13). (From Taylor and Hendess, 1964.)*

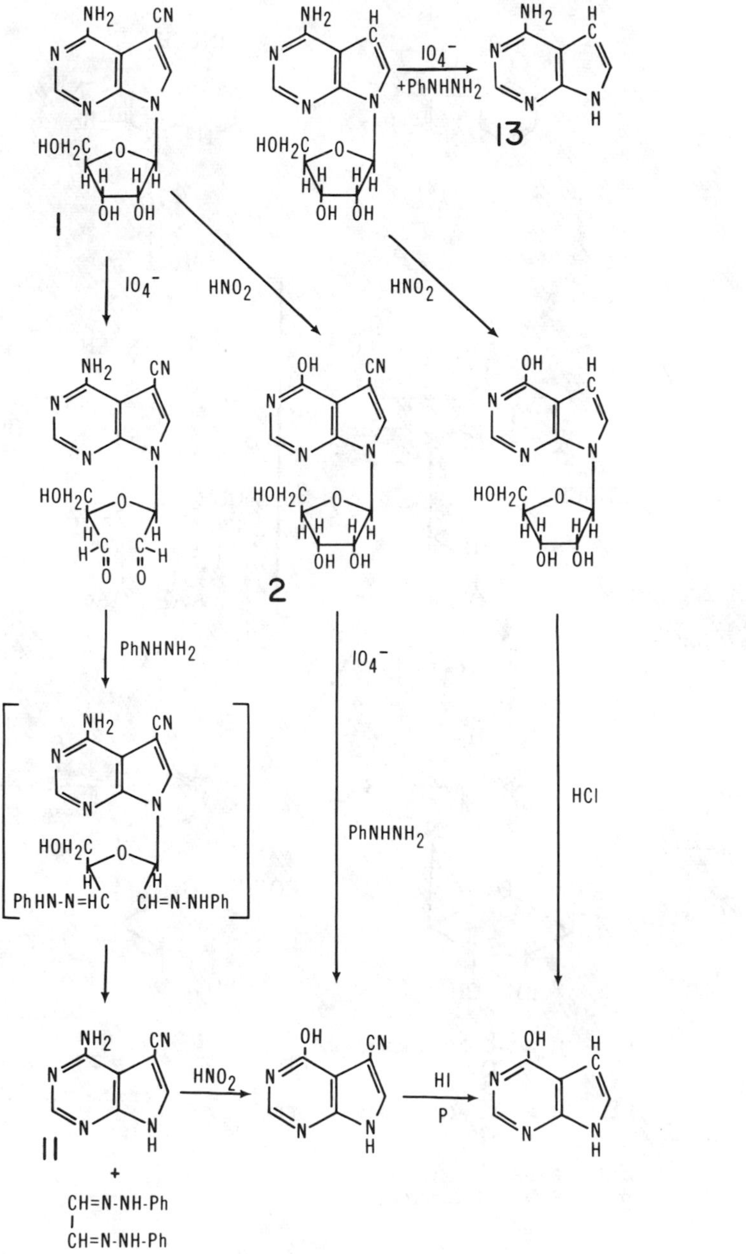

FIGURE 8.4. *Degradation of toyocamycin and tubercidin to isolate carbons* 1′ *and* 2′ *and the pyrrolo*[2,3-*d*]-*pyrimidine base.* (*From Uematsu and Suhadolnik*, 1968.)

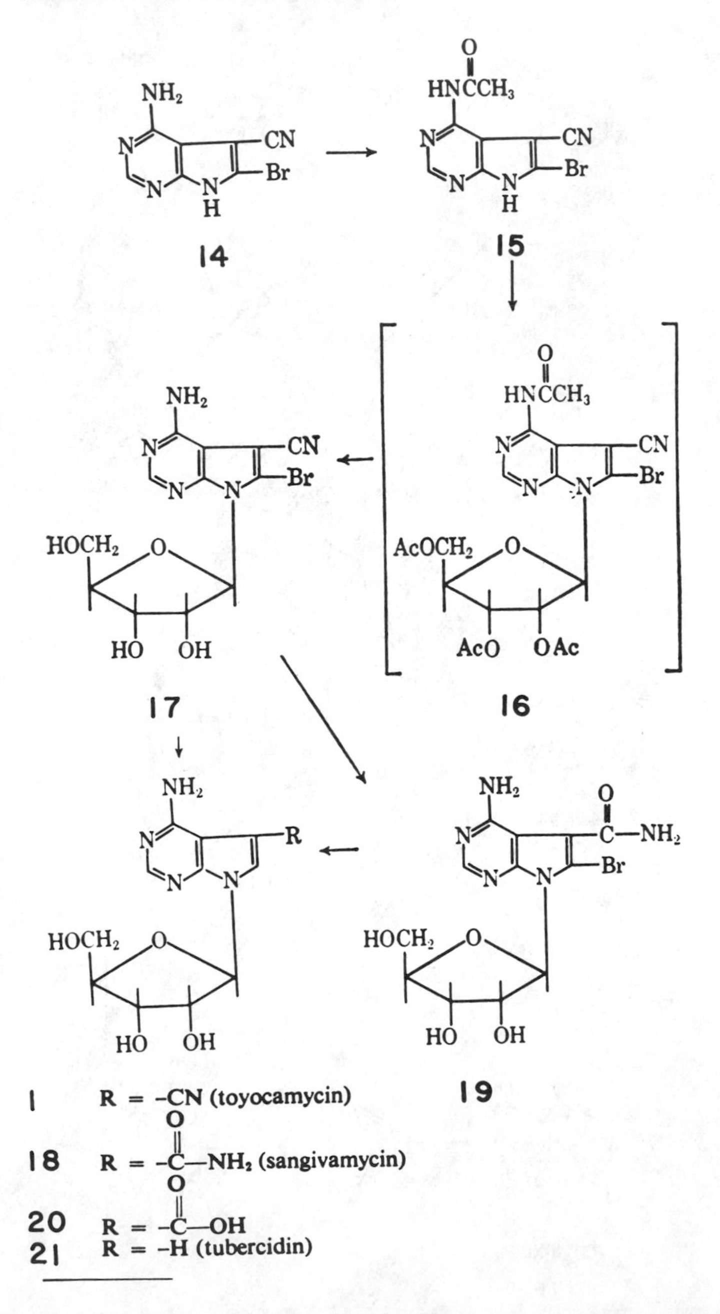


FIGURE 8.5. *Chemical syntheses and structures of toyocamycin* (1), *sangivamycin* (18) *and tubercidin* (21). (*From Tolman et al.*, 1968.)

dine-5′-$N_1$-toyocamycin cyclonucleoside. A rigorous comparison of the physical and chemical properties of unamycin B, antibiotic E-212, and vengicide showed they were identical with those of the chemically synthesized toyocamycin.

The total synthesis and complete structural assignment of sangivamycin (**18**) were also reported. Treatment of compound **17** with 30% hydrogen peroxide in concentrated ammonium hydroxide at room temperature resulted in 65% conversion to 4-amino-6-bromo-5-carboxamido-7-(β-D-ribofuranosyl)pyrrolo[2,3-*d*]pyrimidine (compound **19**). The 6-bromo group from **19** was removed by catalytic hydrogenation to afford 4-amino-5-carboxamido-7-(β-D-ribofuranosyl)pyrrolo[2,3-*d*]pyrimidine (**18**) in an 83% yield. The chemical and physical properties of the chemically synthesized sangivamycin were identical with those of an authentic sample of sangivamycin isolated from the culture filtrates of the *Streptomyces* producing this antibiotic. Final proof for the structural assignment of toyocamycin and sangivamycin was obtained by the conversion of sangivamycic acid to the pyrrolopyrimidine nucleoside antibiotic tubercidin (**21**). The structure of tubercidin had been unequivocally established by Mizuno et al. (1963a,b,c). Tolman, Robins, and Townsend (1969) subsequently reported in more detail their total synthesis of toyocamycin, sangivamycin, and tubercidin. Two excellent alternative syntheses of toyocamycin are described. One involves the acid-catalyzed fusion of 4-chloro-5-cyanopyrrolo[2,3-*d*] pyrimidine with 1,2,3,5-tetra-*O*-acetyl-β-D-ribofuranose. Amination of this acetylated nucleoside affords toyocamycin in 78% yield. The second method involves the acid-catalyzed fusion of 6-bromo-4-chloro-5-cyanopyrrolo[2,3-*d*] pyrimidine with 1,2,3,5-tetra-*O*-acetyl-β-D-ribofuranose. The blocking groups are removed with concomitant amination followed by debromination. Toyocamycin is obtained in good yield.

## DERIVATIVES OF TOYOCAMYCIN

As part of a general program involving pyrrolo[2,3-*d*]pyrimidines and pyrrolo[2,3-*d*]pyrimidine nucleosides, Robins and his colleagues have described their work on the synthesis of several new and interesting 4,5-disubstituted-7-(β-D-ribofuranosyl)pyrrolo[2,3-*d*]pyrimidines. The synthesis of these compounds involves the direct introduction of a functional group at position 5 via electrophilic substitution (Hinshaw et al., 1969) (see Fig. 8.1 for numbering of pyrrolo[2,3-*d*] pyrimidine ring). The experimental procedure involved treatment of 4-chloro-7-(2′,3′,5′-tri-*O*-acetyl-β-D-ribofuranosyl) pyrrolo[2,3-*d*]pyrimidine with *N*-bromoacetamide. The substitution of a bromo-, chloro-, or iodo-group for the cyano group of toyocamycin at the 5-position gave rise to compounds with remarkable biological activity.

Although ribosidation of the chloromercury salt of 4-amino-5-cyanopyrrolo[2,3-*d*]pyrimidine was reported by Taylor and Hendess (1965), the yields were too low to complete characterization of the nucleoside. Tolman et al. (1967) reported a new route for the synthesis of pyrrolo [2,3-*d*]pyrimidine nucleosides by treatment of 4-chloro-5-cyano-6-methylthiopyrrolo-[2,3-*d*]pyrimidine with 1,2,3,5-tetra-*O*-acetyl-β-D-ribofuranose via the fusion reaction (Rousseau et al., 1966). Hinshaw et al. (1970) have studied the reactivity of the cyano group in the pyrrole moeity of toyocamycin by changing the functional group at position 4 in the pyrimidine ring. The differences in reactivity appears to be a function of pH at which the reaction is conducted as well as the reagent and conditions employed. The reactivity of the cyano group could be greatly decreased by the introduction of a group (keto) capable of supporting an anion in position 4. The amino group of position 4 allowed the 5-cyano group to undergo nucleophilic attack; however, a strong nucleophile was required for this reaction to take place.

## INHIBITION OF GROWTH

Toyocamycin strongly inhibits *Candida albicans*, *Cryptococcus neoformans*, and *Mycobacterium tuberculosis*, but has little effect on many gram-positive and gram-negative bacteria, fungi, and yeast (Matsuoka, 1960). The antimicrobial spectrum of toyocamycin is shown in Table 8.1. This antibiotic was shown to be a significant antitumor compound (Saneyoshi et al., 1965), with the antineoplastic activity of toyocamycin dependent on the presence of the amino group at the 4 position and the cyano group at the 5 position. Deamination of the 4-amino group markedly diminished the antitumor activity and also decreased toxicity. When the cyano group at the 5 position was reduced to the amino group, there was a further decrease in activity. Toyocamycin was more active against NF-sarcoma cells than was tubercidin. The $LD_{50}$ of toyocamycin administered intraperitoneally to mice was 20 mg/kg (Saneyoshi et al., 1965) and when administered intravenously to mice it was 10 mg/kg (Matsuoka, 1960).

## BIOSYNTHESIS OF TOYOCAMYCIN

Earlier studies by Smulson and Suhadolnik (1967) indicated that the pyrimidine ring of a preformed purine (presumably adenine) was the precursor for the pyrimidine ring of tubercidin.

Although the isotope labeling pattern on the incorporation of adenine-2-$^{14}C$ into the structurally related pyrrolopyrimidine nucleoside antibiotic tubercidin strongly suggested that the preformed pyrimidine ring of a purine was directly involved in the biosynthesis of the pyrimidine ring of tubercidin by *S. tubercidicus*, chemical degradations of the pyrimidine ring of tubercidin to determine the exact location of $^{14}C$ had not been reported (Smulson and

**Table 8.1**

Antimicrobial Spectrum of Toyocamycin

| Test organisms | Strain no. | Minimum inhibitory concentration, μg/ml |
|---|---|---|
| *Candida albicans* | 3147 | 3.13 |
| *Candida albicans* | YU 1002 | 3.13 |
| *Candida pseudotropicalis* | 7494 | >100 |
| *Candida tropicalis* | | 3.13 |
| *Cryptococcus neoformans* | | >100 |
| *Sporotrichum schenkii* | | >100 |
| *Blastmyces dermatidis* | 4202 | 12.5 |
| *Hormodendrum pedrosoi* | 4207 | 50.0 |
| *Torula utilis* | 4001 | >100 |
| *Helminthosporium leptochloae* | | 3.13 |
| *Saccharomyces sake* | | >100 |
| *Aspergillus niger* | | >100 |
| *Penicillium chrysogenum* | | 3.13 |
| *Trichophyton interdigitale* | | 6.25 |
| *Trichophyton rubrum* | | 25.0 |
| *Microsporum audouini* | 4209 | >100 |
| *Bacillus subtilis* | PCI 219 | >100 |
| *Escherichia coli* | | >100 |
| *Staphylococcus aureus* | 209 P | >100 |
| *Bacillus anthracis* | | 25 |
| *Shigella dysenteriae* | | >100 |
| *Pseudomonas aeruginosa* | | 50 |
| *Mycobacterium* | 607 | >100 |
| *Mycobacterium phlei* | | 3.13 |

From Matsuoka, 1960.

Suhadolnik, 1967). More direct evidence for the utilization of the pyrimidine ring of a purine as the precursor for the pyrimidine ring of toyocamycin has now been established (Uematsu and Suhadolnik, 1970). The chemical degradation used to determine the location of carbon-14 in toyocamycin was modeled after the procedure of Baugh and Shaw (1966), in which they showed that inosinic acid could be readily alkylated at *N*-1, followed by opening of the pyrimidine ring to form the amino pyrrole ribotide. Brown (1968) also reported on the lability of the pyrimidine ring of adenine 1-*N*-oxide.

The biosynthesis of carbon-14 labeled toyocamycin from adenine-2-$^{14}$C by *S. rimosus* was accomplished by the addition of adenine-2-$^{14}$C to the fermenta-

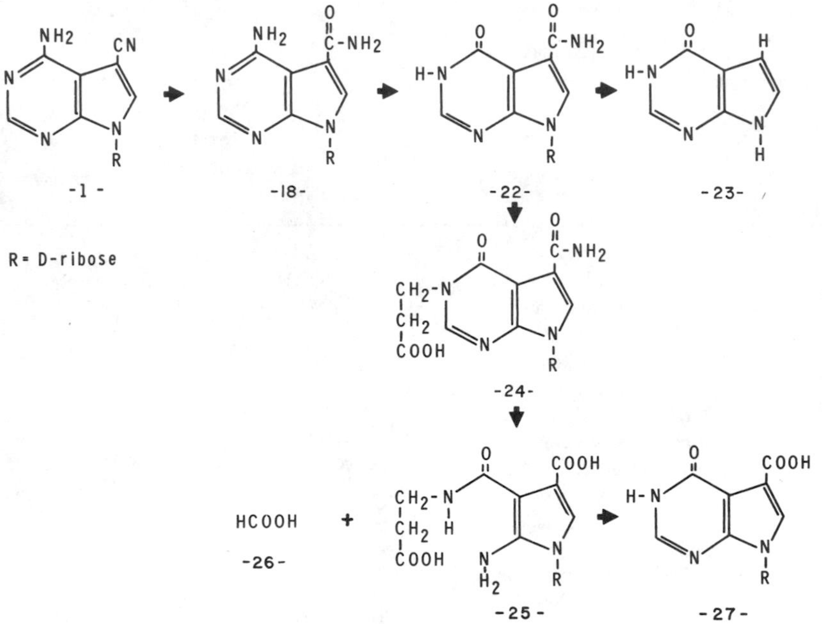

FIGURE 8.6. *Chemical degradation of toyocamycin and isolation of carbon 2 as formic acid. (From Uematsu and Suhadolnik,* 1970.)

tion medium at the time of toyocamycin biosynthesis. Toyocamycin was isolated and degraded as shown in Figure 8.6; the distribution of carbon-14 is shown in Table 8.2. Toyocamycin (**1**) was converted to sangivamycin (**18**) by treatment with Dowex-1-OH⁻. The radioactive sangivamycin was converted to 5-carboxamido-4-hydroxy-7-(β-D-ribofuranosyl)pyrrolo[2,3-*d*]pyrimidine (compound **22**). The distribution of the carbon-14 in the aglycone and the sugar moiety was determined by treatment of compound **22** with red phosphorus and hydriodic acid. The yield of 4-hydroxypyrrolo[2,3-*d*] pyrimidine (compound **23**) was 21%. All the carbon-14 resided in 4-hydroxy-[2,3-*d*]pyrimidine (**23**). The physical and chemical properties of this compound were identical with authentic, chemically synthesized 4-hydroxy-pyrrolo[2,3-*d*]pyrimidine (Davoll, 1960).

Compound **22** was alkylated to form 5-carboxamido-3-(2-carboxyethyl)-4-hydroxy-7-β-D-ribofuranosylpyrrolo[2,3-*d*]pyrimidine (compound **24**). The structure of compound **24** was rigorously established by elemental analysis, nmr, and ir spectroscopy. The pyrimidine ring of the alkylated nucleoside was opened by treatment of compound **24** with 1 *N* NaOH in a sealed tube for 60 min at 130°C. 2-Amino-4-carboxy-3-carbonyl[*N*-(2-carboxyethyl)]-1-β-D-ribofuranosylpyrrole (compound **25**) and formic acid (**26**) were isolated by

**Table 8.2**

Distribution of Carbon-14 in Toyocamycin Isolated from the Adenine-2-$^{14}C$ Experiment[a]

| Compound | Activity, mμc/μmole |
|---|---|
| Sangivamycin (**18**) | 3.85 |
| 5-Carboxamido-4-hydroxy-7-(β-D-ribofuranosyl)pyrrolo[2,3-*d*]pyrimidine (**22**) | 3.80 |
| 4-Hydroxypyrrolo[2,3-*d*]pyrimidine (**23**) | 3.71 |
| Formic acid (**26**) | 3.80 |
| 2-Amino-4-carboxy-3-carbonyl[N-(2-carboxyethyl)]-1-β-D-ribofuranosylpyrrole (**25**) | 0 |
| 5-carboxy-4-hydroxy-7-(β-D-ribofuranosyl)-pyrrolo[2,3-*d*]pyrimidine (**27**) | 0 |

From Uematsu and Suhadolnik, 1970

[a] Seven micromoles of adenine-2-$^{14}C$ (570 mμc/μmole) was added to 3 flasks of *S. rimosus*; after 3 hr, toyocamycin was isolated.

means of a DEAE cellulose column (bicarbonate form). The isolation of formic acid was established by conversion to formaldehyde followed by quantitative analysis. In addition, the $R_f$ of the radioactive formic acid, as determined by paper electrophoresis and paper chromatography, was identical with authentic formic acid (Uematsu and Suhadolnik, 1970). All the carbon-14 resided in the formic acid (Table 8.2). Additional evidence for the structure of compound **25** was obtained by its conversion to 5-carboxy-4-hydroxy-7(β-D-ribofuranosyl)pyrrolo[2,3-*d*]pyrimidine (**27**). Ring closure of compound **25** was accomplished by heating in the presence of 98% formic acid and acetic anhydride. Proof of the structure of compound **27** was obtained from ir and nmr spectroscopy. The mass spectrum of compound **27** had a prominent peak at $m/e$ 267 (due to the loss of $CO_2$ from the parent compound). This peak at 267 is in agreement with the molecular formula, $C_{11}H_{13}N_3O_5$. Final proof for the structure of compound **27** was obtained by the alkaline hydrolysis of compound **22**. The physical and chemical properties of compound **27** synthesized from compound **22** and from compound **25** were identical. The data in Table 8.2 supply unequivocal evidence that the radioactivity in toyocamycin from the adenine-2-$C^{14}$ experiments resides exclusively in carbon 2.

The data in Table 8.3 on the incorporation of adenine-2-$^{14}C$ and adenine-8-$^{14}C$ into toyocamycin by *S. rimosus* show that carbon-14 from adenine-2-$^{14}C$ is incorporated, while carbon-14 from adenine-8-$^{14}C$ is not incorporated. These results suggest that carbon 8 of the imidazole ring of adenine or a purine is lost in the conversion to toyocamycin. The data obtained on the biosynthetic studies with toyocamycin (**1**), sangivamycin (**18**), and tubercidin (**21**) provide experimental evidence showing a close biosynthetic relationship between the pyrimidine ring of a preformed purine and the pyrimidine ring of the pyrrolopyrimidine nucleoside antibiotics (Smulson and Suhadolnik, 1967; Uematsu and Suhadolnik, 1970; Suhadolnik and Uematsu, 1969, 1970). The lack of incorporation of radioactivity into the pyrrolopyrimidine nucleoside antibiotics when adenine-8-$^{14}$C is added to the culture filtrates of the *Streptomyces* producing these antibiotics can be explained by the loss of N-7 and C-8 during the biosynthesis of toyocamycin. The chemical and enzymatic elimination of C-8 of purines has been reported by Burg and Brown (1966), Levenberg and Kaczmarek (1966), Shiota et al. (1967), Krumdieck et al. (1966), and Weygand et al. (1961). Toyocamycin biosynthesis differs from the biosynthesis of the pteridine ring in that nitrogen 7 and carbon 8 of the imidazole ring must be eliminated prior to pyrrole ring formation.

**Table 8.3**

Incorporation of Adenine-2-$^{14}$C and Adenine-8-$^{14}$C into Toyocamycin by *S. rimosus*[a]

| Adenine added | Amount added, μmoles | Toyocamycin isolated, mμc/μmole | Incorporation efficiency |
|---|---|---|---|
| Adenine-2-$^{14}$C | 12 | 2.2 | 0.25 |
| Adenine-8-$^{14}$C | 14 | 0 | 0 |

From Uematsu and Suhadolnik, 1970.
[a] The amount of adenine-2-$^{14}$C added to two flasks: 4.0 μc; 4.6 μmoles; 874 mμc/μmole. Amount of adenine-8-$^{14}$C added to two flasks: 4.0 μc; 4 μmoles; 1000 mμc/μmole. Toyocamycin was isolated 20 min after the addition of carbon-14 labeled adenine.

Evidence for the biosynthetic origin of the two pyrrole carbons and the cyano group of toyocamycin has been obtained by using carbon-14 and tritium-labeled D-ribose and adenosine-U-$^{14}$C. It had been reported earlier by Smulson and Suhadolnik (1967) that the carbon-14 from ribose-1-$^{14}$C was equally distributed in the aglycone and the ribose moieties of tubercidin. These data strongly suggested that ribose was contributing to the biosynthesis of the pyrrole carbon atoms of tubercidin. Since the *N*-riboside bond of tubercidin is extremely resistant to acid hydrolysis, it was not possible to isolate the ribose moiety to determine the exact location of the radioactivity. This experimental difficulty has been overcome by using a chemical degradation method that permits the isolation of the aglycone and carbons 1′ and 2′ of the ribose moiety. Another advantage of using toyocamycin to study the biosynthesis of the aglycone is that the cyano group can be removed as a C-1 unit.

Two pathways were considered for the utilization of ribose in the biosynthesis of pyrrole carbon atoms and the cyano group of toyocamycin (Fig. 8.7). Pathway A in Figure 8.7 proposes that carbons 1, 2, and 3 of ribose become the pyrrole carbons and the cyano group. Pathway B proposes that carbon 1 of ribose becomes the cyano group, while carbons 2 and 3 of ribose contribute to the biosynthesis of the pyrrole carbon atoms. The manner in which the pyrrole ring forms from D-ribose was determined by using ribose-1-$^{3}$H and ribose-3-$^{3}$H. If pathway A in Figure 8.7 were operating and ribose contributed equally to the pyrrole carbon atoms and the cyano group, as well as to the ribose moiety of toyocamycin, the distribution of tritium in the aglycone and the ribose moieties of toyocamycin from ribose-1-$^{3}$H would be 1:1. Using the same reasoning for pathway A with ribose-3-$^{3}$H, all the tritium should reside in the ribose moiety of toyocamycin. On the other hand, if pathway B (Fig. 8.7) were operating in the biosynthesis of toyocamycin, all the tritium in the nucleoside antibiotic would reside in the ribose moiety from ribose-1-$^{3}$H. Ribose-1-$^{3}$H was synthesized as described by Suhadolnik et al. (1967b). Ribose-3-$^{3}$H was synthesized by Hogenkamp (1966–1967).

The data obtained show that the tritium in toyocamycin from the D-ribose-1-$^{3}$H experiments was equally distributed in the aglycone and the ribose moieties. The distribution of the tritium in toyocamycin from the ribose-3-$^{3}$H experiments was such that all the tritium resided in the ribose moiety.

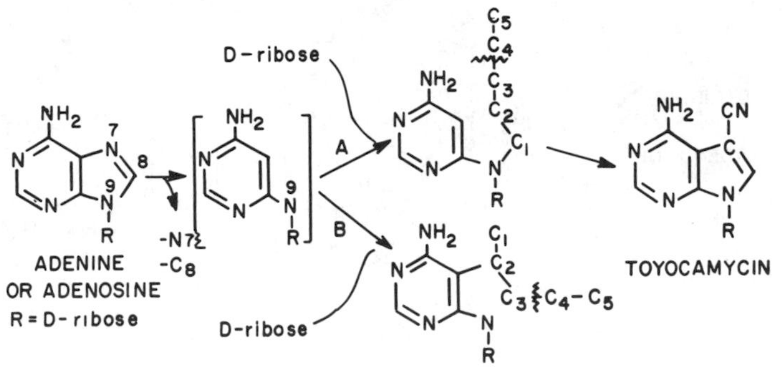

FIGURE 8.7. *Proposed pathways for the utilization of D-ribose in the biosynthesis of the pyrrole carbons and cyano group of toyocamycin.*

These data strongly support the notion that pathway A is functioning in the biosynthesis of toyocamycin (Suhadolnik and Uematsu, 1970). Further evidence for this pathway was obtained from the incorporation of ribose-1-$^{14}$C and ribose-U-$^{14}$C. The data obtained show that the carbon-14 incorporated into the aglycone of toyocamycin from the ribose-1-$^{14}$C experiments resided in the pyrrolopyrimidine ring and no carbon-14 was found in the cyano group. However, 33% of the radioactivity in the aglycone of toyocamycin from the ribose-U-$^{14}$C experiment was shown to reside in the cyano group.

More recent studies by Suhadolnik and Uematsu (1970) have now shown that two ribose units are attached to nitrogen 9 of a purine ring prior to pyrrole ring formation. Carbons 1, 2, and 3 of both ribose units contribute equally to pyrrole ring formation. This evidence was based on the distribution of carbon-14 in toyocamycin from experiments with adenosine labeled with carbon-14 in the adenine and ribose moieties (4 and 96%, respectively) (this $^{14}$C-labeled adenosine was a generous gift of Dr. C. Baugh, Dept. of Medicine, University of Alabama Medical Center). The theoretical distribution of $^{14}$C in the aglycone and ribose moieties of toyocamycin from these experiments was calculated to be 40 and 60%, respectively. The experimental labeling pattern was the same. A biosynthesis scheme to explain these data is shown in Figure 8.8. Elstner and Suhadolnik have now shown that carbon-8 of GTP-8-$^{14}$C (but not $C_8$ of ATP nor $C_8$ of ITP) is lost when incubated with cell-free extracts of *S. rimosus*.

Opening of the imidazole ring might be similar to those studies related to the biosynthesis of folic acid and toxoflavin (Levenberg and Kaczmarek, 1966; Burg and Brown, 1966), in chich carbon 8 of GTP is lost as formic acid. In addition, Shiota et al. (1967) reported on a chemical procedure whereby the imidazole ring of GTP and ATP was opened with $Fe^{2+}$ and mercaptoethanol.

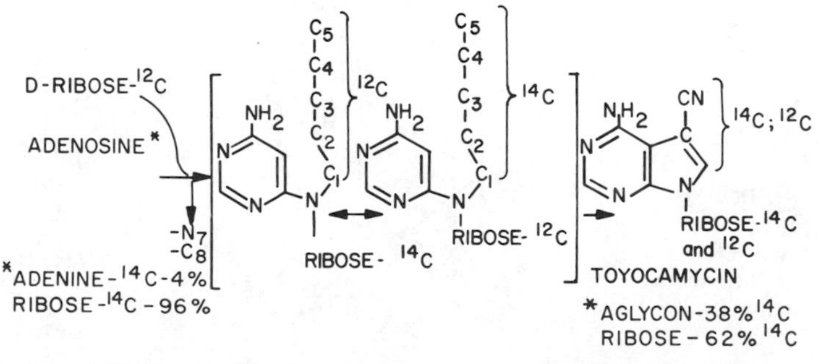

FIGURE 8.8. *Biosynthesis of the pyrrole ring of toyocamycin from a preformed purine nucleoside or nucleotide.*

Townsend and Robins (1963) and Ponnamperuma et al. (1963) have also shown that 7-methylguanosine in the presence of ammonia, or $Co^{60}$ radiation of adenosine results in cleavage of the imidazole ring to form pyrimidine derivatives. Ponnamperuma et al. (1963) reported that 8-hydroxyadenine was one of the products isolated following radiation. The chemistry of guanine and opening of the imidazole ring have recently been reviewed by Shapiro (1968). The ultraviolet absorption spectra of aminopyrimidine, following the opening of the imidazole ring of purines, was reported earlier by Cavalieri et al. (1948) and Cavalieri and Bendich (1950).

In conclusion, data has been presented to show that the carbons of D-ribose serve in the biosynthesis of the pyrrole ring of tryptophan (Yanofsky, 1956), the pyrrolopyrimidine nucleosides (Smulson and Suhadolnik, 1967; Suhadolnik and Uematsu, 1969), the benzimidazole ring of the 5,6-dimethylbenzimidazole moiety of vitamin $B_{12}$ (Alworth et al., 1969), and the pteridine ring of folic acid (Krumdieck et al., 1966; Burg and Brown, 1966). An entirely different biosynthetic mechanism is operative in the for mation of the pyrrole ring of the porphyrin (Shemin et al., 1970). Mathis and Brown (1970) have purified dihydroneopterin aldolase, the enzyme that catalyzes removal of a two-carbon compound from dihydroneopterin.

## 8.2. TUBERCIDIN

### DISCOVERY, PRODUCTION AND ISOLATION

Tubercidin was the second pyrrolopyrimidine nucleoside antibiotic isolated. It is a nucleoside-analog of adenosine in which nitrogen 7 of the imidazole ring of the purine ring has been replaced by a carbon atom. It was first isolated from the culture filtrates of *Streptomyces tubercidicus* from a soil sample obtained from Chiba Prefecture in Japan (Anzai and Marumo, 1957). Nakamura (1961) described the morphological and cultural characteristics of *S. tubercidicus*. Tubercidin and sparsomycin have been isolated from *S. sparsogenes* var. *sparsogenes* (Wechter and Hanze, 1967; tubercidin is referred to in this paper as sparsomycin A). Finally, tubercidin has also been isolated by Higashide et al. (1966) from *S. cuspidosporus* from soil collected in Kyoto, Japan. They also described the isolation of sparsomycin from this same culture. The structure of tubercidin is 4-amino-7-(β-D-ribofuranosyl) pyrrolo[2,3-*d*]pyrimidine (Fig. 8.1).

The culture medium for the production of tubercidin as described by Anzai et al. (1957) is as follows: 1.5% glucose, 1.0% soybean meal, 0.5% peptone, 0.5% meat extract, 0.5% sodium chloride, 0.5% dipotassium hydrogen phosphate, and water. Maximum tubercidin production occurred 2 days after inoculation and shaking of the culture flasks at 27°C. The culture filtrates were adsorbed on charcoal (pH 8) and eluted with 80% acidic aqueous acetone. Following evaporation of the acetone, the residual water solution was adjusted to pH 8 and the tubercidin extracted with 1-butanol. The butanol was removed by evaporation *in vacuo* and the tubercidin crystallized from water. Shirato et al. (1967) described the production and isolation of tubercidin from a *Streptomyces* mutant (T-2-17).

Tubercidin can also be isolated by use of the method described by Dekker (1965). The procedure involves the addition of the culture filtrate to a Dowex-1-$OH^-$ column and washed with water-methanol (85:15). Tubercidin is then eluted quantitatively with water-methanol (40:60). Kaker (Japanese patent—27039, 1969) described the isolation of tubercidin by a combination of ion exchange resins and acetone.

## PHYSICAL AND CHEMICAL PROPERTIES

Tubercidin is a basic crystalline compound that decomposes at 247–248°C. The molecular formula is $C_{11}H_{14}N_4O_4$ and the molecular weight is 266. It is insoluble in acetone, ethyl acetate, chloroform, benzene, and petroleum ether. It is slightly soluble in water (1g/330 ml), methanol (1 g/200 ml), and ethanol (1 g/2000 ml) and readily soluble in acid or alkali (Stecher et al., 1968). The *N*-riboside bond is stable to acid or alkali; there was no loss of activity on boiling for 5 hr between pH 2 and 10. The ultraviolet absorption spectra is as follows: $\lambda_{max}$ 227 m$\mu$, $\lambda_{max}$ 272 m$\mu$ ($\epsilon$ = 12,200 in 0.01 *N* HCl and $\lambda_{max}$ 270 m$\mu$ ($\epsilon$ = 12,100) in 0.01 *N* NaOH (Suzuki and Marumo, 1961). Tubercidin is optically active, $[\alpha]_D^{17} = -67°$ ($c$ = 1 in 50% acetic acid). The p$K_a$ of tubercidin is 5.3. (Suzuki and Marumo, 1960, 1961).

## STRUCTURAL ELUCIDATION

The structure of tubercidin, based on chemical degradation, was reported by Suzuki and Marumo (1960, 1961). When tubercidin was refluxed in water with Dowex 50-$H^+$ for 48 hr, D-ribose was isolated and identified. Treatment of tubercidin with periodate showed that tubercidin consumed 1 mole of periodate per mole of nucleoside. Formic acid was not a product of periodate oxidation. The degradation scheme described by Suzuki and Marumo is shown in Figure 8.9. Desamino-hydroxytubercidin (**28**) was formed by treatment of tubercidin (**21**) with nitrous acid. Treatment of **28** with HCl gave rise to 6-hydroxy-7-deazapurine or 4-hydroxypyrrolo[2,3-*d*]pyrimidine (**23**). Compound **23** was also formed by the fusion of tubercidin with sodium hydroxide. 6-Amino-7-deazapurine or 4-aminopyrrolo[2,3-*d*]pyrimidine (**13**) was formed by refluxing tubercidin in HCl. The yield was about 10%. Compound **23** was converted to 7-deazapurine (**29**). Suzuki and Marumo (1961) concluded that the D-ribofuranosyl moiety is on nitrogen 9 of 7-deazaadenine. This conclusion was supported by the similarity of the ultraviolet absorption spectrum of tubercidin with that of 9-methyl-7-deazaade-

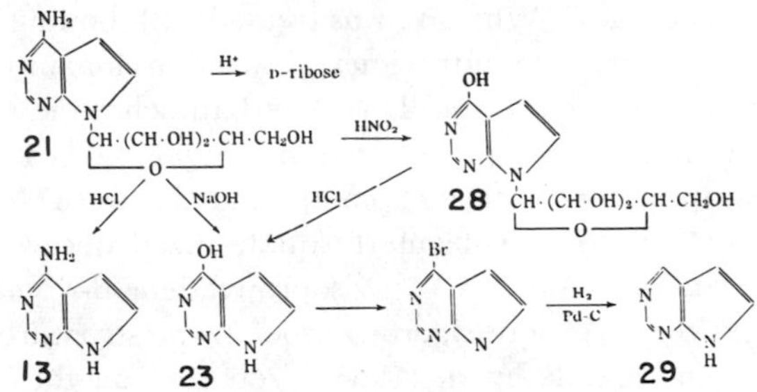

FIGURE 8.9. *Chemical degradation of tubercidin.* (*From Suzuki and Marumo*, 1961.)

nine. On the basis of the data accumulated, Suzuki and Marumo (1961) established the structure of tubercidin as 4-amino-7-(D-ribofuranosyl)-7H-pyrrolo-[2,3-*d*]pyrimidine. The confirmation of the anomeric configuration of tubercidin as $\beta$ was established by Mizuno et al. (1963a, 1963b). The proof of the anomeric configuration of tubercidin was based on the intramolecular quaternization of the 5-*O*-*p*-tolylsulfonate a derivative of tubercidin (Mizuno et al., 1963a) (Fig. 8.10). This type reaction of tubercidin is feasible only if this nucleoside antibiotic possesses the $\beta$-D-configuration. Tubercidin (**21**) was converted to the 2,3-*O*-isopropylidine acetal (**30**) (Hampton and Magrath, 1957). Compound **30** was converted to 5-*O*-*p*-tolylsulfonate (**31**). The ultraviolet absorption maximum of compound **31** was very similar to that of compound **30,** which strongly indicates that this conversion does not affect the

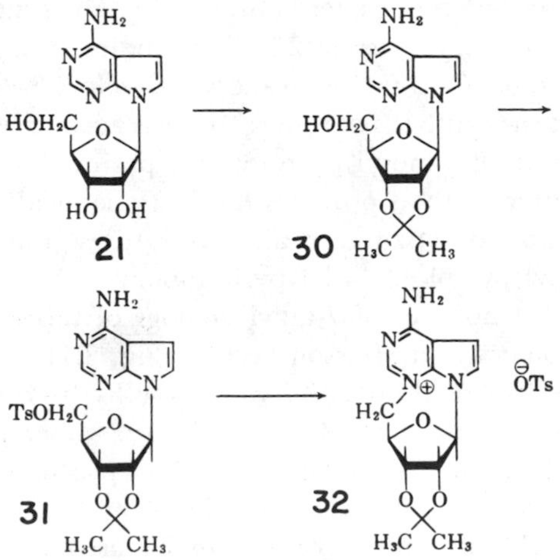

FIGURE 8.10. *Proof of the anomeric configuration of tubercidin.* (*From Mizuno et al.*, 1963*a*.)

chromophoric moiety of **30**. When **31** was treated with boiling acetone, compound **32** was isolated. The ultraviolet absorption maximum of **32** was shifted bathochromically by 9 mμ. This type bathochromic shift is characteristic of an intramolecular quaternized nucleoside (Clark et al., 1951). The chromatographic properties of compound **32** were very similar to those of the same type intramolecularly quaternized derivatives of 2,3-*O*-isopropylideneadenosine or 3-(2,3-*O*-isopropylidene-β-ribofuranosyl)-3H-imidazo[4,5-*b*]pyridine. A more rigorous proof of the structure of tubercidin was obtained when Mizuno et al. (1963b) reported on the synthesis of 4-hydroxy-7-β-D-ribofuranosyl-7H-pyrrolo[2,3-*d*]pyrimidine (**28**). This was achieved by the condensation of 4-amino-5-(2,2-diethoxyethyl)-6(1H)-pyrimidinone with 2,3,4-tri-*O*-acetyl-5-*O*-trityl-D-ribose. Finally, the total chemical synthesis of the three known pyrrolopyrimidine nucleoside antibiotics (toyocamycin, sangivamycin, and tubercidin) (Fig. 8.5), as reported by Tolman et al. (1967), for the first time provided unequivocal evidence for the structural similarity of these three nucleoside antibiotics. The tubercidin (**21**) synthesized by Tolman et al. (1967) had the same properties as the tubercidin isolated from *S. tubercidicus*.

## SYNTHESIS OF ANALOGS OF TUBERCIDIN

Since Bloch et al. (1964) reported that 7-(β-D-ribofuranosyl)pyrrolo[2,3-*d*]-4-pyrimidone (7-deazainosine) (**28**) possesses significant antitumor activity, it was of interest to synthesize a number of 4-substituted 7-(β-D-ribofuranosyl)-pyrrolo[2,3-*d*]pyrimidines to study the structure–activity relationship with lymphoid leukemia L1210. The synthesis of these 4-substituted derivatives of tubercidin was described by Gerster et al. (1967). The synthesis of 4-methylamino-7-(β-D-ribofuranosyl)pyrrolo[2,3-*d*]pyrimidine was accomplished by treatment of 4-chloro-7-(β-D-ribofuranosyl)pyrrolo[2,3-*d*]pyrimidine with methylamine, and the synthesis of tubercidin was accomplished by treatment of 4-chloro-7-(β-D-ribofuranosyl)pyrrolo[2,3-*d*]pyrimidine with methanolic ammonia. This latter method provides the first chemical synthesis of tubercidin since the report of Mizuno et al. (1963b) describing the synthesis of 7-(β-D-ribofuranosyl)pyrrolo[2,3-*d*]-4-pyrimidone.

The synthesis of a number of 7-alkyl analogs of tubercidin has been reported by Montgomery and Hewson (1967) (Fig. 8.11). This synthesis was accomplished via 4,6-dichloro-5-(2,2-diethoxyethyl)pyrimidine. The purpose of the synthesis of these alkylated analogs of tubercidin was to evaluate their biological activity. Since Davoll's (1960) procedure for the synthesis of 4-amino-7H-pyrrolo[2,3-*d*]pyrimidine involved ring closure to a 4-aminopyrimidine, this procedure could not be used to make the 4-alkylaminopyrimidines that were needed by Montgomery and Hewson (1967) to prepare the 7-alkylpyrrolo[2,3-*d*]pyrimidines. The synthesis of the 7-alkyl-4-amino-7H-pyrrolo[2,3-*d*]pyrimidines (**33**) is shown in Figure 8.11. Amination of the chloroalkylpyrrolo[2,3-*d*]pyrimidines was accomplished by

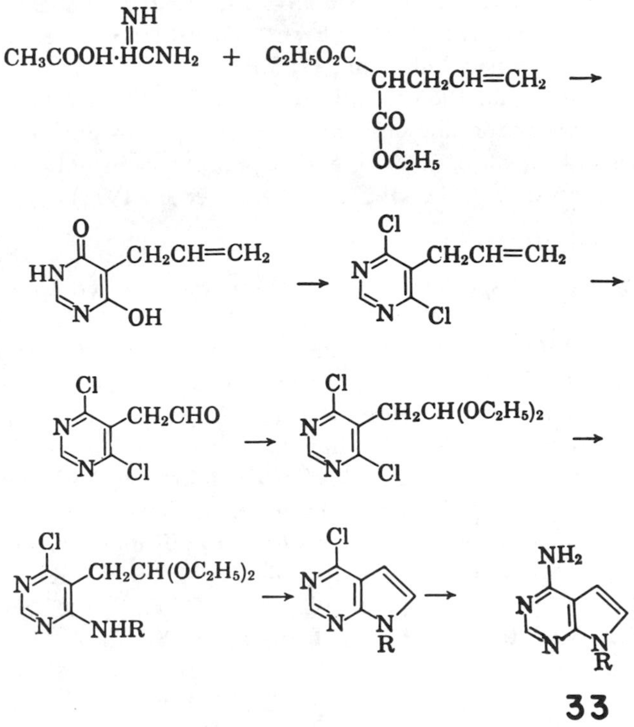

$CH_3COOH \cdot HN{=}CNH_2 + C_2H_5O_2C{-}CH(CO{-}OC_2H_5)CH_2CH{=}CH_2 \rightarrow$

O, HN, N, OH, $CH_2CH{=}CH_2$ → Cl, N, N, Cl, $CH_2CH{=}CH_2$ →

Cl, N, N, Cl, $CH_2CHO$ → Cl, N, N, Cl, $CH_2CH(OC_2H_5)_2$ →

Cl, N, N, NHR, $CH_2CH(OC_2H_5)_2$ → Cl, N, N, N, R → $NH_2$, N, N, N, R

33

R= alkyl

FIGURE 8.11. *Synthesis of 7-alkyl-4-amino-*7H-*pyrrolo*[2,3-*d*]—*pyrimidines.* (*From Montgomery and Hewson,* 1967.)

treatment with ammonia–ethanol in a sealed Parr bomb at 100–110°C for 18 hr. As had been expected, the amination of the chloropyrrolo[2,3-*d*] pyrimidine was much more difficult than the amination of the corresponding 9-alkyl-6-chloropurines. None of these compounds showed any significant activity against leukemia L1210. Iwamura and Hashizume (1968) have also synthesized several compounds related to tubercidin. One of these compounds, 4-amino-5-cyano-6-methylthio-7-β-D-ribofuranosylpyrrolo (2,3-d) pyrimidine, when administered to mice with one day old Ehrlich ascites carcinoma, showed a 33% cure. Synthesis of the 4-thiol derivative of tubercidin was reported by Pike et al. (1964). Interest in the synthesis of the 4-thiol derivative of tubercidin was based on the report that mammalian and microbial cells are able to convert 6-mercaptopurine to the nucleotide by the action of pyrophosphorylase. Those cells resistant to the inhibition by 6-mercaptopurine exhibit an impaired capacity for this conversion.

The chemical synthesis of the 5′-mono-, di-, and triphosphates of tubercidin, with and without tritium, have been reported from several laboratories (Pike et al., 1964; Suhadolnik et al., 1968b; Acs et al., 1964; Uretsky et al., 1968; Nishimura et al., 1966; Smith et al., 1967). The synthesis of certain nucleotide derivatives of tubercidin has been reported by Smith et al. (1967). The synthesis of the 5′-mono-, di-, and triphosphates in red blood cells has also been described (Smith et al., 1966; Smith et al., 1970).

## INHIBITION OF GROWTH

The antimicrobial spectrum of tubercidin was reported by Anzai et al. (1957). Tubercidin inhibited *Mycobacterium tuberculosis* BCG and *Candida albicans*, but had little or no activity against gram-positive bacteria, yeast, or fungi (Table 8.4). Bloch et al. (1967) reported that tubercidin inhibited the growth of *S. faecalis* (8043). Smith et al. (1967) compared the antimicrobial activities of tubercidin, 7-deazainosine, and nucleotide derivatives against a number of microorganisms. Tu and TuMP inhibited *Penicillium oxalicum*, *Glomerella sp.*, and *M. phlei*. The methyl ester of TuMP and 7-deazainosine were not inhibitors. Tubercidin is very toxic to NF mouse sarcoma (Anzai et al., 1957), mouse fibroblasts (Acs et al., 1964), human KB cells (Owen and Smith, 1964), human tumors (Wolberg, 1965), H. Ep. #2 cells (Bennett et al., 1966a; Bennett et al., 1966b; Bennett and Smithers, 1964), and experimental animal tumors (Duvall, 1963; Owen and Smith, 1964; Saneyoshi et al., 1965; Smith et al., 1959). Reproducible inhibition was observed in sarcoma 180 (ascitic form), Ehrlich ascites tumor, and Jensen sarcoma (Owen and Smith, 1964). The cytotoxicity of tubercidin was not reversed by purines, pyrimidines, nucleosides, or nucleotides (Owen and Smith, 1964). Kohls et al. (1966) reported that tubercidin completely inhibited egg laying in the house fly. Acs et al. (1964) reported that the growth of vaccinia (a DNA-virus), Reovirus III (RNA is double-stranded), and Mengovirus (RNA is single-stranded) is inhibited by tubercidin. Reich (1968) reported that 5-bromotubercidin inhibited cells, but stimulated growth of Mengovirus. 7-Deazainosine was also reported by Bloch et al. (1969) to inhibit sarcoma 180 cells *in vitro* and Ehrlich ascites and leukemia P388 cells *in vivo*. This nucleoside did not inhibit *S. faecalis*.

The $LD_{50}$ of tubercidin to mice (intravenous injection) was 35 mg/kg, 1.5 mg/kg in rats, and greater than 25 mg/kg in dogs after a single intravenous injection (Owen and Smith, 1964). Mihich et al. (1969) reported on the $LD_{50}$ of tubercidin and 7-deazainosine in a number of mice.

## BIOSYNTHESIS OF TUBERCIDIN

The natural occurrence of the pyrrolo[2,3-*d*]pyrimidine ribonucleosides and their close structural similarity to adenosine makes the biosynthetic

**Table 8.4**

Antimicrobial Spectrum of Tubercidin

| Test organisms | Minimal inhibitory concentration, μg/ml |
|---|---|
| *E. coli* | >100 |
| *B. subtilis* | >100 |
| *B. pyocianeus* | >100 |
| *Staph. aureus*, 309 P | >100 |
| *B. agri* | >100 |
| *Micrococcus flavus* | >100 |
| *Mycob. tuberculosis* BCG | < 1 |
| *Pen. chrysogenum* | >100 |
| *Sacch. cerevisiae* | >100 |
| *Candida albicans* | >100 (after 24 hr) |
| *Candida albicans* | >100 (after 48 hr) |

From Anzai et al., 1957.

studies of this group of antibiotics most interesting. Five pathways were considered for the biosynthesis of the pyrrolopyrimidine ring of tubercidin (Smulson and Suhadolnik, 1967). The first two pathways assumed that a $C_3$ or $C_4$ acid (propionate or succinate) served as the $C_3$ or $C_4$ skeleton of the pyrrole ring of tubercidin. This pathway would be parallel to the known pathway for the biosynthesis of the purine ring (Buchanan and Hartman, 1959). A third pathway assumed that tubercidin biosynthesis took place via the condensation of the aglycone (7-deazadenine) and ribose-1-phosphate. A fourth pathway assumed that the pyrimidine ring of tubercidin would arise from preformed orotic acid. The final pathway assumed that the pyrimidine ring of a preformed purine would serve as the precursor for the pyrimidine ring of the pyrrolopyrimidine nucleoside antibiotic tubercidin. In this pathway the pyrrole carbon atoms would arise from ribose in a manner analogous to the incorporation of ribose into the pteridine ring of folic acid (Reynolds and Brown, 1964; Krumdieck et al., 1966) and the indole ring of tryptophan (Yanofsky, 1956).

The possibility that a $C_3$ or a $C_4$ acid might serve as a carbon skeleton for the pyrrole ring of the aglycone of tubercidin in a manner similar to the incorporation of formate and glycine into the imidazole ring of the purines was eliminated when it was shown that there was no incorporation of radioactivity into tubercidin. All the radioactive compounds added to the culture medium were taken up by *S. tubercidicus*. The failure of *S. tubercidicus* to convert the aglycone to the nucleoside can probably be attributed to the absence of nucleoside or nucleotide synthesizing enzymes. Since formate-$^{14}C$ and glycine-1-$^{14}C$ were incorporated specifically into the aglycone of tubercidin, it appeared that these two known precursors of purine biosynthesis were first incorporated into a purine and this purine was then the precursor for the pyrrolopyrimidine ring of tubercidin. The results of experiments on the incorporation of adenine-2-$^{14}C$, adenine-U-$^{14}C$, and adenine-8-$^{14}C$ are shown in Table 8.5. Adenine-2-$^{14}C$ and adenine-U-$^{14}C$ were incorporated into the aglycone of tubercidin at a rate about 100 times as great as that of adenine-8-$^{14}C$. While there was about a 100-fold difference in the incorporation of carbon-14 from adenine-2-$^{14}C$ into the aglycone of tuberdicin compared with that of adenine-8-$^{14}C$, there was no difference in the specific activities of the adenine isolated from the RNA of *S. tubercidicus*. The data strongly suggested that carbon 8 of adenine-8-$^{14}C$ was lost in the conversion of the imidazole ring of the purine to the pyrrole ring of tubercidin. The results presented in Table 8.6 show that carbon 8 of the imidazole ring of adenine-8-$^{14}C$ is converted to $CO_2$ with no incorporation of radioactivity into the aglycone of tubercidin. However, there was little $^{14}CO_2$ isolated when adenine-2-$^{14}C$ was added to the culture filtrates. In contrast to the lack of incorporation of carbon-14 into the aglycone from the adenine-8-$^{14}C$ experiment, there was incorporation of carbon-14 into the aglycone from the adenine-2-$^{14}C$ experiment. Studies reported in the section on the biosynthesis of toyocamycin (p. 308) show that carbon 2 of adenine-2-$^{14}C$ is found exclusively in carbon 2 of toyocamycin.

**Table 8.5**

Metabolic Conversion of $^{14}C$-Labeled Adenine to 7-Deazadenine of Tubercidin

| Compound tested | Amount added | | Specific activity added $\mu$C/mmole | Specific activity of radioactive tubercidin formed, $\mu$C/mmole | Specific radioactivity in 7-deazaadenine from hydrolyzed tubercidin, $\mu$C/mmole | Incorporation efficiency of labeled compound into 7-deazaadenine |
|---|---|---|---|---|---|---|
| | $\mu$C | mmole | | | | |
| Adenine-2-$^{14}C$ | 1.82 | 0.0087 | 210 | 6.7 | 6.8 | 3.3 |
| Adenine-8-$^{14}C$ | 1.82 | 0.0087 | 210 | 0.089 | 0.075 | 0.036 |
| Adenine-U-$^{14}C$ | 1.7 | 0.037 | 46 | 2.1 | 2.4 | 5.2 |
| Adenine-8-$^{14}C$ | 20 | 0.037 | 540 | 1.0 | 0.17 | 0.032 |

From Smulson and Suhadolnik, 1967.

**Table 8.6**

Metabolic Fate of Adenine Carbon[a] Atoms 2 and 8

| Labeled substrate | Acid-stable $^{14}C$ | | Acid-volatile $^{14}C$ determined as $CO_2$ | 7-Deazaadenine isolated after hydrolysis of tubercidin, |
|---|---|---|---|---|
| | Time Zero, cpm | After 3 hr, cpm | cpm | cpm |
| Adenine-2-$^{14}C$ | 3,300,000 | 3,100,000 | 440 | 2,300 |
| Adenine-8-$^{14}C$ | 3,300,000 | 2,900,000 | 62,000 | 0 |

From Smulson and Suhadolnik, 1967.

[a] Sterile solutions (5 ml) of adenine-2-$^{14}C$ and adenine-8-$^{14}C$ (10 μmoles) were added to the cultures 42 hr after inoculation. Three hours later the incubation was stopped by the addition of sulfuric acid. The amounts of tubercidin isolated from the adenine-2-$^{14}C$ and adenine-8-$^{14}C$ experiments were 2.2 μmoles and 2.6 μmoles, respectively.

When ribose-1-$^{14}C$ was incorporated into tubercidin, it was observed that the distribution of radioactivity in the aglycone and ribose of tubercidin was 1:1. Although the origins of carbons 5 and 6 of the aglycone have not been definitely established, the equal incorporation in the ribose-1-$^{14}C$ experiments of radioactivity into the aglycone and ribose of tubercidin suggests the possibility of a metabolic conversion of carbons 1 and 2 of ribose to the pyrrole carbons of the aglycone of tubercidin. This would be analogous to the incorporation of ribose into the pteridine ring of folic acid or the indole ring of tryptophane. Additional experiments showing that ribose does contribute carbon atoms 1 and 2 to the pyrrole carbon atoms of the structurally related antibiotic toyocamycin have been presented (p. 313).

## 8.3. SANGIVAMYCIN

### DISCOVERY, PRODUCTION, AND ISOLATION

Rao and Renn (1963) first isolated sangivamycin from the culture filtrates of a species of *Streptomyces*, which was subsequently identified as a strain of *S. rimosus* (ATCC 14,673) by Rao et al. (1969). The medium reported for the production of sangivamycin is as follows: 1% glucose, 1.5% soybean meal, 0.5% distillers' solubles, 0.2% dibasic potassium phosphate, and 0.1% sodium chloride (Rao and Renn, 1963). Sangivamycin was isolated 3–4 days following inoculation and incubation in shake flasks at 25–30°C. The nucleoside antibiotic was adsorbed on charcoal (Darco, G-60, 1.5%) and eluted with 0.05 *N* methanol–HCl. The eluants were concentrated without neutralization and sangivamycin hydrochloride crystallized on standing at

0°C. It was recrystallized from hot water or by dissolving in methanol and adding one part of water to three parts of methanol. About 300 mg of sangivamycin can be isolated and crystallized per liter of medium.

## PHYSICAL AND CHEMICAL PROPERTIES

The molecular formula for sangivamycin is $C_{12}H_{15}N_5O_5$ and the molecular weight is 309. The melting point of sangivamycin is 260°C; $[\alpha]_D^{26}$ $-45.7 \pm 1.9°$ ($c$ = 1.0 in 0.1 *N* HCl) (Tolman et al., 1968). Sangivamycin is slightly soluble in water, soluble in methanol, and insoluble in acetone, ethyl acetate, chloroform, and ether. Sangivamycin is stable in the pH range 2.8–8.0 for 24 hr at room temperature; however, it is unstable in alkaline solutions. The ultraviolet absorption spectra are as follows: $\lambda_{max}$ 278 m$\mu$ ($\epsilon$ = 15,100) and 229 m$\mu$ ($\epsilon$ = 8,200) in ethanol; $\lambda_{max}$ 273 m$\mu$ ($\epsilon$ = 12,800) and 228 m$\mu$ ($\epsilon$ = 9,500), pH 1; $\lambda_{max}$ 277 m$\mu$ ($\epsilon$ = 14,400) and 227 m$\mu$ ($\epsilon$ = 14,100), pH 11 (Tolman et al., 1969). The infrared spectrum has prominent peaks that suggest an amino or hydroxyl group (2.9 and 3.05 $\mu$), as well as a conjugated carbonyl, amide, or C=N system (6.10 $\mu$) (Rao, 1968). When the nmr spectrum of sangivamycin is run in trifluoroacetic acid, there are two broad based singlets at $\tau$ 1.42 and 1.53 that are equal to two to four protons. A doublet is observed at $\tau$ 3.59 and 3.68 and is equal to one proton. Several broad bands are observed in the region $\tau$ 4.87–5.75. The spectrum shows a resemblance to the spectra of purine nucleosides (Rao, 1965, 1968).

## STRUCTURAL ELUCIDATION

The uv and nmr spectra, elementary composition, and chemical reactivities of sangivamycin are closely related to those of toyocamycin. This strongly suggested that sangivamycin and toyocamycin may have the same carbon skeleton. Rao (1968) recently reported on the structure of sangivamycin (Fig. 8.12). Alkaline hydrolysis of toyocamycin (**1**) yielded an acid (**20**) that is the same as the product isolated following alkaline hydrolysis of sangivamycin (**18**). The tetraacetyl derivative of sangivamycin (**34**) on dehydration with $POCl_3$ resulted in the isolation of the tetraacetyl derivative of toyocamycin (**35**). The infrared spectra of both compounds show a characteristic peak at 2230 $cm^{-1}$ (CN). When toyocamycin was hydrolyzed in 2 *N* HCl at 100°C, sangivamycin (**18**) was isolated. On the basis of these data, Rao assigned the structure of sangivamycin as shown in compound **18** (Fig. 8.12).

## CHEMICAL SYNTHESIS

The total synthesis of sangivamycin as reported by Tolman et al. (1968) provided the unequivocal evidence for the structure of sangivamycin and the anomeric configuration. The complete chemical synthesis of sangivamycin is described on p. 306.

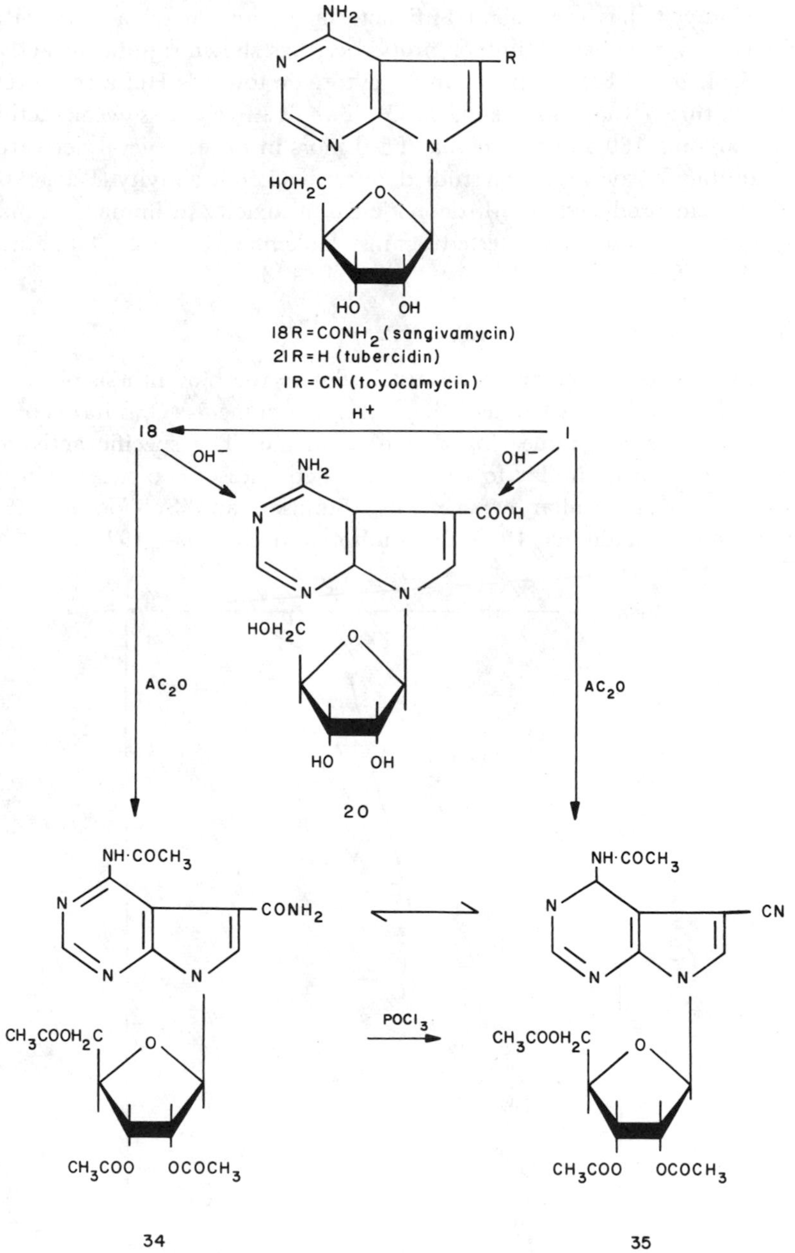

FIGURE 8.12. *Proof of structure of sangivamycin.* (*From Rao*, 1968.)

## INHIBITION OF GROWTH

Sangivamycin has very slight antibacterial or antifungal activity (Rao, 1968). However, it has antitumor properties, has shown significant activity against leukemia L1210 in mice, and is cytotoxic towards HeLa cells grown in cell culture (Rao and Renn, 1963). Sangivamycin has weak activity against sarcoma 180 and carcinoma 755 tumors in mice. Several derivatives of this antibiotic have also been studied for antileukemic activity (Rao, 1968). Sangivamycin produced very little evidence of toxicity in humans at maximally tolerated doses when tested against leukemia (Cavins, 1966; Cavins et al., 1967; Zubrod et al., 1966).

## BIOSYNTHESIS OF SANGIVAMYCIN

The biosynthesis of sangivamycin is similar to the biosynthesis of toyocamycin and tubercidin. Adenine-2-$^{14}$C, but not adenine-8-$^{14}$C, is incorporated into the pyrrolopyrimidine ring of sangivamycin. The specific activity of the sangivamycin is similar to the specific activities of the tubercidin and toyocamycin from similar experiments (Smulson and Suhadolnik, 1967; Uematsu and Suhadolnik, 1970; Suhadolnik and Uematsu, 1970). Rao and

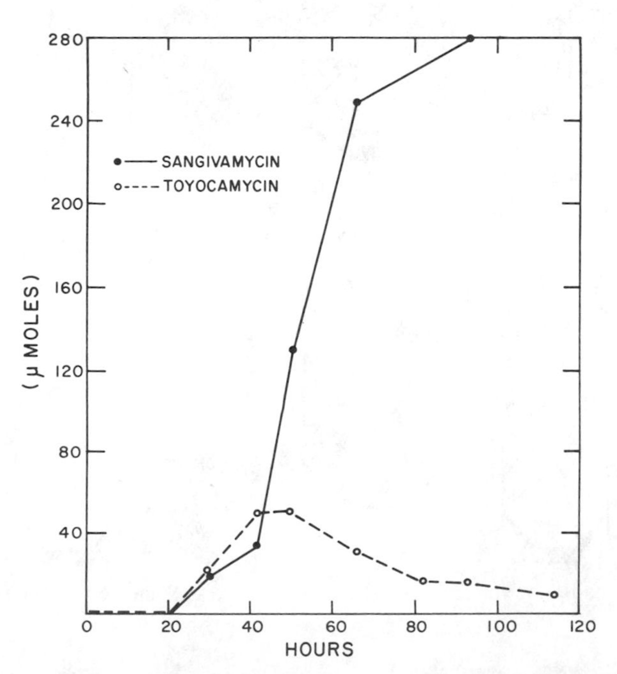


FIGURE 8.13. *Production of toyocamycin and sangivamycin by the sangivamycin-producing Streptomyces. (From Suhadolnik and Uematsu, unpublished results.)*

Renn (1963) and Rao (1968) reported on the presence of sangivamycin in the culture filtrates of the sangivamycin-producing *Streptomyces*. Suhadolnik and Uematsu (1969) have shown that both toyocamycin and sangivamycin can be isolated from the culture filtrates of the same organism (Fig. 8.13). Toyocamycin and sangivamycin appear in the culture filtrates at the same concentrations 30 hr after inoculation. Toyocamycin production reaches a maximum at 42 hr. At this time, sangivamycin production increases markedly. Cell growth approaches the stationary phase at 30 hr.

To study the possibility that toyocamycin was converted to sangivamycin, toyocamycin-$^3$H was added to the culture filtrates of a sangivamycin-producing culture 60 hr after inoculation. Following a 60-min incubation, there was no apparent uptake of the tritium-labeled toyocamycin by the *Streptomyces*. However, there was a 20% conversion of toyocamycin to sangivamycin. Toyocamycin was not converted to sangivamycin by the culture filtrate (free of mycelium). However, when cell-free extracts were made of the *Streptomyces*, there was a rapid conversion to sangivamycin. The enzyme responsible for this conversion has been partially purified. It is stable when stored in 50% glycerol at −20°C and does not lose activity following dialysis; no co-factors appear necessary for the conversion of the nitrile to the carboxamide group. The pH optimum is 6.5. Enzyme activity, with crude extracts, is rapidly lost when incubations are performed at 40°C. The effect of toyocamycin concentration on the formation of sangivamycin is shown in Figure 8.14. The $K_m$ for toyocamycin nitrile hydrolase is $5 \times 10^{-4}$ *M* (Fig. 8.15). *p*-Hydroxybenzonitrile is a competitive inhibitor. The $K_i$ is $5.7 \times 10^{-3}$ *M*.

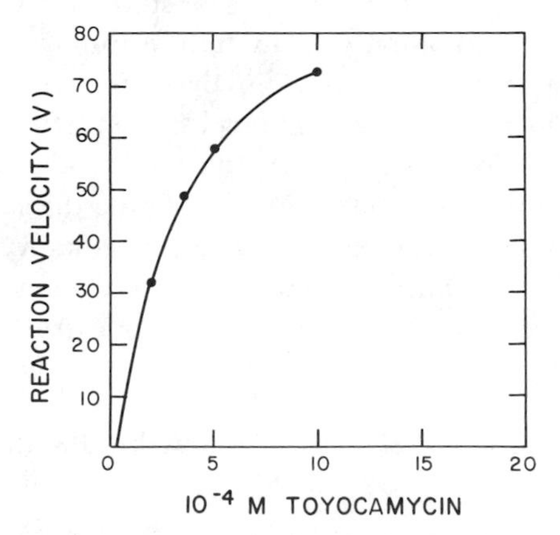


FIGURE 8.14. *The effect of increasing concentrations of toyocamycin on product formation. (From Suhadolnik and Uematsu, unpublished results.)*

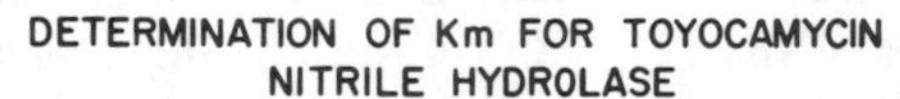


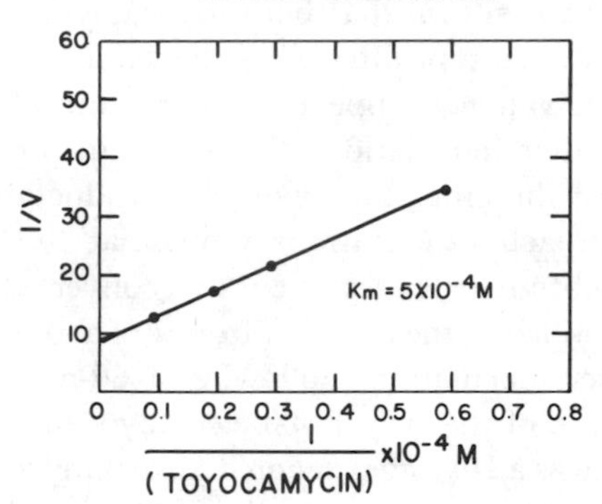


FIGURE 8.15. *Determination of $K_m$ for toyocamycin nitrile hydrolase (From Suhadolnik and Uematsu, unpublished results.)*

The cations, $Hg^{2+}$, $Cu^{2+}$, $Zn^{2+}$, $Fe^{3+}$, and $Fe^{2+}$ are inhibitors of this enzyme reaction.

These data suggest that toyocamycin is the precursor for the biosynthesis of sangivamycin.

## 8.4. BIOCHEMICAL PROPERTIES OF THE PYRROLOPYRIMIDINE NUCLEOSIDES

The three naturally occurring pyrrolo[2,3-*d*]pyrimidine nucleoside antibiotics have been excellent biochemical tools for studying numerous reactions involving microorganisms, animal cells in culture, tumor cells, and viruses. The uptake of tubercidin and tubercidin derivatives by blood cells, the distribution in the whole animal, and clinical studies have been reported in considerable detail. In addition, the unusual biological properties of these nucleosides have also made them useful in elucidating the structural requirements for the interaction at the catalytic sites and regulatory sites of partially purified enzymes. All three antibiotics are analogs of adenosine. Since these three nucleosides have similar structures, the studies related to their biochemical properties will be discussed under one general heading.

### UPTAKE OF TUBERCIDIN BY BLOOD CELLS AND DISTRIBUTION IN THE WHOLE ANIMAL

The biochemical properties of Tu and derivatives of Tu in the blood cell and in whole animals have been studied in much detail by Smith and his co-workers (Smith et al., 1966; Smith et al., 1967; Smith et al., 1970). Unique differences in the biochemical effects of Tu and the methyl ester of TuMP were reported.

More than 95% of the Tu and 92% of TuMP is rapidly absorbed by mammalian blood cells (Smith et al., 1966; Smith et al., 1967) when incubated with them *in vitro*. The methyl ester of TuMP and certain other substituted derivatives of Tu, on the contrary, are neither absorbed nor retained by human blood cells. Most of the Tu is converted inside the cell to the 5′-triphosphate and this process occurs with cells from humans, rat, rabbit, and monkey. On the other hand, the deaminated analog (7-deazainosine) is not concentrated inside red cells, but is found in the serum as the nucleoside (Smith et al., 1970). Since the half-life of red blood cells is not changed after Tu is absorbed therein, Smith et al. (1970) suggests that TuTP does not inhibit those biochemical reactions that require energy. Therefore, the presence of TuTp in the red blood cell does not offer *a priori* an explanation for the high cytotoxicity of Tu to mammalian cells. This finding is contrary to the data reported by Bloch et al. (1967) with *S. faecalis*. Smith et al. (1970) suggest that Tu kills only those mammalian cells in which macromolecular synthesis is occurring. However, Uretsky et al. (1968) have shown that the pyrrolopyrimidine nucleotides could not function as an energy source for amino acid activation or esterification of tRNA by mammalian and bacterial enzymes. They concluded that the rapid and profound inhibition of protein synthesis in intact cells following exposure to Tu could not be explained by the effect on protein synthesis *in vivo*.

Smith et al. (1967, 1970) studied the distribution of Tu in the whole animal by two techniques. The first procedure involved the direct intravenous injection into the animal and the second procedure involved the uptake of tubercidin by whole blood cells (done *in vitro* followed by retransfusion into the animal). When tritium-labeled Tu was administered rapidly, 25% of the radioactivity was found in the urine in 24 hr and only 0.29% could be identified as Tu. When Tu is first incorporated into red blood cells, which in turn are administered to dogs slowly, an entirely different result is observed. Only 5% of the radioactivity is found in the urine after 24 hr and about 20% after 21 days. Paper chromatography of the 24-hr urine sample showed that the radioactivity had the same $R_f$ as authentic Tu. These differences led the authors to conclude that a two-compartment system exists when Tu is administered intravenously. It appears that the mechanism of excretion for the Tu contained in the red blood cell differs from that when Tu is injected directly into the vein.

Considerable radioactivity is found in the urine of mice and rats following intravenous or oral administration of Tu (Smith et al., 1970). No volatile tritium was found in the urine of mice that received Tu-$^3$H. Considerably more 7-deazainosine and the methyl ester of TuMP were recovered from the urine of dogs as compared to the amount of Tu recovered (Smith et al., 1967) when these compounds were administered by the intravenous route.

## INHIBITION OF rRNA SYNTHESIS BY TOYOCAMYCIN

The selective inhibition of rRNA synthesis by toyocamycin in mouse fibroblasts (strain L-929) has been recently studied by Tavitian et al. (1968). Toyocamycin (0.002 $\mu$g/ml) inhibited growth of L cells. At 0.1 $\mu$g/ml the cells developed nuclear alteration and cytoplasmic vacuolation. Although toyocamycin inhibited the synthesis of RNA, DNA, and protein, the inhibition of RNA synthesis was much more pronounced. The sucrose gradient centrifugation profile of RNA extracted from the intact cells showed that low concentrations of toyocamycin completely inhibited the synthesis of 28-S and 18-S RNA. The synthesis of the 4-S and 45-S RNA was not inhibited (Fig. 8.16).

The methyl group of methionine was incorporated into the 45-S RNA in the cells' nucleoli of the toyocamycin-treated cells; the 28-S and 18-S RNAs were not methylated. The 45-S RNA that accumulated in the presence of toyocamycin could not be chased to 28-S and 18-S RNA, even after the removal of the nucleoside antibiotic. Acid-precipitable counts from toyocamycin-$^{3}$H were only found in the 45-S and 4-S region of the sucrose gradient. This radioactivity from the tritium-labeled toyocamycin experiment could not be chased into the 28-S and 18-S RNA. Although 45-S RNA could not be converted to 28-S and 18-S RNA, the cells were able to resume RNA synthesis when toyocamycin was removed. Therefore, the 45-S RNA that accumulated in the presence of toyocamycin that could not be converted to 18-S and 28-S RNA (rRNA) resembled the pulse-labeled 45-S RNA with respect to sedimentation coefficient, cellular localization, and methyl-acceptor activity. Toyocamycin did not inhibit the conversion of normally labeled 45-S RNA to rRNA. Tavitian et al. (1968) attributed the lack of conversion of 45-S RNA to rRNA to the altered structure of the 45-S RNA. Although toyocamycin (0.1 $\mu$g/ml) completely abolished 28-S and 18-S RNA synthesis, guanosine or the methyl group of methionine was incorporated into the 4-S RNA. Additional evidence to show that toyocamycin did not inhibit tRNA synthesis was obtained from two experiments. First, the ratio of uridylic acid to pseudouridylic acid was the same in the tRNA for the treated and untreated cells. Second, the ratio of tritium-labeled methyl groups from methionine in the tRNA to the guanine-$^{14}$C incorporated in the treated and untreated cells was the same. These data suggested that the selective inhibition of 28-S and 18-S RNA could be attributed to the altered structure of its nucleolar precursor. Apparently the transcription of various cellular RNAs on their DNA template is not affected by low concentrations

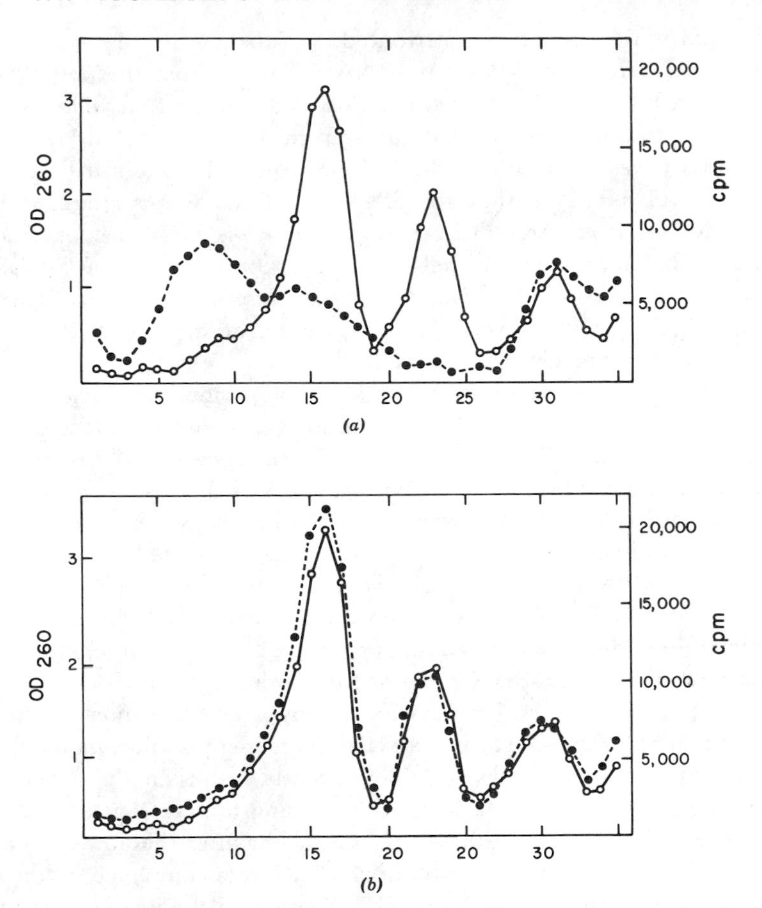


FIGURE 8.16. *Sucrose gradient analysis of RNA. (a) Cells were exposed to* 0.1 *μg of toyocamycin plus guanosine-*$^3$*H, (b) control.* (○), *Absorbance;* (●), *radioactivity.* (*From Tavitian et al.*, 1968.)

of toyocamycin. In a subsequent study, Tavitian et al. (1969) studied the base composition of nucleolar 45-S RNA synthesized in the presence or absence of toyocamycin. They also studied the effect of toyocamycin on the synthesis of mRNA and 5-S RNA. They showed that the base composition of the 45-S RNA that accumulated in the presence of toyocamycin was the same as the base composition of normal 45-S RNA. As reported earlier by Tavitian et al. (1968), this toyocamycin-treated 45-S RNA could not be chased into rRNA. This abnormal 45-S RNA appeared to be contaminated with 32-S RNA. Although RNA synthesis was blocked by toyocamycin, evidence was presented that mRNA was synthesized. A substantial amount of guanine-$^3$H was incorporated into the polyribosomes of toyocamycin-treated cells. This radioactivity was essentially attributed to labeling in the mRNA. The distribution of radioactivity of the RNA extracted from the polyribosomes is shown in Figure 8.17. The radioactivity sedimented as a wide band corresponding to sedimentation constants from 4-S to 30-S. Although some radioactivity was found in the 18-S to 22-S region, this was attributed to the association of mRNA with 18-S rRNA. 5-S RNA was synthesized in the course of toyocamycin treatment. However, this 5-S RNA was not associated with the ribosomes. The intracellular localization of this 5-S RNA is being studied. In conclusion, 4-S, 5-S, 45-S and messenger RNA synthesis were not inhibited by toyocamycin. However, maturation of 45-S RNA in the presence of toyocamycin was inhibited.

Tubercidin, the structurally related pyrrolopyrimidine nucleoside antibiotic, at a concentration of 1–2 μg/ml showed a similar effect to that of toyocamycin on RNA synthesis. However, when tubercidin was removed, the cells did not resume normal RNA synthesis as was the case with the toyocamycin-treated cells. Toyocamycin affected Ehrlich ascites cells and HeLa $S_3$ cells grown in suspension in the same way as it affected L-cells.

Heine (1969) has recently studied the changes in the cellular ultramorphology when HeLa-$S_3$ cells were exposed to toyocamycin in low and high concentrations. She reported that changes were observed in the fine structure of the nucleoli of HeLa-$S_3$ cells exposed to toyocamycin (0.1 μg/ml). There is a loss of many nucleolar granules and many irregularly intertwined cords (nucleolomena). From 6–12 hr there is an increase of fibrillar material and a diminution of the granular part. This process results in the formation of large round nucleoli of high density. According to Tavitian et al. (1968), there is a 4-fold increase in the 45-S rRNA 6 hr after the addition of toyocamycin. The maturation of 28-S and 18-S RNA is completely interrupted. There is no effect on 4-S RNA. Heine suggests that the increase in material

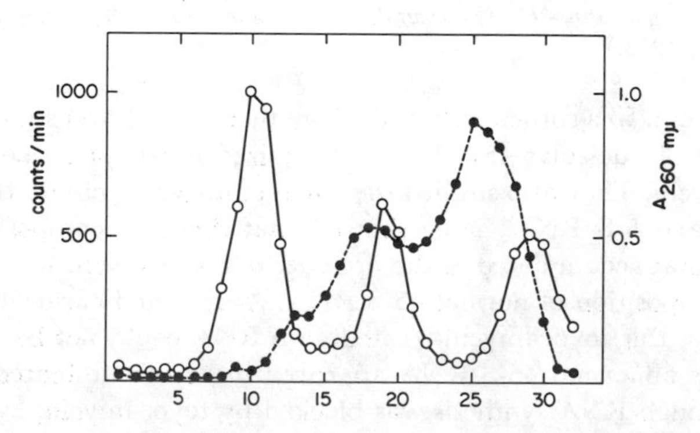


FIGURE 8.17. *RNA associated with polyribosomes.* (○), *Absorbance;* (●), *radioactivity.* (*From Tavitian et al.*, 1969.)

resembling pars fibrosa may contain the 45-S rRNA. At high concentrations of toyocamycin (20 μg/ml), there are marked changes in the morphology of the nucleolus. These changes are similar to those observed for actinomycin, mitomycin, anthramycin, aflatoxin and 4-nitroquinolin-*N*-oxide. However, the action of toyocamycin differs in that the large masses of the granular component (pars granulosa) persist in the nuclei following 24 hour exposure to toyocamycin. This observed persistence of the granules until cell death is in agreement with the earlier findings of Tavitian et al. (1968) which indicate that once toyocamycin is incorporated into the precursor rRNA, cleavage into 28-S and 18-S RNA substructures is not possible.

Truman and Frederiksen (1969) have recently reported their observations on the differential inhibition pattern of nucleochromosomal and cytoplasmic RNA by cordycepin and 3′-amino-3′-deoxyadenosine. They showed that 28-S and 18-S cytoplasmic RNA were very strongly inhibited, whereas 50–60-S nuclear RNA is only slightly inhibited. Cordycepin had the least effect on the inhibition of the 4–5-S peak of the cytoplasmic RNA. These findings, by Truman and Frederiksen with the 3′-deoxy analogs are similar to those of Tavitian et al. (1968) with the pyrrolopyrimidine nucleoside, toyocamycin.

## EFFECT OF THE PYRROLOPYRIMIDINE NUCLEOTIDES ON PROTEIN SYNTHESIS

Since Acs et al. (1964) reported that protein synthesis and nucleic acid synthesis in growing mouse fibroblasts were rapidly inhibited by tubercidin, the enzymatic steps involved in protein synthesis in extracts of mammalian cells were studied in more detail by Uretsky et al. (1968). Although TuTP could not replace ATP as an energy source for the formation of aminoacyl-tRNA with aminoacyl-tRNA synthetases from rabbit liver, rat liver, or *E. coli*, TuTP was not a competitive inhibitor of ATP for the aminoacyl synthetases. This strongly suggests that binding of the nucleotide to the aminoacyl synthetases involves nitrogen 7 of the imidazole ring of ATP. Likewise, ToTP and SaTP could not replace ATP as the energy source for amino acid activation. However, unlike TuTP, ToTP and SaTP function as competitive inhibitors of ATP. It appears that the nitrile group and the carboxamide nitrogen of ToTP and SaTP favor a complex with aminoacyl synthetases at the ATP binding site.

Studies were also reported on the incorporation of TuTP, ToTP, and SaTP into the terminal sequence of tRNA. The data obtained showed that TuMP, ToMP, and SaMP could be incorporated into the terminal sequence of pyrophosphorylyzed tRNA. TuTP did not block the incorporation of CMP into the pyrophosphorylyzed tRNA (Table 8.7). Additional proof that tubercidin was at the 3′-terminal end of tRNA was provided by hydrolysis of tRNA. All the radioactivity of the hydrolyzed RNA was located in the tubercidin. Final proof that the TuMP, ToMP, and SaMP were incorporated adjacent to CMP was provided by incubating $\alpha^{32}$P-TuTP, $\alpha^{32}$P-ToTP, or $\alpha^{32}$P-SaTP with tRNA-C-C-A pyrophosphorylase and tRNA. Alkaline hydrolysis of the $^{32}$P-labeled tRNA resulted in the isolation of 2′(3′)-CM$^{32}$P. These data prove conclusively that tubercidin, toyocamycin, and sangivamycin are substrates for mammalian terminal tRNA-C-C-A pyrophosphorylase. When tubercidin was located at the nucleoside terminus of tRNA, it was able to esterify seven amino acids. Phenylalanine was an exception. When a preparation of tRNA-$^3$H-Tu$^{14}$C-valine was treated with RNase, one radioactive area was observed following electrophoresis. Elution of the radioactive region followed by treatment at pH 10 yielded two distinct radioactive spots. One spot corresponded to valine-$^{14}$C and the other to tubercidin-$^3$H. The acceptor activities of tRNA with sangivamycin and toyocamycin at the 3′ terminal end were greatly reduced.

**Table 8.7**

Incorporation of Tubercidin, Adenosine, and Cytidine Triphosphates into Pyrophosphorylyzed tRNA[a]

| Radioactive nucleotide added | Nonradioactive nucleotide added | Nucleotide incorporated, mμmoles/100 mμmoles pyrophosphorylyzed tRNA |
|---|---|---|
| TuTP-$^3$H | CTP | 25.5 |
| TuTP-$^3$H | None | 6.4 |
| ATP-$^3$H | CTP | 26.6 |
| ATP-$^3$H | None | 6.8 |
| CTP-$^{14}$C | TuTP | 55.3 |
| CTP-$^{14}$C | ATP | 64.5 |
| CTP-$^{14}$C | None | 66.5 |

From Uretsky et al., 1968.

[a] The reaction mixture (0.5 ml) contained 20 μmoles of glycine buffer, pH 9.5, 3 μmoles of $MgCl_2$, 5 μmoles of P-enolpyruvate, 0.2 mg of P-enolpyruvate kinase, 0.4 μmole of TuTP (1.93 × $10^6$ cpm per μmole) or ATP (1.99 × $10^6$ cpm per μmole), 0.4 μmole of CTP, where indicated (3.06 × $10^6$ cpm per μmole), 24.4 mμmoles of pyrophosphorylyzed tRNA, and 0.4 mg of pyrophosphorylase. The reaction mixture was incubated for 45 min at 37°C and the reaction was then terminated by the addition of 0.5 *N* PCA. The precipitates were washed with cold 0.25 *N* PCA, alcohol–ether (3:1, v/v), and ether. They were dissolved in concentrated formic acid, and aliquots were counted in a liquid scintillation counter.

The transfer of amino acids into polypeptides from aminoacyl-tRNA-Tu was studied. Valine-$^{14}$C was incorporated into polypeptide; therefore, aminoacyl-tRNA-Tu can function in peptide bond formation, but at a reduced rate.

Homopolymers of Tu and copolymers of TuMP and UMP, like polyadenylic acid and poly (U,A), were shown to function as templates for the enzymatic formation of polylysine. In view of the results reported above, Uretsky et al. (1968) stated that the rapid and profound inhibition of protein synthesis in intact cells when exposed to tubercidin could not be explained by the effect of these ATP analogs on protein synthesis *in vivo*. They suggested that the action of tubercidin may be an indirect one in that tubercidin may interfere with a critical reaction in energy metabolism or some other process necessary for cellular function. Experimental evidence for this conclusion may be the findings of Bloch et al. (1967). They showed that pyruvate prevented the inhibition of growth of *Streptococcus faecalis* caused by tubercidin. They suggested that this impairment of growth is due primarily to the interference of tubercidin with utilization of glucose (see p. 338 for the discussion of this work by Bloch et al., 1967).

## EFFECT AS FEEDBACK INHIBITORS

Since adenine nucleotides have been shown to be feedback regulators in cell-free systems, it appeared likely that feedback inhibition might explain the effect of analogs of adenine such as tubercidin. Bennett and Smithers (1964) studied the effect of tubercidin as a feedback inhibitor of purine biosynthesis in H. Ep. #2 cells. Tubercidin was as active as adenine as a feedback inhibitor of purine synthesis. Bennett and Smithers suggested that this inhibition probably takes place at the level of the nucleotide. Additional proof that a purine nucleoside kinase is a critical enzyme in the activation of these cytotoxic analogs was supplied by Bennett et al. (1966a). They described experiments showing that tubercidin was as toxic to a subline of H. Ep. #2 cells devoid of AMP pyrophosphorylase as were the parent cells. In a subsequent study, Bennett et al. (1966b) showed that a subline of H. Ep. #2 cells, devoid of adenosine kinase, was not inhibited by tubercidin. More recently, Hill and Bennett (1969) studied the inhibition of partially purified PP-ribose-P amidotransferase from a mouse tumor, adenocarcinoma 755, maintained in cell culture with numerous ribo- and deoxyribonucleotides for this study. The reaction catalyzed by this enzyme is as follows:

$$\text{L-glutamine} + \text{PRPP} + H_2O \xrightarrow{Mg^{2+}} \beta\text{-D-ribosylamine-}$$

$$\text{5-phosphate} + \text{L-glutamate} + \text{PPi}$$

By using their partially purified enzyme, Hill and Bennett (1969) reported that all nucleotides inhibited this enzyme reaction to some extent. However, increased concentrations of $Mg^{2+}$ reversed the inhibition caused by ribo- and deoxyribonucleotide triphosphates. The inhibition by the diphosphates was reversed by $Mg^{2+}$ to a lesser extent, while the monophosphates were generally less affected by $Mg^{2+}$. Apparently the inhibition caused by the triphosphates is due to the removal of $Mg^{2+}$ from the substrate. The inhibition by tubercidin 5′-monophosphate was the same as that of AMP. The most potent inhibitor of this enzyme reaction is 6-methylthiopurine ribonucleotide.

An additional mechanism for resistance of mammalian cells to nucleotide analogs was reported by Bennett et al. (1969). They showed that the resistance of H. Ep. #2 cells to 4-aminopyrazolo[3-4,*d*]pyrimidine could also be explained by an increased rate of degradation of the nucleotide. Apparently the degradative capacity in resistant cells could account for the failure to accumulate nucleotides when grown in the presence of this analog. Therefore, changes in permeability or a loss of nucleotide pyrophosphorylase or a kinase must be considered when evaluating the resistance of cells to purine and pyrimidine analogs.

Henderson and Khoo (1965) studied the forms of purines that act as active feedback inhibitors of purine biosynthesis *de novo* and the site of this inhibition in intact Ehrlich ascites tumor cells *in vitro*. Of the purine analogs studied, only 6-benzylthiopurine and tubercidin were found to stimulate feedback inhibition by blocking PRPP synthesis. These two compounds, together with cordycepin triphosphate and adenine xylonucleoside 5′-triphosphate, can be added to the list of inhibitors for this reaction in tumor cells. Decoyinine and psicofuranine also acted as feedback inhibitors in *S. faecalis*, but did not inhibit Ehrlich ascites tumor cells (Bloch and Nichol, 1964b).

## EFFECT OF TUBERCIDIN ON MOUSE FIBROBLASTS

Acs et al. (1964) studied the effect of tubercidin in mouse fibroblasts (strain L-929) and the effects of tubercidin on cellular and viral functions. When cells were exposed to 1 $\mu$g/ml of tubercidin, colony-forming ability was irreversibly lost in 1 hr. Adenosine, inosine, or deoxyadenosine did not reverse reproductive viability. Actinomycin prevented the cytologic changes attributed to tubercidin. Desaminotubercidin was less toxic than tubercidin. Inhibition of growth was observed only after a lag period of 24 hr. 2′-Deoxytubercidin also inhibited cell growth irreversibly, but was less effective than tubercidin. The aglycone of tubercidin was not toxic to L-cells nor did it inhibit cell division. Acs et al. (1964) attributed this lack of inhibition by the aglycone to the inability of the L-cells to convert it to the nucleoside or nucleotide. Tubercidin inhibited the growth of the DNA-virus, vaccinia. Reovirus III (a double-stranded RNA) and Mengovirus (single-stranded RNA) were also inhibited by tubercidin. Tubercidin, unlike toyocamycin, was shown to inhibit the synthesis of RNA, DNA, and protein equally in cultures of L-cells (Fig. 8.18). Although the rapid inhibition of RNA and

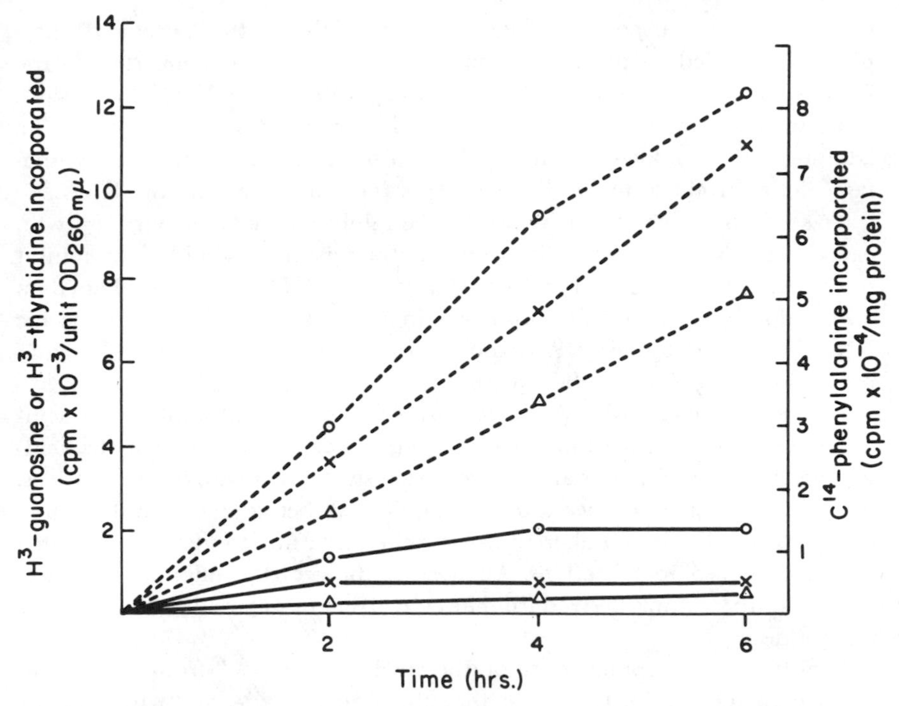


FIGURE 8.18. *Effect of tubercidin (1 μg/ml) on biopolymer formation in suspension cultures of L-cells. Synthesis of DNA, RNA, and protein was measured by the incorporation of $H^3$-thymidine, $H^3$-uridine, and $C^{14}$-L-phenylalanine, respectively. Tubercidin was added 15 min before the respective radioactive precursors. Thymidine incorporation into DNA* (△—△) *with and* (△ - - - △) *without Tu. Uridine incorporation into RNA* (○—○) *with and* (○ - - - ○) *without Tu. Phenylalanine incorporation into protein* (×—×) *with and* (× - - - ×) *without Tu.* (*From Acs et al., 1964*).

DNA synthesis might have been expected, the rapid inhibition of protein synthesis was unexpected. Using another system (KB cells), Smith et al. (1967) also showed that tubercidin inhibited RNA, DNA, and protein synthesis. Although the methyl ester of TuMP is cytotoxic to KB cells, this analog did not inhibit the synthesis of macromolecules during short term incubation (2 hr). Actinomycin, an inhibitor of RNA synthesis, allowed protein synthesis to continue for longer periods of time than did tubercidin. Actinomycin does not inhibit (1) the cellular uptake of Tu, (2) the phosphorylation of Tu, or (3) the incorporation of Tu into DNA. It appears that the effects of actinomycin are related to the inhibition of Tu incorporation into RNA. Hydrolysis of the RNA from cells treated with tritium-labeled tubercidin showed that tubercidin was incorporated into RNA as the ribonucleotide. Only a small amount of the tubercidin incorporated into the RNA was in the nucleoside terminal position. Digestion of the DNA with DNase and phosphodiesterase led to the conclusion that tubercidin was converted to the 2′-deoxynucleotide intracellularly and incorporated into DNA. The presence of an intracellular pool of tubercidin or its nucleotides was shown by the addition of 5-fluorodeoxyuridine, a known inhibitor of thymidylate synthetase. Actinomycin suppressed the incorporation of tubercidin into RNA.

To study the effect of tubercidin on the inhibition of Mengovirus growth, experiments were reported in which actinomycin was added to inhibit cellular RNA synthesis, but not Mengovirus RNA synthesis. The data obtained showed that tubercidin was incorporated into the virus-specific RNA. Therefore, the virus RNA-polymerizing system can utilize tubercidin nucleotides for polynucleotide synthesis.

Desaminotubercidin-$^{3}$H was incorporated into the acid-soluble and acid-insoluble fractions. Following degradation, it was established that the acid-soluble, radioactive material consisted exclusively of tubercidin derivatives. Apparently desaminotubercidin is reaminated before incorporation into nucleic acids. Desaminotubercidin and tubercidin nucleotides were isolated from the acid-soluble fractions. Apparently the growth-inhibitory action of desaminotubercidin is derived from its conversion to tubercidin or tubercidin nucleotides.

Analysis of the components of the acid-soluble pool from those cells exposed to tubercidin-$^{3}$H showed that the 5′-mono-, di-, and triphosphates of tubercidin were present. Neither tubercidin nor its derivatives were converted to guanine analogs.

Tubercidin triphosphate could replace ATP for enzymatic RNA synthesis with *E. coli* RNA polymerase. TuTP could serve efficiently for heteropolymer, as well as homopolymer, formation. On the basis of the data presented, Acs et al. (1964) concluded that tubercidin and its derivatives can substitute in a number of cellular and enzymatic reactions.

## EFFECT OF TUBERCIDIN ON *S. FAECALIS*

Bloch, Nichol, and co-workers have reported their studies on the effect of tubercidin and deaminotubercidin (7-deazainosine) with *S. faecalis* (ATCC 8043) (Bloch and Nichol, 1964a; Bloch et al., 1967; Bloch et al., 1969). Whereas tubercidin inhibited the growth of *S. faecalis*, deaminotubercidin was inactive against this organism. The lack of activity of the deaminotubercidin against *S. faecalis* is related to the inability of this organism to convert the deaminotubercidin to tubercidin. These findings with bacteria are in contrast to the report of Acs et al. (1964) in which they showed that a mammalian cell line (L-cells) can carry out the amination reaction. Bloch et al. (1969) reported that the antitumor activity of tubercidin (0.25–0.5 mg/kg) produced the same effect as 4–8 mg/kg of deaminotubercidin. This difference was attributed to the rate at which deaminotubercidin was converted to tubercidin nucleotides in sensitive cells. Deaminotubercidin and toyocamycin did not inhibit the growth of *S. faecalis*. Another striking difference between the effect of tubercidin in mammalian cells compared with bacterial cells (*S. faecalis*) is that a mixture of amino acids, nucleosides, ribose-5-phosphate, and pyruvate can protect *S. faecalis* against the toxic action of tubercidin (Bloch et al., 1967) while the mammalian cell system does not exhibit such a protective effect. Growing cultures of *S. faecalis* used pyruvate instead of glucose in the presence of tubercidin. However, in the absence of tubercidin, glucose was the main source of energy (Fig. 8.19). The authors concluded that the impairment of growth was attributed to the interference of tubercidin with the metabolism of glucose. Of the purine or pyrimidine nucleosides studied to prevent the inhibition of growth of *S. faecalis* by tubercidin, only adenosine, 2′-deoxyadenosine, and uridine were effective. The purine and pyrimidine bases did not reverse the inhibitory action of tubercidin. Although adenine was readily converted to AMP by AMP pyrophosphorylase, the aglycone of tubercidin was not a substrate. It was also not a substrate for adenosine phosphorylase, since the aglycone was not converted to the nucleoside (tubercidin) or nucleotide (tubercidin 5′-phosphate). This would tend to explain the lack of growth-inhibitory activity

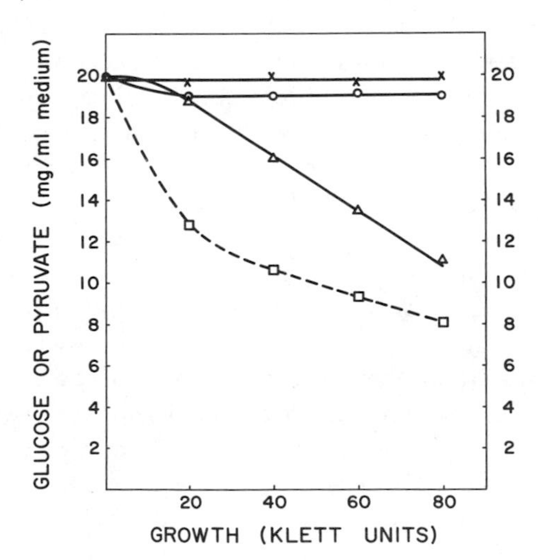


FIGURE 8.19. *Utilization of glucose and pyruvate by S. faecalis in the presence or absence of 7-deazaadenosine.* (○) *Glucose remaining after growth in the presence of pyruvate and deazaadenosine;* (△), *glucose remaining after growth in the presence of pyruvate and absence of deazaadenosine;* (□) *pyruvate remaining after growth in the presence of glucose and deazaadenosine;* (×) *pyruvate remaining after growth in the presence of glucose and absence of deazaadenosine.* (*From Bloch et al., 1967.*)

of the aglycone. Deaminotubercidin was not cleaved to the free base nor was the deaminated aglycone subject to phosphoribosyl transferase activity with cell-free extracts from *S. faecalis* (Bloch et al., 1969). However, incubation of S-180 cells with deaminotubercidin resulted in the isolation of deaminotubercidin 5′-monophosphate and tubercidin nucleotides. In addition, tubercidin and 2′-deoxytubercidin formed from deaminotubercidin were found in the respective nucleic acid fractions of the tumor, but not in the bacterial cells. The deaza analog of hypoxanthine or adenine was not converted to the nucleoside or nucleotides by mammalian extracts. The presence of deaminotubercidin 5′-monophosphate in the acid-soluble extracts requires further examination to determine how deaminotubercidin is converted to tubercidin nucleotides. The authors speculated that the formation of the nucleotide must occur by the phosphorylation of deaminotubercidin, by transphosphorylation, or by amination to tubercidin prior to phosphorylation. If this last reaction were operating in the cell, the isolation of deaminotubercidin-5′-monophosphate would suggest the presence of nucleotide deaminase. This is not impossible since tubercidin 5′-phosphate is a substrate for 5′-adenylic acid deaminase (Bloch et al., 1969). Tubercidin was not a substrate for adenosine deaminase from *S. faecalis* nor was evidence for deamination observed in red blood cells, which are known to deaminate adenosine readily (Smith et al., 1966). In addition, tubercidin did not interfere with deaminase activity (Bloch et al., 1967). Tubercidin was also shown to interfere with the formation of ATP with extracts of sarcoma-180 ascites cells by inhibiting nucleoside diphosphokinase. The same observation was made with cell-free extracts of *S. faecalis*. Tubercidin was a weak inhibitor of adenosine phosphorylase. When cells were grown in the presence of tubercidin-$^{3}$H, tubercidin and 2′-deoxytubercidin were isolated from the RNA and DNA. It was also observed that tubercidin 5′-monophosphate and nicotinamide mononucleotide, when incubated with crude extracts of *S. faecalis*, resulted in the formation of the nicotinamideadenine nucleotide analog.

Although ribose 5-phosphate reversed the inhibition of tubercidin with *S. faecalis*, neither ribose nor deoxyribose-1-phosphate was effective in preventing the growth-inhibitory effect of tubercidin. Since pyruvate reversed the inhibition of tubercidin, it was suggested that the reversal of inhibition by certain amino acids could be attributed to their ability to be converted to pyruvate or by their direct participation in the generation of energy. The reversal of inhibition of tubercidin by *S. faecalis* with ribose 5-phosphate and pyruvate might be attributed to their ability to be utilized for energy production via the nonglycolytic pathway. Since mammalian cells lack the fermentive capacity of bacteria, they cannot utilize ribose-5-phosphate to reverse the inhibitory effect of tubercidin. Of extreme interest was the observation that high levels of tubercidin are incorporated into the nucleic acids in the presence of the reversing metabolites without any effect on the growth of the organism.

## EFFECT ON HYDROCORTISONE AND TRYPTOPHAN-INDUCED TRYPTOPHAN PYRROLASE SYNTHESIS IN RAT LIVER

Tu, the methyl ester of TuMP, and 7-deazainosine exert a marked inhibition of hydrocortisone-induced tryptophan pyrrolase in adrenalectomized rats with Tu being the most potent of these inhibitors (Smith et al., 1967). When Tu was injected 24 hr before hydrocortisone, half of the adrenalectomized animals died. No tryptophan pyrrolase activity could be detected in the surviving animals even following hydrocortisone injection. If the analogs were injected 4 hr after hydrocortisone, the enzyme activity in the rats receiving the analogs was the same as in the control animals. However, the injection of puromycin under identical conditions caused an immediate, marked depression of enzyme activity. From the data obtained with tubercidin and its analogs compared with those obtained from puromycin, Smith et al. (1967) suggested that these abnormal nucleosides affect protein synthesis by acting at the transcriptional level, i.e., inhibiting RNA synthesis, while puromycin affects RNA translation. To determine if these three analogs inhibited substrate-induced enzyme synthesis, Smith et al. (1967) tested their effect on the tryptophan stimulation of tryptophan pyrrolase. Only 7-deazainosine inhibited substrate-induced enzyme synthesis. This finding is in contrast to the inhibition of steroid-induced tryptophan pyrrolase by all three compounds. It is of interest that while 5-azacytidine inhibits RNA, DNA, and protein synthesis, as well as the hormonal induction of tryptophan pyrrolase, it does not inhibit the induction of tyrosine transaminase. These data show that 5-azacytidine has a selective effect on protein synthesis and that the inhibition of tyrosine transaminase occurs at the translational level (see Chap. 7, Sect. 7.1).

## EFFECT OF BINDING OF NUCLEOTIDES CONTAINING TUBERCIDIN TO RIBOSOMES

The stimulation of binding of aminoacyl-tRNA to ribosomes by trinucleoside diphosphate analogs containing tubercidin has been reported by Ikehara and Ohtsuka (1965). This study was based on the findings that ribotrinucleotides can serve as templates for the binding of aminoacyl-tRNAs to ribosomes (Nirenberg et al., 1965). When the binding of the tubercidin analog TupApA was compared with ApApA, the tubercidin trinucleoside diphosphate stimulated the binding of lysyl-$^{14}$C-tRNA to ribosomes and showed that tubercidin can substitute for adenine. Tubercidin did not substitute for cytosine or guanine. TupCpC stimulated the binding of threonyl-$^{14}$C-tRNA, but not the binding of alanyl-tRNA, seryl-tRNA, or pro-tRNA. These data provide strong evidence that tubercidin substitutes only for adenine and have been interpreted in terms of Watson-Crick type hydrogen bond formation between the codon triplet and the anticodon site of tRNA.

## INCORPORATION INTO RNA

The utilization of TuTP, ToTP or SaTP as an ATP analog in the DNA-dependent RNA polymerase system from *E. coli* and *Micrococcus lysodeikticus* has been studied (Acs et al., 1964; Suhadolnik et al., 1968a; Ward et al., 1970; Nishimura et al. 1966). Using poly d(T-C)·d(A-G) as a template, Nishimura et al. (1966) proposed that the ribopolynucleotide synthesized should be poly (Tu-G) in strictly alternating sequence. Tu could then be checked by measuring the incorporation of radioactive GTP into acid-insoluble product. The data obtained showed that the rate of formation of poly (Tu-G) directed by poly d(T-C)·d(A-G) was much lower than that of poly (A-G). However, after long incubation, the final yield of poly (Tu-G) almost reached the same level as that of poly (A-G). Additional proof for the formation for poly (Tu-G) synthesis was shown by treatment of the acid-insoluble product with ribonuclease $T_1$. The acid-insoluble products synthesized in the presence of GTP-$^{14}$C plus TuTP were hydrolyzed by ribonuclease $T_1$. A product of this hydrolysis was the dinucleotide TupGp. TuTP could not replace ATP when other deoxypolynucleotides as poly d(T-G)·d (A-C), poly d(A-T), or poly T was the template. TuTP did not replace GTP in RNA synthesis.

When natural DNAs were used with the RNA polymerase and TuTP, there was essentially no incorporation of CTP-$^{14}$C into the acid-precipitable material. Nishimura et al. (1966) concluded that the replacement of ATP by TuTP resulted in the loss of RNA synthesis by the RNA polymerase from *E. coli* when directed to natural DNAs. Ward et al. (1970) have now reported that TuTP, ToTP, and SaTP are all incorporated into alternating (A-U)-like ribonucleotide polymers with *E. Coli* B RNA polymerase. These three nucleotides serve as analogs of ATP (Table 8.8). Similarly, Suhadolnik et al. (1968a) reported that SaTP can also substitute efficiently for ATP with *M. lysodeikticus* RNA polymerase. The reasons for the discrepancy of the data of Nishimura et al. (1966), with those of Ward et al. (1970) and Suhadolnik et al. (1968a) are not clear. It could be that the RNA polymerase preparations were different. However, in view of the recent incorporation studies of Ward et al. (1970), the nucleotide sequence requirement observed by Nishimura et al. (1966) with TuTP and the interpretation of the nearest-neighbor effects as a cause of cytotoxicity can no longer be maintained.

**TABLE 8.8**

Relative efficiency of pyrrolopyrimidine nucleoside triphosphates as substitutes for ATP with *E coli*. B RNA Polymerase

| Template | ATP | Nucleoside Triphosphate | | | | |
|---|---|---|---|---|---|---|
| | | TuTP | ToTP | SaTP | BrTuTP | TSTP |
| | | Incorporation Relative to ATP (%) | | | | |
| Native calf thymus DNA | 100 | 56 | 61 | 17 | 28 | 3.7 |
| Denatured calf thymus DNA | 100 | 53 | 69 | 26 | 22 | 2.0 |
| Poly d(A-T) | 100 | 27 | 79 | 1.3 | 21 | 0.5 |
| Poly d(A-C)·d(T-G) | 100 | 32 | 50 | — | — | — |
| Poly d(A-G)·d(T-C) | 100 | 53 | 75 | — | — | — |
| Poly dT | 100 | | | | | |

From Ward et al. (1970).

More recently, Ikehara and Fukui (1968) described the physical properties of poly Tu. Poly Tu was synthesized from TuDP using polynucleotide phosphorylase. The chain length was shown to be at least 100 nucleotide units. Poly Tu was shown to have hypochromicity of 25% at 270 m$\mu$ in 0.05 *M* cacodylate when compared with TuDP. The melting profile of poly Tu is broad and similar to that of poly A. The results obtained show that poly Tu did not form the "acid structure" which occurs in acidic solutions of poly A and involves nitrogen 7 of the adenine ring for hydrogen bonding. The formation of a poly Tu and poly U homopolymer complex in a ratio of 1:1 would be expected since $N_3$ and the 4-amino protons of the pyrrolopyrimidine ring of tubercidin are equivalent to $N_1$ and the 6-amino protons of the purine ring of adenosine. The lowering of the $T_m$ by 24°C in the poly Tu–poly U compared with that in the poly A–poly U was attributed by the authors to a weaker stacking of the pyrrolopyrimidine ring compared with that of the purine ring. However, Ward, Reich, and co-workers suggest that lower $Tm$ and the inverted ORD spectra for the homopolymers of complex poly rTu·rU are due to the preference of repeating sequences of Tu to be "non-*anti*" in conformation. Similar reasoning was used for the experimental data obtained for polyformycin (*syn* conformation) (Chapter 9 page 373). Although polynucleotides of tubercidin and formycin have abnormal conformations, tubercidin differs from formycin in that (1) poly Tu codes for polylysine whereas poly F does not and (2) the enzymatic synthesis of poly Tu with either polynucleotide phosphorylase or RNA polymerase proceeds efficiently while poly F synthesis is not efficient. Although poly F and poly Tu assume an abnormal "non-*anti*" conformation, Ward et al. (1970) have shown that poly To differs from poly Tu in that poly To retains the normal *anti* nucleotide conformation. The data obtained on the *Tm* and ORD led Ward et al. (1970) to explain the factors that affect helix stability of poly To and poly Tu. First, the helix-coil transition might be viewed as the partition of individual bases between the aqueous phase and the nonaqueous hydrophobic phase (center of helix). Second, the consistent pattern of the thermostabilities of several families of alternating copolymers indicate that the conformation at the glycosyl bond, and other factors, are important determinants of the physical parameters of ordered polynucleotides. All of the physicochemical data suggest that toyocamycin residues retain the *anti* conformation in both the helical and single-stranded state. Therefore, the cytotoxicity of To must be related to other physical or biological factors. Ward et al. (1970) concluded that the biological properties of Tu, F, and 8-azaguanine might be explained by the conformational abnormalities. Ikehara and Fukui (1968) concluded that the results obtained were consistent with the notion that no Hoogsteen-type base pairing would be involved in the codon-anticodon recognition process and in the protein-synthesizing systems.

Kapuler et al. (1969a; 1969b) have recently studied the utilization of 16 ribonucleoside triphosphate analogs by Mengovirus-induced RNA polymerase. The 5′-triphosphates of toyocamycin, tubercidin, sangivamycin, and 5-bromotubercidin were capable of replacing ATP to the extent of 41, 62, 81, and 22%, respectively. Apparently these purine analogs had normal H-binding sites. Therefore, the substituents in the pyrrolopyrimidine ring equivalent to the "7" position of the purine ring of ATP do not interfere with the enzyme-substrate complex formation. Although the base pairing of these ATP analogs is qualitatively the same as ATP, their effectiveness as replacements for ATP differs (Table 8.8) (Ward et al., 1970). Except for thiosangivamycin, the Km for these analogs is similar to that found for ATP.

The incorporation of SaTP into RNA by partially purified RNA polymerase from *M. lysodeikticus* has been reported by Suhadolnik et al. (1968a). When native calf-thymus DNA or the copolymer of deoxyadenosine and deoxythymidine (poly-d(A-T)) were the primers, SaTP was incorporated into the acid-precipitable material. Homopolymer formation did not take place. Poly (Sa-U) [formed with poly d(A-T) primer] consisted of regularly alternating residues of SaMP and UMP. SaMP and SaDP were not inhibitors of RNA polymerase, nor were these compounds incorporated into acid-precipitable material. 3′-dATP inhibited the incorporation of SaTP. The data show that SaTP competes for ATP in various polymerization reactions. There appear to be marked differences on the efficiency of pyrrolopyrimidine nucleoside triphosphates for incorporation into RNA. For example, SaTP is only 26% as efficient in replacing ATP with *E. coli* B RNA polymerase, with denatured calf thymus DNA primer, but SaTP is 80% as efficient for the replacement of ATP with *M. lysodeikticus* RNA polymerase with native calf thymus DNA primer (Ward et al., 1970; Suhadolnik et al., 1968a). Nitrogen 7 of the purine ring of ATP is not essential for enzyme-substrate complex formation.

## PHOSPHORYLATION

An extensive study of nucleosides as substrates for adenosine kinase with cell-free extracts of H. Ep. #2 cells has been described by Schnebli et al. (1967). They reported that replacement of the imidazole ring of adenosine with the pyrazole ring (formycin) or the pyrrole ring (tubercidin) produced analogs that were good substrates for adenosine kinase. Replacement of the ribofuranosyl group by arabinofuranosyl, xylofuranosyl, or 3-deoxyribosyl resulted in nucleosides that were either poor substrates or not substrates at all. Phosphorylation of tubercidin by acetone powders of *Serratia marcescens* was reported by Shirato (1968). The phosphorylation of sangivamycin to SaMP by liver extracts has been reported by Hardesty et al. (1969). When sangivamycin was added along with adenosine, the formation of AMP was completely inhibited.

## INTERACTION AT REGULATORY AND CATALYTIC SITES OF ENZYMES

The nucleoside antibiotics have also been used as biochemical tools for studying the structural requirements for interaction at the catalytic and regulatory sites of the ribonucleotide reductase (Suhadolnik et al., 1968b; Chassy and Suhadolnik, 1968) and threonine dehydrase (Rabinowitz et al., 1968; Nakazawa et al., 1967). In studies on the structural requirements for interaction at the catalytic site of ribonucleotide reductase from *L. leichmannii*, Suhadolnik et al. (1968b) reported that the 5′-triphosphates of tubercidin, toyocamycin, and sangivamycin were reduced to the 2′-deoxynucleoside triphosphates (Table 8.9). 3′-dATP was not a substrate. The reduction of TuTP, ToTP, and SaTP was greatly increased by the addition of the prime effector dGTP. Without $Mg^{2+}$, TuTP was a better substrate than ATP. Apparently the imidazole ring of ATP can be replaced by the pyrrole ring, but a hydroxyl group at position 3′ is especially important for reduction since, 3′-dATP is not a substrate for ribonucleotide reductase from *L. leichmannii*. 3′-dATP did not bind to the catalytic site since it did not inhibit the reduction of UTP, GTP, or CTP. Although 3′-dATP was not a substrate, it was 50% as good an effector as 2′-dATP for stimulating the reduction of CTP. The pyrrolopyrimidine ribonucleotide antibiotics were less effective than 3′-dATP. The pyrazolopyrimidine ribonucleotide formycin 5′-triphosphate could not be substituted for ATP. The data strongly suggest that C-8 and N-9 of the base portion of ATP, but not N-7, are important for interaction at the regulatory site of ribonucleotide reductase, while the hydroxyl position at carbon 3′ of ribose is not essential. The reduction of nucleotide antibiotics with the partially purified ribonucleotide reductase from *L. leichmannii* might well provide an explanation for the occurrence of these compounds in the DNA of cells treated with these nucleoside antibiotics. Concentrations of the 5′-mono-, di-, and triphosphates of the nucleoside antibiotics toyocamycin and cordycepin have been reported to approach 1 m*M* in the acid-soluble pool of Ehrlich ascites tumor cells. The concentration of ToTP in the acid-soluble pool of Ehrlich ascites tumor cells, as reported by Suhadolnik et al. (1967a), might well explain the formation of the corresponding 2′-deoxyribonucleotide in the cell and the subsequent interaction at the regulatory site of the ribonucleotide reductase. Chassy and Suhadolnik (1968) extended their studies to include the ribonucleotide reductase from *E. coli*. They found that the pyrrolopyrimidine antibiotics TuDP and ToDP were reduced to their corresponding 2′-deoxyribonucleotides, but that SaTP was not a substrate. The reduction of TuDP and ToDP was stimulated by the "prime effector," dGTP. However, TTP, which stimulates the reduction of the purine and pyrimidine nucleoside diphosphates, did not stimulate the reduction of the nucleotide antibiotics. This finding is in contrast to the observation that the reduction of the nucleotide antibiotics was stimulated when the prime effector, dGTP, was added to the ribonucleotide reductase isolated from *L. leichmannii*. The data indicate that nitrogen atom 7 of the imidazole ring of ADP does not play a significant role in binding at the catalytic site.

**Table 8.9**

Nucleoside Triphosphates as Substrates for Ribonucleotide Reductase from *Lactobacillus leichmannii*[a]

| | Deoxyribonucleotide added | | |
|---|---|---|---|
| Substrate | None | dGTP | 2′-dATP |
| | mμmoles | | |
| ATP | 4.8 | 86.4 | 1.4 |
| TuTP | 3.6 | 70.1 | 5.7 |
| ToTP | 0.1 | 47.2 | 0.6 |
| SaTP | 0.1 | 7.5 | 0.1 |
| GTP | 5.2 | 2.8 | 10.7 |
| 3′-dATP | 0 | 0 | 0 |

From Suhadolnik et al., 1968b.
[a] Substrates were 1 m*M* and the effectors, dGTP and 2′-dATP, were 0.5 m*M*.

Replacement of the cyano group with a carboxamide group (SaDP) is sufficient to prevent reduction of the nucleoside diphosphate to the 2′-deoxyribonucleotide. The presence of a hydroxyl group at position 3′ of the ribonucleotides does not appear to be an important structural requirement for interaction at the regulatory site of the ribonucleotide reductase from *E. coli* or *L. leichmannii*.

Rabinowitz et al. (1968) and Nakazawa et al. (1967) used the pyrrolopyrimidine nucleoside monophosphates to determine the structural requirements for interaction at the regulatory site of threonine dehydrase. Analogs differing in the adenine moiety had essentially the same allosteric activity as AMP. The 5′-phosphoryl and the 2′- and 3′-hydroxyl groups of the ribosyl moiety are of prime importance in functioning as allosteric effectors for threonine dehydrase. It appears that the ribosyl moiety of the nucleotide is of primary importance for enzyme regulation, whereas changes in the adenine ring can be tolerated without abolishing allosteric activity.

## 8.5. CLINICAL STUDIES WITH TUBERCIDIN

Tubercidin was selected for clinical testing on the basis of its unique structure, type of biochemical activity, inhibition of mammalian cells, and antitumor activity in animals (Bisel et al., 1970). The first clinical study was concerned with the toxicity of Tu in human beings (Bisel et al., 1970). Ninety-three patients with various types of advanced neoplastic diseases were selected in a Phase I study to determine human toxicity. Patients entering the study had a life expectancy of more than 60 days. Tubercidin is an irritating compound and venous thromboses were observed in 12 cases. In addition, renal toxicity, manifest by proteinuria, azotemia, or both, was encountered. Hematopoietic toxicity and leukopenia were also found. Neither thrombocytopenia nor anemia was observed. Anorexia, nausea, and vomiting also occurred. Although tumor response for a Phase I study was of secondary interest, attempts were made to evaluate the effect of Tu, and tumor response was suggested in three cases. All three were cases of primary pancreatic carcinoma of the islet cell type.

Since nephrotoxicity, venous thrombosis, and necrosis of tissue at the site of injection limit the usefulness of Tu, a new mode of administration was sought to reduce the toxicity. The procedure used was similar to that decribed by Smith et al. (1970), in which Tu was absorbed by red blood cells *in vitro* following which it was then retransfused into the patient (Grage et al., 1970). By using this technique, the problems encountered by direct intravenous administration of Tu have been eliminated. Toxicity to Tu administered in this manner was infrequent, mild, and reversible. Significant regression of malignancy in 4 patients with evaluable lesions was reported between 4 and 12 weeks after onset of therapy. Three of these patients also had islet cell carcinomas of the pancreas. A fourth patient had a carcinoid of the stomach which spread to the cervical lymph nodes.

## Summary

The structures and chemical syntheses of the pyrrolopyrimidine nucleoside antibiotics, tubercidin, toyocamycin, and sangivamycin, have been reported. These three nucleosides are analogs of adenosine. They are excellent substrates for adenosine kinase, but are not subject to phosphorolysis or deamination. The aglycone of tubercidin is not toxic. Sublines of mammalian cells resistant to tubercidin and other adenosine analogs that were devoid of adenosine kinase were resistant to these cytotoxic nucleosides. These results indicate that adenosine kinase is an essential enzyme for the activation of many cytotoxic nucleosides. Tubercidin has been shown to inhibit several organisms, including *M. tuberculosis* and *S. faecalis*. Toyocamycin is a strong inhibitor of *Candida albicans*, *Cryptococcus neoformans*, and *M. tuberculosis*, but has little effect on other bacteria, fungi, or yeast. Sangivamycin does not appear to have antibacterial properties. All three nucleoside antibiotics are highly cytotoxic to vertebrate cell lines in culture. Sangivamycin is not acutely toxic in humans. Tubercidin inhibited RNA and DNA viruses; RNA, DNA, and protein synthesis in mouse fibroblasts and KB cells; and glycolysis in *S. faecalis*. It is absorbed by animal red blood cells and rapidly converted to the 5′-triphosphate. Clinical studies show that tubercidin is less toxic when mixed with blood and then retransfused into the patient. Tubercidin was also incorporated into RNA and DNA of mouse fibroblasts and may inhibit protein synthesis at the transcriptional level rather than at the level of RNA translation. Toyocamycin completely inhibited rRNA synthesis; neither 4-S, 5-S, 45-S nor mRNA was inhibited. The pyrrolopyrimidine nucleotides could not function as an energy source for activation of amino acids or esterification of tRNA by mammalian or bacterial enzymes. All three nucleosides were incorporated at the 3′ termini of tRNA. The coding properties of several tubercidin-containing polyribonucleotides were the same as those of the corresponding polymers containing adenosine. The 5′-triphosphates of tubercidin, toyocamycin, and sangivamycin can replace ATP with Mengovirus-induced RNA polymerase, *E. coli* RNA polymerase, and *M. lysodeikticus* RNA polymerase. Ribonucleotide reductase from *L. leichmannii* and *E. coli* reduced these nucleotide antibiotics to their corresponding 2′-deoxynucleotides. These nucleotide analogs also interact at the regulatory sites of several enzymes. Physical measurements ($T_m$) of poly (Tu-U)-poly U compared with those of poly A-poly U were different. The $T_m$ and

ORD spectra of poly Tu, To and F have been compared with similar polymers of adenine nucleotides. The data obtained permit the identification and show the importance of the *syn* or *anti* conformation of nucleotides of Tu, To and F in helical homopolymer complexes, alternating complexes and single-stranded homopolymers. To, in poly To, appears to retain the normal *anti* conformation; F in poly F has the *syn* conformation; and Tu in poly Tu appears to be intermediate between poly F (*syn*) and poly A (*anti*) and, therefore, assumes an abnormal "non-*anti*" conformation. Although poly Tu and poly F have abnormal conformations, their biological properties differ in that (1) poly Tu codes effectively for polylysine synthesis; poly F does not and (2) the enzymatic polymerization of Tu proceeds efficiently with RNA polymerase or polynucleotide phosphorylase; polymerization of F does not. Since poly To retains the *anti* conformation, the cytotoxicity of To must be related to factors other than the conformational abnormalties.

The biosyntheses of tubercidin, toyocamycin, and sangivamycin have been studied *in vivo* and *in vitro*. The pyrimidine ring of a purine appears to be the precursor for the pyrimidine ring of the pyrrolopyrimidine nucleoside antibiotics; carbons 1′,2′ and 3′ of the ribose moity of adenosine serves as the carbon source for pyrrole carbons 6,5 and the cyano group of toyocamycin. The nitrile group of toyocamycin can be converted to the carboxamide group to form sangivamycin with a partially purified enzyme from the sangivamycin-producing *Streptomyces*.

## References


Acs, G., E. Reich, and M. Mori, *Proc. Natl. Acad. Sci. (U.S.)*, **52**, 493 (1964).

Alworth, W. L., H. N. Baker, D. A. Lee, and B. A. Martin, *J. Amer. Chem. Soc.*, **91**, 5662 (1969).

Anzai, K., and S. Marumo, *J. Antibiotics (Tokyo)*, **10A**, 20 (1957).

Anzai, K., G. Nakamura, and S. Suzuki, *J. Antibiotics (Tokyo)*, **10A**, 201 (1957).

Arcamone, F., G. Cassinelli, G. D'Amico, and P. Orezzi, *Experientia*, **24**, 441 (1968).

Aszalos, A., P. Lemanski, R. Robison, S. Davis, and B. Berk, *J. Antibiotics (Tokyo)*, **19A**, 285 (1966).

Baugh, C. M., and E. N. Shaw, *Biochim. Biophys. Acta*, **114**, 213 (1966).

Bennett, L. L., Jr., and D. Smithers, *Biochem. Pharmacol.*, **13**, 1331 (1964).

Bennett, L. L., Jr., M. H. Vail, S. Chumley, and J. A. Montgomery, *Biochem. Pharmacol.*, **15**, 1719 (1966a).

Bennett, L. L., Jr., H. P. Schnebli, M. H. Vail, P. W. Allan, and J. A. Montgomery, *Mol. Pharmacol.*, **2**, 432 (1966b).

Bennett, L. L., P. W. Allan, D. Smithers, and M. H. Vail, *Biochem Pharmacol.*, **18**, 725 (1969).

Bisel, H. F., F. J. Ansfield, J. H. Mason, and W. L. Wilson, *Cancer Res.*, **30**, 76 (1970).

Bloch, A., and C. A. Nichol, *Antimicrobial Agents Chemotherapy*, 1964a, 530.

Bloch, A., and C. A. Nichol, *Biochem. Biophys. Res. Commun.*, **16**, 400 (1964b).

Bloch, A., M. T. Hakala, E. Mihich, and C. A. Nichol, *Proc. Amer. Assoc. Cancer Res.*, **5**, 6 (1964).

Bloch, A., R. J. Leonard, and C. A. Nichol, *Biochim. Biophys. Acta*, **138**, 10 (1967).

Bloch, A., E. Mihich, R. J. Leonard, and C. A. Nichol, *Cancer Res.*, **29**, 110 (1969).

Brown, G. B., Progress in Nucleic Acid Research and Molecular Biology, Vol. 8, J. N. Davidson and W. E. Cohn Eds., Academic Press, New York, 1968, P. 212.

Buchanan, J. M., and S. C. Hartman, *Advan. in Enzymology*, Vol. 21, F. F. Nord, Ed., Interscience, New York, 1959.

Burg, A. W., and G. M. Brown, *Biochim. Biophys. Acta*, **117**, 275 (1966).

Cavalieri, L. F., A. Bendich, J. F. Tinker, and G. B. Brown, *J. Amer. Chem. Soc.*, **70**, 3875 (1948).

Cavalieri, L. F., and A. Bendich, *J. Amer. Chem. Soc.*, **72**, 2587 (1950).

Cavins, J. A., *Proc. Amer. Assoc. Cancer Res.*, **7**, 12 (1966).

Cavins, J. A., T. C. Hall, K. B. Olson, C. L. Khung, J. Horton, J. Colsky, and R. K. Shadduck, *Cancer Chemotherapy Rept.*, **51**, 197 (1967).

Chassy, B. M., and R. J. Suhadolnik, *J. Biol. Chem.*, **243**, 3538 (1968).

Clark, V. M., A. R. Todd, and J. Zussman, *J. Chem. Soc.*, 1951, 2952.

Davoll, J., *J. Chem. Soc.*, 1960, 131.

Dekker, C. A., *J. Am. Chem. Soc.*, **87**, 4027 (1965).

Duvall, L. R., *Cancer Chemotherapy Rept.*, **30**, 61 (1963).

Fujii, S., H. Hitomi, M. Imanishi, and K. Nakazawa, *Ann. Rept. Takeda Res. Lab.*, **14**, 8 (1955).

Gerster, J. F., B. Carpenter, R. K. Robins, and L. B. Townsend, *J. Med. Chem.*, **10**, 326 (1967).

Grage, T. B., D. B. Rochlin, A. J. Weiss, and W. L. Wilson, *Cancer Res.*, **30**, 79 (1970).

Hampton, A., and D. I. Magrath, *J. Amer. Chem. Soc.*, **79**, 3250 (1957).

Hardesty, C. T., N. A. Chaney, V. S. Waravdekar, and J. A. R. Mead, *Biochem. Biophys. Res. Commun.*, **195**, 581 (1969).

Heine, U., *Cancer Res.*, **29**, 1875 (1969).

Henderson, J. F., and M. K. Y. Khoo, *J. Biol. Chem.*, **240**, 2349 (1965).

Higashide, E., T. Hasegawa, M. Shibata, K. Mizuno, and H. Akaike, *Takeda Kenkyusho Nempo*, **25**, 1 (1966).

Hill, D. L., and L. L. Bennett, Jr., *Biochemistry*, **8**, 122 (1969).

Hinshaw, B. C., J. F. Gerster, R. K. Robins, and L. B. Townsend, *J. Heterocyclic Chem.*, **6**, 215 (1969).

Hinshaw, B. C., J. F. Gerster, R. K. Robins, and L. B. Townsend, *J. Org. Chem.*, **35**, 236 (1970).

Hogenkamp, H. P. C., *Carbohydrate Res.*, **3**, 239 (1966–67).

Ikehara, M., and T. Fukui, *J. Mol. Biol.*, **38**, 437, (1968).

Ikehara, M., and E. Ohtsuka, *Biochem. Biophys. Res. Commun.*, **21**, 257 (1965).

Iwamura, H. and T. Hashizume; *Agr. Biol. Chem.*, **32**, 1010 (1968).

Kapuler, A. M., D. C. Ward, N. Mendelsohn, H. Klett, G. Acs, and S. Spiegelman, *Federation Proc.*, **28**, 731 (1969a).

Kapuler, A. M., D. C. Ward, N. Mendelsohn, H. Klett, and G. Acs, *Virology*, **37**, 701 (1969b).
Katagiri, K., K. Sato, and S. Nishiyama, *Shionogi Kenkyusho Nempo*, **7**, 715 (1957).
Kikuchi, K., *J. Antibiotics (Tokyo)*, **8A**, 145 (1955).
Kohls, R. E., A. J. Lemin, and P. W. O'Connell, *J. Econ. Entomol.*, **59**, 745 (1966).
Krumdieck, C. L., E. Shaw, and C. M. Baugh, *J. Biol. Chem.*, **241**, 383 (1966).
Levenberg, B., and D. K. Kaczmarek, *Biochim. Biophys. Acta*, **117**, 272 (1966).
Mathis, J. B., and G. M. Brown, *J. Biol. Chem.*, **245**, 3015 (1970).
Matsuoka, M., *J. Antibiotics (Tokyo)*, **13A**, 121 (1960).
Matsuoka, M., and H. Umezawa, *J. Antibiotics (Tokyo)*, **13A**, 114 (1960).
Mihich, E., C. L. Simpson, and A. I. Mulhern, *Cancer Res.*, **29**, 116 (1969).
Mizuno, Y., M. Ikehara, K. A. Watanabe, S. Suzaki, and T. Itoh, *J. Org. Chem.*, **28**, 3329 (1963a).
Mizuno, Y., M. Ikehara, K. A. Watanabe, and S. Suzaki, *J. Org. Chem.*, **28**, 3331 (1963b).
Mizuno, Y., M. Ikehara, K. A. Watanabe, and S. Suzaki, *Chem. Pharm. Bull. (Tokyo)*, **11**, 1091 (1963c).
Montgomery, J. A., and K. Hewson, *J. Med. Chem.*, **10**, 665 (1967).
Nakamura, G., *J. Antibiotics (Tokyo)*, **14A**, 90 (1961).
Nakazawa, A., M. Tokushige, and O. Hayaishi, *J. Biol. Chem.*, **242**, 3868 (1967).
Nirenberg, M., P. Leder, M. Bernfield, R. Brimacombe, J. Trupin, F. Rottman, and C. O'Neal, *Proc. Natl. Acad. Sci. (U.S.)*, **53**, 1161 (1965).
Nishimura, H., K. Katagiri, K. Satō, M. Mayama, and N. Shimaoka, *J. Antibiotics (Tokyo)*, **9A**, 60 (1956).
Nishimura, S., F. Harada, and M. Ikehara, *Biochim. Biophys. Acta*, **129**, 301 (1966).
Ohkuma, K., *J. Antibiotics (Tokyo)*, **13A**, 361 (1960).
Ohkuma, K., *J. Antibiotics (Tokyo)*, **14A**, 343 (1961).
Owen, S. P., and C. G. Smith, *Cancer Chemotherapy Rept*, *No.* **36**, 19 (1964).
Pike, J. E., L. Slechta, and P. F. Wiley, *J. Heterocyclic Chem.*, **1**, 159 (1964).
Ponnamperuma, C., R. M. Lemmon, and M. Calvin, *Radiation Res.*, **18**, 540 (1963).
Rabinowitz, K. W., J. D. Shada, and W. A. Wood, *J. Biol. Chem.*, **243**, 3214 (1968).
Rao, K. V., 150th National Meeting, American Chemical Society, Atlantic City, N.J., September, 1965, Abstracts 24P.
Rao, K. V., *J. Med. Chem.*, **11**, 939 (1968).
Rao, K. V., and D. W. Renn, *Antimicrobial Agents Chemotherapy*, 1963, 77.
Rao, K. V., W. S. Marsh, and D. W. Renn, U.S. Pat. No. 3,116,222 (1963).
Rao, K. V., W. S. Marsh, and D. W. Renn, U.S. Pat. No. 3,423,398, 1969; *Chem. Abstr.*, **70**, 86268y, 1969.
Reich, E., 156th National Meeting, American Chemical Society, Atlantic City, N. J., 1968, Medi 30.
Reynolds, J. J., and G. M. Brown, *J. Biol. Chem*, **239**, 317 (1964)
Rousseau, R. J., L. B. Townsend, and R. K. Robins, *Chem. Commun.*, 1966, 265
Sanėyoshi, M., R. Tokuzen, and F. Fukuoka, *Gann*, **56**, 219 (1965).
Schnebli, H. P., D. L. Hill, and L. L. Bennett, Jr., *J. Biol. Chem.*, **242**, 1997 (1967).
Shapiro, R., in Progress in Nucleic Acid Research and Molecular Biology, Vol. 8., J. N. Davidson and W. E. Cohn, Eds., Academic Press, 1968, p. 93.

Shemin, D., *Naturwissenschaften*, **57**, 185 (1970).
Shiota, T., M. P. Palumbo, and L. Tsai, *J. Biol. Chem.*, **242**, 1961 (1967).
Shirato, S., *J. Fermentation Technol.*, **46**, 233 (1968).
Shirato, S., Y., Miyazaki, and I. Suzuki, *J. Fermentation Technol.*, **45**, 60 (1967).
Smith, C. G., W. L. Lummis, and J. E. Grady, *Cancer Res.*, **19**, 847 (1959).
Smith, C. G., L. M. Reineke, H. Harpootlian, *Proc. Amer. Assoc. Cancer Res.*, **7**, 259 (1966).
Smith, C. G., G. D. Gray, R. G. Carlson, and A. R. Hanze, *Advan. Enzyme Reg.*, **5**, 121 (1967).
Smith, C. G., L. M. Reineke, M. R. Bruch, A. M. Shefner, and E. C. Muirhead, *Cancer Res.*, **30**, 69 (1970).
Smulson, M. E., and R. J. Suhadolnik, *J. Biol. Chem.*, **242**, 2872 (1967).
Stecher, P. G., M. Windholz, D. S. Leahy, D. M. Bolton, and L. G. Eaton, Eds., The Merck Index, Eighth Edition, Merck & Co., Inc., 1968, Pg. 1087.
Stheeman, A. A., and A. P. Struyk, Switzerland Pat. No. 331988 (1953).
Struyk, A. P., and A. A. Stheeman, *Chem. Abstr.*, **51**, 10009a, (1957); British Pat. No. 764,198 (1956).
Struyk, A. P., and A. A. Stheeman, *Chem. Abstr.*, **62**, 11114g (1965); Netherlands Pat. No. 109,006 (1964).
Suhadolnik, R. J., T. Uematsu, and H. Uematsu, *Biochim. Biophys. Acta*, **149**, 41 (1967a).
Suhadolnik, R. J., T. Uematsu, and R. M. Ramer, *Carbohydrate Res.*, **5**, 479 (1967b).
Suhadolnik R. J., and T. Uematsu, 69th Annual Meeting, American Society for Microbiology, Miami Beach, Fla., May, 1969, Abstracts, p. 135.
Suhadolnik, R. J., T. Uematsu, H. Uematsu and R. G. Wilson, *J. Biol. Chem.*, **243**, 2761 (1968a).
Suhadolnik, R. J., S. I. Finkel, and B. M. Chassy, *J. Biol. Chem.*, **243**, 3532 (1968b).
Suhadolnik, R. J., and T. Uematsu, *J. Biol. Chem.*, **245**, No. 16, *in press* (1970).
Suzuki, S., and S. Marumo, *J. Antibiotics (Tokyo)*, **13A**, 360 (1960).
Suzuki, S., and S. Marumo, *J. Antibiotics (Tokyo)*, **14A**, 34 (1961).
Tavitian, A., S. C. Uretsky, and G. Acs, *Biochim. Biophys. Acta*, **157**, 33 (1968).
Tavitian, A., S. C. Uretsky, and G. Acs, *Biochim. Biophys. Acta*, **179**, 50 (1969).
Taylor, E. C., and R. W. Hendess, *J. Amer. Chem. Soc.*, **86**, 951 (1964).
Taylor, E. C., and R. W. Hendess, *J. Amer. Chem. Soc.*, **87**, 1995 (1965).
Tolman, R. L., R. K. Robins, and L. B. Townsend, *J. Heterocyclic Chem.*, **4**, 230 (1967).
Tolman, R. L., R. K. Robins, and L. B. Townsend, *J. Amer. Chem. Soc.*, **90**, 524 (1968).
Tolman, R. L., R. K. Robins, and L. B. Townsend, *J. Amer. Chem. Soc.*, **91**, 2102 (1969).
Townsend, L. B., and R. K. Robins, *J. Amer. Chem. Soc.*, **85**, 242 (1963).
Truman, J. T., and S. Frederiksen, *Biochim. Biophys. Acta*, **182**, 36 (1969).
Uematsu, T., and R. J. Suhadolnik, *J. Org. Chem.*, **33**, 726 (1968).
Uematsu, T., and R. J. Suhadolnik, *Biochemistry*, **9**, 1260 (1970).
Uretsky, S. C., G. Acs, E. Reich, M. Mori, and L. Altwerger, *J. Biol. Chem.*, **243**, 306 (1968).
Ward, D. C., A. Cerami, E. Reich, N. Mendelsohn and G. Acs, *J. Biol. Chem.*, **245**, *in press* (1070).

Wechter, W. J., and A. R. Hanze, U. S. Pat. No. 3,336,289 (1967).
Wellcome Foundation Ltd., Brit. Pat. 812,366, 1959; *Chem. Abstr.*, **54**, 592 (1960).
Weygand, F., H. Simon, G. Dahms, M. Waldschmidt, H. J. Schliep, and H. Wacker, *Angew. Chem.*, **73**, 402 (1961).
Wolberg, W. H., *Biochem. Pharmacol.*, **14**, 1921 (1965).
Yamamoto, H., S. Fujii, K. Nakazawa, A. Miyake, H. Hitomi, and M. Imanishi, *Takeda Kenkyusho Nempo*, **16**, 26 (1957).
Yanofsky, C., *Biochim. Biophys. Acta*, **20**, 438 (1956)
Zubrod, C. G., S. Shepartz, J. Leiter, K. M. Endicott, L. M. Carrese, and C. G. Baker, *Cancer Chemotherapy Rept.*, **50**. 496 (1966).

CHAPTER 9

# Pyrazolopyrimidine Nucleosides and Coformycin



## ABBREVIATIONS

Adenosine 5′-triphosphate, ATP; 2-formamido-*N*-ribosylacetamide 5′-phosphate, FGAR; formycin, F; tubercidin, Tu; toyocamycin, To; formycin 5′-monophosphate, FMP; formycin 5′-diphosphate, FDP; formycin 5′-triphosphate, FTP; 5-phosphoribosyl-1-pyrophosphate, PRPP; homopolymers of adenosine, formycin, and uridine, poly A, poly F, and poly U; alternating ribocopolymers of formycin, guanosine, cytidine, uridine, and adenosine, poly (F-U), poly (A-U), poly (F-G), poly (F-C), poly (A-C), poly (A-G); 1:1 mixtures of homopolymers of formycin with the homopolymer of uridine, poly rF·rU; tRNA molecules containing formycin (F) termini, tRNA-CCF; poly d(T-C)·d(A-G) refers to the DNA-like polymer that contains alternating deoxythymidylate and deoxycytidylate units in one strand and alternating deoxyadenylate and deoxguanylate units in the complementary strand; poly d(T-G)·d(A-C) refers to alternating deoxythymidylate and deoxyguanylate units in one strand and deoxyadenylate and deoxycytidylate units in the complementary strand; poly d(A-T) refers to a DNA-like polymer that contains alternating deoxyadenylate and deoxycytidylate units.

The pyrazolopyrimidine antibiotics represent a new class of naturally occurring nucleosides that contain the unusual *C*-riboside linkage. They are formycin (formycin A), formycin B (laurusin) and oxoformycin B (Fig. 9.1). These three nucleosides are structural analogs of adenosine, inosine, and xanthosine, respectively. The structure assigned to formycin is 7-amino-3-(β-D-ribofuranosyl)-pyrazolo[4,3-*d*]pyrimidine; formycin B is 3-(β-D-ribofuranosyl)-pyrazolo[4,3-*d*]-6(H)-7-pyrimidone; oxoformycin B is 3-(β-D-ribofuranosyl)-pyrazolo[4,3-*d*]-4(H), 6(H)-5,7-pyrimidione (Koyama et al., (1966). Formycin is a cytoxic analog of adenosine (Hori et al., 1964; Umezawa et al., 1965). Recent biochemical studies by Ward, Fuller, and Reich (1969b) show that formycin is the most effective analog to replace adenosine. FTP is the first nucleotide analog, with a modified base, that can substitute for ATP as a substrate for aminoacyl-tRNA synthetase. It was originally isolated from the culture filtrates of the actinomycetes, *Nocardia interforma n. sp.*, by Hori et al. (1964). Aizawa et al. (1965) also isolated formycin from *Streptomyces lavendulae*. More recently, formycin has been isolated from the culture filtrates of *S. gunmaences* (Nippon Kayaku Co., Ltd., 1968).

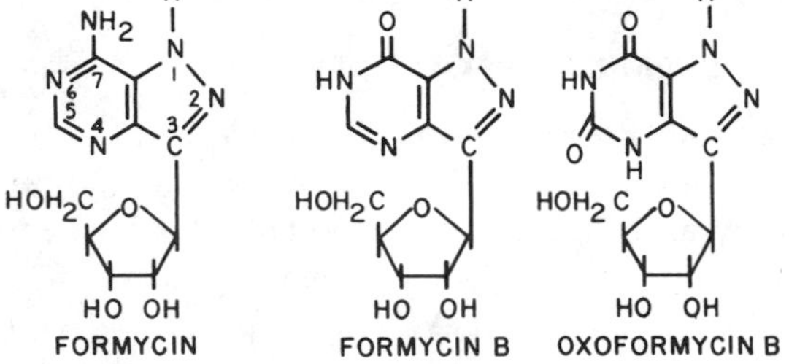

FORMYCIN FORMYCIN B OXOFORMYCIN B

FIGURE 9.1. *Structures for* (*A*) *formycin*, (*B*) *formycin B*, *and* (*C*) *oxoformycin B*.

Koyama and Umezawa (1965) and Aizawa et al. (1965) reported on the isolation of the deaminated product, formycin B or laurusin from *N. interforma* and *S. lavendulae*. Finally, coformycin, a substance that appears to be a ribonucleoside, has also been isolated from the culture filtrates of *N. interforma* (Niida et al., 1967; Sawa et al., 1967a, 1967b). Coformycin does not appear to be either a pyrazolopyrimidine nucleoside nor the C-C pyrazoloriboside, pyrazomycin, that has recently been isolated by Williams et al. (1969). (See chapter 10, Pyrazomycin). Oxoformycin B has been isolated from the culture filtrates of *N. inferforma*, and from the urine of mice, and rabbits (Ishizuka et al., 1968a, 1968b; Sheen et al., 1968b, 1970).

Formycin inhibits tumor cells, bacteria, fungi, and viruses. Formycin B is less toxic but inhibits *X. oryzae* and influenza virus. In contrast, oxoformycin B does not inhibit the growth of any organism that has been checked to date. The antitumor activity for formycin appears to require its conversion to the 5′-mono-, di-, and triphosphates.

## 9.1. FORMYCIN

### DISCOVERY, PRODUCTION, AND ISOLATION

The fermentation medium for the production of formycin by *N. interforma* is as follows: 2% glucose, 1% N-Z amine, 0.2% yeast extract, and 0.3% sodium chloride (Hori et al., 1964). Formycin is produced by shake culture at 27°C. At the time of maximum formycin production, the pH of the medium rises to 8.2 (4 or 5 days). Formycin is isolated from the culture filtrate (pH 7.0) by adsorption on charcoal followed by washing with water, 50% aqueous methanol, and acetone. The nucleoside is eluted with 50% aqueous acetone. The acetone is removed *in vacuo* followed by lyophilization. Further purification is attained by Sephadex G-25. The formycin is eluted with water, concentrated *in vacuo*, and lyophilized. The dried residue is dissolved in HCl–water (pH 3.5) and added to a strong cation-exchange resin saturated with pyridine-acetate buffer (pH 5.5). Pure formycin is obtained by lyophilization. The yield of the nucleoside antibiotic is about 6 mg/100 ml.

### PHYSICAL AND CHEMICAL PROPERTIES

Formycin is an amphoteric compound with p$K_a$'s at 4.5 and 9.5 (Hori et al., 1964). The p$K_a$ of formycin at 9.5 is readily explained by the removal of a proton from the pyrazole nitrogen (Robins et al., 1966). The titration equivalent is 280 (Hori et al., 1964). The molecular formula is $C_{10}H_{13}N_5O_4 \cdot H_2O$; $[\alpha]_D^{20} = -35.5°$ ($c = 1\%$ in 0.1 $N$ HCl); mp 141–144°C. (decomp. above 252°C) (Koyama and Umezama, 1965). The chemical syntheses of FMP and FTP have been reported (Ward et al., 1969c). The ultraviolet absorption spectra are as follows: $\lambda_{max}$ 295 m$\mu$ ($E^{1\%}_{1\,cm}$ 380) (neutral aqueous solution); $\lambda_{max}$ 234 m$\mu$ ($E^{1\%}_{1\,cm}$ 280) and 295 m$\mu$ ($E^{1\%}_{1\,cm}$ 340) (in acidic solution); $\lambda_{max}$ 235 m$\mu$ ($E^{1\%}_{1\,cm}$ 500) and 305 m$\mu$ ($E^{1\%}_{1\,cm}$ 260) (in alkaline solution) (Hori et al., 1964). Formycin is soluble in water and methanol and moderately soluble in ethanol, but insoluble in acetone and ether. The *C*-riboside bond is stable in acid. The use of thin-layer chromatography, paper chromatography, and paper electrophoresis describing the migration of formycin has been reported by Hori et al. (1964).

## PROOF OF STRUCTURE

The proof of structure for formycin was established by X-ray analysis, degradation studies, and nmr spectra (Koyama et al., 1966; Kawamura et al., 1966; Watanabe et al., 1966; Robins et al., 1966). The X-ray analysis showed that formycin was a *C*-riboside and had the $\beta$ configuration. The pyrazolopyrimidine base and its exocyclic amino group were shown to be coplanar. The ribose ring was puckered in such a way that carbon 2′ is displaced from the plane of carbon 1′, oxygen 1′, and carbon 4′. The displacements of carbon 2′ and carbon 3′ indicated that the ribofuranose ring of formycin (as the hydrobromide monohydrate) exists in the C(2′)-*endo*-C(3′)-*exo*-conformation (Koyama et al., 1966). The plane of the sugar is at a dihedral angle of 64.2° to the base. This is slightly smaller than the reported value for some nucleosides and nucleotides. Koyama et al. (1966) also studied the conformation about the glycosidic bond. The value of the torsion angle was reported to be +148.33°. The known reported value for normal *syn* orientation is known to be $+150 \pm 45°$, while the *anti* orientation is $-30 \pm 45°$. These data indicate that formycin exists in the *syn* orientation. The importance of the orientation of formycin and nucleosides in biological reactions will be discussed in more detail in the section related to the mechanism of action of formycin (Ward and Reich, 1968; Ward, Reich and Stryer, 1969a; Ward, Fuller and Reich, 1969b; Ward et al; 1970).

Simultaneously with the structural elucidation of formycin by the X-ray method, Kawamura et al. (1966) reported on the isolation of the aglycone of formycin, which was shown to be 7-aminopyrazolo[4,3-*d*]pyrimidine. The chemical degradation of formycin to the aglycone is shown in Figure 9.2. Robins and his co-workers (1966) also reported on the structure of formycin.

A comparison of the ultraviolet absorption spectrum of formycin with 3-methyl-7-aminopyrazolo[4,3-*d*]pyrimidine at pH 7 and 13 revealed spectra that were almost identical (Robins et al., 1966, 1956). The nmr spectra of formycin in $D_6$-dimethylsulfoxide (DSS as an internal standard) showed a very sharp singlet at 8.3 $\delta$. The broad peak (2H) at 7.5 $\delta$ is due to $NH_2$. A very broad peak of low absorption (1H) was observed at 12.8 $\delta$. This peak is typical of "NH" absorption of pyrazoles. The nmr data also indicated the

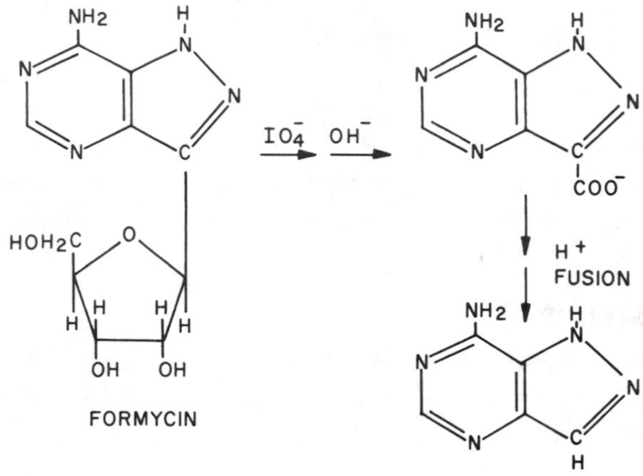

FIGURE 9.2. *Degradation of formycin to the aglycone (7-amino-pyrazolo[4,3-d]pyrimidine). (From Kawamura et al., 1966.)*

presence of 3 hydroxyl protons. The nmr spectrum for the carbohydrate moiety of adenosine was essentially the same as that of formycin, except for the position of the anomeric proton of adenosine. The position of the anomeric proton of adenosine is at 6.15 $\delta$ as compared to the anomeric proton of formycin at 5.14 $\delta$. This peak at 5.14 $\delta$ is consistent with the notion that the sugar moiety is attached to a carbon instead of a nitrogen. The presence of the anomeric proton at 5.14 $\delta$ is similar to that reported for pseudouridine (5.08 $\delta$). Therefore, the anomeric protons of *C*-glycosides or their derivatives appear at higher field than the values for *N*-glycosides.

Since formycin is readily deaminated to formycin B, Robins et al. (1966) provided proof that these two nucleoside antibiotics were *C*-nucleosides with attachment of the sugar at position 3 in both nucleosides. The enzymatic deamination of formycin by adenosine deaminase is additional proof for the $\beta$ assignment, since LePage and Junga (1965) reported that adenosine deaminase does not attack the $\alpha$-anomer of adenosine analogs, but does deaminate the corresponding $\beta$-anomers. Robins et al. (1966) concluded that formycin showed a striking steric resemblance to adenosine and predicted that the pyrazolopyrimidine antibiotics should resemble the purine nucleosides in the enzymatic reactions of various biochemical systems.

Most recently, Townsend and Robins (1969) have reported the mass spectra of formycin, formycin B, and showdomycin. Although the structural elucidation of formycin and formycin B was accomplished without the use of mass spectrometry, Townsend and Robins have shown that the initial structural elucidation could have been facilitated by mass spectrometry. The fragmentation pattern for formycin is shown in Figure 9.3. A marked dif-

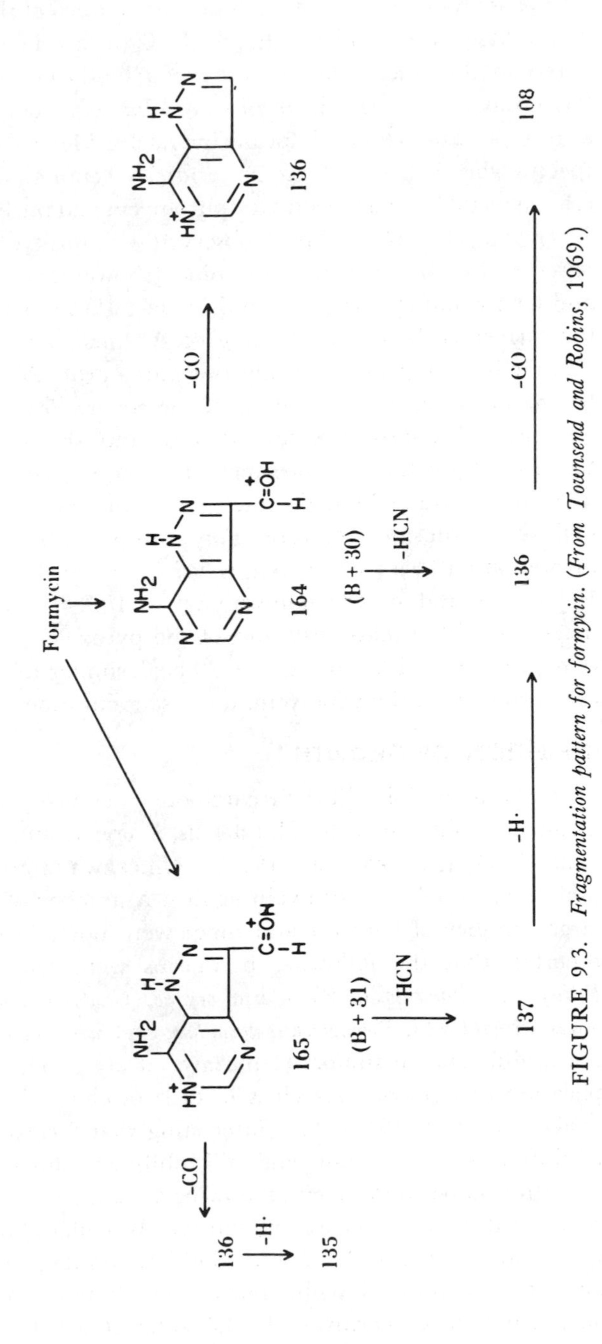


FIGURE 9.3. *Fragmentation pattern for formycin.* (*From Townsend and Robins*, 1969.)

ference is observed in the fragmentation pattern of the nucleoside antibiotics with a base-ribosyl linkage that is C—C (formycin and formycin B) as compared to the nucleosides that have a base-ribosyl linkage that is C—N (adenosine). The base peak observed for formycin and formycin B occurs at B + 30. The base peak for purine nucleosides is B + 1 or B + 2 (see mass spectra for 3′-deoxyadenosine and 3′-acetamido-3′-deoxyadenosine and references to the mass spectra for puromycin and purine nucleosides, pp. 5, 55, 77 and 88). The B + 30 peak observed with formycin and formycin B is also observed for the nucleoside antibiotic showdomycin (Chap. 11, Page 395) and for pseudouridine (Rice and Dudek, 1969). These results indicate that the carbon–carbon bond of the glycosyl linkage is not ruptured. It appears that the B + 30 peak is the aglycone plus a protonated formyl group which is the result of the fragmentation of the sugar. From the mass spectral data obtained with formycin, formycin B, and showdomycin, Townsend and Robins suggest that the presence of a major M-103 peak for an unknown nucleoside (where M is the molecular ion) might be strong evidence of a C-linked pentoside structure. For example, they reported that the fragmentation pattern for formycin is 267 − 103 = 164; for formycin B, 288 − 103 = 165, and for showdomycin 229 − 103 = 126. It will be of interest to determine if the mass spectrum of the pyrazolo *C*-nucleoside pyrazomycin (Chap. 10, Sect. 10.1) has a B + 30 peak similar to the *C*-nucleosides formycin, formycin B, showdomycin, and pseudouridine.

## INHIBITION OF GROWTH

Formycin inhibits Ehrlich carcinoma in mice, mouse leukemia L-1210, Yoshida rat sarcoma cells, HeLa cells, *X. oryzae*, and *Mycobacterium* 607 (Hori et al., 1964; Ishizuka et al., 1968b). Umezawa et al. (1965) reported on the inhibitory activity of formycin against a number of organisms. Although a large number of bacteria and fungi were not inhibited by formycin, they reported that the following organisms were sensitive to this antibiotic: *Pellicularia filamentosa*, *Piricularia oryzae*, *Cladospolium sphaerospemum*, *Blastomyces dermantitidis*, *Pseudomonas dacunhae*, and *Sclerotium rolfsii*. Formycin B does not inhibit animal tumors (Umezawa et al., 1965), but does inhibit multiplication of influenza $A_1$ virus in cells of chick chorioallantoic membrane (Takeuchi et al., 1966). It is interesting that formycin B, which showed no inhibition of mammalian cells, still inhibited influenza virus.

Tritium-labeled formycin, following subcutaneous injection into mice, was concentrated in the spleen and kidney (Ishizuka et al., 1968b). Formycin B was found in the peritoneum, bladder, kidney, and spleen. It caused a marked inhibition of white blood cells. There was a rapid oxidation of formycin B to oxoformycin B (Ishizuka et al., 1968b). The inhibition of growth following modification of the functional groups on the pyrazolo-pyrimidine nucleoside was also reported (Kunimoto et al., 1968).

Coformycin, an inhibitor of adenosine deaminase and a compound isolated from the culture filtrates of *N. interforma* (Niida et al., 1967), increased the toxicity of formycin (Sawa et al., 1967b; Umezawa et al., 1967). Coformycin shows a synergistic effect with formycin. Since coformycin inhibits the deamination of formycin in the medium, the effect of formycin on Yoshida rat sarcoma cells is markedly increased (Sawa et al., 1967a). The $LD_{50}$ for intravenous or intraperitoneal injection into mice was 250–500 mg/kg.

## BIOSYNTHESIS

Sawa et al. (1968) reported on the amination of formycin B to form formycin. These workers have utilized several interesting and useful biochemical techniques in their studies on the conversion of formycin B to formycin. First, they compared the conversion of formycin B to formycin by low- and high-producing formycin strains of *N. interforma*. The second technique utilized was based on the report that sodium fluoride inhibits adenylosuccinate synthetase. Copper sulfate was also used because of its reported stimulation of the phosphorylating enzymes. A combination of sodium fluoride and copper sulfate was used to inhibit phosphatase activity and permit the isolation of 5′-phosphorylated intermediates. The data on the amination of formycin B to yield formycin with low-producing and high-producing strains of *N. interforma* are shown in Table 9.1. The amination reaction predominated in the metabolism of Strain b. With Strain a, the major product was oxoformycin B. These data strongly suggested that the biosynthesis of formycin proceeded via formycin B. The addition of hadacidin to Strain b completely blocked the conversion of formycin B to formycin. Since hadacidin is a known inhibitor of adenylosuccinate synthetase, these data strongly suggested that the biosynthesis of formycin proceeded via formycin B.5′-monophosphate. When tritium-labeled formycin B was added to Strain b of *N. interforma* in the presence of sodium fluoride and copper sulfate, formycin B 5′-monophosphate was isolated following paper electrophoresis. Very little formycin 5′-monophosphate was detected. These data demonstrated that the conversion of radioactive formycin B to radioactive formycin was severely inhibited. When sodium fluoride and copper sulfate were added 60 min after the addition of tritium-labeled formycin B, the accumulation of formycin 5′-monophosphate was observed in the acid-soluble pool.

The formation of oxoformycin B was inhibited by allopurinol (4-hydroxypyrazolo[3,4-*d*]pyrimidine), a known inhibitor of xanthine oxidase. Although these data suggested that the conversion of formycin B to oxoformycin B required xanthine oxidase, formycin B was not oxidized when a purified

**Table 9.1**

Metabolic Conversion of Formycin B to Formycin and Oxoformycin B

| | | $\frac{{}^{3}\text{H as Formycin A or Oxoformycin B}}{{}^{3}\text{H recovered in the medium}} \times 100\,\%$ | | | | |
|---|---|---|---|---|---|---|
| Strain | Conversion to | 3 days | 4 days | 5 days | 6 days | 7 days |
| Low productivity | Formycin A | 6.7 | 3.0 | 1.1 | — | 0 |
| (a) | Oxoformycin B | 65.4 | 73.0 | 73.7 | — | 28.4 |
| High productivity | Formycin A | 5.7 | 42.1 | 73.0 | 17.6 | — |
| (b) | Oxoformycin B | 2.1 | 11.6 | 16.1 | 8.7 | — |

From Sawa et al., 1968.

preparation of xanthine oxidase was used. This observation is consistent with the studies of Lettre et al. (1967), in which they found that 9-ribosylpurines were not substrates for xanthine oxidase. The mechanism by which formycin B is enzymatically converted to oxoformycin B has been shown to require aldehyde oxidase from liver (Tsukada et al., 1969; Sheen, Martin and Parks, 1970).

## 9.2. FORMYCIN B

### DISCOVERY, PRODUCTION, AND ISOLATION OF FORMYCIN B OR LAURUSIN

This nucleoside antibiotic was isolated from the culture filtrates of *N. interforma* (Koyama and Umezawa, 1965) and *Streptomyces lavendulae* (Aizawa et al., 1965). The culture medium for the production of formycin B from *N. interforma* is the same as that described for the production of formycin. The procedure described here for the isolation of formycin B is that described by Koyama and Umezawa (1965). The crude powder containing formycin and formycin B were concentrated and added to a Sephadex G-25 column. The effluent containing formycin B was concentrated to a small volume and added to a Dowex-50-$H^+$ column. Formycin B appeared in the aqueous effluent and yielded a crystalline product.

### PHYSICAL AND CHEMICAL PROPERTIES

Formycin B melts at 254–255°C (Aizawa et al,.1965). The molecular formula is $C_{10}H_{12}N_4O_5$. The molecular weight is 268. It is soluble in water and insoluble in organic solvents. The p$K_a$ is 8.8 (acidic); $[\alpha]_D^{20} = -51.5°$ ($c = 1\%$ in water). Formycin B shows the following ultraviolet maxima: 221 m$\mu$ ($E_{1\,\text{cm}}^{1\%} = 543$)

and 276 mμ ($E^{1\%}_{1\,cm}$ = 300) in 0.1 *N* HCl; 219 mμ ($E^{1\%}_{1\,cm}$ = 348) and 280 mμ($E^{1\%}_{1\,cm}$ = 294) in water; 230 mμ ($E^{1\%}_{1\,cm}$ = 643) and 292 mμ ($E^{1\%}_{1\,cm}$ = 338) in 0.1 *N* NaOH (Koyama and Umezawa, 1965).

### PROOF OF STRUCTURE

The formula for formycin B strongly suggests that this nucleoside antibiotic might be a deaminated product of formycin. Umezawa et al. (1965) reported that treatment of formycin with nitrite yielded formycin B. Laurusin and formycin B are also known to be identical. When formycin was added to the culture filtrates of the formycin-producing strain (*N. interforma*), about 75% of the added formycin was converted to formycin B after 10 days.

The structural elucidation of formycin B or laurusin was determined by the use of X-ray analysis, chemical degradation, ir, and nmr (Koyama et al., 1966; Kawamura et al., 1966; Watanabe et al., 1966; Robins et al., 1966). Kawamura et al., (1966) degraded formycin B and isolated the aglycone using the procedure as described for the degradation of formycin. These studies provided unequivocal evidence that the structure of formycin B or laurusin is 3-(β-D-ribofuranosyl)-pyrazolo[4,3-*d*]-6(H)-7-pyrimidone.

### INHIBITION OF GROWTH

Formycin B inhibited the pathogenic bacterium for rice plant disease called "Shirahagare" caused by *X. oryzae* (Aizawa et al., 1965). To inhibit *X. oryzae* in rice plants, 3.12 μg/ml of formycin B was required; for formycin 0.78 μg/ml was required (Koyama and Umezawa, 1965). Similarly, formycin was much more toxic than formycin B in inhibiting the growth of *P. filamentosa*. Formycin B did not inhibit Ehrlich carcinoma or L-1210 (Ishizuka et al., 1968b). Umezawa et al. (1965) studied the inhibitory action of formycin B. It did not inhibit bacteria, fungi, HeLa cells, or Yoshida rat sarcoma cells. Except for the protective effect against infection of *X. oryzae* in rice plants, formycin B was not an inhibitor. However, it has shown antiviral properties against influenza virus (Takeuchi et al., 1967; Kunimoto et al., 1968). Apparently the type substituent on position 7 of the pyrazolo [4,3-*d*]pyrimidine ring is in large part responsible for the biological activities of these nucleoside antibiotics.

The $LD_{50}$ in mice following the intravenous injection of formycin B was about 1000 mg/kg (Takeuchi et al., 1967). Thus, formycin B is 2–4 times less toxic than formycin.

## 9.3. OXOFORMYCIN B

### DISCOVERY, PRODUCTION, AND ISOLATION

Oxoformycin B, 3-(β-D-ribofuranosyl)-pyrazolo[4,3-*d*]-4(H),6(H)-5,7-pyrimidione, is the xanthosine analog of the formycin family. Formycin B is converted to oxoformycin B by *N. interforma*, *Pseudomonas fluorescens*, and *Streptomyces kasugaensis* (Sawa et al., 1968). Oxoformycin B has been found in the urine of mice and rabbits following the injection of formycin or formycin B (Sheen et al., 1968b; Ishizuka et al., 1968a). Tsukada et al. (1969) and Sheen, Martin, and Parks (1970) have now shown that hepatic aldehyde oxidase is responsible for the oxidation of formycin B to oxoformycin B. Since no other metabolic product appeared in the urine, oxoformycin B was assumed to be an end product of detoxification.

The medium for the production of oxoformycin B from *N. interforma* was the same as that reported for the production of formycin (Hori et al., 1964).

## CONVERSION OF FORMYCIN B TO OXOFORMYCIN B

Oxoformycin B has been isolated from two strains of *N. interforma* and from the urine of mice and rabbits (Sawa et al., 1968; Sheen et al., 1968b; Ishizuka et al., 1968a). The conversion of formycin B to oxoformycin B was not immediately obvious since there is no known direct oxidation of inosine to xanthosine. The possibility that xanthine oxidase was oxidizing formycin B to oxoformycin B seemed unlikely for two reasons. First, formycin B is not cleaved by purine nucleoside phosphorylase to the hypoxanthine analog, and, second, nucleosides are not substrates for this enzyme (Lettre et al., 1967). A possible role for xanthine oxidase was eliminated when allopurinol, 1H-pyrazolo[3,4-*d*]pyrimidine-4-ol, a known xanthine oxidase inhibitor, was administered simultaneously with formycin B to mice. Oxoformycin B was isolated from the urine (Sheen, Martin, and Parks, 1970). Sawa et al. (1968) also showed that a purified preparation of xanthine oxidase did not oxidize formycin B to oxoformycin B. Tsukada et al. (1969) and Sheen, Martin, and Parks (1970) subsequently reasoned that oxoformycin B might be formed by aldehyde oxidase from liver. By using a partially purified preparation of rabbit liver hepatic oxidase, they were able to show the enzymatic conversion of formycin B to oxoformycin B. The $K_m$ reported by Sheen, Martin, and Parks (1970) is $2.0 \times 10^{-4}$ *M*. Formycin B is a competitive inhibitor of *N*′-methyl nicotinamide. The $K_i$ for formycin B is $1.6 \times 10^{-4}$ *M*. Inosine is neither a substrate nor an inhibitor for aldehyde oxidase (Sheen, Martin, and Parks, 1970). However, these same investigators showed that xanthine oxidase is competitively inhibited by formycin and formycin B with a $K_i$ of $1.3 \times 10^{-5}$ *M*. The aglycone of formycin B (7-hydroxypyrazolo[4,3-*d*]pyrimidine) and 7-hydroxy-3-methylpyrazolo[4,3-*d*]pyrimidine were poor substrates for aldehyde oxidase and were not substrates for xanthine oxidase (Sheen, Martin, and Parks, 1970). Using the same enzymes, Johns, Spector, and Robins (1969) reported that the unsubstituted isomer, pyrazolo[3,4-*d*]pyrimidine was a rapidly oxidized substrate for both enzymes. The enzymatic oxidation *in vitro* of 4-hydroxypyrazolo[3,4-*d*]pyrimidine to 4,6-dihydroxypyrazolo

[3,4-*d*]pyrimidine by aldehyde oxidase and xanthine oxidase was very slow, whereas the *in vivo* oxidation of pyrazolo[3,4-*d*]pyrimidine is very rapid. The present evidence indicates that 6-hydroxylation occurs by xanthine oxidase (Elion et al., 1966). Sheen, Martin, and Parks (1970) compared the subtle differences in the biochemical properties of the formycins with those of the structurally similar purine ribonucleosides, such as adenosine, inosine and xanthosine. The C—C riboside bond of formycin is 1.55 Å (Ward and Reich, 1968) whereas the C—N riboside bond of adenosine is 1.47 Å. This longer bond distance between the aglycone and the ribose moieties of the formycins allows an unhindered rotation of the ribofuranosyl group of almost 360°. In contrast, the rotation of the ribofuranosyl moiety of inosine is markedly restricted. The differences in conformation between inosine and formycin B appear to allow formycin B to assume a conformation that can adapt itself to the active sites of aldehyde oxidase and xanthine oxidase, while the natural nucleosides are not capable of such changes (Sheen, Martin, and Parks, 1970; Parks, private communication). In addition to these conformational differences, additional factors, such as the electronic properties of the pyrazolopyrimidine ring system or the purine ring system, must also be important.

To study the conversion of tritium-labeled formycin B to oxoformycin B by *N. interforma*, the cells were centrifuged, washed and suspended in glucose-Czapek media and shaken for 3 hr at 27°C (Sawa et al., 1968). Two strains of *N. interforma* were used for these studies. One strain (Strain a) showed poor production of formycin. Strain b was a good producer of formycin. The low-producing Strain a of *N. interforma* could be obtained by repeated passages or prolonged storage of the producing organism in agar slant cultures. When tritium-labeled formycin B was added to Strain b, the amination reaction predominated and formycin was the main product (Sawa et al., 1968). The low producing formycin culture (Strain a) had a limited rate of amination. The major product isolated was oxoformycin B. The pathway proposed by Sawa et al. (1968) for the biosynthesis of oxoformycin B by *N. interforma* is as follows:

Formycin B 5′-monophosphate → formycin A 5′-monophosphate (1)

Formycin A 5′-monophosphate → formycin A (2)

Formycin A → formycin B (3)

Formycin B → oxoformycin B (4)

## INHIBITION OF GROWTH

Oxoformycin B was not inhibitory against Yoshida sarcoma, *X. oryzae*, or influenza virus (Kunimoto et al., 1968). Following the subcutaneous injection of formycin B, oxidation to oxoformycin B occurred rapidly in all organs except the peritoneum (Ishizuka et al., 1968b).

## 9.4. COFORMYCIN

### DISCOVERY, ISOLATION, AND PRODUCTION

Coformycin was isolated together with formycin from the culture filtrates of *N. interforma* and *S. kaniharaensis* SF-557. The latter culture was isolated from a soil sample in Hiroshima, Japan (Niida et al., 1967; Sawa et al., 1967a; Tsuruoka et al., 1967). Coformycin was isolated from the culture broth by passing the filtrate through a column of Amberlite IR-120-$NH_4^+$. The column was washed with water, and coformycin was eluted with 0.01 *N* $NH_4OH$. The fractions containing the compound were collected, concentrated, and stored at 5°C. Formycin crystallized. The mother liquor was added to a carbon column and coformycin was eluted with 50% aqueous acetone. Formycin crystallized on concentration of the fractions. DEAE-sephadex A-25 ($OH^-$) was used to separate coformycin and formycin. Coformycin was eluted with water, while formycin remained on the column. Following concentration, purification on a Dowex 50-$NH_4^+$ column, and concentration of the eluant, white needle crystals were obtained (Tsuruoka et al., 1967).

### PHYSICAL AND CHEMICAL PROPERTIES

The molecular formula for coformycin is $C_{11}H_{16}N_4O_5$; mp 182–184°C; $[\alpha]_D^{24°} = +34°$ ($c = 1\%$ in water). The $\lambda_{max}$ in water is 282 m$\mu$ ($E_{1\,cm}^{1\%}$ 290); $\lambda_{max}$ in 0.1 *N* HCl is 264 m$\mu$ ($E_{1\,cm}^{1\%}$ 257); $\lambda_{max}$ in 0.1 *N* NaOH is 284 m$\mu$ ($E_{1\,cm}^{1\%}$ 290); p$K_a$ is 5.3 (Tsuruoka et al., 1967). Coformycin is stable at pH 7–9, but unstable in acid solution. Acid hydrolysis of coformycin yields ribose (Niida et al., 1967).

The data suggest that coformycin has a close structural relationship to a ribofuranoside. Apparently, coformycin does not have the same structure as pyrazomycin, the pyrazolo nucleoside recently isolated by Williams et al. (1969) (see Chapter 10.1, Pyrazomycin).

### INHIBITION OF GROWTH

Coformycin inhibits the deamination of adenosine and formycin by adenosine deaminase (Sawa et al., 1967a). Coformycin showed a strong synergistic effect with formycin on inhibiting the growth of bacteria, except *X. oryzae*. Coformycin alone does not exhibit antibacterial activity (Shomura et al., 1967). In the presence of formycin it is a strong inhibitor of gram-negative bacteria.

### MODE OF ACTION

Coformycin with formycin shows a synergistic effect on the inhibition of growth of Yoshida rat sarcoma cells. This synergistic effect of coformycin is attributed to the inhibition of adenosine deaminase. When coformycin and formycin were added to cells of *E. coli*, there was no deamination of formycin. Without coformycin, formycin was rapidly deaminated to form formycin B. Similar observations were made with Ehrlich carcinoma cells. Therefore, it appears that coformycin blocks the deamination of formycin by bacteria and tumor cells, which may indeed explain the synergism with formycin on inhibition of deamination in these cells. Coformycin also inhibited the deamination of adenosine and formycin when purified adenosine deaminase was used. Sawa et al. (1967a) studied the mode of inhibition of coformycin with purified adenosine deaminase. The mode of inhibition of this enzyme by coformycin was the mixed type. The $K_i$ value of coformycin for adenosine was $6.5 \times 10^{-8}$ *M* and for formycin, $1.9 \times 10^{-8}$ *M*. It will be of interest to see if coformycin is a 6-substituted derivative of a purine nucleoside.

## 9.5. BIOCHEMICAL PROPERTIES OF FORMYCIN AND FORMYCIN B

Formycin is a cytotoxic analog of adenosine. The experimental evidence obtained strongly suggests that the action of the pyrazolopyrimidine nucleoside on the neoplastic cell occurs at the nucleotide level. Formycin 5′-triphosphate has been proposed by Caldwell et al. (1967), Henderson et al. (1967), Acs et al. (1967), Ward et al. (1969c) and Ward, Fuller, and Reich (1969b) as the form of this nucleoside antibiotic which interferes with either purine nucleotide metabolism or causes its cytotoxic effect by intefering with RNA and protein synthesis. Since the conversion of formycin to formycin B by adenosine deaminase results in a nucleoside antibiotic that is less toxic or even inactive against tumors, deamination may explain the resistance of organisms to this compound (Sawa et al., 1967b). In addition, since the 5′-mono-, di-, and triphosphates of formycin appear to be essential for inhibition, the absence of a kinase in the cell should play a very important role in determining the level of toxicity of this adenosine analog. Caldwell et al. (1969) have shown that formycin is rapidly phosphorylated by Ehrlich ascites tumor cells *in vitro*. The 5′-triphosphate is the major product when adequate energy is supplied. The 5′-mono- and diphosphates also form. The concentration of formycin phosphates approached that of adenosine phosphates. The ratio of the deaminated analog (formycin B) to phosphorylation was about 5 to 1. Caldwell et al. (1966, 1967) have reported studies on a formycin-resistant subline of Ehrlich ascites tumor cells that lacked adenosine kinase. The ability of the nucleoside analogs to inhibit purine synthesis *de novo* by Ehrlich ascites carcinoma cells and the resistant subline is shown in Table 9.2. The data show that the resistant cells (EAC-R2 cells) demonstrate less inhibition of purine synthesis *de novo* with formycin than the parent line of sensitive cells (EAC cells). The degree of inhibition in both tumor lines with adenine was essentially the same. Adenosine kinase and adenosine deaminase appear to be two important enzymes in bacteria and mammalian tumors responsible for resistance to formycin. Indeed, Umezawa et al. (1967) reported that the simultaneous addition of coformycin, a potent inhibitor of adenosine deaminase, broadened the antibiotic spectrum of formycin.

**Table 9.2**

Inhibition of Purine Synthesis *de novo* in Tumor Cells

| Compound (1 m*M*) | Inhibition (%) of FGAR accumulation | |
|---|---|---|
| | In EAC cells[a] | In EAC-R2 cells[c] |
| Adenine | 88 | 86 |
| Me6MPR[b] | 88 | 9 |
| 2′-Deoxyadenosine | 82 | 68 |
| 6-Methoxypurine ribonucleoside | 92 | 33 |
| Purine ribonucleoside | 84 | 14 |
| 6-Methylaminopurine ribonucleoside | 88 | 33 |
| Formycin | 98 | 10 |

From Caldwell et al., 1967.

[a] EAC cells: parent sensitive Ehrlich ascitic cells; parent tumor line sensitive to nucleoside analogs.

[b] 6-(Methylmercapto)purine ribonucleoside.

[c] EAC-R2 cells: Ehrlich ascites cells resistant to formycin.

Bennett and his co-workers reported that replacement of the imidazole ring of purines with the pyrazole ring resulted in nucleosides that were good substrates for adenosine kinase from human tumor cells (Schnebli et al., 1967). Apparently these two enzymes cannot distinguish the *C*-riboside bond from the normal *N*-riboside bond. In attempts to find more potent chemotherapeutic agents, Kunimoto et al. (1968) studied the biological activity of structurally modified formycin. Assuming that adenosine kinase and adenosine deaminase play key roles in the toxicity of formycin in tumor cells, they synthesized and studied the 7-*N*-methyl derivative of formycin. If toxicity were dependent on phosphorylation by adenosine kinase, the 7-*N*-methylated formycin should have been a good inhibitor, since these type derivatives are known to be readily phosphorylated, but very poorly deaminated. Contrary to expectations, the 7-*N*-methyl formycin was inactive against Yoshida sarcoma cells and influenza virus. A possible explanation for the inactivity of the *N*-methyl derivatives of formycin may be that they can be phosphorylated to the 5′-monophosphate, but not to the di- and triphosphates. Modification on the ribose moiety of formycin, except the phosphorylated derivatives, resulted in marked loss of activity. The inhibition of the phosphorylated formycin is probably attributed to the removal of the phosphate by phosphatase before this nucleoside entered the cells.

Since cordycepin 5′-monophosphate had been reported by Overgaard-Hansen (1964) to inhibit the formation of 5-phosphoribosyl-1-pyrophosphate, Sawa et al. (1965) suggested that formycin might have a similar mode of action. They studied the enzymatic phosphorylation of formycin and formycin B.

Formycin and formycin B were added to washed cells of *Serratia marcescens.* The phosphate donor was *p*-nitrophenylphosphate. Formycin was phosphorylated to the 5′-monophosphate. Formycin B was not phosphorylated. A similar reaction involved the transfer of phosphate from *p*-nitrophenylphosphate to the 2′- or 3′-hydroxyl group of formycin or formycin B by a strain of *Proteus mirabilis.* Both these nucleosides were phosphorylated. 5′-Nucleotidase and 3′-nucleotidase hydrolyzed the 5′- and 3′-monophosphates of formycin and formycin B. The biological activities of the formycin phosphates against tumor cells, bacteria, and fungi were also reported.

The importance of the phosphorylation of formycin and/or its deamination to formycin B was reported by Umezawa et al. (1967) and Ishizuka et al. (1968b). The presence of glucose increased the incorporation of formycin into intact cells of Ehrlich carcinoma. These results indicated that phosphorylation of formycin stimulated the transport of this nucleoside into the cells. The synergism of coformycin with formycin was attributed to the inhibition of the deamination of formycin. With *E. coli*, formycin was rapidly deaminated to formycin B. When coformycin was added, formycin became effective as an inhibitor to *E. coli* because deamination of formycin to formycin B was inhibited.

The effect of formycin on the synthesis of protein, RNA, and DNA on HeLa $S_3$ cells in synchronized culture has been studied by Kunimoto et al. (1967). The inhibition observed for formycin was compared with that of phleomycin and bleomycin. At high concentrations (10 μg/ml), formycin inhibited DNA synthesis. At low concentrations (0.1 μg/ml), formycin inhibited protein synthesis by 25%, but did not inhibit RNA synthesis. Cell division was inhibited about 40% by formycin at a concentration of 0.1 μg/ml. In order to dissociate the effects on cell division by formycin from the effects on DNA synthesis, the antibiotic was added at various stages of DNA synthesis. At 0.1 μg/ml, formycin caused an appreciable inhibition of cell division when added at 3 hr, but little inhibition of cell division when added at 9 hr. At this low concentration, formycin did not inhibit DNA synthesis. Kunimoto et al. (1967) stated that two factors are indispensable for cells to enter mitosis. Formycin appears to inhibit the one factor that is important (between 3 and 9 hr) in the early stage of the DNA synthesis phase. Formycin does not appear to inhibit the second factor which acts just before or during the mitotic phase.

The biochemical effects of formycin on purine metabolism in Ehrlich ascites tumor cells were studied by Henderson et al. (1967). They reported that the doses of formycin used to inhibit the tumor cells did not cause a loss of weight of the animals studied. Formycin was shown to interfere with nucleotide synthesis by depressing the synthesis of FGAR and PRPP. Formycin probably acts at the triphosphate level by competing for ATP to inhibit 5-phosphoribose pyrophosphokinase. A similar type inhibition has been reported for 3′-deoxyadenosine 5′-triphosphate, tubercidin, psicofuranine, and decoyinine. When a subline of Ehrlich ascites tumor cells, unable to phosphorylate formycin, was used, the inhibition of FGAR synthesis dropped from 78.8% in the parent line of cells to 7.8% in the resistant subline. This was taken as presumptive evidence that formycin acts as an inhibitor at the nucleotide level.

The effects of formycin on the purine utilization are shown in Table 9.3. Formycin did not affect the conversion of adenine to the adenine and guanine nucleotides. However, nucleotide synthesis from hypoxanthine and guanine were both inhibited. Henderson et al. (1967) suggested that the inhibition of nucleotide synthesis from guanine and hypoxanthine may be attributed either to PRPP deficiency or to the direct inhibition of phosphoribosyltransferase. Since guanine-hypoxanthine phosphoribosyltransferase is inhibited by ATP, this enzyme may also be sensitive to FTP. Parallel experiments showed that formycin did not alter the distribution of radioactivity into any one fraction. Formycin was not able to serve as a ribosyl donor for

**Table 9.3**

Effect of Formycin on Purine Utilization

| Purine | Formycin | Total acid-soluble nucleotide, cpm | Acid-soluble | | Nucleic acid | |
|---|---|---|---|---|---|---|
| | | | Adenine, cpm/μg | Guanine, cpm/μg | Adenine, cpm/μg | Guanine, cpm/μg |
| Adenine | − | 10,720 | 16,690 | 1090 | 1470 | 131 |
| Adenine | + | 9,620 | 15,340 | 945 | 1140 | 101 |
| Guanine | − | 7,190 | 84.9 | 447 | 7.54 | 83.4 |
| Guanine | + | 3,080 | 29.6 | 205 | 4.41 | 49.8 |
| Hypoxanthine | − | 12,070 | 10,830 | 1140 | 110 | 28 |
| Hypoxanthine | + | 4,930 | 3,530 | 657 | 61 | 21 |

From Henderson et al., 1967.

inosine synthesis, nor did it inhibit the guanosine-supported synthesis of inosine or the hypoxanthine-inosine exchange. Formycin (0.1 m*M*) inhibited the incorporation of methionine and lysine into protein by about 38%. The incorporation of glycine was not inhibited. The reason for the selective inhibition is not known.

Formycin 5′-monophosphate was a very potent inhibitor of adenylate synthesis by adenine phosphoribosyltransferase (Table 9.4). Other phosphorylated derivatives of formycin were not inhibitors. Formycin B 5′-phosphate was about as active as IMP in this system. The purine nucleoside kinase activity of Ehrlich ascites tumor cell extracts was not inhibited by formycin or formycin B. The corresponding 5′- or mixed 2′- and 3′-phosphates were not inhibitors. ATP concentrations in control and formycin-treated cells were the same, which is in agreement with the report of Hori et al. (1964) in which they stated that formycin did not inhibit glycolysis in Yoshida ascites sarcoma cells. Apparently formycin does not act by causing changes in energy metabolism.

**Table 9.4**

Effects of Formycin and Its Derivatives on Adenine Phosphoribosyltransferase[a]

| Inhibition | Concentration, m*M* | Inhibition, % |
|---|---|---|
| Formycin | 2 | 15 |
| | 0.1 | 5 |
| Formycin 5′-phosphate | 1 | 93 |
| | 0.1 | 78 |
| Formycin 2′- and 3′-phosphate | 1 | 41 |
| | 0.1 | 30 |
| Formycin B 5′-phosphate | 1 | 46 |
| | 0.1 | 10 |
| Formycin B 2′-and 3′-phosphate | 1 | 34 |
| | 0.1 | 17 |

From Henderson et al., 1967.

[a] One microgram of enzyme protein was incubated for 12 min at 30°C with 1 m*M* $MgSO_4$, 1 μ*M* adenine-$^{14}C$, 0.01 m*M* PRPP, 0.12 *M* Tris buffer (pH 7.4), and formycin derivatives in a total volume of 0.50 ml. Figures are means of results of two experiments.

More recently, Ikehara et al. (1968, 1969a) reported on the synthesis of formycin triphosphate and its incorporation into ribopolynucleotide by DNA-dependent RNA polymerase from *E. coli* strain B. It was shown that formycin 5′-triphosphate could replace ATP when DNAs from calf thymus, salmon sperm, $T_2$, and *E. coli* were the templates. The rate of synthesis of the formycin-containing RNA was about 47–68% that of RNA with the normal nucleoside triphosphates. When poly d(A-T), poly d(T-C)·d(A-G), and poly d(T-G)·d(A-C) were the templates, the amounts of poly (F-U), poly (F-G), and poly (F-C) synthesized were 56, 94, and 67%, respectively, of the amounts of the corresponding poly(A-U), poly(A-G), and poly(A-C) synthesized. Proof for the incorporation of FTP in place of ATP was confirmed by digestion of the ribopolynucleotide with bovine pancreatic or $T_1$ ribonuclease. The authors stated that the incorporation of tubercidin 5′-triphosphate and formycin 5′-triphosphate into ribopolynucleotides (as ATP analogs) with the DNA-dependent RNA polymerase system was not only directed by the Watson-Crick type base-pairing, but was also dependent on the neighboring base sequences. With formycin-5′-triphosphate the selectivity was not as strict as in the case of tubercidin 5′-triphosphate. Poly d(T-C)·d(A-G) was the most efficient template for tubercidin and formycin. It was suggested that the incorporation of formycin is regulated by the sequence -CTC- in the template and that the neighboring C residue exerts a stabilizing effect for the formation of enzyme-template-substrate, while the G residue does not stabilize this complex. An alternative interpretation for the "regulatory sequences" as proposed by Ikehara et al. (1968, 1969a) would be that the F,C repeating sequence prefers (more than F,G) the *anti* conformation. This would be supported by the fact that pyrimidines (e.g., C) are conformatively much more restricted in rotation at the glycosidic bond than purines (e.g., G) (Ward, private communication).

The binding of histidine-tRNA and threonine-tRNA by poly (F-C) was 40–80% that observed with poly (A-C) (Ikehara et al., 1968). It appears that the coding properties of ribopolynucleotides containing formycin are the same as ribopolynucleotides containing adenosine. When tubercidin replaced adenosine in chemically synthesized trinucleotides, Ikehara and Ohtsuka (1965) found that the trinucleotides containing tubercidin were as effective as stimulators of lysine-tRNA and threonine-tRNA binding to the ribosomes as were the corresponding normal trinucleotides.

Although FTP is an excellent substrate for *E. coli* RNA polymerase by supporting rates of RNA synthesis 47–94% that reported for ATP, FTP was not a substrate for Mengovirus polymerase. Similarly, Kapuler and Spiegelman (1970) reported that with $Q\beta$ replicase, FTP could only replace ATP by 1–2%. Kapuler et al. (1969) suggested that the RNA virus-induced polymerases cannot utilize FTP in place of ATP, while the DNA-dependent RNA polymerase freely replaces ATP by FTP. Apparently there are fundamental differences between the two polymerases.

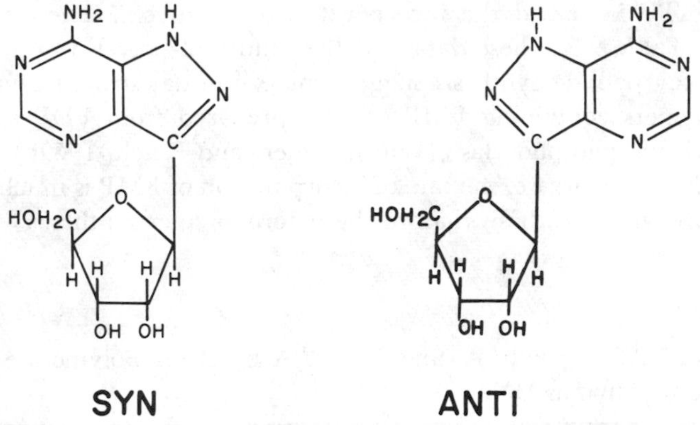

FIGURE 9.4. *The structure of formycin in the syn and anti conformation.*

While studying the formation of formycin polymers, Ward and Reich (1968) observed that these polymers had numerous abnormal properties. They concluded that the anomalous behavior of formycin polymers may be attributed to the conformational properties that are unique to formycin residues in enzymatically synthesized polyribonucleotides. They proposed that formycin residues in ribopolymers exist in either the *syn* or the *anti* conformation (Fig. 9.4). On the basis of the data obtained, they concluded that the formycin units are: "(*1*) *anti* in ordered structures of the Watson-Crick-type; (*2*) probably a mixture of *syn* and *anti* in single-stranded copolymers; (*3*) entirely *syn* in neutral polyformycin." The behavior of the alternating copolymer of formycin and pseudouridine suggested that the pseudouridine residues in polymers may also undergo analogous conformational transitions.

While copolymers containing formycin behave normally, polyformycin failed to code for the synthesis of any polypeptide. Another abnormal property is that formycin 5′-triphosphate completely and efficiently replaces ATP for RNA synthesis when directed by different naturally occurring and synthetic DNA templates, provided at least one additional nucleotide is being polymerized. The nearest-neighbor sequence of the products formed with formycin 5′-triphosphate is identical with that synthesized from ATP. The reactions with formycin 5′-triphosphate proceed at velocities equal to 30–80% those with ATP. Homopolymer (polyformycin) synthesis, when directed by denatured DNA, proceeded at a rate 0.5% that of poly A synthesis. Ward and Reich (1968) and Ward et al. (1969c) reported that poly F was not formed with poly U or poly dT as the template. Under these same conditions, poly A would be formed. FTP was effectively used by RNA polymerase when both poly U and poly dT or denatured DNA were the templates, provided some ATP was present. Increasing initial concentrations of ATP increased the incorporation of formycin. These results are shown in Table 9.5. These data and the failure of poly F to function as a template for peptide synthesis suggest some unusual structure for poly F. Homopolymers containing FMP can be prepared from FDP and *E. coli* polynucleotide phosphorylase (Ward, Fuller, and Reich, 1969b). As in the case of RNA polymerase, the rate of incorporation of FMP is much slower in the homopolymer synthesis than in the heteropolymer synthesis.

**Table 9.5**

Synthesis of Poly F, Poly A, and Poly F,A by RNA Polymerase with Denatured Calf Thymus DNA Template[a]

| Substrate conc., mμmoles/ml | | Input ratio | Mμmoles NMP Incorporated/ml[b] | | Product ratio |
|---|---|---|---|---|---|
| ATP | FTP | ATP/FTP | AMP | FMP | AMP/FMP |
| — | 400 | — | — | 0.40 | — |
| 10 | 390 | 1:39 | 0.98 | 0.76 | 1.29 |
| 40 | 360 | 1:9 | 1.25 | 2.20 | 0.57 |
| 200 | 200 | 1:1 | 3.29 | 2.62 | 1.26 |
| 360 | 40 | 9:1 | 23.2 | 2.88 | 8.05 |
| 390 | 10 | 39:1 | 58.2 | 1.88 | 30.6 |
| 400 | — | — | 84.0 | — | — |

From Ward and Reich, 1968.

[a] Synthesis was monitored by following the incorporation of $H^3$-FMP ($3.7 \times 10^4$ cpm/μmole) and $^{14}C$-AMP ($9 \times 10^5$ cpm/μmole) into acid-insoluble form.

[b] The reaction mixtures (0.1 ml) containing 20 mμmoles of heat-denatured DNA template were incubated at 37° C for 30 min. Poly F,A, random copolymer.

Another abnormal reaction of poly F compared to poly A is its interaction with nucleases. Compared to poly A, the depolymerization of poly F proceeds 100 times as slowly as with snake venom phosphodiesterase, 1000 times as slowly as with micrococcal nuclease, and 10,000 times as slowly as with spleen phosphodiesterase. Ward and Reich (1968) stated that the behavior of poly F with those enzymes known not to have a base specificity is indeed unique among polyribonucleotides, this was taken as additional evidence that the polymer structure must indeed be unusual. Another unexpected finding concerned the interaction of poly F with bovine pancreatic RNase (Ward, Fuller and Reich, 1969b; Ikehara, 1969c). While poly A is not a substrate for RNase, poly F is degraded quantitatively to 2′,3′-cyclic-FMP at a rate close to that reported for poly C (Ward, Fuller, and Reich, 1969b).

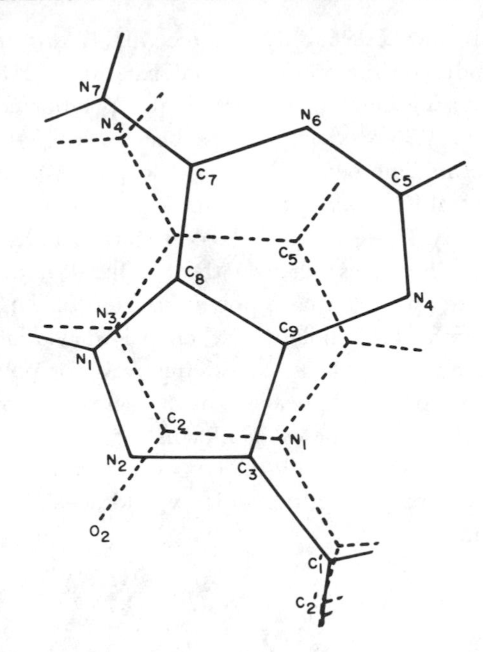


FIGURE 9.5. *Stereochemical comparison of* (——) *3′-FMP* (*syn*) *and* ( - - - ) *an H-bonding hybrid of* 3′-*CMP and* 3′-*UMP* (*anti*). *The phosphorus atoms of both nucleotides have been superimposed, but only the bases and glycosyl bonds are illustrated. The torsion angles about the glycosyl bonds differ by* 180° *in the two nucleotides. All other torsion angles are identical and are like those in ribose nucleotides of helical RNA. This permits a comparison of atomic positions using torsion angles which are stereochemically acceptable. The identity in the H-bonding groups* ($N^7$ *and* $N^1$ *of F with the* $N^4$ *of C and* $N^3$ *of U*) *can be improved, although the* 3′-*phosphate groups are then no longer in exactly the same positions.* (*From Ward, Fuller, and Reich,* 1969*b.*)

A comparison of the three-dimensional structure of FMP with CMP and UMP demonstrates that FMP can be considered as a hybrid of CMP and UMP with respect to H-bonding groups (Fig. 9.5) (Ward, Fuller, and Reich, 1969b). This notion requires that CMP and UMP have the *anti* conformation, while the formycin residues in FMP have the *syn* conformation. Evidence for the *anti* and *syn* conformations of nucleosides and nucleotides has been presented by Emerson et al. (1967), Ts'o et al. (1969), Ward and Reich (1968), Hart and Davis (1969a,b), Koyama (1966) and Miles et al. (1969a, 1969b).

FTP was not incorporated into polymer with highly purified *E. coli* DNA polymerase (Ward et al., 1969c).

While the coding properties of a copolymer of FMP and UMP are analogous to the corresponding AU polymer, poly F fails to code as a homopolymer

(Table 9.6) (Ward et al., 1969c). More recently, Ward et al. (1970) have extended their studies on the abnormal conformation of F and poly F to the physical and conformational properties of the polynucleotides containing pyrrolopyrimidines. Whereas poly F has been identified as *syn*, poly Tu appears to be intermediate between poly F and poly A (*anti*) and, therefore, assumes an abnormal "non-*anti*" conformation. Although the conformation of poly Tu and poly F are abnormal, Ward et al. (1970) have observed several important differences between the biochemical properties of these two polymers. First, the enzymatic polymerization of Tu, but not F, proceeds efficiently with RNA polymerase or polynucleotide phosphorylase. Second, poly Tu, but not poly F, codes effectively for polylysine synthesis. While poly Tu and poly F appear to have abnormal conformations, the conformation of poly To has the *anti* conformation. Since poly To retains the *anti* conformation, the cytotoxicity of To must be related to factors other than the conformation abnormalties which would account for the biological properties of F and Tu.

**Table 9.6**

Incorporation of Amino Acids into Polypeptides: Stimulation by Various Synthetic RNA Polymers

| Amino acid[a] | $\mu\mu$moles $^{14}$C-amino acid incorporated per ml |
|---|---|
| Lysine (AAA) | |
| Poly A | 1520 |
| Poly F | 0 |
| Poly (A-F) (1.9:1) | 468 |
| Poly A + poly F | 1478 |
| Poly (F-U) (1.0:1) | 0 |
| Isoleucine (AUU) | |
| Poly (A-U) | 725 |
| Poly (F-U) | 345 |
| Tyrosine (UAU) | |
| Poly (A-U) | 152 |
| Poly (F-U) | 115 |
| Phenylalanine (UUU) | |
| Poly U | 2640 |
| Poly (A-U) | 682 |
| Poly (F-U) | 472 |

From Ward et al., 1969c.

[a] The polymer concentrations (m*M*) used were: poly A, 34.2; poly F, 39.5; poly U, 33.1; poly (A-F), 31.0; poly (F-U), 29.0.

Added information concerning the unique conformational properties of formycin polymers was obtained from optical and thermal properties, fluorescence, and optical rotatory dispersion curves. The data show that poly F exists in the single-stranded state. The double-stranded polymers containing formycin residues resemble polymers containing adenosine. While the thermostability of alternating polymers of formycin is similar to that of the adenosine alternating polymers, the $T_m$ values for the homopolymer pair are substantially different (Table 9.7). Ikehara et al. (1969a) reported that the $T_m$ of poly (F-U) was the same or only slightly lower than that of poly (A-U).

**Table 9.7**

$T_m$ Values of Comparable Adenosine and Formycin Polymers

| Alternating polymers | | | | Homopolymer Pairs | |
|---|---|---|---|---|---|
| Polymer | $T_m$, °C | Polymer | $T_m$, °C | Polymer | $T_m$, °C |
| r(A-U) | 32 | r(F-U) | 33 | rA:rU (1:1) | 57 |
| r(A-FU) | 32 | r(F-FU) | 35 | rF:rU (1:1) | 22 |
| r(A-T) | 46 | r(F-T) | 48 | rA:dT (1:1) | 65 |
| r(A-BU) | 53 | r(F-BU) | 63 | rF:dT (1:1) | 37 |
| r(A-ψ) | 58 | r(F-ψ) | 32 | | |

From Ward and Reich, 1968.
[a] The $T_m$ of alternating polymers was determined in 0.001 $M$ Na citrate, and that of the homopolymer pairs in 0.01 $M$ Tris (pH 7.9)–0.1 $M$ KCl.

Evidence that poly F exists as a single-stranded structure at neutral pH was determined by fluorescent studies. Final proof for the anomalous behavior of poly F in enzymatic systems was obtained from the ORD spectrum. Ward and Reich (1968) concluded that the biochemical properties of poly F could be explained by the individual formycin residues in the polymer existing in the *syn* conformation (Fig. 9.3). By assigning the *syn* conformation, Ward and Reich (1968), Ward, Fuller, and Reich (1969b) and Ward et al. (1970) accounted for all the anomalies of the poly F structure, as well as for its function. They stated that the conformational determinants for the individual formycin residues included the nature of the adjacent nucleotide sequence, the presence or absence of base-pairing, helical structure, and the ionic environment.

Until recently, conformational analysis of the glycosidic bond in the purine ribosides has been difficult to determine in view of the apparently low energy barriers to rotation about the $N$-glycoside bond. Recently, Ts'o et al.

(1969) and Hart and Davis (1969a) reported their NMR studies and the application of the nuclear Overhauser effect as a more direct approach to determining the conformation of purine and pyrimidine ribosides. The two extreme conformations in the purine series are designated *anti* and *syn*. The NMR and double-resonance experiments reported indicated that adenosine is mainly *anti* and guanosine is mainly *syn*. The data of Hart and Davis (1969a) suggest that both *syn* and *anti* conformations are allowed for adenosine and guanosine under their experimental conditions but that quantitative differences exist. Using the 2′, 3′-isopropylidine derivatives of adenosine and guanosine, Hart and Davis concluded that, "the *anti* range of conformations is less favored in the adenosine case than in the guanosine case, a difference that was not anticipated." Hart (private communication) is currently studying the conformation of several of the nucleoside antibiotics. On the basis of the data reported by Ward and Reich (1968), it would appear that application of the nuclear Overhauser effect would favor the *syn*-like conformation for formycin.

In an equally elegant study, Ward, Reich, and Stryer (1969a) have reported on the fluorescent properties under physiological conditions of the chromophore of formycin and its phosphorylated derivatives. Since formycin functions as an analog of adenosine in base-pairing in a wide variety of enzyme reactions, including the nucleotide polymerizing enzymes, this nucleotide can be a very useful biochemical tool for studies of nucleic acid structure and function. The ultraviolet absorption band of formycin is at a wavelength longer than that of the normal purine and pyrimidine bases. It is also fluorescent at ambient temperature, neutral pH, and a wide range of ionic environments. Therefore, the fluorescence properties of polymers that contain formycin can be studied under physiological conditions. By using fluorescence techniques, the binding properties and catalytic mechanisms of numerous enzymes can be studied. In addition, Ward, Reich, and Stryer (1969a) suggested that the substitution of a fluorescent analog for the AMP moiety of coenzymes could also be used to study structure and function. The authors suggested the use of this technique as a sensitive assay for nucleases since the incorporation of formycin into polymers resulted in a 100-fold reduction in the quantum yield of fluorescence. When the double-stranded helical polymer containing a fluorescent nucleotide was denatured by heating, there was a three- to ten-fold increase in fluorescence. They also reported on the fluorescence properties of a mixture of uncharged tRNA species containing terminal formycin residues. Formycin can be incorporated into the nucleoside terminus of tRNA since FTP is a substrate for rabbit liver tRNA-CCA pyrophosphorylase (Ward et al., 1969c). These workers concluded that the terminal formycin residue is involved in interactions with other parts of the tRNA molecule. The interactions of formycin are sensitive to temperature

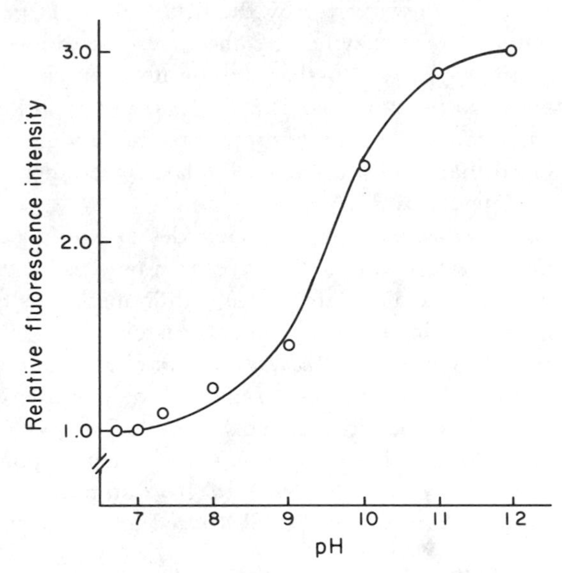


FIGURE 9.6. *The relative fluorescence intensity of tRNA-CCF excited at 295 mμ as a function of pH.* (*From Ward, Reich and Stryer*, 1969*a*).

and pH changes. This is based on the observation that there is a marked shift in the p*K* of the dissociable proton on the pyrazole ring. The anion fluorescence of the free nucleoside had a p*K* of 9.5. In tRNA-CCF, the p*K* value is 12. It has also been observed that the relative fluorescence intensity of tRNA-CCF increased about 3 times between pH 9.0 and 11.0 (Fig. 9.6). The fluorescence intensity of free formycin actually decreases through this pH range. Another significant difference in the fluorescence behavior of free formycin and tRNA-CCF is the change in fluorescence intensity with increased temperature. Formycin fluorescence decreases with increased temperature, while the intensity of tRNA-CCF increases with temperature. The fluorometric $T_m$ of tRNA-CCF is 55°C while the absorbance $T_m$ is 62°C. Since the dissociable proton on the pyrazole ring of formycin in tRNA–CCF is changed, it may be that this proton is involved in hydrogen bonding. The authors stated that the use of this technique should permit a careful analysis of the conformational changes in charged and uncharged tRNA–CCF.

Ikehara et al. (1969b) have also reported recently that the polymerization of 8-substituted purine ribonucleoside diphosphates did not form homopolymers with polynucleotide phosphorylase. Copolynucleotides were obtained from these 8-substituted analogs when ADP or GDP was added. Ikehara et al. stated that the inability of these analogs to form homopolymers could be explained in two ways: (*1*) by the distortion of large substituents located at position 8 with the enzyme sites and (*2*) by the different conformation of the analog diphosphates from that of naturally occurring diphosphates. They further stated that the slow incorporation encountered in the polymerization of formycin diphosphate tends to support the latter view. Kapuler et al. (1969) also reported that 8-substituted GTP analogs were neither substrates nor inhibitors of Mengovirus RNA polymerase.

Since a number of enzymes require nucleotides for the regulation of the allosteric site, the nucleoside antibiotics have been used as biochemical tools to study the interaction and the structural requirements for regulation of enzymes. One such study involved the use of formycin triphosphate with the ribonucleotide reductase from *Lactobacillus leichmannii*. In this work, Suhadolnik et al. (1968) studied the structural requirements for interaction at the allosteric site of the ribonucleotide reductase (Table 9.8). FTP had little or no effect as a stimulator of this enzyme, while the 5′-triphosphates of cordycepin, tubercidin, and toyocamycin were effective stimulators. Apparently the pyrazolopyrimidine ring cannot be substituted for the purine ring.

**Table 9.8**

Allosteric Stimulation of CTP Reduction by Nucleotide Antibiotics

| Addition | Deoxycytidine formed | Stimulation (ratio of activity) |
|---|---|---|
| | mμmoles | |
| None | 0.93 | 1.0 |
| 2′-dATP | 12.70 | 13.5 |
| 3′-Deoxy-5′-ATP | 6.75 | 7.2 |
| TuTP | 6.02 | 6.5 |
| ToTP | 3.79 | 4.1 |
| FTP | 1.19 | 1.3 |

From Suhadolnik et al., 1968.

Hayaishi and his co-workers and Wood and his co-workers also used formycin 5′-monophosphate to study the requirements for activation of the enzyme threonine dehydrase (Nakazawa et al., 1967; Rabinowitz et al., 1968). Substitution of the pyrazole ring for the imidazole ring in AMP resulted in a nucleotide analog (FMP) that could replace AMP as an allosteric activator. Thus, it appears that there is little binding function or steric restriction in the imidazole portion of the adenine ring for this enzyme.

Sheen et al. (1968a) have examined formycin as a biochemical tool to study the molecular properties and the mechanism of erythrocytic purine nucleoside phosphorylase. Formycin B, a *C*-riboside analog of inosine, is neither a substrate for nucleoside phosphorylase from human erythrocytes nor inosine kinase in animal tissue (Sheen et al., 1968a, 1970). However, formycin B is a potent competitive inhibitor of the purified erythrocytic purine nucleoside phosphorylase. The inhibition constant ($K_i = 1 \times 10^{-4}$ $M$) is of the same order of magnitude as the $K_m$ for inosine ($5 \times 10^{-5}$ $M$). Since formycin B has a low toxicity in mice, the inhibition constant as reported was unexpected. Parks suggested that this relatively nontoxic nucleoside might be extremely useful for studying those disorders of purine metabolism, such as gout, or it might be used to alter the toxicity and/or antitumor behavior of other nucleoside analogs. When formycin B was added to assay systems containing the purified purine nucleoside phosphorylase and 6-mercaptopurine ribonucleoside, it was shown to be a competitive inhibitor (Fig. 9.7). Formycin B was also a competitive inhibitor for the hydrolysis of inosine by purine nucleoside phosphorylase and inhibited the degradation of 6-mercaptopurine ribonucleoside in intact sarcoma 180 cells. However, whereas the Dixon plot had a concave upward curvature in those inhibitions studied with intact human erythrocytes, this plot was linear in the sarcoma 180 ascites cell studies. The possibility that formycin B interfered with a transport mechanism in those experiments with intact human erythrocytes was eliminated when the same type concave upward curvature was obtained with cells that were pre-incubated for 30 min or where no preincubation was employed. These data indicate that the principal effect of formycin B is not on the nucleoside transport mechanism, but on some system

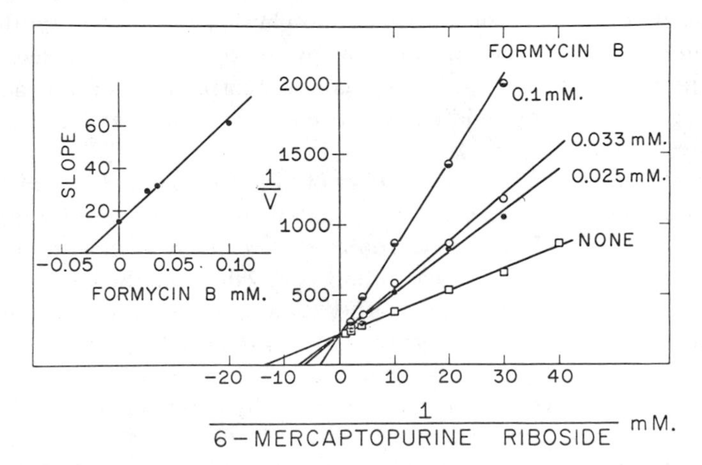


FIGURE 9.7. *Inhibition by formycin B of purified purine nucleoside phosphorylase with 6-mercaptopurine ribonucleoside as substrate. (From Sheen et al.,* 1968*a*).

that does not depend on the integrity of the cell wall. When hemolyzed cells and xanthine oxidase were added to the assay mixture containing formycin B along with mercaptopurine ribonucleoside, linear reciprocal velocity versus formycin B concentration plots were obtained. Apparently xanthine oxidase removes 6-mercaptopurine as soon as it is formed in the assay medium. 6-Mercaptopurine is either being metabolized to the nucleotide or is acting as a product inhibitor. Sheen et al. (1968a) concluded that formycin B acts on the intact cells by inhibiting purine nucleoside phosphorylase.

On the basis of these findings, it seems highly likely that formycin B might be an extremely useful biochemical tool in evaluating the importance of nucleoside analogs such that their antitumor activity could be markedly increased. The authors speculated that formycin B might inhibit normal degradation of nucleosides and thereby decrease the production of uric acid in conditions such as gout. Another consideration proposed by Parks and co-workers is the possibility that the lack of a nucleoside kinase in animal tissue would render nucleosides as formycin B metabolically inert. In turn, these nucleoside analogs may selectively inhibit the growth of invading organisms or tissues that have the ability to phosphorylate formycin B.

Simultaneously with the report by Sheen et al. (1968a), Hori et al. (1968) reported their findings on the biochemical effects of formycin B on *X. oryzae*. They discovered that the inhibition of *X. oryzae* by formycin B is not due to the amination to form formycin. Formycin B appears to interfere with purine and pyrimidine nucleoside metabolism in *X. oryzae* by blocking the entry of exogenously supplied nucleosides into the cell. Sheen et al. were able to show that formycin B inhibited the incorporation of thymidine, uridine, and adenosine into nucleic acids; however, it did not inhibit the uptake of orotic acid, formic acid, or inorganic phosphoric acid. Protein synthesis was not inhibited by formycin B. The dioxopyrazolopyrimidine nucleoside, oxoformycin B, had no effect on the uptake or incorporation of radioactive leucine or purine and pyrimidine precursors into protein and nucleic acids, respectively.

To determine if the inhibitory effect of formycin B could be attributed to an interference with some process essential to the uptake of exogenously supplied nucleosides into the cell, data was presented to show that the suppression of nucleoside uptake was almost porportional to the logarithmic concentrations of formycin B. The incorporation of leucine was not inhibited to any significant extent. When formycin B was removed by washing the cells, the uptake of adenosine-$^{14}C$ was the same as in control cells. Formycin B did not have any effect on adenosine kinase, thymidine kinase, nucleosidase, or nucleoside phosphorylase. In fact, formycin B appeared to be somewhat stimulatory to the activity of adenosine kinase and thymidine kinase. A kinetic study was reported in which the external concentration of adenosine-

$^{14}C$ was varied and the concentration of formycin B was kept constant. It was observed that both formycin and formycin B noncompetitively inhibited the incorporation of adenosine into the TCA-insoluble material. Similar results were obtained with thymidine as the substrate. Although formycin B interferes with some process that is essential to the entry of exogenous nucleosides into the cell, *X. oryzae* does retain the *de novo* synthesis pathway for purine and pyrimidine nucleotide biosynthesis. Hori et al. (1968) stated that it is difficult to explain why the inhibition of uptake of exogenous nucleosides would be fatal to this organism. They suggested that the *de novo* pathway for nucleotides in this organism may not be active enough to maintain cell division. The antagonistic effect of formycin B was shown to be reversed by the addition of purine and pyrimidine nucleosides. The corresponding bases were not effective in reversing this inhibition. It is noteworthy that Sheen et al. (1968a) reported that formycin B was a potent competitive inhibitor of erythrocytic purine nucleoside phosphorylase. However, this enzyme did not seem to be inhibited by formycin B in *X. oryzae*.

The biochemical effects of formycin B in purine biosynthesis *de novo* with Ehrlich ascites tumor cells were studied by Henderson et al. (1967). While formycin was a potent inhibitor of purine biosynthesis, as determined by the accumulation of FGAR, formycin B caused no inhibition of FGAR synthesis under the same conditions.

## ANTIVIRAL EFFECT

The first report on the antiviral effect of formycin and formycin B was by Takeuchi et al., (1966). They reported that formycin and formycin B were equally active as inhibitors of influenza $A_1$ virus in chorioallantoic membrane. This report is extremely interesting since formycin B did not inhibit mammalian cells, nor did it inhibit the growth of *Mycobacterium* 607. It did inhibit the growth of *X. oryzae* (Sawa et al., 1965; Hori et al., 1964; Koyama and Umezawa, 1965). Since *S. marcescens* and intact Ehrlich carcinoma can phosphorylate formycin, but cannot phosphorylate formycin B, this suggests that these two nucleoside antibiotics may exhibit inhibition of viruses by a similar mode of action. (Sawa et al., 1965). In a subsequent paper, Takeuchi et al. (1967) studied the influence of coformycin on the antiviral effect of formycin and formycin B. The study was undertaken since coformycin had no synergistic effect with formycin against influenza $A_1$ virus. The minimum effective concentrations for antiviral activity for formycin were from 6 to 12.5 $\mu g/ml$ and from 10 to 20 $\mu g/ml$ for formycin B. The minimal effective concentrations needed for antiviral activity for deaminoformycin, deaminochloroformycin, deaminomethylthioformycin, deaminoiodoformycin, and deaminomercaptoformycin were also reported. In a more recent paper, Kunimoto et al. (1968) reported on the biological activity of formycin derivatives as inhibitors of influenza virus. When the 7-amino group was replaced with a hydrogen, chloride, or iodide, or the amino group substituted with a methyl or dimethyl group, there was no inhibition of influenza virus. Replacement of the 7-amino group with a —SH or —$SCH_3$ group resulted in a nucleoside that was only slightly inhibitory to influenza virus. Modification on the ribose moiety resulted in complete loss of activity. At present there is no clear understanding of the antiviral action of formycin.

A summary of the cellular systems and enzymes affected by the pyrazolopyrimidine nucleoside antibiotics is presented in Table 9.9.

**Table 9.9**

Summary of Cellular Systems and Enzymes Affected by the Pyrazolopyrimidine Nucleoside Antibiotics

| System or enzyme | References |
|---|---|
| Adenosine kinase | Schnebli et al., 1967; Hori et al., 1968; Acs et al., 1967; Ishizuka et al., 1968b; Sawa et al., 1968; Umezawa et al., 1967; Caldwell et al., 1966; Ward et al., 1969c. |
| 5′-Nucleotidase | Ward, Fuller, and Reich, 1969b |
| 3′-Nucleotidase | Sawa et al., 1965 |
| Conversion of formycin to Formycin B | Ishizuka et al., 1968a; Koyama and Umezawa, 1965 |
| Adenosine deaminase | Umezawa et al., 1967; Fukagawa et al., 1965; Sawa et al., 1967a; Watanabe et al., 1966; Kunimoto et al., 1968; Ward, Fuller and Reich, 1969b |
| Nucleosidase | Hori et al., 1968 |
| Nucleoside phosphorylase (in human erythrocytes) | Sheen et al., 1968a |
| Inhibition of synthesis of protein, DNA, and RNA | Kunimoto et al., 1967 |
| Inhibition of phosphoribosyl-pyrophosphate synthesis | Henderson et al., 1967 |
| Resistance of animal cells to formycin | Caldwell et al., 1966, 1967 |
| Adenine phosphoribosyl transferase | Henderson et al., 1967 |
| Purine nucleoside kinase | Henderson et al., 1967 |
| Myokinase | Acs et al., 1967; Ward et al., 1969c |
| Phosphoenolpyruvate kinase (rabbit) | Acs et al., 1967; Ward et al., 1969c. |
| Polynucleotide phosphorylase (*E. coli*) | Acs et al., 1967; Ward et al., 1969c; Ikehara et al., 1969b. |
| RNA polymerase (*E. coli*) | Acs et al., 1967; Ward et al., 1969c; Ikehara et al., 1968, 1969a |

**Table 9.9 continued**

| System or enzyme | References |
|---|---|
| tRNA-CCA pyrophosphorylase | Acs et al., 1967; Ward et al., 1969c |
| Aminoacyl-tRNA synthetases of ATP by FTP | Acs et al., 1967; Ward et al., 1969c |
| Yeast hexokinase | Acs et al., 1967; Ward et al., 1969c |
| Phosphodiesterase (snake venom) | Ward and Reich, 1968 |
| Micrococcal nuclease | Ward and Reich, 1968 |
| Phosphodiesterase (spleen) | Ward and Reich, 1968 |
| RNase (Bovine, pancreatic) | Ward, Fuller, and Reich, 1969b; Ikehara et al., 1969c |
| Ribopolynucleotides containing formycin and coding properties for protein synthesis | Ward and Reich, 1968; Ward, Fuller and Reich, 1969b; Ikehara et al., 1968, 1969a |
| *syn* and *anti* conformation of formycin | Koyama et al., 1966; Ward and Reich, 1968; Ward, Fuller and Reich, 1969b |
| Fluorescence properties of formycin | Ward, Reich, and Stryer, 1969a |
| Nucleotide pool of Ehrlich ascites cells | Caldwell et al., 1966 |
| NAD pyrophosphatase | Ward et al., 1969c |
| NAD synthetase | Ward et al., 1969c |

## Summary

Formycin is a nucleoside antibiotic in which the pyrazolo [4,3-*d*]pyrimidine ring has replaced the purine ring. This nucleoside has a carbon–carbon bond instead of a carbon-nitrogen nucleosidic linkage. Formycin is structurally related to adenosine and effectively replaces adenosine nucleotides in a number of reactions. It appears that adenosine kinase plays a key role in the mode of action of formycin since a formycin-resistant strain of Ehrlich ascites tumor cells lacks this enzyme. Evidence was also presented to show that formycin can be converted to the deaminated nucleoside, formycin B, by adenosine deaminase. Further metabolism of formycin B to oxoformycin B results in a nucleoside that is less active or entirely inactive toward tumors. For example, formycin B is not a substrate for nucleoside phosphorylase and does not appear to be phosphorylated by an inosine kinase in animal tissue. The reason for the limited antibiotic spectrum of formycin may be that those organisms that have a very active adenosine deaminase or lack an adenosine kinase. Coformycin, a potent inhibitor of adenosine deaminase, broadens the antibiotic spectrum of formycin. Although formycin and formycin B both inhibit influenza $A_1$ virus, the mode of action is not clear.

Since formycin is a cytotoxic analog of adenosine, the formycin nucleotides have been effectively substituted for adenosine nucleotides in many enzymatic reactions. The enzymes reported in this chapter in which formycin has been utilized include adenosine kinase, adenosine deaminase, hexokinase, myokinase, polynucleotide phosphorylase, phosphoenolpyruvate kinase, DNA-dependent RNA polymerase, aminoacyl-tRNA synthetases, tRNA-CCA pyrophosphorylase, phosphodiesterase, NAD pyrophosphatase, and NAD synthetase. Formycin can replace adenosine as the terminal nucleoside in tRNA and still transfer amino acids into peptides.

The formycins have great freedom of rotation about the C—C bond. This permits a greater ease of conformational transitions in the formycin polymers. This physical property may explain why polyformycin resembles polycytosine rather than polyadenosine in its reaction with bovine pancreatic ribonuclease A. It may also explain why formycin codes like adenosine in copolymers for peptide synthesis, but in the homopolymer, polyformycin does not mimic polyadenosine in coding for polylysine synthesis. The chromophore of formycin, in contrast to adenine and other purine and pyrimidine bases, is fluorescent under physiological conditions. Numerous potential applications of formycin in studies related to structure and function of nucleic acids and proteins are now available. The biological properties and crystal structure of formycin and formycin polymers indicate that this nucleoside has the *syn* conformation. The $T_m$ and ORD data of the nucleotides of purines, pyrimidines, pyrrolopyrimidines and the pyrazolo pyrimidines in helical homopolymer complexes, alternating copolymers and single-stranded homopolymers have been studied. There is a considerable destabilization of the homopolymer pair structure of poly rF·rU. This denaturation in poly F coincides with the transition from the *anti* to *syn* conformation. This abnormal conformation of formycin would explain (1) the difficulty observed in the enzymatic synthesis of poly F,(2) the susceptibility of poly F to attack by RNase and (3) the inability of poly F to code for polylysine synthesis. Although poly F and poly Tu have abnormal conformations, their biological properties differ in that (1) poly Tu, but not poly F, can code effectively for polylysine and (2) polymerization of Tu, but not F, occurs efficiently with polynucleotide phosphorylase or RNA polymerase. The information available to date does not provide a clear understanding of the cytotoxicity and antiviral activity of formycin. At high concentrations, formycin inhibited DNA synthesis in synchronized cell cultures of HeLa $S_3$ strain. It did not inhibit RNA synthesis and only had a slight effect on the inhibition of protein synthesis. Several studies have been reported in which formycin has been used as a biochemical tool to study the structural requirements for interaction at the catalytic and regulatory sites of enzymes.

Marked differences in the biological properties of formycin and formycin B have been reported. For example, formycin is readily phosphorylated, but formycin B is not. The resistance of tumor cells to formycin has been shown to be attributed to the lack of adenosine kinase. This observation strongly suggests that adenosine kinase may play a key role in the mode of action of formycin. Since the deaminated nucleoside, formycin B, is less toxic, this may be another mechanism by which cells become resistant to formycin.

The biosynthesis of formycin appears to proceed via formycin B 5′-monophosphate. Formycin 5′-monophosphate and formycin B 5′-monophosphate were both detected in the nucleotide pool of *N. interforma*. This amination reaction appears to require adenylosuccinate synthetase. Formycin B is converted to oxoformycin, the xanthosine analog, by partially purified aldehyde oxidase isolated from rabbit liver.

## References


Acs, G., D. C. Ward, A. Cerami, E. Reich, and S. Uretsky, *Abstracts, Seventh International Congress of Biochemistry*, 1967; p. 647, IV-B-121.

Aizawa, S., T. Hidaka, N. Ōtake, H. Yonehara, K. Isono, N. Igarashi, and S. Suzuki, *Agr. Biol. Chem. (Tokyo)*, **29**, 375 (1965).

Caldwell, I. C., J. F. Henderson, and A. R. P. Paterson, *Can. J. Biochem.*, **45**, 735 (1967).

Caldwell, I. C., J. F. Henderson, and A. R. P. Paterson, *Proc. Amer. Assoc. Cancer Res.*, **7**, 11 (1966).

Caldwell, I. C., J. F. Henderson, and A. R. P. Paterson, *Can. J. Biochem.*, **47**, 901 (1969).

Elion, G. B., A. Kovensky, G. H. Hitchings, E. Metz, and R. W. Rundles, *Biochem. Pharmac.*, **15**, 863 (1966).

Emerson, T. R., R. J. Swan, and T. L. V. Ulbricht, *Biochemistry*, **6**, 843 (1967).

Fukagawa, Y., T. Sawa, T. Takeuchi, and H. Umezawa, *J. Antibiotics (Tokyo)*, **18A**, 191 (1965).

Hart, P. A., and J. P. Davis, *J. Amer. Chem. Soc.*, **91**, 512 (1969a).

Hart, P. A., and J. P. Davis, *Biochem. Biophys. Res. Commun.*, **34**, 733 (1969b).

Henderson, J. F., A. R. P. Paterson, I. C. Caldwell, and M. Hori, *Cancer Res.*, **27**, 715 (1967).

Hori, M., E. Ito, T. Takida, G. Koyama, T. Takeuchi, and H. Umezawa, *J. Antibiotics (Tokyo)*, **17A**, 96 (1964).

Hori, M., T. Wakashiro, E. Ito, T. Sawa, T. Takeuchi, and H. Umezawa, *J. Antibiotics (Tokyo)*, **21A**, 264 (1968).

Ikehara, M., and E. Ohtsuka, *Biochem. Biophys. Res. Commun.*, **21**, 257 (1965).

Ikehara, M., K. Murao, F. Harada, and S. Nishimura, *Biochim. Biophys. Acta*, **155**, 82 (1968).

Ikehara, M., K. Murao, F. Harada and S. Nishimura, *Biochim. Biophys. Acta*, **174**, 696 (1969a).

Ikehara, M., Q. Tazawa, and T. Fukui, *Biochemistry*, **8**, 736 (1969b).

Ikehara, M., K. Murao, and S. Nishimura, *Biochim. Biophys. Acta*, **182**, 276 (1969c).
Ishizuka, M., T. Sawa, G. Koyama, T. Takeuchi, and H. Umezawa, *J. Antibiotics (Tokyo)*, **21A**, 1 (1968a).
Ishizuka, M., T. Sawa, S. Hori, H. Takayama, T. Takeuchi, and H. Umezawa, *J. Antibiotics (Tokyo)*, **21A**, 5 (1968b).
Johns, D. G., T. Spector, and R. K. Robins, *Biochem. Pharmacol.* **18**, 2371 (1969).
Kapuler, A., D. C. Ward, N. Mendelsohn, and G. Acs, *Virology*, **37**, 701 (1969).
Kapuler, A. M., and S. Spiegleman, *Proc. Natl. Acad. Ssi.* (U.S.), **66**, 539 (1970).
Kawamura, K., S. Fukatsu, M. Murase, G. Koyama, K. Maeda, and H. Umezawa, *J. Antibiotics (Tokyo)*, **19A**, 91 (1966).
Koyama, G., and H. Umezawa, *J. Antibiotics (Tokyo)*, **18A**, 175 (1965).
Koyama, G., K. Maeda, H. Umezawa, and Y. Iitaka, *Tetrahedron Letters*, 1966, 597.
Kunimoto, T., M. Hori, and H. Umezawa, *J. Antibiotics (Tokyo)*, **20A**, 277 (1967).
Kunimoto, T., T. Wakashiro, I. Okamura, T. Asajima, and M. Hori, *J. Antibiotics (Tokyo)*, **21A**, 468 (1968).
LePage, G. A., and I. G. Junga, *Cancer Res.*, **25**, 46 (1965).
Lettre, H., N. K. Kapoor, and D. Werner, *Biochem. Pharmacol.*, **16**, 1747 (1967).
Miles, D. W., M. J. Robins, R. K. Robins, M. W. Winkley, H. Eyring, *J. Am. Chem. Soc.*, **91**, 824 (1969a); 831, (1969b)
Nakazawa, A., M. Tokushige, O. Hayaishi, M. Ikehara, and Y. Mizuno, *J. Biol. Chem.*, **242**, 3868 (1967).
Niida, T., T. Niwa, T. Tsuruoka, N. Ezaki, T. Shomura, and H. Umezawa, 153rd Scientific Meeting of Japan Antibiotics Research. Association, January, 1967.
Nippon Kayaku Co., Ltd., Japan Pat. No. 10,928, 1967; Chem. Abstr., **68**, 24552M (1968).
Overgaard-Hansen, K., *Biochim. Biophys. Acta*, **80**, 504 (1964)
Rabinowitz, K. W., J. D. Shada, and W. A. Wood, *J. Biol. Chem.*, **243**, 3214 (1968).
Rice, J. M. and G. O. Dudek, *Biochem. Biophys. Res. Commun.*, **35**, 383 (1969).
Robins, R. K., L. B. Holum, and F. W. Furcht, *J. Org. Chem.*, **21**, 833 (1956).
Robins, R. K., L. B. Townsend, F. Cassidy, J. F. Gerster, A. F. Lewis, and R. L. Miller, *J. Heterocyclic Chem.*, **3**, 110 (1966).
Sawa, T., Y. Fukagawa, Y. Shimauchi, K. Ito, M. Hamada, T. Takeuchi, and H. Umezawa, *J. Amtibiotics (Tokyo)*, **18A**, 259 (1965).
Sawa, T., Y. Fukagawa, I. Homma, T. Takeuchi, and H. Umezawa, *J. Antibiotics (Tokyo)*, **20A**, 227 (1967a).
Sawa, T., Y. Fukagawa, I. Homma, T. Takeuchi, and H. Umezawa, *J. Antibiotics (Tokyo)*, **20A**, 317 (1967b).
Sawa, T., Y. Fukagawa, I. Homma, T. Wakashiro, T. Takeuchi, and M. Hori, *J. Antibiotics (Tokyo)*, **21A**, 334 (1968).
Schnebli, H. P , D. L. Hill, and L. L. Bennett, Jr., *J. Biol. Chem.*, **242**, 1997 (1967).
Sheen, M. R., B. K. Kim, and R. E. Parks, Jr., *Mol. Pharmacol.* **4**, 293 (1968a).
Sheen, M. R., B. K. Kim, H. Martin, and R. E. Parks, Jr., *Proc. Am. Assoc. Cancer Res.*, **9**, 63 (1968b).
Sheen, M. R., H. F. Martin, and R. E. Parks, Jr., *Mol. Pharmacol.*, **6**, 99 (1970).
Shomura, T., T Niwa, J. Yoshida, C. Moriyama, and T. Niida, *Meiji Shika Kenkyu Nempo*, **9**, 21 (1967).

Suhadolnik, R. J., S. I. Finkel, and B. M. Chassy, *J. Biol. Chem.*, **243**, 3532 (1968).
Takeuchi, T., J. Iwanaga, T. Aoyagi, M. Murase, T. Sawa, and H. Umezawa, *J. Antibiotics* (*Tokyo*), **20A**, 297 (1967).
Takeuchi, T., J. Iwanaga, T. Aoyagi, and H. Umezawa, *J. Antibiotics* (*Tokyo*), **19A**, 286 (1966).
Townsend, L. B. and R. K. Robins, *J. Heterocyclic Chem.*, **6**, 459 (1969).
Tsukada, I., T. Kunimoto, M. Hori, and T. Komai, *J. Antibiotics* (*Tokyo*), **22A**, 36 (1969).
Tsuruoka, T., N. Ezaki, S. Amano, C. Uchida, and T. Niida, *Meiji Shika Kenkyu Nempo*, **9**, 17, (1967).
Ts'o, P. O. P., N. S. Kondo, M. P. Schweizer, and D. P. Hollis, *Biochemistry*, **8**, 997 (1969).
Umezawa, H., T. Sawa, Y. Fukagawa, G. Koyama, M. Murase, M. Hamada, and T. Takeuchi, *J. Antibiotics* (*Tokyo*), **18A**, 178 (1965).
Umezawa, H., T. Sawa, Y. Fukagawa, I. Homma, M. Ishizuka, and T. Takeuchi, *J. Antibiotics* (*Tokyo*), **20A**, 308 (1967).
Ward, D. C., and E. Reich, *Proc. Natl. Acad. Sci.* (*U.S.*), **61**, 1494 (1968).
Ward, D. C., E. Reich, and L. Stryer, *J. Biol. Chem.*, **244**, 1228 (1969a).
Ward, D. C., W. Fuller, and E. Reich, *Proc. Natl. Acad. Sci.* (*U.S.*), **62**, 581 (1969b).
Ward, D. C., A. Cerami, E. Reich, G. Acs, and L. Altwerger, *J. Biol. Chem.*, **244**, 3243 (1969c).
Ward, D. C., A. Cerami, E. Reich, N. Mendelsohn, and G. Acs, *J. Biol. Chem.*, **245**, *in press* (1970).
Watanabe, S., G. Matsuhashi, S. Fukatsu, G. Koyama, K. Maeda, and H. Umezawa, *J. Antibiotics* (*Tokyo*), **19A**, 93 (1966).
Williams, R. H., K. Gerzon, M. Hoehn, and D. C. DeLong, 158th National Meeting American Chemical Society, New York, 1969, Abstracts, Micr. 38.

CHAPTER 10

# Pyrazole Nucleosides

## 10.1. PYRAZOMYCIN



## 10.1. PYRAZOMYCIN

### INTRODUCTION

The fourth and most recent *C*-nucleoside to be isolated is pyrazomycin (Fig. 10.1). This nucleoside has been isolated at the Lilly Research Laboratories from the culture filtrates of a strain of *Streptomyces candidus* (Williams et al., 1969).

The structure of pyrazomycin has been reported to be a 3(5)-ribofuranosyl-4-hydroxypyrazole-5(3)-carboxamide (Gerzon et al., 1969). This novel nucleoside is thus structurally related to the *C*-ribosides formycin and formycin B. While these two latter nucleosides have the pyrazolopyrimidine ring system, pyrazomycin is a substituted pyrazole.

The biosynthesis of the formycin-pyrazomycin group of *C*-nucleosides presents an interesting problem for investigation. Pyrazomycin is a potent antiviral agent (Streightoff et al., 1969).

### DISCOVERY, ISOLATION AND PRODUCTION

Pyrazomycin was discovered and isolated from the fermentation broth of *S. candidus*.

The nucleoside is isolated from the culture medium by adsorption on carbon and subsequently eluted with acetone-water (1:1). The eluant is passed through a Dowex-50-$H^+$ column and a Sephadex G-10 column and is finally absorbed onto a cellulose column. It is eluted from the cellulose column with 70% *n*-propanol, concentrated, and allowed to crystallize from water at 4°C.

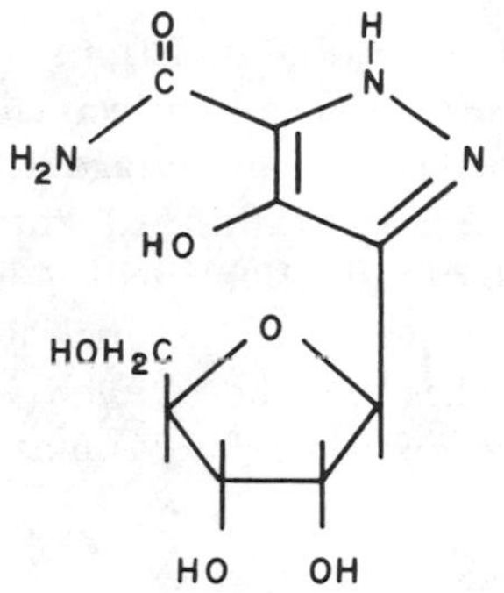

FIGURE 10.1. *Structure of pyrazomycin* [3(5)-*ribofuranosyl*-4-*hydroxypyrazole*-5(3)-*carboxamide*].

## PHYSICAL AND CHEMICAL PROPERTIES AND STRUCTURAL ELUCIDATION

The molecular formula, as established by elemental analyses and the mass spectral molecular ion at $m/e$ 259, is $C_9H_{13}N_3O_6$; mp 108–113°C; $[\alpha]_D$ −47° ($c$ = 1.25 in water). The observed $pK_a$ ($H_2O$) is 6.7; the ultraviolet absorption spectrum: $\lambda_{max}$ ($\epsilon \times 10^{-3}$) = 263 m$\mu$ (6.2) at pH 7.0; 307 m$\mu$ (8.1) at pH 12.0. The bathochromic shift in alkali is compatible with the 4-hydroxy function. The carboxamide group was inferred by bands in the infrared spectrum at 1650 and 1610 $cm^{-1}$. The nmr spectrum ($D_2O$) exhibits proton resonances of a *C*-riboside. There is a doublet at 5.0 $\delta$ for the C-1′ proton, a characteristic feature of other *C*-ribosides (see formycin p. 357; showdomycin, p. 396). The structure of the pyrazole aglycone is confirmed through the laboratory synthesis of *O,N*-dimethyl-5-carbamoyl-4-hydroxy-pyrazole-3-carboxylic acid, which was found to be identical with the corresponding amide–acid obtained by conversion of pyrazomycin with diazomethane to its *O,N*-dimethyl derivative and subsequent oxidation with aqueous permanganate.

## VIRUS INHIBITION

Pyrazomycin prevents the proliferation of rhinovirus, measles, herpes simplex, and vaccinia virus in monolayers of culture cells. The minimal inhibitory concentration for vaccinia virus is less than 2 $\mu$g/ml. Vaccinia virus lesions in mice are reduced when pyrazomycin is given orally, subcutaneously, or intraperitoneally and when it is applied topically at a concentration of 0.25% as an ointment.

## BIOCHEMICAL PROPERTIES

Pyrazomycin competes for uridine and is a strong inhibitor of orotidylic acid decarboxylase. Reversal studies in vaccinia virus-infected cell monolayers show that pyrazomycin is an antagonist of uridine metabolism. Pyrazomycin 5′-phosphate was synthesized enzymatically by nucleoside phosphotransferase (Strider et al., 1968) with *p*-nitrophenylphosphate as a phosphate donor. The nucleotide was hydrolyzed by 5′-nucleotidase. Pyrazomycin 5′-phosphate has been shown to be twenty times as potent as an inhibitor of orotidylic acid decarboxylase as 6-azauridylic acid (Streightoff et al., 1969).

## Summary

Pyrazomycin is the fourth *C*-nucleoside antibiotic that has been isolated from the culture filtrates of the *Streptomyces*. Pyrazomycin is an inhibitor of virus multiplication. The mode of action appears to be related to a competition with uridine metabolism.

## References

Gerzon, K., R. H. Williams, M. Hoehn, M. Gorman, and D. C. DeLong, 2nd Intern. Cong. Heterocyclic Chemistry, Montpellier, France, July 10, 1969, Abstract C-30.

Streightoff, F., J. A. Nelson, J. C. Cline, K. Gerzon, R. H. Williams, and D. C. DeLong, 9th Conference on Antimicrobial Agents and Chemotherapy, Washington, D.C., 1969, Abstract No. 18.

Strider, W., C. Harvey and A. L. Nussbaum, *J. Med. Chem.*, **11**, 524 ((1968).

Williams, R. H., K. Gerzon, M. Hoehn, and D. C. DeLong, 158th National Meeting, American Chemical Society, New York, 1969, Abstract, Micr. 38.

CHAPTER 11

# Maleimide Nucleosides



## 11.1. SHOWDOMYCIN

### INTRODUCTION

The nucleoside antibiotic showdomycin represents a new class of *C*-substituted nucleoside antibiotics. This nucleoside was discovered by the research group at the Shionogi Research Laboratory (Nishimura et al., 1964). It is elaborated by *Streptomyces showdoensis*. The structure is closely related to pseudouridine and uridine (Fig. 11.1). The structure of showdomycin has been established as 2-(β-D-ribofuranosyl)maleimide (Darnall, Townsend, and Robins, 1967; Nakagawa et al., 1967). It is moderately active against gram-positive and gram-negative bacteria and cytotoxic to tumor cells. Recent studies on the biosynthesis and mechanism of action of showdomycin will be discussed in this chapter. A complete review of showdomycin can be found in the Annual Report of the Shionogi Research Laboratory (1968) and Roy-Burman (1970).

### DISCOVERY ISOLATION AND PRODUCTION

Showdomycin was first isolated from the culture filtrates of *S. showdoensis n. sp.* by Nishimura et al. (1964). Showdomycin is produced in a fermentation medium containing 5% glycerol, 5% potato starch, 5% glucose, 5% polypeptone, 4% potato juice, and 3% sodium chloride. The flasks are shaken

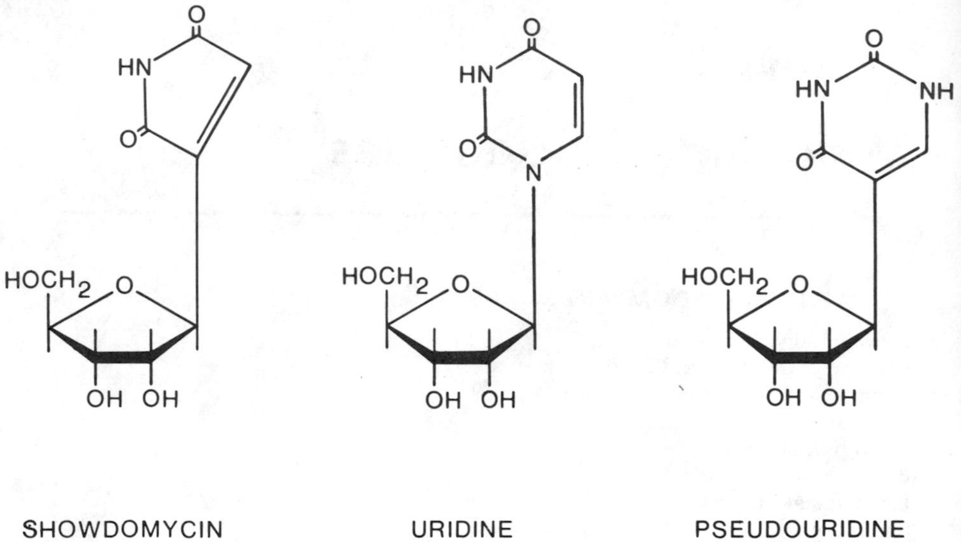

SHOWDOMYCIN URIDINE PSEUDOURIDINE

FIGURE 11.1. *Formula of showdomycin, uridine, and pseudouridine* (*From Roy-Burman, Roy-Burman and Visser*, 1968.)

on an incubator shaker at 28°C. The yield of the nucleoside antibiotic from 28 to 49 hr cultures is about 400 μg/ml (Nishimura et al., 1964).

Since the original strain of *S. showdoensis* only produced 170 μg of showdomycin per milliliter of synthetic medium, Mayama et al. (1968) subsequently isolated a mutant (*S. showdoensis* $N_2$-209-56) from the parent strain (C-224) following treatment with *N*-nitroquinoline 1-oxide. This mutant is a much better producer of showdomycin in synthetic medium (1000 μg/ml). When this mutant was grown in organic medium, the yield of showdomycin was only 10% that of the nucleoside produced in the synthetic medium. An additional mutant ($U_5$-166) was also isolated following uv irradiation of the original parent, *S. showdoensis* C-224. This strain was a better producer of showdomycin in organic medium than either the parent strain or mutant $N_2$-209-56. The production of showdomycin by submerged fermentation with the parent strain and mutants of *S. showdoensis* in synthetic medium or organic medium has been described in much detail by Kimura et al. (1968).

The medium as supplied by Dr. Arnold Demain (private correspondence) for the production of showdomycin is as follows: 1% dextrose, 0.1% asparagine, 0.01% $K_2HPO_4$, 0.05% $MgSO_4 \cdot 7H_2O$, 0.001% $FeSO_4 \cdot 7H_2O$, and 0.05% yeast extract in distilled water (pH 7.2). Inoculation is carried out by adding 5 ml of sterile medium to one agar slant and adding 0.5 ml of a suspension of *S. showdoensis* to each flask containing 200 ml medium (Suhadolnik and Ramer, manuscript in preparation). The 2-liter baffled flasks were maintained on an incubator shaker at 29°C. Growth and production of showdomycin was complete 36 hr after inoculation.

Showdomycin was isolated from the fermentation medium as follows. The culture filtrate was acidified to pH 5.0 (10% HCl) and filtered (Nishimura et al., 1964). Charcoal (2 g/100 ml) was added to the filtrate and the solution was stirred and filtered. Showdomycin was eluted from the charcoal by stirring for 30 min with acetone–water (4:1). This procedure was repeated three times. The acetone was removed by distillation. The aqueous portion was then extracted with 1-butanol. The butanol extract was concentrated to one-fourth the original volume and added to a silica gel column prepared with benzene. The column was developed with benzene–acetone (2:8) to elute showdomycin. The solvent was removed by vacuum distillation to a small volume. Showdomycin crystallized on standing. The nucleoside was recrystallized from a mixture of hot acetone–benzene and allowed to recrystallize overnight at 0°C.

## PHYSICAL AND CHEMICAL PROPERTIES

The molecular formula of showdomycin is $C_9H_{11}NO_6$; mp 160–161°C $[\alpha]_D^{22.5} = +49.9°$ ($c$ = 1% in water) (Nakagawa et al., 1967). The ultraviolet absorption spectrum has a $\lambda_{max}$ at 220 mμ ($E_{1\,cm}^{1\%} = 442$). It is soluble in water, but insoluble in ether and benzene. It is unstable in alkali and loses its $\lambda_{max}$ at 220 mμ in dilute aqueous ammonia in 5 min (Darnall et al., 1967; Nishimura et al., 1964).

The mass spectrum for showdomycin has recently been reported by Townsend and Robins (1969). The parent peak was observed at B + 30 ($m/e$ = 126). This same peak was observed for formycin, formycin B, and pseudouridine (see Chap. 9, Fig. 9.3 and page 358 for discussion of fragmentation pattern for the C—C nucleoside antibiotics formycin and formycin B as reported by Townsend and Robins, 1969). The B + 30 peak for showdomycin may also be considered as a M-103 peak, where $M$ is the molecular ion (showdomycin 229 − 103 = 126). It appears that the mass spectra of carbon-linked nucleosides do lend themselves to patterns that may prove to be extremely useful for determining the structure of C—C linked nucleosides or C—C linked pentoses (Townsend and Robins, 1969).

## STRUCTURAL ELUCIDATION AND SYNTHESIS

The structural elucidation of showdomycin is in large part attributed to the research group at the Shionogi Research Laboratories, Japan, and to Robins and his co-workers. The $pK_a$ of showdomycin is 9.29. The $pK_a$ for uridine is 9.17 and pseudouridine 9.1. These data strongly support the notion of an imide structure and the presence of an acidic "NH"-type proton. In addition, the infrared spectrum of showdomycin has a very strong carbonyl band at 1704 cm$^{-1}$ (Nishimura et al., 1964; Nakagawa et al., 1967). This absorption band is similar to that exhibited by maleimide.

Darnall et al. (1967) reported that showdomycin consumed 1.1 moles of hydrogen with a concomitant loss of ultraviolet absorption. Maleimide under similar conditions absorbed 1.0 moles of hydrogen. Similarly, Nakagawa et al. (1967) reduced the triacetate and acetonide of showdomycin to their respective dihydro derivatives. They also studied the proton magnetic resonance spectra of showdomycin. A definite single absorption peak (1 proton) was noted at 10.78 $\delta$ in dry deuterated dimethylsulfoxide-$d_6$. This peak is typical of the "NH" proton of a cyclic amide. Showdomycin showed a sharp doublet (1 proton) at 6.74 $\delta$. This is typical of an aromatic or vinylic proton. The pmr of showdomycin in deuterium oxide and deuterato acetic acid-$d_4$) was also studied. The "NH" proton at 10.78 $\delta$ was absent due to deuterium exchange. The three protons found in the 3.2–5.3 $\delta$ region also exchanged with deuterium. Therefore, showdomycin has 4 exchangeable protons. From these data, the optical rotation, and the molecular formula, Darnall et al. (1967) concluded that a carbohydrate must be present. Attempts to demonstrate the presence of a sugar by acid hydrolysis were unsuccessful. When showdomycin was treated with aqueous hydrazine at 100°C, D-ribose was detected by paper chromatography. This procedure was similar to that employed by Davis and Allen (1957) for the detection of D-ribose in the C-nucleoside, pseudouridine. One mole of periodate was consumed per mole of showdomycin. This firmly established the furanose configuration. Additional proof that showdomycin was a C-riboside was obtained by the anomeric proton centered at 4.82 $\delta$. Pseudouridine under similar conditions shows the anomeric proton at 4.72 $\delta$. This compares with 6.0 $\delta$ for the anomeric proton of uridine. The carbon-substituted nucleoside antibiotics formycin and laurusin exhibit the anomeric proton (Hb) at 5.14 $\delta$ in deuterated dimethylsulfoxide-$d_6$ and at 5.37 $\delta$ in deuterium oxide and deuteroacetic acid-$d_4$ (Robins et al., 1966). Showdomycin was assigned the $\beta$ configuration since the pmr spectrum of the C-3′, C-4′, C-5′ proton region (3.7–4.8 $\delta$) shows that this region has an absorption pattern that is identical to those for pseudouridine and formycin. The pmr spectrum of the $\alpha$-anomer of pseudouridine shows the anomeric proton 0.33 ppm $\delta$ downfield from that of the $\beta$-isomer (Cohn, 1960). The structural similarity of showdomycin and pseudouridine is shown in Figure 11.1. On the basis of their chemical and physical data, Darnall et al. (1967) reported that showdomycin is a *C*-substituted nucleoside antibiotic that is structurally related to pseudouridine. They assigned showdomycin the structure, 2-($\beta$-D-ribofuranosyl)maleimide (Fig. 11.1).

Tsukuda et al. (1967) also reported their studies on the structure of showdomycin and its derivatives. Based on the results of X-ray structure analysis of *N*-methyl-bisdeoxocycloshowdomycin acetonide hydrobromide, they were able to show that showdomycin must be represented by the structure, 2-($\beta$-D-ribofuranosyl)maleimide (Fig. 11.1). Nakagawa et al. (1967) also used a combination of chemical and physical methods to show that the structure of showdomycin was the *C*-maleimide riboside. This is now the fifth *C*-substituted ribonucleoside isolated from natural sources. The other nucleosides are pseudouridine (Cohn, 1960), formycin (Hori et al., 1964), laurusin (Aizawa, 1965), and pyrazomycin (Williams et al., 1969).

Kalvoda, Farkas, and Sorm (1970) have utilized several elegant procedures for the first total chemical synthesis of showdomycin.

## INHIBITION OF GROWTH

Showdomycin is a broad spectrum antibiotic and also shows remarkable activity against Ehrlich ascites tumor in mice and HeLa cells (Matsuura et al., 1964; Nishimura et al., 1964; Shionogi and Co., Ltd., 1964). It is moderately active against gram-positive and gram-negative bacteria and is extremely active against *Streptococcus hemolyticus* and *S. pyogènes* (Nishimura et al., 1964). The antibacterial spectrum of showdomycin is shown in Table 11.1. The $LD_{50}$ for mice was 25 mg/kg when given intraperitoneally, 18 mg/kg when given subcutaneously, and 110 mg/kg when given intravenously (Nishimura et al., 1964). Matsuura et al. (1964) studied the effect of showdomycin against HeLa cells. The *in vitro* activity against HeLa cells was relatively low as compared with other known antitumor antibiotics. Showdomycin was most effective in inhibiting tumor growth when treatment was started within short periods following tumor inoculation. Matsuura et al. (1964) reported that showdomycin caused metaphase arrest in HeLa cells. Only consecutive daily doses of 5–20 mg/kg/day by intraperitoneal injections were effective in

**Table 11.1**
Antibacterial Spectrum of Showdomycin

| Test organisms | Minimal inhibitory concentration, mcg/ml |
|---|---|
| 1. *Shigella dysenteriae* | 10 |
| 2. *Shigella paradysenteriae*, Ohara | 20 |
| 3. *Salmonella typhosa* | 50 |
| 4. *Salmonella paratyphi* A | 50 |
| 5. *Escherichia coli*, Umezawa | 50 |
| 6. *Klebsiella pneumoniae* | 50 |
| 7. *Stapholycoccus aureus*, 209P | 50 |
| 8. *Salutina lutea* | 50 |
| 9. *Diplococcus pneumoniae*, type II | 50 |
| 10. *Streptococcus hemolyticus*, D | 2 |
| 11. *Streptococcus hemolyticus*, HA | 2 |
| 12. *Corynebacterium diphtheriae*, S | 10 |
| 13. *Corynebacterium diphtheriae*, Tront | 10 |
| 14. *Staphylococcus aureus*, 209P | 50 |
| 15. *Mycobacterium tuberculosis* var. *hominis*, H37Rv | >200 |

From Nishimura et al., 1964.

inhibiting the growth of Ehrlich ascites tumor cells (100 μg/ml) in Swiss albino rats. Showdomycin was less effective against ascites hepatoma AH-130 in Wistar rats. Showdomycin was more toxic in mice when administered intraperitoneally or subcutaneously than when administered intravenously. At concentrations of 50–100 mg/ml, showdomycin was cytotoxic to cultured HeLa cells. Ascites hepatoma AH-130 in rats was not markedly inhibited by showdomycin (Matsuura et al., 1964). The toxicity of showdomycin against *E. coli* B/r is lower than that observed with other thiolbinding reagents (Titani and Katsube, 1969a). The reversal of toxicity by nucleosides (Komatsu and Tanaka, 1969) indicates a low toxicity of showdomycin to *E. coli* B/r.

## BIOSYNTHESIS

Since showdomycin represents a new class of *C*-substituted ribonucleoside antibiotics, it was of interest to study the biosynthesis of this nucleoside antibiotic and to compare the biosynthetic pathway for showdomycin with that of the previously studied nucleoside antibiotics (Suhadolnik, 1967). Suhadolnik and Ramer (1968) have recently reported on the biosynthesis of showdomycin. Six pathways were considered as possibilities in the biosynthesis of this antibiotic (Fig. 11.2). The data suggest that acetate serves as the precursor for either two or four of the carbons of the maleimide ring of showdomycin. The ribose moiety arises from either ribose-1-$^{14}$C or glucose-6-$^{14}$C. The radioactive showdomycin from these experiments was degraded.

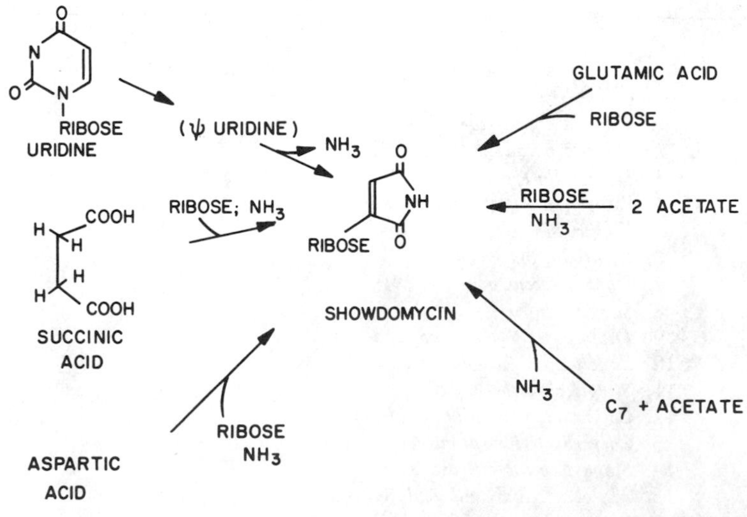

FIGURE 11.2. *Proposed pathways for the biosynthesis of showdomycin.*

All the carbon-14 in the radioactive showdomycin from the ribose-1-$^{14}C$ experiment resided exclusively in the pentose moiety. Degradation of the maleimide ring of showdomycin from the acetate-2-$^{14}C$ experiment showed that all the carbon-14 was in the maleimide ring. There was no carbon-14 in the carbonyl groups (Table 11.2). All the radioactivity from the acetate-1-$^{14}C$ experiments resided in the maleimide moiety of showdomycin. However, only 50% of the carbon-14 was isolated as $^{14}CO_2$ by the Hofmann hypobromite degradation method. If acetate were contributing the four carbons of the maleimide ring, one would expect that all the carbon-14 from acetate-1-$^{14}C$ would reside in the carbonyl carbons of showdomycin and should be isolated as $CO_2$ by the Hofmann hypobromite degradation method. The data obtained imply that either this degradation procedure is not applicable to a 2-substituted maleimide or else the $^{14}C$ from acetate-1-$^{14}C$ does not reside exclusively in the carbonyl carbons. Carbon-14 labeled pyruvate, asparate, glutamate, succinate, formate, and glyoxalate were not incorporated into showdomycin. Fluoroacetate did not inhibit the incorporation of carbon-14 labeled acetate into the maleimide moiety of showdomycin, but did inhibit the incorporation of radioactive acetate into the pyrimidine bases in the RNA of *S. showdoensis*. These findings suggest that two molecules of acetate are not converted to a $C_4$ dicarboxylic acid via the tricarboxylic acid cycle which then becomes the maleimide ring of showdomycin. If this were happening, the incorporation of radioactivity into the maleimide of show-

**Table 11.2**

Distribution of Carbon-14 in Showdomycin from Acetate, D-Glucose and D-Ribose Experiments

| | | Showdomycin | | |
|---|---|---|---|---|
| | | | Distribution of $^{14}C$ | |
| | mμc/μmole[a] | mμc/μmole | maleimide | ribose |
| Acetate-1-$^{14}C$ | 630 | 1.4 | 100 | 0 |
| Acetate-1-$^{14}C$ (plus $1 \times 10^{-3}$ *M* fluoroacetate) | 630 | 2.0 | 100 | 0 |
| Acetate-2-$^{14}C$ | 630 | 2.8 | 100 | 0 |
| Fluoroacetate | | | 100 | 0 |
| D-Glucose-U-$^{14}C$ | 0.11 | 0.05 | | 96 |
| D-Ribose-1-$^{14}C$ | 0.51 | 0.19 | | 90 |

From Suhadolnik and Ramer, manuscript in preparation.
[a] Specific activities of compounds added.

domycin from the carbon-14 labeled acetate would have been inhibited in a manner similar to the decrease in incorporation of radioactivity into the pyrimidine bases when fluoroacetate was added. When fluoroacetate-1-$^{14}C$ was added to showdomycin-producing cultures, the incorporation of $^{14}C$ into the maleimide ring was equal to the incorporation of labeled acetate. If fluoroacetate does replace acetate in the biosynthesis of the maleimide ring, fluoroshowdomycin could be the product.

## BIOCHEMICAL PROPERTIES

Several major contributions have appeared that are related to the biochemical properties of showdomycin. Komatsu and Tanaka (1968) studied the action of showdomycin on the incorporation of amino acids and purine and pyrimidine bases into protein and nucleic acids in *E. coli* K-12 cells. The reversal of inhibition of showdomycin by 2-mercaptoethanol was also studied. Showdomycin inhibited nucleic acid synthesis. This inhibition was cancelled by the addition of nucleosides. As seen in Table 11.3, the incorporation of adenine-$^{14}C$ and uracil-$^{14}C$ was inhibited more at lower concentrations of showdomycin than was the incorporation of amino acids into protein.

**Table 11.3**

Inhibitory Effects of Showdomycin on the Incorporation of Amino Acids and Purine and Pyrimidine Bases in *E. coli* K-12 Cells

| | | Incorporation of labeled precursors into the corresponding macromolecules in *E. coli* K-12 cells (cpm/ml cells)[a] | | | | | |
|---|---|---|---|---|---|---|---|
| SHM, μM | Cell growth (Δ O.D. at 660 mμ) | Adenine-$^{14}C$ | Uracil-$^{14}C$ | Leu-$^{14}C$ | Phe-$^{14}C$ | Lys-$^{3}H$ | Val-$^{14}C$ |
| 0 | 0.357 | 3134 | 5533 | 3771 | 2994 | 24060 | 4192 |
| 1 | 0.334 | 2421 | 4062 | 3704 | 2815 | | |
| 2 | 0.314 | 1752 | 2683 | 3455 | 2273 | 16629 | 3811 |
| 4 | 0.196 | 915 | 1362 | 1878 | 1556 | 9938 | 3193 |
| 8 | −0.021 | 369 | 468 | 796 | 892 | 3707 | 1782 |

From Komatsu and Tanaka, 1968.
[a] Showdomycin, SHM; Leucine, Leu; Phenylalanine, Phe; Lysine, Lys; Valine, Val.

The inhibition of incorporation of carbon-14 labeled purine or pyrimidine base was shown to be reversed by the simultaneous addition of a purine or pyrimidine nucleoside. Similarly, 2-mercaptoethanol reversed the inhibition of incorporation of adenine-$^{14}C$ by showdomycin. The sulfhydryl compounds or the nucleosides must be added early in the stage of showdomycin inhibition in order to block the action of this nucleoside antibiotic on the synthesis of nucleic acids or protein in the cell. Pseudouridine did not reverse the inhibitory action of showdomycin on the incorporation of amino acids or purine or pyrimidine bases into macromolecules. This finding is in contrast to the finding that the ribo- and deoxyribonucleosides did reverse the inhibitory reaction of showdomycin. The authors concluded that the *N*-glycosyl linkage between the base and sugar in the *N*-riboside nucleosides may be a structural requirement for the reversal of inhibition by showdomycin. The inhibitory action of the structurally related analog, *N*-ethylmaleimide, is not reversed by the addition of nucleosides. This is in contrast to the reversal of inhibition of showdomycin when nucleosides are added. Apparently a marked difference exists in the mechanism of action of showdomycin and *N*-ethylmaleimide.

Roy-Burman et al. (1968) studied the effect of showdomycin with cell-free preparations of Ehrlich ascites cells. They studied the sensitivity of showdomycin to uridine kinase, UMP kinase, and UDP kinase. Showdomycin did not inhibit uridine kinase, but specifically inhibited UMP kinase and uridine phosphorylase. Showdomycin is not a substrate for nucleoside kinase nor uridine phosphorylase. Showdomycin would not be an expected substrate for enzymatic phosphorolysis since the base-ribosyl linkage is C—C and not N—C. Similarly, pseudouridine, which is also a *C*-nucleoside, is not hydrolyzed by nucleosidase (Cohn, 1960) or nucleoside phosphorylase (Adler and Gutman, 1959). High concentrations of showdomycin blocked the utilization of orotic acid and caused a marked accumulation of UMP. This may probably be attributed to the inhibition of the UMP kinase step by showdomycin. The reduction in orotic acid utilization may be attributed to the direct inhibition of orotidylic acid pyrophosphorylase. Showdomycin strongly inhibited bovine liver uridine-5′-diphosphate-α-D-glucose dehydrogenase, but had no effect on rabbit muscle lactic acid dehydrogenase. To determine if showdomycin exerted its inhibitory effects by acting as an alkylating agent, Roy-Burman et al. (1968) pre-incubated showdomycin with cysteine. This pre-incubation completely removed the inhibitory effect of showdomycin on UDP-glucose dehydrogenase. Kinetic data showed that the uncompetitive nature of UDP-glucose dehydrogenase inhibition by showdomycin is attributed to its alkylating action on the enzyme (Fig. 11.3). The inhibitory effect of showdomycin on UDP-glucose dehydrogenase is completely removed by preincubation of showdomycin with cysteine. The authors also reported that adenosine phosphorylase is completely resistant to showdomycin, Tsai, Holmberg and Ebner (1970) reported that bovine mammary UDP-galactose 4-epimerase is slightly inhibited by showdomycin. This selective inhibition of certain enzymes make this nucleoside a useful sulfhydryl reagent. Ribonucleosides, deoxyribonucleosides, and thiols (L-cysteine,

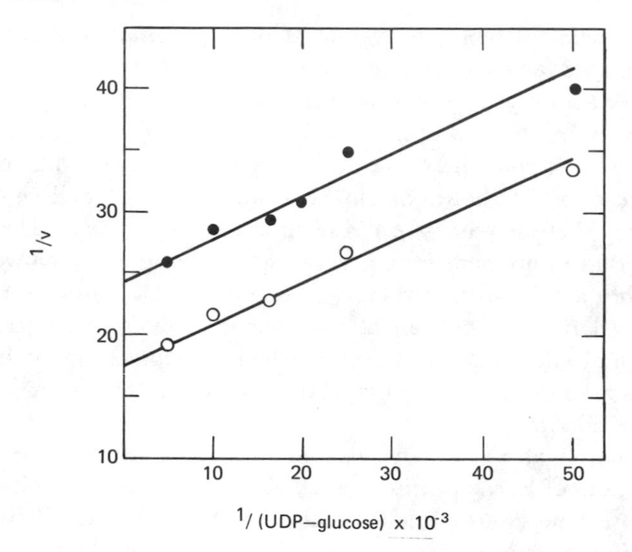


FIGURE 11.3. *Inhibition of UDP-glucose dehydrogenase activity by showdomycin with respect to UDP-glucose. Reaction mixtures, pH* 8.0, *contained in* 1 *ml:* $2 \times 10^{-3}$ *M NAD*$^+$, $5 \times 10^{-2}$ *M tris-(hydroxymethyl)aminomethane acetate, 35 units of enzyme, and the indicated amount of UDP-glucose. Initial reaction velocity was determined in the* (○) *absence and* (●) *presence of* 27 $\times$ $10^{-6}$ *M showdomycin. Velocity (V) is expressed in terms of change in* $A_{340}$ *mμ per minute. UDP, uridine-5'-diphosphate; NAD*$^+$, *nicotinamide adenine dinucleotide.* (*From Roy-Burman, Roy-Burman and Visser*, 1968.)

glutathione, and β-mercaptoethanol) reverse the inhibition of showdomycin in *E. coli* while the purine or pyrimidine bases or nucleoside phosphates had no reversal effect (Nishimura and Komatsu, 1968). Darnall et al. (1967) first suggested that showdomycin acts as an alkylating agent. Whereas Roy-Burman (1970) showed that showdomycin did not inhibit DNA polymerase and weakly inhibited *E. coli* RNA polymerase, Maryanka and Johnston (1970) observed a 50% inhibition of RNA polymerase at 2.5 mM. CNA provides protection against showdomycin.

Komatsu (private communication) has recently shown that showdomycin is a strong inhibitor of DNA synthesis in *E. coli*, but does not inhibit RNA synthesis. The increased toxicity of showdomycin when injected subcutaneously compared to intravenous administration suggests that skin irritation may contribute to the acute systemic toxicity of this nucleoside. Harada et al. (1967) studied the effect of showdomycin on the permeability of minute vessels in the skin. They found that showdomycin elicited or increased vascular permeability in the skin of the rat. Formycin and toyocamycin produced a similar effect, but much less than showdomycin.

The radiosensitization of showdomycin by *E. coli* B/r as reported by Titani and Katsube (1969a) indicates that showdomycin might be a useful compound to study the mechanism of radiosensitization, since the toxicity of showdomycin against *E. coli* B/*r* is lower than other known thiol-binding reagents. In addition, while showdomycin toxicity can be cancelled by the addition of nucleosides, N-ethylmaleimide toxicity is not reversed (Komatsu and Tonaka, 1968).

Beljanski, et al. (1970) have studied the effect of showdomycin on macromolecular synthesis in *E. coli*. They reported that showdomycin causes a massive biosynthesis of rapidly labeled RNAs which are not complementary to DNA. Showdomycin also changes the base ratio in ribosomal 23 S RNA. Polynucleotide phosphorylase from the mutant has new properties. Titani and Katsube (1969b) also investigated the radiosensitivity property of showdomycin, since it is a known thiol-binding agent (Hadler et al., 1969; Roy-Burman et al., 1968). The experiment was done to determine the enhancement of the lethal effect of alkylation to *E. coli* B/r when combined with other alkylating agents. Although showdomycin alone was not very toxic to the cells, the lethal effects of two alkylating agents were considerably enhanced by showdomycin. This effect was called "radiomimetics sensitization".

## Summary

Showdomycin is a *C*-substituted riboside antibiotic with the structure 2-(β-D-ribofuranosyl)maleimide. It is active against gram-negative and gram-positive bacteria and inhibits Erhlich ascites tumor cells. Biosynthetic studies indicate that the $C_4$ moiety of showdomycin is formed from two acetates. This condensation is not inhibited by fluoroacetate, which suggests that the formation of maleimide does not proceed through the tricarboxylic acid cycle intermediates.

Although the biological activity of showdomycin has not been established, it does possess antitumor activity against Ehrlich mouse ascites tumor cells and HeLa cells. Showdomycin inhibits nucleic acid synthesis. Purines, pyrimidines, and sulfhydryl agents reverse the inhibition of showdomycin. Pseudouridine does not reverse this inhibition. DNA, but not RNA, synthesis is inhibited by showdomycin in *E. coli*. Showdomycin is not phosphorylated by nucleoside kinase nor is the *C*-riboside bond cleaved. UMP kinase, uridine phosphorylase, uridine-5′-diphosphate-α-D-glucose dehydrogenase are inhibited by showdomycin. Cysteine reverses this inhibition. The chemical reactivity of the maleimide moiety of showdomycin with sulhydryl groups may be responsible for the selective inhibition of enzymes. The radiosensitizing effect of showdomycin with *E. coli* B/r has been reported. The dose-survival curve is similar to that reported with N-ethylmaleimide. Showdomycin enhances with lethal alkylation of *E. coli* B/r when combined with other alkylating agents.

## References

Aizawa, S., T. Hidaka, N. Ōtake, H. Yonehara, K. Isono, N. Igarashi, and S. Suzuki, *Agr. Biol. Chem. (Tokyo)*, **29**, 375 (1965).
Adler, M., and A. B. Gutman, *Science*, **130**, 862 (1959).
Beljanski, M., P. Bourgavel, and M. M. Beljanski, *Annales L'Institut Pasteur*, **118**, 19 (1970).
Cohn, W. E., *J. Biol. Chem.*, **235**, 1488 (1960).
Darnall, K. R., L. B. Townsend, and R. K. Robins, *Proc. Natl. Acad. Sci. (U.S.)*, **57**, 548 (1967).
Davis, F. F., and F. W. Allen, *J. Biol. Chem.*, **227**, 807 (1957).
Hadler, H. I., B. E. Claybourn, T. P. Tschang and T. L. Moneau, *J. Antibiotics*, **22**, 183 (1969).
Harada, M., M. Takeuchi, and K. Katagiri, *J. Antibiotics (Tokyo)*, **20A**, 369 (1967).
Hori, M., E. Ito, T. Takida, G. Koyama, T. Takeuchi, and H. Umezawa, *J. Antibiotics (Tokyo)*, **17A**, 96 (1964).
Kalvoda, L., J. Faraks, and F. Sorm, *Tetrahedron Letters*, 2297 (1970).
Kimura, T., H. Kyotani, and M. Ozaki, *Ann. Rept. Shionogi Res. Lab.*, **18**, 23 (1968).
Komatsu, Y., and K. Tanaka, *Agr. Biol. Chem. (Tokyo)*, **32**, 1021 (1968).
Maryanka, D., and I. R. Johnston, *FEBS Letters*, **7**, 125 (1970).
Matsuura, S., O. Shiratori, and K. Katagiri, *J. Antibiotics (Tokyo)*, **17A**, 234 (1964).
Mayama, M., H. Nagata, K. Motokawa, *Ann. Rept. Shionogi Res. Lab.*, **18**, 13 (1968).
Nakagawa, Y., H. Kanō, Y. Tsukuda, and H. Koyama, *Tetrahedron Letters*, 1967, 4105.
Nishimura, H., M. Mayama, Y. Komatsu, H. Katō, N. Shimaoka, and Y. Tanaka, *J. Antibiotics (Tokyo)*, **17A**, 148 (1964).
Nishimura, H., and Y. Komatsu, *J. Antibiotics (Tokyo)*, **21A**, 250 (1968).
Robins, R. K., L. B. Townsend, F. Cassidy, J. F. Gerster, A. F. Lewis, and R. L. Miller, *J. Heterocyclic Chem.*, **3**, 110 (1966).
Roy-Burman, S., P. Roy-Burman, and D. W. Visser, *Cancer Res.*, **28**, 1605 (1968).
Roy-Burman, P., *Recent Results in Cancer Research*, **25**, 80 (1970).
Shionagi and Co., Ltd. (by H. Nishimura), Fr. M2751, September 21, 1964; *Chem. Abstr.*, **62**, 2675b (1967).
Suhadolnik, R. J., *Antibiotics*, II, 400 (1967).
Suhadolnik, R. J., and R. M. Ramer, 156th National Meeting American Chemical Society, Atlantic City, N.J., September 1968, Abstract 26.
Titani, Y. and Y. Katsube, *Biochim. Biophys. Acta*, **192**, 367 (1969a).
Titani, Y. and Y. Katsube, *J. Antibiotics*, **23**, 43 (1969b).
Townsend, L. B., and R. K. Robins, *J. Heterocyclic Chem.*, **6**, 459 (1969).
Tsai, C. M., N. Holmberg, and K. E. Ebner, *Arch. Biochem. Biophys.*, **136**, 233 (1970).
Tsukuda, Y., Y. Nakagawa, H. Kano, T. Sato, M. Shiro, and H. Koyama, *Chem. Commun.*, 1967, 975
Williams, R. H., K. Gerzon, M. Hoehn, and D. C. LeDong, 158th National Meeting, American Chemical Society, N.Y., 1969, Abstract, Micr. 38.

# Author Index

Numbers in roman numerals indicate the page on which an author's work is cited. Numbers in *italics* show the page on which the complete reference is listed.

# Subject Index

Page numbers in *italics* indicate illustrations. Page numbers followed by the letter "t" indicate tabular information.